Biochemistry of Alkaloids

Edited by
K. Mothes, H. R. Schütte and M. Luckner

Distribution:
VCH Verlagsgesellschaft, P. O. Box 1260/1280, D-6940 Weinheim (Federal Republic of Germany)
USA and Canada: VCH Publishers, 303 N.W. 12th Avenue, Deerfield Beach, FL 33442-1705 (USA)

ISBN 3-527-26079-X (VCH Verlagsgesellschaft)
ISBN 0-89573-072-3 (VCH Publishers)

Biochemistry of Alkaloids

Edited by
K. Mothes, H. R. Schütte and M. Luckner

With contributions by
H. Böhm, D. Gröger, D. Gross, S. Johne, H.-W. Liebisch, M. Luckner, J. Miersch, K. Mothes (deceased), D. Neumann, H. Reinbothe (deceased), W. Roos, D. Schlee, K. Schreiber, H. R. Schütte

Authors:
H. Böhm Chapter 3
D. Gröger Chapter 15
D. Gross Chapter 13, 19.1
S. Johne Chapter 16.4—16.5
H.-W. Liebisch Chapter 9, 11.1—11.2, 12.1, 14.14
M. Luckner Chapter 1, 2, 4, 16.1—16.3, 18
J. Miersch Chapter 8.2
K. Mothes (†) Chapter 1
D. Neumann Chapter 6
H. Reinbothe (†) Chapter 8.1
W. Roos Chapter 6
D. Schlee Chapter 7, 17
K. Schreiber Chapter 19.5
H. R. Schütte Chapter 2, 10, 11.3, 12.2, 14.1—14.13, 19.2—19.4

CIP-Kurztitelaufnahme der Deutschen Bibliothek

Biochemistry of alkaloids / ed. by K. Mothes ...
With contributions by H. Böhm ... — Weinheim ;
Deerfield Beach, FL : VCH, 1985.
ISBN 3-527-26079-X (Weinheim)
ISBN 0-89573-072-3 (Deerfield Beach)

NE: Mothes, Kurt [Hrsg.]; Böhm, Hartmut [Mitverf.]

Printed in the German Democratic Republic
Satz und Druck: VEB Druckhaus „Maxim Gorki", DDR - 7400 Altenburg

Preface

Alkaloids are an outstanding group of secondary natural products. Their toxicity has been known since ancient times and their use in medication caused interest to be focussed first on the chemistry and later on the fascinating biochemistry of these substances. Nearly all facts known today about alkaloid biochemistry were established in the last decade, i.e. their biosynthesis and its regulation, the genetics and enzymatics of alkaloid metabolism, the accumulation of alkaloids and their turnover and the significance of alkaloid formation for the producer organisms. It is the aim of this monograph to give a comprehensive survey of these fields.

The first part of the book contains chapters of general nature (methods used in the investigation of alkaloid biosynthesis, genetics of alkaloid metabolism, principles regulating alkaloid metabolism, ecological significance of alkaloid biosynthesis). In the second part the biosynthetic pathways leading to the most important groups of alkaloids are discussed. The alkaloids are grouped according to their biosynthesis and some natural products related biosynthetically to true alkaloids are included (e.g. protoalkaloids, heterocyclic non-protein amino acids and antibiotics, cyclic peptides etc.). Due to the limitation of space the book does not deal with the metabolic fate of alkaloids in organisms not producing them. However, in appropriate cases there are indications on the physiological action of alkaloids in humans, the use of alkaloid formation in taxonomy etc.

The book is based on the monograph *Biosynthese der Alkaloide* (Deutscher Verlag der Wissenschaften, Berlin 1969) edited by two of us (K. M. and H. R. S.), which covers the relevant literature until 1968. Older publications are therefore discussed only briefly and emphasis is laid on the results obtained in recent years. We have tried to review these recent results critically and hope that no important information has been overlooked. Two series were particularly useful: *The Alkaloids* and *Biosynthesis,* which are published annually as part of the *Specialist Periodical Report* by the Chemical Society in London and which cover most of the known alkaloid groups. The book is written by a small number of authors, all working in Hallensian research groups. We thank them for their readiness to accept our ideas and for their discipline in the preparation of the manuscripts. We thank Dr. M. Blackburn, Department of Chemistry, Sheffield, U. K., and other colleagues for the stylistic revision of the manuscript and many of our colleagues for fruitful discussions.

We acknowledge the assistance of Mr. W. Schwarzkopf in the preparation of the drawings, Mr. A. Unverricht and Mrs. J. Leipnitz in editing the manuscripts, and thank our publishing house, the VEB Deutscher Verlag der Wissenschaften, especially Dr. B. Fichte and Dr. I. Schwiebs, for close and intelligent cooperation.

Halle, March 1985

K. Mothes †
H. R. Schütte
M. Luckner

Contents

1 Historical Introduction

K. Mothes and M. Luckner

1.1 Alkaloid Chemistry

Modern alkaloid research was initiated by Friedrich Wilhelm Sertürner who was the first able to isolate an alkaloid from a crude natural product. Sertürner was a German pharmacist with a special interest in opium. In 1806 he published a paper in which he described that the "prinzipium somniferum" of opium is an "alkali" [34], which was later called morphine (*cf.* chapter 14.6), although until that time the active principles of crude drugs were thought to be acids. After more than 10 years Sertürner's discovery led to the isolation of a series of further basic substances from other physiologically active natural products:

- Narcotine (chapter 14.10) 1817 by Robiquet [28],
- Emetine (chapter 14.2) 1817 by Pelletier and Magendie [22],
- Strychnine (chapter 15.3) 1818 by Pelletier and Caventou [20],
- Veratrine (chapter 19.5) 1818 by Meißner [11],
- Brucine (chapter 15.3) 1819 by Caventou and Pelletier [2, 3],
- Piperine (chapter 12.1) 1819 by Oersted [18],
- Caffeine (chapter 17) 1820 by Runge [29, 30],
- Cinchonine and Quinine (chapter 15.3) 1820 by Pelletier and Caventou [21].

Dr. W. Meißner, a pharmacist living in Halle, created the collective term "alkaloids" for these compounds: "Überhaupt scheint es mir auch angemessen, die bis jetzt bekannten alkalischen Pflanzenstoffe nicht mit dem Namen Alkalien, sondern Alkaloide zu belegen, da sie doch in manchen Eigenschaften von den Alkalien sehr abweichen" [11]. Later it was demonstrated that the alkalinity is due to the presence of a basic nitrogen atom. During the next decades the meaning of the term "alkaloids" changed repeatedly and even today it is not clearly defined; in this book therefore we use the term in a broad sense. In addition to the most important groups of basic *N*-heterocyclic compounds (true alkaloids), we also describe biogenetically related nitrogen-containing substances, such as amines (protoalkaloids), *N*-heterocyclic non-protein amino acids and antibiotics, cyclic peptides etc., even when these substances in some cases are not really basic.

Until the middle of this century the determination of the chemical structure of isolated alkaloids proved difficult. The chemical methods used frequently resulted in the determination of partial structures only, which then had to be combined by suitable experiments. The final step in structure determination was generally the synthesis of the alkaloid in question and successful syntheses were important milestones in the history of alkaloid research.

The first alkaloid to be synthesized was coniine, in 1886 by Ladenburg (*cf.* chapter 12) [8], which had already been isolated in 1827 [6].

Hence it was more than half a century before the structure of this relatively simple alkaloid was determined. The same happened with other alkaloids. Nicotine (*cf.* chapter 13.4) discovered in 1828 [25] was not synthesized until 1904 [24], quinine (*cf.* chapter 15.3) isolated in 1820 (see above) was synthesized in 1944 [40, 41], morphine (*cf.* chapter 14.6) isolated in 1806 was synthesized in 1952 [5], and strychnine (*cf.* chapter 15.3) isolated in 1818 was synthesized in 1954 [38, 39]. Hence in many cases it took more than 100 years of intensive chemical research for the structure of alkaloids isolated at the birth of the chemistry of natural products to become known and be accepted in the field of organic chemistry.

In the middle of this century a new period of alkaloid research began and the methods of alkaloid isolation became increasingly sophisticated. New physico-chemical procedures were introduced into the chemistry of natural products resulting in a rapid decrease in the time between the isolation and structure determination of alkaloids. Since then the number of known alkaloids has greatly increased as shown by the following figures:

1900: Picted [23] reported about 100 alkaloids, whose chemical structures were all more or less unknown.

1908: Euler [4] described about 100 alkaloids including about 30 with unknown chemical structure.

1922: Wolffenstein [37] reported about 150 alkaloids at least partially known and about 100 chemically unknown or almost unknown alkaloids.

1960: Boit [1] described about 2000 alkaloids including about 600 with unknown chemical structure.

1970: Raffauf [26] listed about 4000 alkaloids.

1978: Waller and Nowacki [36] estimated the number of known alkaloids to be about 6000, 800 of which had been discovered in the two preceding decades. Today about 7000 more or less well-characterized alkaloids may be known.

1.2 Distribution of Alkaloids

Alkaloids were originally isolated exclusively from higher plants and until the middle of this century alkaloids were recorded to be typical plant products (*cf. e.g.* [19]). In the meantime, however, several alkaloids have also been found in cultures of microorganisms, *e.g.* bacteria, *Actinomycetes,* fungi and algae, and have been isolated from animals, *e.g.* many insects and isoquinolines (*cf.* chapter 14.2) from the human body. Hence, even if some of the "animal" alkaloids are absorbed with the diet (*e.g.* the pyrrolizidines found in caterpillars and imagines of certain butterflies, chapter 7), there is no doubt that certain animals are able to synthesize alkaloids *de novo.*

It is a general rule that the alkaloids with complicated chemical structures are less widespread in nature than simple compounds. Nicotine (*cf.* chapter 13.4), for instance, an alkaloid with a relatively uncomplicated structure, occurs in many different plant species, whereas strychnine, a highly complicated compound (*cf.* chapter 15.3) is restricted to some genera of *Loganiaceae.* It is of special interest, however, that some complicated types of alkaloids are found in obviously unrelated groups of organisms. For instance

— ergolines (chapter 15.4) in moulds and some higher plants
— quinazolines (chapter 16.5) in plant species and a few insects
— peptides of the maytansine type in cultures of *Actinomycetes* and in higher plants and
— benzodiazepines (chapter 16.2) in procaryotic as well as in eucaryotic microorganisms.

It is unknown whether these compounds are formed on identical pathways in the different groups of organisms. Nevertheless, it seems extremely unlikely that the sets of complicated reactions involved in their biosynthesis developed in duplicate during evolution. It may therefore be speculated that during evolution an exchange of genetic material took place between different groups of organisms, as it may occur even today between *Agrobacterium tumefaciens,* the crown gall producing organism, and its host plants [7] and that in this way genes of the alkaloid metabolism may be transferred from one organism to another [15, 16].

1.3 Biosynthesis

At the beginning of this century there were speculations on the biogenetic origin of alkaloids [35]. These were later substantiated by formation of alkaloids under "physiological" conditions, *e.g.* room temperature, nearly neutral pH [27, 31—33]. However, only the use of the isotopically labelled compounds that became available in the 1950s provided the experimental basis for a real understanding of alkaloid biosynthesis and metabolism.

At the end of the 1960s the use of isotopically labelled "tracers" made it clear that the overwhelming majority of alkaloids is derived from amino acids and some amino acid-like compounds, such as nicotinic acid and anthranilic acid (*cf.* chapters 10—18). In only a few instances it was found that alkaloid skeletons are formed from nitrogen-free precursors and the alkaloid-nitrogen is introduced at a late stage of biosynthesis (*cf.* coniine, chapter 12, and the isoprenoid alkaloids, chapter 19). Hence most alkaloids can now be grouped according to precursors and pathways of biosynthesis in "biogenetic families" that indicate their "natural" (evolutionary) relationship.

Recently, interest in alkaloid formation has been concentrated on the identification of the biosynthetic intermediates and on the mechanistic and stereochemical details of the formation and conversion of these compounds. Facilitated tracer experiments (*cf.* chapter 2) brought about a breakthrough in this field. Application of enzymology, however, is still restricted. In most experimental systems the difficulties in the isolation of biologically active enzyme preparations from alkaloid-producing cells are enormous and as yet the enzymatics of alkaloid metabolism have only been investigated in detail in a few microbial systems and cell cultures of higher plants (*cf.* the biosynthesis of benzodiazepine and quinoline alkaloids in cultures of *Penicillium cyclopium*, chapter 16.2, and of indole alkaloids in cell cultures of *Catharanthus roseus*, chapter 15.3).

1.4 Regulation of Alkaloid Biosynthesis

Alkaloid metabolism is not equally expressed in the cells of the producer organisms. In most organisms alkaloids are synthesized in certain groups of specialized cells during certain stages of their differentiation. Expression of the facilities necessary for alkaloid formation and accumulation is an aspect of cell differentiation, *i.e.* the expression of selected parts of the genetic material present in all cells of the producer (*cf.* chapter 4). Experiments with alkaloid producing microorganisms and plants were very important in the elucidation of this basic feature, which provides a molecular biological rationale for many aspects of alkaloid metabolism [9, 10].

In addition to the regulation of enzyme synthesis by differential gene expression alkaloid formation, storage and breakdown are controlled by the *in vivo* activity of the enzymes present. Alkaloid metabolism proceeds at the same time as the reactions of primary metabolism in the producer cells. Compartmentation and channelling of enzymes, precursors, intermediates and products are therefore of great significance in its regulation (*cf.* chapter 5). Already at the beginning of this century it was shown by light microscopy that alkaloids are accumulated in vacuoles and the periplasmic space of plants. Later these results were substantiated by electron microscopy and kinetic *in vivo* techniques (*cf.* chapters 5 and 6), which demonstrated the existence of complex patterns of pools and fluxes, of sources and sinks controlling the formation, accumulation and degradation of alkaloids in individual and cooperating cells. Coordination of the different fields of metabolism during differentiation and development of the producer organisms is brought about by integration in programmes of differential gene expression [17], which are also of prime importance for the regulation of alkaloid metabolism (*cf.* chapter 4).

1.5 Significance of Alkaloid Formation for the Producer Organism

Most alkaloids are formed in plants, probably because of the special mode of excretion of secondary products in this group of organisms [14, 14a]. From the animal body most excess and waste products are removed with the urine, but they are also released easily with bile and sweat and from microbial cells. In contrast, plants have a "metabolic excretion" [12, 13] into intracellular or intraorganismic storage spaces (vacuoles, cell walls, intercellulars) which allow the accumulation of high concentrations of secondary products in the plant body. This facilitates the detection of minor products as well as the further transformation of basic structures resulting in large families of related compounds. Substances with a general high toxicity, *e.g.* alkaloids with antitumour activity, which inhibit DNA and protein biosynthesis, only occur in very small concentrations. Obviously higher amounts are poisonous to the producing plant itself, although these compounds, like most other secondary products, may accumulate in non-plasmic compartments as well as in the metabolically active regions of the cells. At all we have no idea how many secondary products were formed during evolution but became extinct together with their producer organisms because of their unbearable toxicity.

The multiplicity of secondary products (*cf.* [9a]) is caused by the random processes of mutation, *i.e.* it reflects the gambling of nature rather than a sophisticated strategy. Hence the often expected significance of secondary products must be regarded with caution in many cases. Several plant alkaloids obviously deter and repel potential predators or attacking microorganisms (*cf.* chapter 7). They possess a high physiological activity for animals and human beings, which is the basis of their use in medicine. However, there are many examples to demonstrate that alkaloids toxic for most animals are tolerated or even used by others. One example is the pyrrolizidine alkaloids formed in plants, which are poisonous to most animals but are tolerated by some specialized insects that may even metabolize these substances in the production of pheromones (*cf.* chapter 7). Another example is the bulbous agaric, which contains highly poisonous peptides that cause the death of many human beings every year. Roe deer, however, may feed on this mushroom without being affected.

It would be useful to study these ecological relations under purely scientific conditions. This, however, may be impossible without a carefully investigated background of biochemistry, enzymology and genetics. It is the aim of this book to stimulate further work in these exciting fields of research with respect to the alkaloids.

1.6 References

[1] Boit, H.-H.: Ergebnisse der Alkaloidchemie bis 1960. Akademie-Verlag, Berlin 1961.
[2] Caventou, E.; Pelletier, J.: Ann. Chim. Phys. (2) **12** (1819) 113.
[3] Caventou, E.; Pelletier, J.: J. Pharm. Chim. **5** (1819) 529.
[4] Euler, H.: Grundlagen und Ergebnisse der Pflanzenchemie, part 1. Friedr. Vieweg & Sohn, Braunschweig 1908.
[5] Gates, M.; Tschudi, G.: J. Am. Chem. Soc. **74** (1952) 1109.
[6] Giseke, A. L.: Arch. Pharm. **20** (1827) 97.
[7] Krauspe, R.: In: Nover, L.; Luckner, M.; Parthier, B. (Eds.): Cell Differentiation. Gustav Fischer Verlag, Jena, and Springer-Verlag, Berlin 1982, p. 569.
[8] Ladenburg, A.: Ber. dtsch. chem. Ges. **19** (1886) 439.
[9] Luckner, M.: Pharmazie **26** (1971) 717.

[9a] Luckner, M.: Secondary Metabolism in Microorganisms, Plants and Animals. Gustav Fischer Verlag, Jena, and Springer-Verlag, Berlin 1984.
[10] Luckner, M.; Nover, L.; Böhm, H.: Secondary Metabolism and Cell Differentiation. Springer-Verlag, Berlin 1977.
[11] Meißner, W.: J. Chem. Phys. **25** (1818) 379.
[12] Mothes, K.: In: Flaschenträger, B.; Lehnartz, E. (Eds.): Physiologische Chemie, Vol. II/ 2d/β, p. 971. Springer-Verlag, Berlin 1966.
[13] Mothes, K.: Naturwissenschaften **53** (1966) 317.
[14] Mothes, K.: Abhandl. Sächs. Akad. Wiss., Math.-Naturwiss. Klasse **52** (1972) 1.
[14a] Mothes, K.: Österr. Akad. Wiss., Math.-Naturwiss. Klasse, Sb. Abt. I, Vol. **181** (1973) 1.
[15] Mothes, K.: Science and Scientists (Tokyo) **1981**, 323.
[16] Mothes, K.: Pharmazie **36** (1981) 199.
[17] Nover, L.; Luckner, M.; Parthier, B. (Eds.): Cell Differentiation. Gustav Fischer Verlag, Jena, and Springer-Verlag, Berlin 1982.
[18] Oested, H. C.: J. Chem. Phys. **29** (1819) 80.
[19] Paech, K.: Biochemie und Physiologie der sekundären Pflanzenstoffe. Springer-Verlag, Berlin 1950.
[20] Pelletier, J.; Caventou, E.: Ann. Chim. Phys. (2) **8** (1818) 323.
[21] Pelletier, J.; Caventou, E.: Ann. Chim. Phys. (2) **15** (1820) 291.
[22] Pelletier, J.; Magendie, F.: Ann. Chim. Phys. (2) **4** (1817) 172.
[23] Pictet, A.: Die Pflanzenalkaloide und ihre chemische Konstitution, 2nd ed. Springer-Verlag, Berlin 1900.
[24] Pictet, A.; Rotschy, A.: Ber. dtsch. chem. Ges. **37** (1904) 1225.
[25] Posselt, W.; Reimann, L.: Geigers Mag. Pharmac. **24** (1828) 138.
[26] Raffauf, R. F.: A Handbook of Alkaloids and Alkaloid-Containing Plants. Wiley-Interscience, New York 1970.
[27] Robinson, R.: J. Chem. Soc. (London) **111** (1917) 762.
[28] Robiquet, P. J.: Ann. Chim. Phys. (2) **5** (1817) 275.
[29] Runge, F. F.: Phytochemische Entdeckungen. Berlin 1820, p. 144.
[30] Runge, F. F.: J. Chem. Phys. **31** (1820) 308.
[31] Schöpf, C.: Angew. Chem. **50** (1937) 779, 797.
[32] Schöpf, C.: Angew. Chem. **61** (1949) 31.
[33] Schöpf, C.; Lehmann, G.: Liebigs Ann. Chem. **518** (1935) 1.
[34] Sertürner, F. W.: Trommsdorffs J. Pharmaz. **14** (1806) 47.
[35] Trier, G.: Über einfache Pflanzenbasen und ihre Beziehungen zum Aufbau der Eiweißstoffe und Lecithin. Bornträger-Verlag, Berlin 1912.
[36] Waller, G. R.; Nowacki, E. K.: Alkaloid Biology and Metabolism in Plants. Plenum Press, New York 1978.
[37] Wolffenstein, R.: Die Pflanzenalkaloide und ihre chemische Konstitution, 3rd ed. Springer-Verlag, Berlin 1922.
[38] Woodward, R. B.: Experientia, Suppl. II **1955** 213.
[39] Woodward, R. B.; Cava, M. P.; Ollis, W. D.; Hunger, A.; Daeniker, H. U.; Schenker, K.: J. Am. Chem. Soc. **76** (1954) 4749.
[40] Woodward, R. B.; von Doering, W. E.: J. Am. Chem. Soc. **66** (1944) 849.
[41] Woodward, R. B.; von Doering, W. E.: J. Am. Chem. Soc. **67** (1945) 860.

2 Methods in Biosynthetical Research

M. Luckner and H. R. Schütte

2.1 Use of Isotopically Labelled Compounds (Tracer Techniques)

To investigate natural product biosynthesis several experimental methods can be used. The most important for the investigation of metabolic pathways is the so called tracer technique [1—4, 8, 9, 11]. This involves the administration ("feeding") of isotopically labelled compounds and has greatly increased our knowledge in the field of biosynthesis. The method can be used with stable (*e.g.* ^{2}H = deuterium, ^{13}C, ^{15}N, ^{18}O) or radioactive (*e.g.* ^{3}H = tritium, ^{14}C, ^{32}P, ^{35}S) isotopes. The principle is to feed putative precursors labelled at one or more specific positions with an isotope, to isolate the natural product and, after a suitable period, to determine whether it contains any of the isotope that was in the administered precursor.

The application of the precursor can be achieved by injection, with higher plants by feeding via the roots (in hydroculture) or severed shoots, by spraying or smearing on the leaves, by vacuum infiltration, or by means of a cotton wick passing through the shoot, which sucks up the solution of the precursor into which it dips. But such precursor feeding subjects the plant to an unnatural situation in that the site of feeding and the concentration of substrates are not normal, and the plant may alter its metabolic behaviour as a consequence. Growth in an isotopically labelled "natural" environment circumvents some of these problems. For plants this is conveniently obtained by exposure to $^{14}CO_2$ [7].

At the end of the feeding experiment it should be determined to what extent the incorporation of the precursor has proceeded, *i.e.* the so-called incorporation rate should be measured, which is the relationship between the radioactivity in the compound fed and in the product formed. The results may be calculated either as absolute or as specific incorporation rates. In the absolute incorporation rate, the quantity of the precursor administered is correlated to the portion incorporated into the particular product under investigation. It is usually less suited to the assessment of biochemical investigations than is the specific incorporation rate, which measures the dilution of the precursor. It depends to a great extent on the quantity of the product synthesized during the period of the experiment and is therefore usually subject to great variation. In addition, the exact determination of the total quantity of the product formed is a prerequisite for the calculation of the absolute incorporation rate, a condition that usually cannot be satisfied.

The specific incorporation rate (*i.e.* the dilution of the precursor) permits the calculation of the portion of the compound formed that originates from the precursor fed and the portion that originates from the pool of the endogenous precursor present in the organism. Thus, in the case of a specific incorporation rate of 0.1% (or a dilution of 1:1000), one molecule in a thousand of the compound under investigation is formed from the isotopically labelled precursor. The specific incorporation rate is therefore dependent on the ratio of incorporation of endogenous to administered precursor molecules. It is influenced by the absolute rate of synthesis during the interval of the experiment only as far as this ratio is altered.

The specific incorporation rate permits conclusions to be drawn regarding the relationship between the product formed and the precursor fed, since it is usually higher if the added precursor can be changed to the product in one, or a few, steps. The size of the incorporation rate must not, however, be over-rated, since the size of the endogenous pools of precursors and of intermediates plus the limits of permeability for the precursor entering the cell from outside play a considerable role. The specific incorporation rates, preferably in experiments with higher plants are usually small (frequently of the order of 0.01%) because in the organisms are rather big pools of unlabelled precursors and products that dilute the labelled compound fed. Furthermore the precursor may be used in other competing reactions during the necessary transport to the sites of biosynthesis. In microorganisms the specific incorporation rates may be much higher because the presence of an unlabelled product at the beginning of the experiment may easily be avoided and precursor transport does not play a significant rôle. The experimental difficulties faced in biochemical experiments on higher plants and animals may be reduced by the use of cell cultures, which may be fed in the same way as cultures of microorganisms.

In any case it has to be found out whether the formation of the substance under investigation actually takes place during the experimental period. This can easily be detected in control experiments by feeding [^{14}C] glucose, $^{14}CO_2$ or similar substances that enter primary metabolism easily.

If labelled alkaloids are obtained after feeding an isotopically labelled precursor, it is usually necessary to determine the location of the isotope to substantiate a preconceived biosynthetic scheme. If there is a direct biogenetical relationship, *i.e.* if the administered compound is a direct precursor of the product built, then in the alkaloid only those atoms will be labelled that directly correspond to the labelled positions of the precursor. If, however, the substance fed first enters the pathways of general metabolism and the alkaloid is then synthesized from its conversion products, smearing (randomization) of labelling is observed which will be the greater the longer the time between feeding and extracting the alkaloid. Hence biosynthetic experiments with isotopically labelled compounds in which the pertinent isotope distribution is not localized do not usually stand up to rigorous criticism.

Quite sophisticated degradations are often necessary to determine the location of the isotopes in the ultimate natural products [10]; many examples in the field of alkaloid biosynthesis are cited in this book. It is possible to avoid such degradations by using the stable isotope carbon-13 as tracer [5, 6]. Its location in a natural product can be determined by nuclear magnetic resonance. The method is dependent on the low natural abundance of ^{13}C (1.11%). If a ^{13}C-labelled precursor is incorporated into an alkaloid, the enriched carbon in this compound will have an enhanced signal in the ^{13}C-NMR spectrum. However, this method only works when the specific incorporation of the administered compound leads to an enhancement of the NMR signal of at least 30 to 40%. If, for example, a natural product was found to have 1.6% ^{13}C at a certain carbon atom and was derived from a precursor enriched 90% at the analogous carbon, the specific incorporation would be 0.55% and the NMR signal for this carbon would be enhanced by 45%. The ^{13}C method has been used extensively in the study of microbial natural products since one can usually expect high specific incorporation in such a system (see above). Greater dilution of the administered precursor can be tolerated if two contiguous ^{13}C atoms are labelled. The natural abundance of such contiguous carbons is only $1.11 \times 1.11\% = 0.0123\%$. Contiguous ^{13}C atoms give rise to satellite peaks (due to spin-spin coupling) in the ^{13}C-NMR, located near the central singlet peaks that arise from naturally abundant ^{13}C.

For certain investigations multiple labelling of the precursors is necessary. Either different atoms of the same element or different elements participating in the structure may be labelled with isotopes. For example, if one has to determine whether the whole molecule of an amino acid including the amino group is incorporated into an alkaloid, it is necessary to label the nitrogen of the amino group as well as a carbon atom as an internal standard. In the case of direct incorporation, the ratio of label in both these atom species in the compound fed and the compounds isolated must be the same. Multiple labelling often also reduces the need to determine the isotope distribution.

To determine biosynthetic pathways it is necessary to test the incorporation of possible intermediates. If these are converted into the compound under investigation, it may be assumed that they also play a part in normal metabolism. It has been shown in certain cases, however, that "unusual" molecules may also be incorporated into secondary products. This is due, *inter alia,* to the inadequate substrate specificity of the enzymes involved or, more seldom, to the induction of additional enzymes after exposure of the cells to a high level of an "unnatural substrate".

Further information about possible intermediates may be obtained by "competition experiments" or "trapping experiments". In the first case the specific incorporation rate of a labelled precursor is determined with and without administration of greater amounts of an unlabelled supposed intermediate. The unlabelled intermediate incorporated into the product will reduce the specific incorporation rate of the labelled precursor. If the unlabelled compound is not an intermediate, then the specific incorporation rate will remain unchanged. In "trapping" experiments the labelled precursor is fed together with an excess of the unlabelled intermediate. If it is a real intermediate then it must become labelled by synthesis from the labelled precursor feeding into the pool of the unlabelled compound. This can be proved by extraction and purification of the given carrier material. Competition and trapping experiments enable the study of hypothetical intermediates even if they are not available in an isotopically labelled form. The experiments fail, however, if intermediates of biosynthetic chains are strictly channelled and do not mix with pools of compounds administered from outside to the cell or to the site of biosynthesis (*cf.* chapter 5).

2.2 Investigations with Mutants

The use of mutants stems from the hypothesis that each structural gene is responsible for a specific step of a biosynthetic pathway. The blocking of a structural gene encoding a secondary metabolic enzyme depresses synthesis of the product and frequently causes accumulation of the substrate of this enzyme. Since in most metabolic pathways the starting material is converted into the end product through intermediary stages, the reaction chain may be blocked by mutation at different places. The number and location of blocks within the metabolic chain can be determined by "co-synthesis", *i.e.* normalization of product formation if mutants with different blocks are grown together. This method gives an idea of the number of structural genes and enzymes involved in a metabolic chain.

By feeding compounds that "normalize" synthesis of secondary products in the blocked mutants and by isolation of the substances accumulated in the particular mutants, it is possible to elucidate the metabolic pathway in question. Difficulties may arise, however, because the intermediates themselves may fail to accumulate, whereas more or less modified derivatives may do so, and also because the accumulated compounds may be transformed by pathways that are not observed in the wild type strain. Biosynthetic research by use of mutants is of special importance with microorganisms.

2.3 References

[1] Brown, S. A.; Fetter, L. R.: Progr. Phytochem. **3** (1972) 1.
[2] Evans, E. A.: Tritium and its Compounds. Butterworth, London 1974.
[3] Evans, E. A.; Muramatsu, M. (Eds.): Radiotracer Techniques and Application. Marcel Dekker, New York/Basel 1977.
[4] Floss, H. G.: Lloydia **35** (1972) 399.
[5] Grutzner, J. B.: Lloydia **35** (1972) 375.
[6] Leete, E.: Rev. Latinoamer. Quim. **11** (1980) 8.
[7] Parker, H. I.; Blaschke, G.; Rapoport, H.: J. Am. Chem. Soc. **94** (1972) 1276.
[8] Schütte, H. R.: Radioaktive Isotope in der organischen Chemie und Biochemie. Deutscher Verlag der Wissenschaften, Berlin 1966.
[9] Schütte, H. R.: Isotopenpraxis **16** (1980) 313; **20** (1984) 8.
[10] Simon, H.; Floss, H. G.: Bestimmung der Isotopenverteilung in markierten Verbindungen. Springer-Verlag, Berlin 1967.
[11] Simon, H. (Ed.): Messung von radioaktiven und stabilen Isotopen. Springer-Verlag, Berlin 1974.

3 The Biochemical Genetics of Alkaloids

H. Böhm

In the first version of this chapter [7] a comprehensive review was given of the "Genetik des Alkaloidmerkmals", arranged according to areas of experimental genetic research (*e.g.* intraspecific and interspecific crosses, different types of mutations). The subject is now discussed with reference to certain biochemical phenomena. Therefore only relevant data are taken from the previous paper and results published during the last 15 years have been added.

Higher plants have remained the most common objects of experiments concerning the genetic aspects of alkaloid metabolism. Nearly all results are still based on cross-pollination of different individuals, while work on molecular genetics in this field is rare. Experiments with plant cell cultures have recently contributed new data to the discussion on the biochemical genetics of alkaloids.

In 1978 Waller and Nowacki [109] reviewed the "Genetic control of alkaloid production". The research activities of the second author have made this article a source of information on the genetics of lupine alkaloids. In the same year in a review of ergot alkaloid formation, genetic aspects were given particular consideration [98].

3.1 The Genetic Control of Basic Alkaloid Formation

Alkaloid-containing organisms essentially differ from organisms with no alkaloid characteristics [64] by their ability to transform primary metabolites, especially amino acids, into alkaloids. According to an earlier proposal [7] the first alkaloid molecule occurring in the course of a transformation sequence is a basic alkaloid. Its presence alone indicates the alkaloid characteristic of an organism. Moreover, a basic alkaloid determines the skeleton and the amount of alkaloids derived from it.

From this point of view it is reasonable to assume the formation of basic alkaloids under strict genetic control. However, alkaloid-free mutants, which support this assumption, have not been reported even in recent years. Therefore, one should consider whether mutants with very low alkaloid concentrations are significant to this problem.

Plants poor in alkaloids of *Lupinus luteus*, *L. angustifolius*, and *L. albus* possess a maximum alkaloid content of 0.1% of seed dry weight. Their cross-pollinations with alkaloid-rich plants of the same species (about 1.0 to 2.0% alkaloid concentration in dry seeds) result in F_1 progenies rich in alkaloid. Among the F_2 descendants the ratio of bitter and sweet plants is about 3:1. This segregation pattern indicates that the low alkaloid concentration is caused by the recessive mutation in one gene [25–27, 29, 60, 105]. – Because the alkaloid level of seeds corresponds to that of the mother plant, a discrepancy can exist between the genotype and phenotype of lupine seedlings with regard to the alkaloid character. This is the case if a bitter descendant arises from a sweet mother plant and *vice versa* [27, 66, 105, 110a].

Systematic cross-pollinations of sweet lupines revealed the existence of various independent genes responsible for low alkaloid levels within the species investigated (*cf.* 3.4.2) [22–25, 27–29, 48, 58, 61, 75, 105]. Besides the spontaneous mutants selected from lupine populations one new mutation leading to trace amounts of alkaloids could experimentally be induced [114]. Each changed gene causes a characteristic alkaloid residue [21, 26, 59, 75, 94]. A variety of *L. albus* containing two of these genes shows a lower alkaloid concentration than plant material whose genome has only one of the two factors [75].

As demonstrated by qualitative analysis, the alkaloid spectrum of a mutant poor in alkaloid is identical to that in the wild lupine. Even the quantitative relations between the individual components are similar in both genotypes [30a, 30b, 93, 94].

Mutants isolated from the *Papaver somniferum* variety "Indra" generally contain 0.02 to 0.03%, at most 0.10%, morphine in dry mature capsules. Because other alkaloids are detectable only in trace amounts, their morphine concentrations are practically identical with the total alkaloid contents. From cross-pollinations between plants poor in morphine and normal *P. somniferum* types containing about 0.40% morphine in the capsule dry weight, F_1 hybrids were derived which possess an average morphine content of 0.46% in their dry mature capsules. On average F_2 progenies comprise 17.4% and never more than 25% of the descendants poor in morphine. This finding is rated as a 3:1 segregation. Consequently, the low morphine level is caused by one recessive gene [70, 72–74]. After the cross between mutants poor in morphine of *P. somniferum* and *P. setigerum* (*cf.* 3.4.2) F_3 descendants occur which are free from morphine. They probably accumulate two different recessive genes responsible for low morphine concentration [70]. Nyman was unsuccessful in the attempt to combine a low morphine level and a relatively high concentration of papaverine or narcotine in the same individual by crossing the mutant poor in morphine with suitable *P. somniferum* genotypes [70]. The biosynthetic pathways leading to papaverine, narcotine and morphine, respectively, branch out directly or only some steps after norlaudanosoline, the basic alkaloid of benzylisoquinoline derivatives (*cf.* chapter 14). The cited work indicates that the low level of morphine in *P. somniferum* mutants is not due to the inhibition of a specific step of morphine biosynthesis. It is caused by the restricted formation of an alkaloid from which various biosynthetic pathways start.

The experiments performed with alkaloid-poor mutants of *Lupinus* and *Papaver* species show the participation of genes in the biosynthesis of basic alkaloids. Since the mutation of one of these factors produced a certain alkaloid residue, they were predominantly discussed as genes that only quantitatively influence biosynthetic reactions (*cf.* [7]). The present knowledge on the mechanism of genetic control in higher organisms supports the idea that in alkaloid-poor plants the mutation of a gene directly, or qualitatively, responsible for a certain step leads to an incomplete interruption of the formation of the respective basic alkaloid [67, 70].

Recently enzymes catalysing final reactions of basic alkaloid formations were characterized in cell cultures derived from various plants: strictosidine synthase in *Catharanthus roseus* and its relatives [103, 104], 17-oxosparteine synthase in *Lupinus polyphyllus* [112, 113], and norlaudanosoline synthase in different species [83]. There are certainly correlations between the last two enzymes and individual genes changed in *Lupinus* and *Papaver* mutants. Their investigation should produce remarkable progress in the field of the biochemical genetics of alkaloids.

3.2 Genes Responsible for the Transformation of Alkaloids

Most alkaloids occurring in living systems are formed by the conversion of other alkaloids. Some of the reactions involved in these processes were genetically analysed or at least characterized.

3.2.1 Methylation

Dehydrogenation and the complete *O*-methylation of norlaudanosoline result in the papaverine molecule. From reciprocal crosses between *Papaver somniferum* types with and without papaverine, F_1 descendants were obtained that always contained this alkaloid. However, its quantity is generally lower in hybrids than in the papaverine-producing parent [74].

The inheritance of gramine and other alkaloid-like substances showing different methylation levels was studied in canary grass, *Phalaris arundinacea*. The genetic analysis of plants derived from self- and cross pollination of the different genotypes support the following model. One gene, T, is responsible for the *N*-methylation, a second gene, M, controls the *O*-methylation. M is epistatic to T. Gramine can only occur if both genes are represented by recessive alleles [54].

3.2.2 Oxidation

Sparteine, a quinolizidine compound without oxygen function, is the main alkaloid in *Lupinus arboreus* and related species. In contrast, lupanine and hydroxylupanine containing one and two oxygen atoms, respectively, predominate in *L. hartwegii* and its relatives. The latter substances are the main alkaloids in all the F_1 plants obtained from crosses between representatives of both groups of *Lupinus* species [34, 66, 68]. From the segregation patterns found in F_2 progenies it was concluded that the conversion of sparteine into lupanine is controlled by one gene. Intensive conversion is dominant over low conversion activity. A similar condition has been assumed for the step from lupanine to hydroxylupanine. These genetic data agree with the result of tracer experiments on the biosynthesis of hydroxylupanine [66, 68].

The quantitative relation of hyoscyamine and scopolamine in mature plants of *Datura stramonium* differs from that in *D. ferox*. Whereas scopolamine shows a lower concentration than hyoscyamine in the former it predominates in the latter. This situation results from a long term conversion of hyoscyamine to scopolamine in *D. ferox* plants in contrast to a relatively short period of transformation characteristic of *D. stramonium*. The analyses of F_1 to F_4 progenies derived from reciprocal interspecific crosses revealed a monohybrid, dominant inheritance of the high portion of scopolamine in the hyoscyamine/scopolamine amount [76, 77]. This finding indicates that at least one of the reaction steps involved in the epoxidation of hyoscyamine is genetically governed.

3.2.3 Demethylation

All the F_1 plants obtained from crosses between nicotine-rich and nicotine-poor *Nicotiana* species or *N. tabacum* varieties are poor in nicotine [3, 12, 18, 19, 37, 41, 49, 52, 90, 91, 96, 97, 110]. The nicotine concentration essentially depends on the extent of nicotine "degradation", *i.e.* the intensity of the transformation of nicotine into nornicotine [4, 12, 19, 38, 51, 53, 108]. *Nicotiana* plants with a high nicotine level contain small amounts of nornicotine and *vice versa*. Consequently, the F_1 character described above indicates the dominance of intensive overweak alkaloid conversion. The number of genes that participate in the nicotine demethylation has varied, concluded from the segregation patterns of F_2 to F_4 progenies of intraspecific *N. tabacum* crosses. Obviously, there are at least two genes (*cf.* [7]).

According to Koelle's detailed studies in this field up to four genes — A-factors — responsible for the nicotine demethylation can be accumulated in *N. tabacum* plants. With respect to the quality of their function they are independent of each other [39—43]. The quantitative effect of A-factors on the conversion of nicotine to nornicotine is not always additive, as believed earlier [45], but it can decrease with the increasing number of these genes in a plant genome [46].

The biosynthesis of morphine is known only in *Papaver somniferum* and in the probable progenitor of this species, *P. setigerum* [30]. Its final stage comprises two *O*-demethylations (*cf.* chapter 14). Thebaine is the first molecule in these reactions, and among the intermediates codeine accumulates more or less. Morphine is normally the main alkaloid of *P. somniferum*. However, a genotype was recently selected from this species in which thebaine occupies the position of morphine [69, 71]. Since the concentration of codeine or codeinone is obviously not increased, a gene responsible for the demethylation of thebaine should be changed by mutation.

Surprisingly, the progenies obtained by self-pollination of the *P. somniferum* mutant and its descendants actually comprise increasing portions of plants showing thebaine as the main alkaloid, but they never uniformly consist of these genotypes. Perhaps the realization of the new alkaloid character in *P. somniferum* is drastically influenced by environmental factors [71].

P. bracteatum forms large amounts of thebaine but no morphine. Traces of codeine were detected by chemical analysis [47, 57], but the biochemical relevance of these findings is not clear [9]. After interspecific crosses between *P. somniferum* and *P. bracteatum*, morphine is present as the main alkaloid in each F_1 plant [5, 10]. This result proves at least the genetic control of the transformation of codeine into morphine if not that of both *O*-demethylations involved in the reactions leading from thebaine to morphine. Also the F_1 plants obtained from crosses between *P. somniferum* and other thebaine-containing *Papaver* species that are free from morphine, *e.g. P. orientale* and *P. pseudo-orientale* exhibit morphine throughout [10].

Plants of the species *P. orientale* are able to demethylate thebaine at the aromatic ring. In this way the alkaloid oripavine is formed. All the F_1 descendants of the interspecific cross between *P. orientale* and *P. bracteatum,* in which oripavine is not detectable, contain this alkaloid. Obviously, the transformation of thebaine into oripavine is also genetically controlled [10].

3.3 Alkaloid Quantity as a Genetic Character

Because many alkaloids are of practical significance the quantitative aspects of their occurrence have been frequently investigated. Unfortunately, such experiments often contribute only to the economic improvement of the objects and not to the analysis of the genetic background of quantitative phenomena.

As discussed in section 3.1, the total amount of alkaloids in an organism cannot be higher than the quantity of the basic alkaloid(s) formed. In other words, the amount is determined by the number of molecules that change from primary metabolism to alkaloid biosynthesis as part of secondary metabolism. Certainly, the final amount of alkaloid depends not only on synthetic reactions but also on degradation processes. However, genes responsible for the degradation of alkaloids, which result in structures without alkaloidal properties, are not known.

3.3.1 Wild Type Genes and their Combination

Within natural populations of an alkaloid bearing organism one generally finds a broad range of alkaloid concentrations. The progenies of selected individuals with low or high alkaloid concentrations normally show decreased or increased mean values of alkaloid contents compared with the respective date of the start population. Using this effect the alkaloid level in various organisms can be remarkably improved by repeated positive selection (*cf.* [7]).

The progenies of diallel crosses between several *Papaver somniferum* varieties were analysed with respect to the hereditary variability of the morphine content in mature capsules. According to the data obtained, additive and dominant components play an important role. Epistasis is of significance only in some cases. Dominant genes participating in the control of morphine quantity can accumulate in certain genotypes, among the varieties investigated were "Hanácký modry" and "Nordster". In the first variety they have a negative, in the second a positive effect [31, 32].

The experiments described before reveal a slight hybrid vigour [31], in agreement with the result of other crosses between *P. somniferum* varieties: The average morphine concentration of an F_1 progeny can be higher than that of either parent [14, 15, 63, 87]. Findings that indicate heterosis with respect to alkaloid concentration were also reported from crosses within and between solanaceous species: *Atropa belladonna* [85, 86], *Datura stramonium* [13, 102], *D. aurea*, *D. candida* [17], *Scopolia tangutica*, *S. stramonifolia* [62], *Duboisia myoporoides*, *D. leichhardtii* [33]. A similar phenomenon was observed in experiments with *Claviceps purpurea* [106]. Cross pollinations between three lines of *Catharanthus roseus* did not result in an increased ajmalicine content of F_1 plants [49a]. — After the somatic hybridization of *Datura stramonium* and *D. innoxia* not only the first progeny of fertile hybrids (= *D. straubii*) shows a scopolamine concentration higher than that of either parent [88a].

Whether or not progenies derived from reciprocal crosses differ in their alkaloid concentrations is not uniformly answered in the literature. Whereas maternal inheritance could be excluded in experiments with *Datura* [100], tobacco [12], and *Papaver* species [10, 32], it had to be included from the result of crosses between genotypes of *Nicotiana tabacum* [44], *Lupinus* spec. [2], and species of the genus *Duboisia* [33].

A further quantitative phenomenon shall be first characterized by a model: The alkaloid D is an intermediate of the biosynthesis of E. Its concentration in a given species surpasses that of E in another species. Under these conditions it is possible that the interspecific hybrid possesses an increased E concentration compared with that of the E producing or accumulating parent. Examples of this "genetic feeding experiment" are the crosses between *Datura stramonium* and *D. ferox* [76—78] (*cf.* 3.2.2) as well as between *Papaver bracteatum* and *P. somniferum* [5, 10] (*cf.* 3.2.3). Although *D. ferox* transforms hyoscyamine into scopolamine to a relatively high extent, the level of the latter alkaloid is low (about 0.15% of leaf dry weight). On the other hand, the hyoscyamine concentration is about 0.30% in dry *D. stramonium* leaves. The F_1 plants derived from the interspecific cross reach an average scopolamine content of 0.28% in leaf dry weight [76]. Representatives of the species *Papaver bracteatum* can accumulate thebaine up to about 4% dry weight (dw) in their capsules and about 1%/dw in their roots (*cf.* [9]). This alkaloid is at least in substantial amount, not transformed into another alkaloid by the perennial poppy. The capsules of normal *P. somniferum* types contain up to 1%/dw morphine besides traces of the intermediate thebaine. From crosses between *P. bracteatum* and *P. somniferum* F_1 descendants were obtained which show up to 2.5%/dw morphine in their mature capsules [5, 10].

Unfortunately, these *Papaver* hybrids are sterile as a result of diverse chromosome numbers of their parents. Therefore, they cannot be directly used as crop plants. This complication may frequently confine not only the theoretical value but also the practical significance of interspecific crosses that correspond to a "genetic feeding experiment". Moreover, although the effect of such crosses is generally detectable it can be very low and does not coincide with the concentration of the intermediate alkaloid symbolized by D in our model. This possibility is exemplified by the cross between *P. bracteatum* and plants of *P. orientale* that contain about 0.3% oripavine in their mature capsules. In the same organs of F_1 hybrids the maximum concentration of oripavine is 0.5%/dw. Thebaine occurs only in trace amounts [10]. These findings show that the quantity of a certain alkaloid from one of the parents of an interspecific cross is not necessarily available under the genetic condition of the respective hybrid.

3.3.2 The Alkaloid Level of Mutants

Plants poor in alkaloids, selected from *Lupinus* species and from *Papaver somniferum*, were described in section 3.1 because their features are probably caused by the mutation of genes responsible for the quality and not — or not only — for the quantity of biosynthetic processes. All reaction steps affected by the mutation are involved in the biosynthesis of basic alkaloids (*cf.* 3.1).

Interesting arguments for a strict quantitative control of the introduction of primary metabolites into the alkaloid metabolism were recently obtained from experiments with plant cell cultures [88, 95]. After the influence of mutagenetic agents on cultivated *Catharanthus roseus* cells several cell culture strains could be selected which are resistant against 5-methyltryptophan. These variants can show a 30-to-40-fold increase of the tryptophan concentration compared with the control. Besides an isoprenoid compound, tryptophan is the characteristic component of *Catharanthus* alkaloids (*cf.* chapter 15). Nevertheless, the alkaloid levels of the *C. roseus* cell culture strains overproducing this amino acid remained unchanged.

The amphidiploid species *Nicotiana tabacum* was probably originally poor in nicotine. In other words, it very intensively transformed nicotine into nornicotine (*cf.* 3.2.3). High concentrations of nicotine in *N. tabacum* became possible only after the mutation of the gene responsible for nicotine demethylation. Low nicotine levels in populations and varieties of *N. tabacum* are explained by the function of wild genes or by the back mutation of genes that had lost the ability to initiate intensive nicotine transformation [41, 53].

The genome of a diploid organism consists of pairs of identical chromosomes that represent two chromosome sets. One additional chromosome belonging to a certain chromosome pair indicates a simple genome mutation designated as trisomy. The more chromosome pairs present, the more trisomic genotypes are possible. The alkaloids of different trisomic ($2n + 1$) *Datura stramonium* plants have repeatedly been analysed. According to these experiments the quantities of individual alkaloids [99], as well as the total alkaloid amount [55], can be influenced by the presence of a certain additional chromosome.

Polyploid organisms are characterized by the possession of more than two chromosome sets. Whether or not the biosynthetically active cells of a polyploid plant produce a higher alkaloid amount than those of a diploid plant is a question of general biological relevance. However, the answer is difficult. As shown by a comparison of diploid and tetraploid *Datura tatula*, polyploidization can lead to a changed quantitative relation of plant parts that synthesize and accumulate alkaloids, respectively. Therefore, the increased alkaloid concentrations found in leaves of several tetraploid *D. tatula* genotypes must not be caused by an increased alkaloid production of the respective roots. The analyses of whole tetraploid plants gave no evidence of a direct influence of polyploidy on alkaloid formation [81, 82]. These results were confirmed by most related experiments (*cf.* [7]). Nevertheless, it is possible to increase by polyploidization the alkaloid concentration of plant parts used as crude drug. The leaves of haploid plants from different *Datura* species have lower alkaloid concentrations than those of comparable diploid plants [35, 56].

3.4 New Qualitative Alkaloid Characters of Hybrids

3.4.1 Compromise Structures

As postulated many years ago [1], hybrids derived from the cross-pollination of plants possessing different alkaloid spectra regularly contain the bases of both parents. Among possible irregularities the occurrence of hybrid alkaloids unknown in the parents, even as natural compounds, is of special interest. In some cases the genetic background of the formation of hybrid specific alkaloids was elucidated. The leaves of a *Solanum dulcamara* chemovariety contain the aglyca solasodine, (*25R, 22R*)-*N*-spirosol-5-en-3β-ol, and soladulcidine, (*25R, 22R*)-*N*-5α-spirosolane-3β-ol, those of another the aglycone tomatidenol, (*25S, 22S*)-*N*-spirosol-5-en-3β-ol, alone. F_1 plants obtained by the cross-pollination of both chemovarieties form tomatidine, (*25S, 22S*)-*N*-5-spirosolane-3β-ol, besides the parental aglyca. This alkaloid had not been isolated from the species *S. dulcamara* before. It obviously results from the co-operation of independent genes responsible for the situation at C-5 (double bond or H) and C-25 (*R* or *S* configuration) [80]. Similar processes may lead to new withanolides. These steroid alkaloids have been identified after their isolation from hybrids obtained by reciprocal crosses between chemovarieties of *Withania somnifera*. In each case, the withanolides of one parent are characterized by different levels of ring saturation and those of the other parent by oxygen functions at certain skeleton positions [16, 36, 65].

3.4.2 Genetic Complementation

The cross pollination of two sweet lupine mutants — S_1 and S_2 — can result in bitter F_1 plants. In the F_2 progeny a ratio of 9 bitter to 7 sweet individuals then occurs [25, 27, 29, 58, 67, 75, 107]. This result indicates that the gene whose mutation causes the lack of alkaloids in the lupine S_1 is different from the factor changed in the S_2 genotype. The first gene may be designated as s_1, the second as s_2. In the hybrid genome one unchanged allele of s_1 is present — contributed by the S_2 parent — as well as one of s_2 — provided by the S_1 mutant. Consequently, active alleles of all genes responsible for alkaloid biosynthesis are combined in hybrid descendants of the cross-pollination between lupines such as S_1 and S_2. Genetic complementation has also been reported from the cross-breeding of *Papaver somniferum* and *P. setigerum* types that are very poor in morphine [70, 74]. In other fields of plant metabolism it recently became highly significant because it can indicate successful somatic hybridization [89].

A certain type of feeding experiment is regarded as the biochemical counterpart of genetic complementation. If a mutant is fed with one of the intermediates located after the step blocked in a biosynthetic pathway, the final product of this sequence can occur. From the biochemical genetics of alkaloids at least two positive examples are known: the complementation of alpinigenine biosynthesis in e^h-types of *Papaver bracteatum* [8] and that of ergotoxin biosynthesis in a strain of *Claviceps purpurea* [50].

3.5 The Inheritance of Temporal and Spatial Patterns

In the species *Lupinus angustifolius* a genotype exists whose alkaloid content considerably increases during flowering after being very low in earlier stages of development. A modifier gene is assumed to be responsible for this peculiarity [20, 94]. The e^h-type of *Papaver bracteatum* forms alpinigenine only up to about the third month of its life. If it is crossed with other *P. bracteatum* plants that either always or never contain alpigenine (e^+ or e^- types), the limited alpinigenine biosynthesis shows a monohybrid segregation [6, 11]. The long term conversion of hyoscyamine into scopolamine reported from *Datura ferox* (*cf.* 3.2.2) only takes place in parts of the plant above ground. Consequently, *D. ferox* differs from other *Datura* species not only in a temporal but also in a spatial aspect of hyoscyamine conversion. These characteristics are hereditary [76, 77]. Mutation experiments with *D. stramonium* resulted in a plant whose leaves were free from alkaloids. The genotype was considered to be an alkaloid-free mutant [101]. However, as long as its root is not analysed, one must assume that the alkaloid transport is blocked in this individual. *Solanum dulcamara* chemovarieties do not differ in the alkaloid spectra of whole plants but in the alkaloids transported from the roots, with more or less identical alkaloid compositions into the leaves [79, 84, 92]. The selective alkaloid transport of the species *S. dulcamara* is a genetic character [111]. Whereas the root of *Papaver somniferum* is poor in alkaloid, that of *P. bracteatum* is rich in alkaloid. All the hybrids derived from cross breeding between *P. somniferum* and *P. bracteatum* (*cf.* 3.3.1) show high alkaloid concentrations in their roots. Obviously, the transport of thebaine from leaves into the root [9] is genetically controlled in *P. bracteatum* [5, 10].

3.6 References

[1] Alston, R. E.; Turner, B. L.: Biochemical Systematics. Prentice-Hall, Englewood Cliffs, New York 1963.
[2] Anokhina, V. S.; Kuptsova, A. G.; Belanovskaya, O. F.: Genetika **17** (1981) 1266.
[3] Arghyroudis, D.: Proc. 2nd Intern. Sci. Tobacco Congr. 1958, 364.
[4] Blaim, K.: Roczniki Nauk rolniczych, Ser. A, **88** (1964) 917.
[5] Böhm, H.: Planta Med. **13** (1965) 234.
[6] Böhm, H.: Planta Med. **15** (1967) 215.
[7] Böhm, H.: In Mothes, K.; Schütte, H. R. (Eds.): Biosynthese der Alkaloide. Deutscher Verlag der Wissenschaften, Berlin 1969, p. 21.
[8] Böhm, H.: Bioch. Physiol. Pflanzen **162** (1971) 474.
[9] Böhm, H.: Pharmazie **36** (1981) 660.
[10] Böhm, H.: Planta Med. **48** (1983) 193.
[11] Böhm, H.: Z. Naturforsch. subm.
[12] Burk, L. G.; Jeffrey, R. N.: Tobacco **147**, No. 25 (1958) 20.
[13] Carlsson, G.: Medd. Gullåkers Växtförädlings Anst. 1959, 169.
[14] Dános, B.: Abh. Dtsch. Akad. Wiss. Berlin, Klasse Chem., Geol., Biol. **1966**, No. 3, p. 363.
[15] Deneva, T.: Genet. Sel. **8** (1975) 419.
[16] Eastwood, F. W.; Kirson, I.; Lavie, D.; Abraham, A.: Phytochem. **19** (1980) 1503.
[17] El-Dabbas, S. W.; Evans, W. C.: Planta Med. **44** (1982) 184.
[18] Gerstel, D. U.; Mann, T. J.: Crop Sci. **4** (1964) 387.
[19] Griffith, R. B.; Valleau, W. D.; Stokes, G. W.: Science **121** (1955) 343.
[20] Hackbarth, J.: Saatgutwirtschaft **6** (1954) 320.
[21] Hackbarth, J.: Abh. Dtsch. Akad. Wiss. Berlin, Klasse Chem., Geol., Biol. **1956**, No. 7, 58.
[22] Hackbarth, J.: Z. Pflanzenzücht. **37** (1957) 1.
[23] Hackbarth, J.: Z. Pflanzenzücht. **37** (1957) 81.
[24] Hackbarth, J.: Z. Pflanzenzücht. **37** (1957) 185.
[25] Hackbarth, J.: Z. Pflanzenzücht. **45** (1961) 334.
[26] Hackbarth, J.: Genetica agraria **15** (1962) 357.
[27] Hackbarth, J.; Sengbusch, R. von: Züchter, **6** (1934) 249.
[28] Hackbarth, J.; Troll, H.-J.: Züchter **13** (1941) 63.
[29] Hackbarth, J.; Troll, H.-J.: Z. Pflanzenzücht. **34** (1955) 409.
[30] Hammer, K.; Fritsch, R.: Kulturpfl. **25** (1977) 113.
[30a] Harrison, J. E. M.; Williams, W.: Euphytica **31** (1982) 357.
[30b] Harrison, J. E. M.; Williams, W.: Z. Pflanzenzücht. **90** (1983) 32.
[31] Hlaváčková, Z.: Genet. šlechtěni **11** (1975) 113.
[32] Hlaváčková, Z.: Genet. šlechtěni **15** (1979) 217.
[33] Ikenaga, T.; Abe, M.; Itakura, A.; Ohashi, H.: Planta Med. **35** (1979) 51.
[34] Kazimierski, T.; Nowacki, E.: Genetica polon. **2** (1961) 93.
[35] Kibler, R.; Neumann, K. H.: Planta Med. **33** (1978) 289.
[36] Kirson, I.; Abraham, A.; Lavie, D.: Israel J. Chem. **16** (1977) 20.
[37] Koelle, G.: Tabak-Forsch. 1955, No. 14, 13.

[38] Koelle, G.: Tabak-Forsch. 1957, No. 19, 46.
[39] Koelle, G.: Tabak-Forsch. 1957, No. 22, 65.
[40] Koelle, G.: Tabak-Forsch. 1960, Coll. Vol. 3, 125.
[41] Koelle, G.: Züchter **31** (1961) 346.
[42] Koelle, G.: Züchter **35** (1965) 222.
[43] Koelle, G.: Z. Pflanzenzücht. **55** (1966) 375.
[44] Koelle, G.: Biol. Zbl. **86** (1967) 745.
[45] Koelle, G.: Z. Pflanzenzücht. **68** (1972) 281.
[46] Koelle, G.: Z. Pflanzenzücht. **75** (1975) 71.
[47] Küppers, F. J. E. M.; Salemink, C. A.; Bastart, M.; Paris, M.: Phytochem. **15** (1976) 444.
[48] Lamberts, H.: Thesis, Wageningen 1955.
[49] Lašuk, G. I.: Izvest. Akad. Nauk SSSR **70** (1950) 265.
[49a] Levy, A.; Milo, J.; Ashri, A.; Palevitch, D.: Euphytica **32** (1983) 557.
[50] Maier, W.; Erge, D.; Gröger, D.: Planta Med. **40** (1980) 104.
[51] Mann, T. J.; Weybrew, J. A.: Tobacco **146**, No. 12 (1958) 20.
[52] Mann, T. J.; Weybrew, J. A.: Tobacco **147**, No. 13 (1958) 20.
[53] Mann, T. J.; Weybrew, J. A.; Matzinger, D. F.; Hall, J. L.: Crop Sci. **4** (1964) 349.
[54] Marum, P.; Hovin, A. W.; Marten, G. C.: Crop Sci. **19** (1979) 539.
[55] Mechler, E.; Haun, N.: Planta Med. **42** (1981) 102.
[56] Mechler, E.; Kohlenbach, H. W.: Planta Med. **33** (1978) 294.
[57] Meshulam, H.; Lavie, D.: Phytochem. **19** (1980) 2633.
[58] Mikołajczyk, J.: Genetica Polon. **2** (1961) 19.
[59] Mikołajczyk, J.: Postepy Nauk rolniczych **13** (1961) 3.
[60] Mikołajczyk, J.: Genetica Polon. **7** (1966) 181.
[61] Mikołajczyk, J.; Nowacki, E.: Genetica Polon. **2** (1961) 55.
[62] Minina, S. A.; Mashkova, L. P.: Rastit. Resur. **12** (1976) 546.
[63] Moldenhawer, K.: Wiss. Z. Karl-Marx-Univ. Leipzig, Math.-Nat. Reihe **1963**, 415.
[64] Mothes, K.; Romeike, A.: In: Ruhland, W. (Ed.): Handbuch der Pflanzenphysiologie. Springer-Verlag, Berlin/Göttingen/Heidelberg 1958, Vol. 8, p. 989.
[65] Nittala, S. S.; Lavie, D.: Phytochem. **20** (1981) 2741.
[66] Nowacki, E.: Genetica Polon. **4** (1963) 161.
[67] Nowacki, E.: Genetica Polon. **5** (1964) 189.
[68] Nowacki, E.; Bragdø, M.; Duda, A.; Kazimierski, T.: Flora **151** (1961) 120.
[69] Nyman, U.: Hereditas **89** (1978) 43.
[70] Nyman, U.: Hereditas **93** (1980) 115.
[71] Nyman, U.: Hereditas **93** (1980) 121.
[72] Nyman, U.; Hall, O.: Hereditas **76** (1974) 49.
[73] Nyman, U.; Hall, O.: Hereditas **84** (1976) 69.
[74] Nyman, U.; Hansson, B.: Hereditas **88** (1978) 17.
[75] Porsche, W.: Züchter **34** (1964) 251.
[76] Romeike, A.: Planta Med. **6** (1958) 426.
[77] Romeike, A.: Kulturpflanze **9** (1961) 171.
[78] Romeike, A.: Kulturpflanze **10** (1962) 140.
[79] Rönsch, H.; Schreiber, K.: Liebigs Ann. Chem. **694** (1966) 169.
[80] Rönsch, H.; Schreiber, K.; Stubbe, H.: Naturwissenschaften **55** (1968) 182.
[81] Rudorf, W.: Jenaische Z. Naturwiss. **77** (1944) 290.
[82] Rudorf, W.; Schwarze, P.: Planta **39** (1951) 36.
[83] Rueffer, M.; El-Shagi, H.; Nagakura, N.; Zenk, M. H.: FEBS-Letters **129** (1981) 5.
[84] Sander, H.: Planta Med. **11** (1963) 303.
[85] San Martin Casamada, R.: Farmacognosia 8 (1948) 213.
[86] San Martin Casamada, R.: Farmacognosia **13** (1953) 357.
[87] Sárkány, S.; Dános, B.; Sárkány-Kiss, I.: Ann. Univ. Scient. Budapest, Sect. Biol. **2** (1959) 211.
[88] Schallenberg, J.; Berlin, J.: Z. Naturforsch. **34c** (1979) 541.

[88a] Schieder, O.: In Czygan, F.-C. (Ed.): Biogene Arzneistoffe, Friedrich Vieweg und Sohn, Braunschweig/Wiesbaden 1984, p. 194.
[89] Schieder, O.; Vasil, I. K.: International Review of Cytology, Supplement 11B. Academic Press, New York 1980, p. 21.
[90] Schmuck, A. A.: Izvest. Akad. Nauk SSSR, Biol. Ser. **1937**, 1693.
[91] Schmuck, A. A.; Khmura, M. I.: Trudy prikladnoy botanike, genetike i selekcii, Ser. A. **15** (1935) 111.
[92] Schreiber, K.; Rönsch, H.: Arch. Pharmaz. **298** (1965) 285.
[93] Schwarze, P.: In: Kappert, H.; Rudorf, W. (Eds.): Handbuch der Pflanzenzüchtung. Verlag Paul Parey, Berlin/Hamburg 1958, Vol. 1, p. 307.
[94] Schwarze, P.; Hackbarth, J.: Züchter **27** (1957) 332.
[95] Scott, A. I.; Mizukami, H.; Lee, S. L.: Phytochemistry **18** (1979) 795.
[96] Smith, H. H.; Abashian, D. V.: Am. J. Bot. **50** (1963) 435.
[97] Smith, H. H.; Smith, C. R.: J. agric. Res. **65** (1942) 347.
[98] Spalla, C.; Marnati, M. P.: Proc. 5th FEMS Symp. Academic Press, London 1978, p. 219.
[99] Starý, F.: Abh. Dtsch. Akad. Wiss. Berlin, Klasse Chem., Geol., Biol. 1963, No. 4, 175.
[100] Steinegger, E.: Pharmac. Acta Helvetiae **27** (1952) 311.
[101] Steinegger, E.: Spillmann, P.: Pharmac. Acta Helvetiae **42** (1967) 183.
[102] Steinegger, E.; Zbinden, F.: Pharmac. Acta Helvetiae **36** (1961) 74.
[103] Stöckigt, J.; Zenk, M. H.: FEBS-Letters **79** (1977) 233.
[104] Treimer, J. F.; Zenk, M. H.: FEBS-Letters **97** (1979) 159.
[105] Troll, H.-J.: Z. Pflanzenzücht. **39** (1958) 35.
[106] Tudzynski, P.; Esser, K.: Theor. Appl. Genet. **61** (1982) 97.
[107] Turbin, N. V.; Anokhina, V. S.: Bull. Mosk. obščestva ispytatelej prirody, biol. Otd. **68** (1963) 116.
[108] Valleau, W. D.: J. agric. Res. **78** (1949) 171.
[109] Waller, G. R.; Nowacki, E. K.: Alkaloid Biology and Metabolism in Plants. Plenum Press, New York/London 1978, p. 49.
[110] Weybrew, J. A.; Mann, T. J.: Tobacco **156**, No. 8 (1963) 30.
[110a] Williams, W.; Harrison, J. E. M.: Phytochemistry **22** (1983) 85.
[111] Willuhn, G.: Planta Med. **14** (1966) 408.
[112] Wink, M.; Hartmann, T.: FEBS-Letters **101** (1979) 343.
[113] Wink, M.; Hartmann, T.; Schiebel, H. M.: Z. Naturforsch. **34c** (1979) 704.
[114] Zachow, F.: Züchter **37** (1967) 35.

4 Regulation of Alkaloid Metabolism

M. Luckner

4.1 Alkaloid Formation and Cell Differentiation

It is well known that the different features of alkaloid metabolism, *i.e.* biosynthesis, accumulation, transformation, degradation, release etc. of alkaloids, are not expressed equally in all cells of the producer organisms and during all of their developmental stages [14]. Examples in this respect are:

- the biosynthesis of hyoscyamine (*cf.* chapter 11.1) which proceeds mainly in the roots of *Solanaceae,* whereas the alkaloid is stored in the leaves and may be epoxidized during transportation in stem tissue [19—21]
- synthesis of benzodiazepine alkaloids (*cf.* chapter 16.2) in hyphae and conidiospores of the mould *Penicillium cyclopium,* whereas only the conidiospores are able to transform these compounds to quinoline derivatives [11] and
- formation of alpinigenine (*cf.* chapter 14.10) in young plants of *Papaver bracteatum* (e^h-type) [1, 12], but not after reaching maturity.

Alkaloid metabolism depends on the state of specialization of the participating cell groups and cells (phase dependence), which is caused by the selective expression of different parts of the genetic material of the producer organism: like secondary metabolism in general, the expression of the different characteristics of alkaloid metabolism is the result of differentiation processes [9, 10, 12].

Cell differentiation is of fundamental significance in the development of all living beings [17]. It includes the cyclic programmes of cell multiplication as well as the programmes by which specialized cells are formed. Cell specialization, especially in multicellular organisms, enables the "sharing of labour" which increases efficiency and fitness of the organism as a whole, whereas the individual cell may loose important vital characteristics. In nearly all cases the formation of alkaloids has no significance for the producer cells themselves but may be of importance for the organism *in toto* (*cf.* chapter 7).

Investigation of the regulation of alkaloid metabolism is still it its infancy. Most results have been obtained with relatively simple systems, *i.e.* with microorganisms and plant cells cultivated *in vitro.* These systems will therefore form the centre of the following discussion.

4.2 Regulation of Enzyme Content and Activity

The principles regulating the amount of the enzymes of alkaloid biosynthesis and their activity *in vivo* were investigated in the formation of benzodiazepine and quinoline alkaloids in *Penicillium cyclopium* (*cf.* chapter 16).

Development of *P. cyclopium*, like that of other organisms, proceeds in a programmed manner, including a large number of steps of differential gene expression [11]. The development of the mould may be divided into three phases:

a) the phase in which dormancy of the propagation units, the conidiospores, is overcome (germination phase),

b) the phase of hyphal growth (trophophase) and

c) the phase of cell specialization (idiophase).

The idiophase is characterized by synthesis of the alkaloids and formation of the specialized cells of the penicilli which detach the conidiospores [5, 15].

During the transition to the idiophase the enzymes of benzodiazepine alkaloid biosynthesis become measurable [24]. The increase of the *in vitro* activities of the enzymes cyclopeptine dehydrogenase, dehydrocyclopeptine epoxidase, and cyclopenin *m*-hydroxylase (*cf.* chapter 16) as well as the increase of the rates of alkaloid formation are inhibited by 5-fluorouracil or cycloheximide indicating the necessity of RNA and protein biosynthesis [3, 16, 18]. Hence an important regulatory feature in the expression of alkaloid biosynthesis is the *de novo* formation of the enzymes involved at the level of transcription.

For two enzymes the control of transcription is, however, superimposed by post transcriptional, in fact post translational, regulation [11]. One of them is cyclopenase, a constituent of the mature conidiospores of *P. cyclopium* [25, 26]. Expression of cyclopenase activity proceeds during spore ripening. It cannot be hindered by cycloheximide in concentrations that prevent protein biosynthesis and therefore is not coupled directly to the formation of the enzyme protein [3].

The signals that trigger the formation of the enzymes of alkaloid biosynthesis are still unknown. In most cases they obviously originate internally and form part of the developmental programmes of the producer organisms (*cf.* chapter 4.3). Probably the interaction of inducers and repressors determines the actual enzyme amount in certain organisms. Product repression was shown for instance in *Nicotiana tabacum.* In this plant the alkaloid nicotine (*cf.* chapter 13) is able to repress certain enzymes necessary for its biosynthesis [13].

In addition to the regulation by enzyme amount, secondary metabolism is controlled by the regulation of the activity of secondary metabolic enzymes [10] (*cf.* chapter 5), as shown by:

— the large discrepancies that exist in many organisms between the relatively high *in vitro* measurable activities of enzymes involved in secondary metabolism and the much lower rates of product formation *in vivo* and

— the differences in the time-course of appearance of the *in vitro* activities of secondary metabolic enzymes and *in vivo* product biosynthesis during the development of the producer organisms.

In vivo activity of the enzymes responsible for alkaloid synthesis may be controlled for instance by limited substrate supply. Hence alkaloid production by older batch cultures of *P. cyclopium* or of mature conidiospores gradually ceases because of the exhaustion of substrates [11]. In other cases compartmentation of enzymes and substrates is the reason for low *in vivo* activity. For instance, in the conidiospores of *P. cyclopium* cyclopenase shows a very limited activity because it is separated by membranes from its substrates cyclopenin and cyclopenol [11, 22].

Furthermore, the activity of secondary metabolic enzymes may be influenced by the accumulation of products (product inhibition by allosteric interaction or by competition with the substrates at the binding sites [4, 10]). Although in the living cell alkaloids and the synthesizing enzymes in most cases are separated by compartmentation (*cf.* chapter 6), in some instances product inhibition can be observed. One example is ergoline alkaloid biosynthesis in *Claviceps* (*cf.* chapter 15.4). The first enzyme of this biosynthetic pathway, DMAT synthetase, undergoes inhibition by certain ergoline alkaloids *in vitro*. Hence its activity in crude homogenates depends on the alkaloid content present in the disintegrated mould cells. It is also interesting that the *in vivo* rate of ergoline biosynthesis depends on the alkaloid concentration in the mycelium and both processes fluctuate during cultivation of the mould. These fluctuations may be explained by

a) accumulation of alkaloids at periods of high enzyme activity where release of the alkaloids in the medium is rate limiting,
b) increasing inhibition of DMAT-synthetase (and probably other enzymes of the pathway) by the accumulating alkaloids and
c) decrease of the concentration of stored alkaloids by release from the cells and again increase of enzyme activity and alkaloid biosynthesis [4, 6].

In many instances, however, the means by which the activity of enzymes of alkaloid metabolism are regulated *in vivo* are not known. Probably the fluxes of precursors, intermediates, cosubstrates, *i.e.* their rates of biosynthesis, their compartmentation, and channelling, play an important role. As yet, however, the rates of these fluxes are known only in a few instances (*cf.* chapter 5).

4.3 The Integration of Alkaloid Metabolism in the Developmental Programmes of the Producer Organisms

There are three arguments indicating that expression of alkaloid biosynthesis (as that of other secondary products) is integrated into developmental programmes [10, 12]:

a) the phase dependence, *i.e.* the fixation of alkaloid biosynthesis in relation to other chemical and morphological parameters expressed during development and the lack of alkaloid formation if development is stopped by physiological means or suitable mutations before the developmental stage of alkaloid synthesis is reached;
b) the indirect influence of signals on alkaloid metabolism, which influence the expression of large parts of the producer's genetic material, *e.g.* by hormones;
c) the determination of the rates of alkaloid synthesis at certain stages of development (determination phases) prior to the real expression of alkaloid formation.

With alkaloid formation in *P. cyclopium*, for instance, it is possible to demonstrate the stepwise expression of its individual features during maturation of hyphae and detachment and ripening of conidiospores, the influence of intracellular and extracellular signals on idiophase development including alkaloid formation and the existence of a determination phase (*cf.* [11]). It was shown, however, that the integration of alkaloid metabolism in the programmes of hyphae and conidiospore formation and maturation is not absolute. Drastic changes in the physiological conditions as well as mutations may alter the time points of expression of the different features of alkaloid metabolism within the developmental programmes. Alkaloid formation may even be avoided by mutation without hindrance of the further progress of the programmes [23]. This flexibility seems to be due to the location of alkaloid metabolism, like that of other secondary products, on side branches of the developmental programmes as investigated in detail with the secondary products (peptide antibiotics, dipicolinic acid, sulfolactic acid and spore pigments) synthesized during endospore formation and maturation in bacilli (for a summary *cf.* [10]).

4.4 Conclusions

The results discussed above demonstrate that several control mechanisms operate in alkaloid metabolism. The most important are:

a) regulation of the amount of the enzymes involved in biosynthesis, transformation, and degradation of alkaloids, in most cases by control of the transcription of the corresponding genes, but also by post-translational control and probably at other levels such as enzyme degradation;
b) regulation of the activity of the enzymes, for instance, by controlling the level of substrates at the site of alkaloid biosynthesis, transformation *etc.*, or by accumulating products.

The significance of these control mechanisms may vary in different organisms, but in all systems investigated so far they result in a thorough regulation of the overall process. Alkaloid metabolism is integrated into the programmes of cell differentiation and development of the producer organisms and in this way is harmonized with other biochemical activities and with morphological features. The molecular organization of this integration is unknown, as is the organization of the programmes. There are, however, many external and endogenous signals, including hormones and growth factors (for a summary *cf.* [8]), and precursors [2, 7] that influence smaller or larger parts of the programmes and in this way also affect the formation of alkaloids. Furthermore, it has been shown that alkaloid formation may be deleted, *e.g.* by mutation, without killing the organism. This indicates that its expression is not really essential for the individual producer cell. In other words alkaloid metabolism has those characteristics also typical for other fields of secondary product biosynthesis and cell specialization.

4.5 References

[1] Böhm, H.: Biochem. Physiol. Pflanzen **162** (1971) 474.

[2] Dunkel, R.; Müller, W.; Nover, L.; Luckner, M.: Nova Acta Leopoldina Suppl. **7** (1976) 281.

[3] El Kousy, S.; Pfeiffer, E.; Ininger, G.; Roos, W.; Nover, L.; Luckner, M.: Biochem. Physiol. Pflanzen **168** (1975) 79.

[4] Floss, H. G.; Robbers, J. E.; Heinstein, P. F.: Recent Adv. Phytochem. **8** (1974) 141.

[5] Framm, J.; Nover, L.; El Azzouny, A.; Richter, H.; Winter, K.; Werner, W.; Luckner, M.: Europ. J. Biochem. **37** (1973) 78.

[6] Heinstein, P. F.; Floss, H. G.: Nova Acta Leopoldina Suppl. **7** (1976) 299.

[7] Krupinsky, V. M.; Roberts, J. E.; Floss, H. G.: J. Bacteriol. **125** (1976) 158.

[8] Lovkova, J. Ja.: Biosynthesis and Metabolism of Alkaloids in Plants. Nauka, Moscow 1981.

[9] Luckner, M.: Pharmazie **26** (1971) 717.

[10] Luckner, M.: In: Encyclopedia of Plant Physiology, New Series, Vol. 8: Bell, E. A.; Charlwood, B. V. (Eds.): Secondary Plant Products. Springer-Verlag, Berlin 1980, p. 23.

[11] Luckner, M.: J. Nat. Prod. **43** (1980) 21.

[12] Luckner, M.; Nover, L.; Böhm, H.: Secondary Metabolism and Cell Differentiation. Springer-Verlag, Berlin 1977.

[13] Mizusaki, S.; Tanabe, Y.; Noguchi, M.; Tamaki, E.: Plant Cell Physiol. **14** (1973) 103.

[14] Mothes, K.: In: Mothes, K.; Schütte, H. R. (Eds.): Biosynthese der Alkaloide. Deutscher Verlag der Wissenschaften, Berlin 1969, p. 1.

[15] Nover, L.; Luckner, M.: Biochem. Physiol. Pflanzen **166** (1974) 293.

[16] Nover, L.; Luckner, M.: Nova Acta Leopoldina Suppl. **7** (1976) 229.

[17] Nover, L.; Luckner, M.; Parthier, B.: Cell Differentiation. Gustav Fischer Verlag, Jena, and Springer-Verlag, Berlin 1982.

[18] Nover, L.; Müller, W.: FEBS-Letters **50** (1975) 17.

[19] Romeike, A.: Flora **143** (1956) 67.

[20] Romeike, A.: Flora **148** (1959) 306.

[21] Romeike, A.: Planta Med. **8** (1960) 491.

[22] Roos, W.; Fürst, W.; Luckner, M.: Nova Acta Leopoldina Suppl. **7** (1976) 175.

[23] Schmidt, I.; Nover, L.; Ininger, G.; Luckner, M.: Z. Allg. Mikrobiol. **18** (1978) 219.

[24] Voigt, S.; El Kousy, S.; Schwelle, N.; Nover, L.; Luckner, M.: Phytochemistry **17** (1978) 1705.

[25] Wilson, S.; Luckner, M.: Z. Allg. Mikrobiol. **15** (1975) 45.

[26] Wilson, S.; Schmidt, I.; Roos, W.; Fürst, W.; Luckner, M.: Z. Allg. Mikrobiol. **14** (1974) 515.

5 Compartmentation and Channelling in Alkaloid Biosynthesis

W. Roos

The integration of secondary metabolic pathways into cellular substructures frequently has been demonstrated [8a]. In the field of alkaloid metabolism the experimental systems characterized have revealed the following basic elements of spatial organization.

5.1 Membrane Association of Enzymes

Several enzymes involved in alkaloid metabolism have been shown to be integrated into or bound to cellular membrane systems. This association may be a structural requirement for the vectorial processes discussed below.

Clavicipitic acid synthetase is part of the microsomal membrane fraction and (to a lesser extent) of the mitochondria in *Claviceps purpurea* [14]. The enzyme system involved in the biosynthesis of mandelonitrile in *Sorghum* seedlings is bound to the ER membranes [13]. In the mould *Penicillium cyclopium* the enzyme cyclopenase, which catalyses the conversion of cyclopenin to viridicatin (*cf.* chapter 16.3), is most probably part of the cytoplasmic membrane of the conidiospores as indicated by its solubilization as a protein-phospholipid complex from a crude cell wall-membrane fraction [19]. In *Bacillus brevis* Vm 4 the enzymes synthesizing the polypeptide edeine are associated with a complex of cytoplasmic membrane and DNA. It has been suggested that this kind of localization may permit the fast excretion of the antibiotic avoiding its inhibitory effects on DNA synthesis and protein synthesis [7].

5.2 Compartmentation and Channelling of Precursors

The branching of secondary pathways from primary metabolism results in a competition for common substrates. Regulation of the competing metabolic chains occurs not only *via* the selective control of the amount and activity of the (key) enzymes (differential gene expression, *cf.* chapter 4.2, allosteric regulation, cofactor supply, *etc.*), but also with the aid of intracellular precursor pools, *i.e.* areas of different substrate concentrations serving selectively as reservoirs of distinct metabolic pathways. The establishment of pools in addition to the vectorial transfer of substrates to the catalytic sites (channelling) is due either to the effects of the microenvironment at protein or membrane surfaces or to the reversible accumulation in membrane-surrounded organelles.

In fungal [2, 3, 16, 18], plant [10], and animal cells [17, 20] amino acids the principle precursors of alkaloids (*cf.* chapters 10—18, 20, 21) are frequently found to be distributed between two pools:

— a cytosolic compartment of low capacity, which maintains the substrate supply of protein synthesis, and
— a pool of high capacity whose content expands largely with the external amino acid concentration. In yeast [3, 18, 18a] and *Neurospora* [2, 16] this pool obviously represents the amino acid content of the vacuole (or of vesicles with a similar function) as shown by kinetic experiments and isolation of the organelles. It absorbs the excess amino acids of the cytoplasm and serves as a reservoir whose content responds to changes in the cytosolic amino acid concentration.

In *Penicillium cyclopium*, kinetic studies suggest a similar compartmentation of phenylalanine, a precursor of the peptide alkaloids cyclopenin and cyclopenol (*cf.* chapter 16.2). As in other fungal cells, more than 90% of phenylalanine taken up from the medium are concentrated several thousand-fold in a centrally located pool (most probably in the vacuole). This pool, which is maintained by a dynamic equilibrium of active uptake and passive efflux, exchanges with the medium only *via* the small cytosolic pool. This thus exerts a regulatory function on the vacuolar pool by determining the rate of efflux from the cell as well as the degree of trans-inhibition of uptake [8]. Both uptake and concentration of extracellular phenylalanine are catalysed by various uphill transporting permeases, some of which are characterized by kinetic methods in intact cells (W. Roos, unpublished results).

Isotopic dilution analysis suggests that the two phenylalanine pools serve different metabolic purposes. The content of the expandable pool may be used for the biosynthesis of alkaloids but not of proteins, whereas the cytosolic phenylalanine supplies both processes. Thus, external phenylalanine may reach the site(s) of alkaloid synthesis *via* two channels:

a) *via* a direct way from a cytosolic pool (giving rise to a "primary labelling" of alkaloids) and
b) an indirect way *via* the expandable pool (causing a "secondary" labelling) (Fig. 5.1).

It must be noted, however, that both pathways under all experimental conditions used contribute no more than 10% to the total alkaloid synthesis. The main precursor supply comes from phenylalanine biosynthesis or degradation of proteins [9].

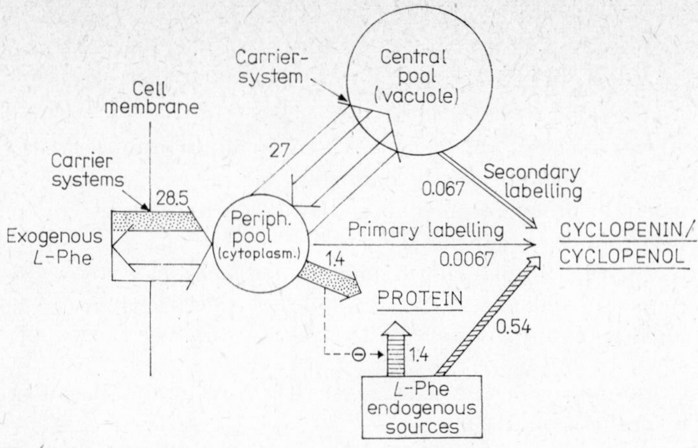

Fig. 5.1
Channelling of phenylalanine in *Penicillium cyclopium* (from [9])

Fluxes of phenylalanine in surface cultures 3 days after inoculation. Full arrows: pathway of exogenous phenylalanine predominating at low concentrations; open arrows: pathway of excess exogenous phenylalanine not incorporated into proteins; hatched arrows: pathways of endogenously produced phenylalanine. The numbers (pmol *L*-phenylalanine/cm^2 culture area · sec) are an estimation of the capacity of the individual channels.

5.3 Channelling of Intermediates

The coexistence of secondary pathways with other (even competing) metabolic chains may often be attributed to the limited diffusion of intermediates. A specific sequential arrangement of the catalytic sites allows the stepwise passage of the intermediates combined with their successive conversion. The reaction sequence of such a "metabolic channel" may be determined by the following substructures:

a) the sequence of catalytic sites within a single, multifunctional polypeptide (polyenzyme), as found with the biosynthesis of the depsihexapeptide antibiotic enniatin B [21];
b) the co-operation of mono- or multi-functional enzymes in a non-covalent complex, which occurs in the biosynthesis of several other antibiotic peptides, *e.g.* gramicidin S, tyrocidine, bacitracin, and alamethicin (*cf.* [4, 6] for recent reviews);
c) the assembly of enzymes in proximity to each other on membrane surfaces, which has been demonstrated with the biosynthesis of mandelonitrile [1].

Whereas in cases a) and b) the intermediates remain bound covalently, in case c) the diffusion of non-covalently bound intermediates is reduced by the microenvironment of the membrane-associated complex [22]. In each case, metabolic channelling is indicated by "catalytic facilitation" [23], *i.e.* the preferred conversion of the initial substrate compared with externally added intermediates. In the next section examples are given of both principles with two well-elaborated systems.

5.3.1 Synthesis of Gramicidin S in *Bacillus brevis* ATCC 9999

The biosynthesis of the ionophoric decapeptide gramicidin S [4–6] (cyclo-[*D*-Phe-Pro-Val-Orn-Leu]$_2$) occurs on a soluble multi-enzyme complex consisting of two multi-functional polypeptide chains: GS 1 (100 K dal) bears the initiation site, where the starter amino acid *L*-phenylalanine is activated as an aminoacyl-adenylate, then it is bound as a thiolester and epimerized to *D*-phenylalanine. Likewise, the other substrates (proline, valine, ornithine and leucine) become AMP-activated and then bind in this sequence to thiol groups of the polyenzyme GS 2 (280 K dal), which serves as a "thiotemplate" for the following steps. Synthesis is initiated by the transfer of *D*-phenylalanine from GS 1 to a specific binding site on GS 2 from where it attacks the NH-group of the proline-thiolester to form the (thiol-activated) dipeptide *D*-Phe-Pro-S-enzyme. This peptide is then transferred to the SH-group a centrally arranged 4′-phosphopantetheine residue of GS 2. The newly formed thiolester undergoes a gradual "intrinsic" trensfar attacking with its NH_2-terminal of the thiol-activated carboxyl of the proximal amino acid. Repetition of this process gives rise to the formation of an enzyme-bound pentapeptide. Finally, head-to-tail-cyclization with a second enzyme bound molecule yields gramicidin S.

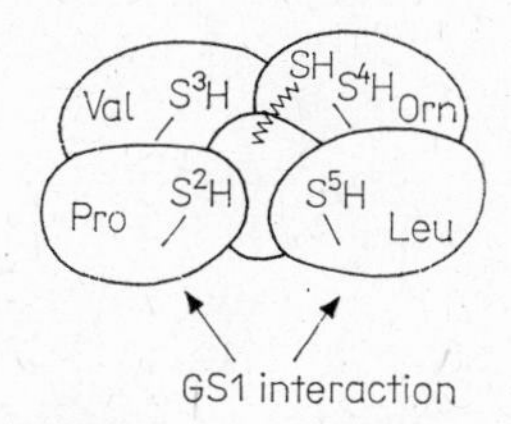

Fig. 5.2
Model of the GS 2 multi-enzyme of gramicidin S-synthetase (from [5a])

The amino acid activating "subunits" with their peripheral SH-groups are arranged around the pantetheine-carrying peptide with the central SH group. The docking of GS 1 (see text) may involve sites of both the proline- and leucine-activating peptides.

Each activation- and elongation-step is considered to be catalyzed by a functional domain containing a binding site for the AMP-amino acid and peripheral SH group. A detailed analysis of proteolytic fragments activating one or more amino acids as well as mutational alterations of the enzyme structure shows that the formation of similar antibiotic tyrocidine [5a, 6] is disintegrated by an (unindentified) proteolytic factor from *Bacillus brevis* to amino acid activating units of each 70 K dal. The structure of the functional peptide synthetase is therefore thought to consist of similar-sized amino acid activating subunits arranged around a 4′-phosphopantetheine-containing peptidyl carrier protein (Fig. 5.2).

The initiation of peptide synthesis requires the formation of a complex of the heavy enzyme GS 2 with GS 1. This complex not only allows the transfer of *D*-phenylalanine (see above), but establishes cooperativity between the enzymes, as shown by the fact that synthesis of gramicidin S on the heavy enzyme loaded with all 5 substrate amino acids requires the presence of the light enzyme [11]. Under *in vitro* conditions the enzyme system was shown to accept tyrosine, tryptophan, and other homologues in place of phenylalanine. This substitution does not occur in intact cells. Thus, channelling of substrates to the activating sites may play an additional regulatory role.

5.3.2 Formation of *p*-Hydroxymandelic Acid Nitrile in *Sorghum bicolor*

A microsomal membrane preparation obtained from germinating seedlings of *Sorghum bicolor* catalyses the conversion of *L*-tyrosine to *p*-hydroxymandelonitrile by the following sequence of reactions (*cf.* [1, 15]):

$$L\text{-Tyrosine (1)} \xrightarrow{(O)} N\text{-Hydroxy-tyrosine (2)} \xrightarrow[-CO_2]{-2H} p\text{-Hydroxyphenyl-acetaldoxime (3)}$$

$$\xrightarrow{-H_2O} p\text{-Hydroxyphenyl-acetonitrile (4)} \xrightarrow{(O)} p\text{-Hydroxy-mandelonitrile (5)}$$

(*p*-Hydroxymandelonitrile (5) is further converted to the cyanogenic glucoside dhurrin by a soluble enzyme.) None of the intermediates is covalently bound to the enzyme proteins catalysing the formation of *p*-hydroxymandelonitrile. Nevertheless, kinetic facilitation has been demonstrated by two types of experiments suggesting that the membrane-bound enzyme system constitutes a metabolic channel:

- although the labelled intermediates are incorporated into the mandelonitrile (5), the rate of its formation is significantly higher with tyrosine (1) and *N*-hydroxytyrosine (2) than with the late intermediates, *i.e.* the aldoxime (3) and the acetonitrile (4);
- [^{3}H]*p*-hydroxyphenyl-aldoxime and [^{3}H]*p*-hydroxyphenyl-acetonitrile produced from ^{3}H-tyrosine do not readily exchange with added [^{14}C]*p*-hydroxyphenyl aldoxime or [^{14}C]*p*-hydroxyphenyl acetonitrile, but instead they are preferentially incorporated into the end product.

5.4 Influence of Membrane Barriers on Alkaloid Biosynthesis

The significance of membrane permeability for compartmentation and channelling of secondary metabolites may be demonstrated with the metabolic fate of the benzodiazepine alkaloids, which are synthesized in the spores and hyphal cells of *Penicillium cyclopium* (*cf.* chapter 16.2) and excreted into the cell walls and the culture medium. These alkaloids are substrates of the enzyme cyclopenase, which is located on the inner side of the cytoplasmic membrane of conidiospores. In intact cells, the low permeability of this membrane towards the alkaloids almost completely protects them from transformation by the enzyme. Artificial increase of membrane permeability, *e.g.* by lowering the cellular ATP level [12a], greatly increases the conversion of the benzodiazepines to quinoline derivatives. The latter remain stored in the cells due to their low water solubility [12].

The uptake of the benzodiazepine alkaloids from the medium by suspended conidiospores is tightly coupled with their conversion by cyclopenase. Derivatives of cyclopenin/cyclopenol, which are not attacked by cyclopenase, do not accumulate in the spores. Hyphal cells, which are free of cyclopenase, are unable to provide any storage capacity for externally added cyclopenin and cyclopenol, either in the native or the permeabilized state, even if the permeability is increased in the presence of these alkaloids (W. Roos, unpublished results).

5.5 Conclusions: Significance of Compartmentation and Channelling

By analogy with other metabolic pathways, the existence of the vectorial elements discussed above effectively promotes the integration of alkaloid biosynthesis into the total cellular metabolism. This spatial control may govern

— the balance between alkaloid formation and other competing pathways,
— the selection of certain metabolic sequences by means of the vectorial transfer of precursors,
— the establishment of high local concentrations of precursors and intermediates,
— the common metabolic regulation of the associated enzymes and
— the accumulation of the often cytotoxic alkaloids, thus avoiding their contact with the cytoplasm (*cf.* next chapter).

5.6 References

[1] Conn, E. E.; McFarlane, I. J.; Moeller, B. L.; Shimada, M.: Proc. 12th FEBS-Meeting, Dresden 1978, Vol. 55, p. 63.
[2] Davis, R. H.; Bowman, B. J.; Weiss, R. L.: J. Supramolec. Struc. **9** (1978) 473.
[3] Huber-Wälchli, V.; Wiemken, A.: Arch. Mikrobiol. **120** (1979) 141.
[4] Kleinkauf, H. et al.: Proc. 12th FEBS-Meeting, Dresden 1978, Vol. 55, p. 37.
[5] Kleinkauf, H.: Planta Med. **35** (1979) 1.
[5a] Kleinkauf, H.; Koischwitz, H.: In: Molecular Biology, Biochemistry and Biophysics, Vol. 32: Chapeville, F.; Haenni, A. L. (Eds.): Chemical Recognition in Biology. Springer-Verlag, Berlin 1980, p. 427.
[6] Kleinkauf, H.; von Döhren, H.: Current Topics Microbiol. Immunol. **91** (1981) 129.
[7] Kurylo-Borowska, Z.: Biochim. Biophys. Acta **399** (1975) 31.
[8] Luckner, M.; Roos, W.: In: Krumphanzl, V.; Sikyta, B.; Vanek, Z. (Eds.): Overproduction of Microbial Products. Academic Press, London 1982, p. 111.

[8a] Luckner, M.; Diettrich, B.; Lerbs, W.: Progress Phytochem. **6** (1980) 103.
[9] Nover, L.; Lerbs, W.; Müller, W.; Luckner, M.: Biochim. Biophys. Acta **584** (1979) 270.
[10] Oaks, A.; Bidwell, R. G. S.: Ann. Rev. Plant Physiol. **21** (1970) 43.
[11] Pass, L.; Zimmer, T. L.; Laland, S. G.: Eur. J. Biochem. **47** (1974) 607.
[12] Roos, W.; Fürst, W.; Luckner, M.: Nova Acta Leopoldina Suppl. **7** (1976) 175.
[12a] Roos, W.; Luckner, M.: Biochem. Physiol. Pflanzen **171** (1977) 127.
[13] Saunders, J. A.; Conn, E. E.; Chin Ho Lin; Mikio Shimada: Plant Physiol. **60** (1977) 629.
[14] Saini, M. S.; Cheng, M.; Anderson, J. A.: Phytochemistry **15** (1976) 1497.
[15] Shimada, M.; Conn, E. E.: Arch. Biochem. Biophys. **180** (1977) 199.
[16] Weiss, R. L.: J. Biol. Chem. **248** (1973) 5409.
[17] Wettenhall, R. E. H.; London, D. R.: Biochem. Biophys. Acta **390** (1975) 363.
[18] Wiemken, A.; Dürr, M.: Arch. Mikrobiol. **101** (1974) 45.
[18a] Wiemken, A.: In: Nover, L.; Lynen, F.; Mothes, K. (Eds.): Cell Compartmentation and Metabolic Channelling. Gustav Fischer Verlag, Jena, Elsevier/North Holland, Amsterdam 1980, p. 225.
[19] Wilson, S.; Luckner, M.: Z. Allg. Mikrobiol. **15** (1975) 45.
[20] van Veenrovij, W. J.; Moonen, H.; van Loon-Klaassen, L.: Eur. J. Biochem. **50** (1974) 297.
[21] Zocher, R.; Kleinkauf, H.: Biochem. Biophys. Res. Commun. **81** (1978) 1162.
[22] Srere, P. A.; Mosbach, K.: Ann. Rev. Microbiol. **28** (1974) 61.
[23] Gaertner, F. H.; Ericson, M. C.; de Moss, J. A.: J. Biol. Chem. **245** (1970) 595.

6 Storage of Alkaloids

D. Neumann

Early work on the localization of alkaloids, based on more or less unspecific precipitation in biological material with $KI \cdot I_2$, $KBiI_4$, HgI_2, $HgCl_2$, picric acid, tannic acid *etc.*, gave rough results regarding the distribution of these compounds in plant organs and tissues [7, 8, 17]. In more recent investigations electron microscopy [29] was used and whole plants [11, 25, 26, 30], callus [31], cell suspension cultures [27] and isolated cell organelles [10, 16, 28, 34] were investigated. Alkaloids were frequently found in actively growing tissues, such as meristems and phloem and xylem initials, and in cells with active metabolism, such as the epidermal cells of leaves and shoots [40], in hairs [39], bundle sheats and laticifers [3, 10, 11, 24, 30, 38]. In *Cinchona*, for instance, alkaloids are present in rather high concentrations (10 to 12% of the dry weight) in parenchyma cells and other living cells of the bark [37]. An exception is the accumulation of alkaloids in dead parts of the seeds of *Datura*, *Atropa*, *Conium* [3] and *Nigella* [23] and in the dead wood cells of *Berberis* [4]. In several cases alkaloids are released from living cells into the surroundings [6, 36].

6.1 Cellular Aspects of Alkaloid Storage

Most alkaloids are toxic for the cytoplasm even of the producing cells [13, 25, 35]. Their accumulation therefore requires not only a well regulated and compartmentalized system of biosynthesis, but they must be stored in a way that avoids damage of the producing organism. Unlike animals higher plants do not have an excretory system and poisonous substances must be stored inside the plant body (*cf.* chapter 1). In many cases the storage of secondary products means the well-regulated accumulation in compartments that allow their controlled re-utilization. Indeed, different types of alkaloids (*e.g.* nicotine, chapter 13; ricinine, chapter 13, and morphine, chapter 14.6) undergo a rapid turnover, which may be demonstrated by feeding radioactive-labelled precursors in pulse-chase experiments. The half-life of the alkaloids depends on the physiological state of the organism in question and varies from a few hours to a couple of days. During degradation substances of primary metabolism and in the last step CO_2 may be formed [10a, 33b]. Dead

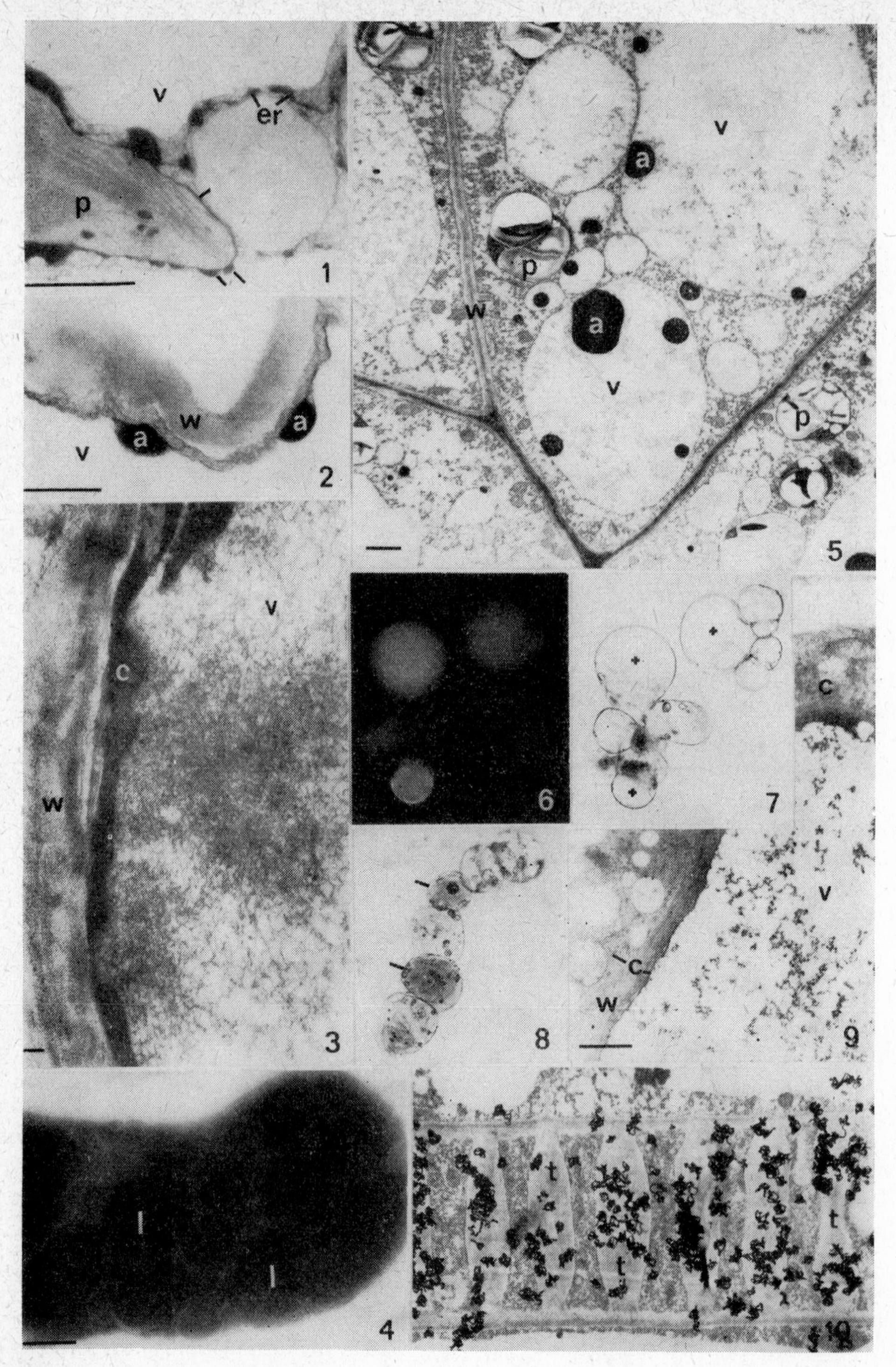

Fig. 6.1–6.10

parts of plants (xylem vessels, sclerenchymal tissues, stone and cork cells), which are able to bind alkaloids non-specifically [21, 22, 30], do not satisfy the prerequisite conditions for a controlled accumulation. Such tissues are therefore used only very rarely for alkaloid storage. Living cells are more suitable for regulated storage and are therefore the main site of alkaloid accumulation.

The resistance of storage cells to accumulated secondary products may, on the one hand, be based on an adaptation of the protoplasmic constituents as was shown for toxic non-protein amino acids in plants [9] and batrachotoxin, a complex steroidal alkaloid in animals [4a], and on the other hand be brought about by compartmentation. It seems that the resistance of plants to their alkaloids is based on the latter principle, *i.e.* on separation of the poisonous substances from the cytoplasm in vacuoles and other non cytoplasmic compartments.

Localization of alkaloids in different plant cells

Fig. 6.1 Part of a cell of a rooted leaf of *Nicotiana rustica* exhibiting dark alkaloid precipitations after treatment with hexachloroplatinate in the endoplasmic reticulum (er) and between the outer and inner membrane of the plastid envelopes (—); chloroplast (p), vacuole (v); bar = 0,5 μm (Magnification 36000 ×)

Fig. 6.2 The same cell as in Fig. 6.1 showing alkaloid precipitations after treatment with hexachloroplatinate (a) in the vacuole (v); cell wall (w); bar = 0,5 μm (Magnification 24000 ×)

Fig. 6.3 Part of an old laticifer of *Chelidonium majus* with diffuse alkaloid precipitates after treatment with hexachloroplatinate in the vacuole (v); degenerated cytoplasm (c); cell wall (w); bar = 1 μm (Magnification 3600 ×)

Fig. 6.4 Ultrathin cryosection of a hyphal cell of *Claviceps purpurea*. Note the strong osmiophilic lipid droplets (l), storage compartments for ergot alkaloids; bar = 1 μm (Magnification 8000 ×)

Fig. 6.5 *In vitro* cultivated cell of *Macleaya cordata* with dark alkaloid precipitations after treatment with hexachloroplatinate (a) inside vacuoles (v); cell wall (w), amyloplast (p); bar = 1 μm (Magnification 5400 ×)

Fig. 6.6 Fluorescence micrograph of a suspension culture of *Catharanthus roseus*. The alkaloid containing cells show a blue fluorescence (Magnification 150 ×)

Fig. 6.7 Light micrograph of the same cell clusters as in Fig. 6.6, indicating that alkaloid storage is restricted to the same cells (+) (Magnification 150 ×)

Fig. 6.8 Cell filament of a *Catharanthus* cell suspension culture after vital staining with neutral red at pH 3. Only two cells of the filament are stained (—), indicating a vacuolar pH of 3. The remaining cells of the filament have a vacuolar pH of 5 (Magnification 150 ×)

Fig. 6.9 Alkaloid storage cell of a suspension culture of *Catharanthus roseus* with diffuse alkaloid precipitations after treatment with OsO_4 inside the vacuole (v); cell wall (w), cytoplasm (c); bar = 1 μm (Magnification 8000 ×)

Fig. 6.10 Autoradiograph of a young xylem vessel of *Chelidonium majus* after labelling with [^{3}H] protopine. Note the exclusive labelling of the lignified cell wall thickenings (t); bar = 1 μm (Magnification 6400 ×)

6.1.1 Alkaloid Storage in Vacuoles

The vacuole is one of the most prominent compartments of the mature plant cell occupying more than 70% of the cell volume. Indications for vacuolar localization, for instance of the alkaloid nicotine, were found by histochemical methods [1, 2, 25] as well as by isolation of vacuoles [34]. However, electron micrographs demonstrated that, in addition to the vacuole, other aqueous compartments participate in the storage of alkaloids. In rooted leaves of *Nicotiana rustica* for instance alkaloid was found in ER cisternae between the outer and inner membrane of the plastid envelope, and in plastid vacuoles that originate from thylakoid membrane dilatations [25] (Fig. 6.1, 6.2). The nicotine content is extremely high in the rooted leaves, which facilitates the localization of the alkaloid [18].

6.1.2 Alkaloid Storage in Laticifers

While in *Solanaceae* and many other plant families the bulk of the cells of a particular organ (*e.g.* of leaves or roots) can store alkaloids, in other plants (*e.g.*, in *Papaveraceae* [10, 11, 24, 30, 38] and *Apocynaceae* [41]) alkaloids are accumulated in particular cells, *e.g.* in laticifers (Fig. 6.5, 6.6, 6.7, 6.9). The existence of these idioblasts is an essential prerequisite for the accumulation of alkaloids and there is a direct connection between alkaloid synthesis and formation of storage cells.

Recently it has been shown for callus cultures of *Macleaya cordata* [31] that the bulk of callus cells is able to produce alkaloids, while the accumulation takes place only in a few storage cells (Fig. 6.5). In callus cultures that lost the ability of storage cell formation, no alkaloid was detected. It is not yet clear whether this fact is caused by the absence of biosynthesis or by the degradation of the alkaloids formed due to the non-existence of a space for accumulation. The data, however, support the latter possibility [31].

When alkaloid storage was investigated by precipitation methods using electron microscopy the *Papaveraceae Macleaya cordata* (Fig. 6.5), *Sanguinaria canadiensis* (Fig. 6.3), *Chelidonium majus*, and *Corydalis cava* showed similar pictures [26, 30]. In all cases alkaloids were accumulated in the vacuoles of particular idioblasts (*Sanguinaria, Corydalis*) or articulated laticifers (*Macleaya, Chelidonium*). Young alkaloid storage cells exhibited an ultrastructure that resembling the normal cells of the surrounding tissue. However, the young cells degenerate rapidly, a process which was frequently accompanied by fusion of the individual alkaloid storage cells. The tonoplast membrane became electron-dense after OsO_4-fixation, indicating that additional lipids may be incorporated, and after a short time no organelles or membranes were visible in the dark unstructured cytoplasm [26, 30] (Fig. 6.3). At the period of the highest alkaloid accumulation the storage cells did not contain a living protoplast, *i.e.* the cells were unable to plasmolyse.

An exception to the rule that alkaloid storage cells degenerate was found in *Macleaya callus* cultures, where no fusion of the single idioblasts took place and no degeneration of the cytoplasm was observed [31] (Fig. 6.5). Nevertheless, considerable amounts of alkaloids were accumulated in the vacuoles of these living single storage cells indicating that degeneration is not a prerequisite of alkaloid storage in *Papaveraceae*.

As for the *Papaveraceae*, the alkaloids of *Apocynaceae* accumulate in special storage cells, non-articulated laticifers [41] (Fig. 6.6, 6.7, 6.9). The blue fluorescence of some *Apocynaceae* alkaloids may be used as a sensitive tool in investigations with these cells.

Cell suspension cultures of *Catharanthus roseus* showed a bright blue fluorescence in a few of the cells present, indicating the accumulation of alkaloids (Figs. 6.6, 6.7). The isolation of vacuoles confirmed the assumption that alkaloids were accumulated in these compartments. Electron micrographs showed alkaloid precipitations after OsO_4-fixation exclusively inside a certain fraction of the vacuole population [27] (Fig. 6.9). As in the cultivated cells of *Macleaya*, the ultrastructure of the storage cells did not differ from that of the normal suspension cells (Fig. 6.9).

6.1.3 Alkaloid Storage in Lipid Droplets

Some alkaloids may be stored in their unprotonated form inside lipid droplets that accumulate in certain cells. Sclerotial cells and particularly submerged grown cells of *Claviceps purpurea* accumulate large amounts of lipids during the phase of alkaloid formation [14, 15] (Fig. 6.4). The alkaloids produced in such cells were stored in lipid droplets according to their lipid solubility [28]. The retention in this compartment was caused by the relatively high pH in the cytoplasm, which prevented loss of alkaloids from the lipid phase. A re-utilization of the alkaloids seemed possible after digestion of the lipid droplets inside the vacuoles [15].

6.2 The Principle of Alkaloid Retention

Although the storage of alkaloids leads to high concentrations in particular compartments, accumulation seems not to be accomplished by an active transport [12a, 16, 20—22, 30, 33a]. The alkaloids are accumulated more likely as a consequence of their physico-chemical properties, *i.e.* the driving force for alkaloid storage is the pH gradient between cytoplasm and vacuoles and accumulation inside the vacuole is caused by an ion trap mechanism.

Alkaloid uptake experiments with [^{14}C] nicotine and [^{14}C] ajmalicine in different cell suspension cultures clearly demonstrated the dependence of alkaloid accumulation on the pH-difference between the solution and the cells [12a, 33a]. The significance of the pH gradient for alkaloid accumulation was also shown by the differences in pH of individual cells in *Catharanthus roseus* cell suspension cultures (Fig. 6.8). Alkaloid storage cells exhibited a vacuolar pH of about 3, *i.e.* at least 2 units lower than in vacuoles of normal cells [27]. In the same order of magnitude were the pH differences found in lysosomes of macrophages accumulating weakly basic substances [33]. While the difference in the vacuolar pH was also measurable in young *Catharanthus* cells with a very low alkaloid content, cells grown in an alkaloid, non-producing medium exhibited no variation in the vacuolar pH. This indicates that the formation of storage cells is necessary for alkaloid production in cell suspension cultures of *Catharanthus roseus,* as shown already for *Macleaya.*

The pH difference is probably brought about by H^+-ion pumps in the tonoplast of alkaloid storage cells. Inhibition of these pumps by KCN exhibited no changes in the vacuolar pH during a period of 2 h, while dinitrophenol led immediately to an efflux of H^+-ions from vacuoles of the storage cells [27].

Recently it was suggested a highly selective active transport of alkaloids from experiments with isolated vacuoles of *Catharanthus roseus* [6 a]. However, the results confirm the characterization of a H^+-translocation ATPase in the tonoplast [34 a] rather than an active transport of the alkaloids.

In addition to the pH difference between cytoplasm and vacuole the uptake and retention of alkaloids is caused by particular properties of the tonoplast. This membrane seems to be highly permeable to neutral forms of alkaloids, but only slightly permeable to their protonated forms. The alkaloids will be trapped by protonation inside the vacuole and accumulated there. This mechanism seems to be a general principle in the storage of weak bases [12, 32, 33]. In plant cells, additional salt formation [5] or binding to organic substances [11] cannot be excluded.

6.3 Long-distance Transport of Alkaloids

Accumulation of alkaloids in some plants includes transport over a long distance because many alkaloids are synthesized far from their storage site. In tobacco and other *Solanaceae,* for example, alkaloids are synthesized in the roots and are translocated *via* xylem vessels to accumulate in the leaves [19]. The passage from the synthesizing root cells to the vessels and to the accumulating leaf cells cannot be explained by pH differences only. Binding of alkaloids to lignified cell wall thickenings of the xylem cells may lead to an accumulation in these structures [30] (Fig. 6.10). Alkaloids are removed from the cell wall thickenings by Ca^{2+}-ions. This process probably plays a role in the regulation of the long-distance transport and in the passage of alkaloids from xylem vessels to the accumulating leaf cells.

6.4 References

[1] Chaze, J.: C. R. **185** (1927) 80.
[2] Chaze, J.: Ann. Sci. Nat. Bot. **14** (1932) 5.
[3] Clautriau, G.: Rec. l'Inst. Bot. Bruxelles **2** (1906) 237.
[4] Cromwell, B. T.: Biochem. J. **27** (1933) 860.
[4a] Daly, J. W.; Myers, C. W.; Warnick, J. E.; Albuquerque, E. X.: Science **208** (1980) 1383.
[5] Dawson, R. F.: Plant Physiol. **21** (1946) 115.

[6] Deinzer, M. L.; Thomson, P. A.; Burgett, D. M.; Isaacson, D. L.: Science **195** (1977) 497.
[6a] Deus-Neumann, B.; Zenk, M. H.: Planta **162** (1984) 250.
[7] Errera, L.: Rept. Brit. Ass. **1904**, 815.
[8] Errera, L.: Rec. l'Inst. Bot. Bruxelles **2** (1906) 185.
[9] Fowden, L.: Nova Acta Leopoldina Suppl. **7** (1976) 117.
[10] Fairbairn, J. W.; Hakim, F.; Kheir, Y. E.: Phytochemistry **13** (1974) 1133.
[10a] Ibraeva, B. Ž.; Lovkova, M. Ja.; Klychev, L. K.: Izvest. Akad. Nauk SSSR, Ser. Biol. **1981**, 97.
[11] Jans, B. P.: Ber. Schweiz. Bot. Ges. **83** (1973) 306.
[12] Johnson, R. G.; Carlson, N. J.; Scappa, A.: J. Biol. Chem. **253** (1978) 1512.
[12a] Kurkdjian, A.: Physiol. Vég. **20** (1982) 73.
[13] Lärz, H.: Flora **135** (1942) 319.
[14] Lösecke, W.; Neumann, D.; Gröger, D.; Schmauder, H.-P.: Arch. Microbiol. **125** (1980) 251.
[15] Lösecke, W.; Neumann, D.; Schmauder, H.-P.; Gröger, D.: Z. Allg. Mikrobiol. **21** (1981) 767.
[16] Matile, Ph.; Jans, B.; Rickenbacher, R.: Biochem. Physiol. Pflanzen **161** (1970) 447.
[17] Molisch, H.: Mikrochemie der Pflanzen. Jena 1923; p. 285.
[18] Mothes, K.; Engelbrecht, L.; Tschöpe, K.-H.; Hutschenreuther-Trefftz, G.: Flora **144** (1957) 518.
[19] Mothes, K.; Romeike, A.: Die Alkaloide. In: Ruhland, W. (Ed.): Handbuch der Pflanzenphysiologie, p. 1008. Springer-Verlag, Berlin/Göttingen/Heidelberg 1958.
[20] Müller, E.: Nova Acta Leopoldina Suppl. **7** (1976) 123.
[21] Müller, E.; Nelles, A.; Neumann, D.: Biochem. Physiol. Pflanzen **162** (1971) 272.
[22] Müller, E.; Neumann, D.; Nelles, A.; Bräutigam, E.: Nova Acta Leopoldina Suppl. **7** (1976) 133.
[23] Munsche, D.: Flora **154** (1964) 317.
[24] Nessler, C. L.; Mahlberg, P. G.: Am. J. Bot. **64** (1977) 541.
[25] Neumann, D.: Biochem. Physiol. Pflanzen **168** (1975) 511.
[26] Neumann, D.: Nova Acta Leopoldina Suppl. **7** (1976) 77.
[27] Neumann, D.; Krauss, G.; Hieke, M.; Gröger, D.: Planta med. **48** (1983) 20.
[28] Neumann, D.; Lösecke, W.; Maier, W.; Gröger, D.: Biochem. Physiol. Pflanzen **174** (1979) 504.
[29] Neumann, D.; Müller, E.: Flora **158** (1967) 479.
[30] Neumann, D.; Müller, E.: Biochem. Physiol. Pflanzen **163** (1972) 375.
[31] Neumann, D.; Müller, E.: Biochem. Physiol. Pflanzen **165** (1974) 271.
[32] Ohkuma, S.; Poole, B.: J. Cell Biol. **90** (1981) 656.
[33] Poole, B.; Ohkuma, S.: J. Cell Biol. **90** (1981) 665.
[33a] Renaudin, J.-P.; Guern, J.: Physiol. Vég. **20** (1982) 533.
[33b] Robinson, T.: Science **184** (1974) 430.
[34] Saunders, J. A.: Plant Physiol. **64** (1979) 74.
[34a] Sce, H.: Physiol. Plant **61** (1984) 683.
[35] Schmidt, H.-H.: Protoplasma **40** (1951) 507.
[36] Thurston, R.; Smith, W. T.; Cooper, B. P.: Entomol. Exp. Appl. **9** (1966) 428.
[37] Tschirsch, A.; Vesterle, O.: Anatomischer Atlas der Pharmakognosie und Nahrungsmittelkunde. Leipzig 1900, p. 35.
[38] Vagujfalvi, D.: Acta Bot. Acad. Sci. Hung. **17** (1971) 217.
[39] Verzar-Petri, G.: Acta Bot. Acad. Sci. Hung. **18** (1973) 257.
[40] White, H. A.; Spencer, M.: Canad. J. Bot. **42** (1964) 1481.
[41] Yoder, L. R.; Mahlberg, P. G.: Am. J. Bot. **63** (1976) 1167.

7 The Ecological Significance of Alkaloids

D. Schlee

Generally, ecosystems are biological systems and functional unities involving the transfer of energy, foods, and inorganic nutrients between the environment and living organisms. In addition to these transfers within an ecosystem, living organisms interact with their immediate biotic environment by means of chemical substances.

In the network of an ecochemical continuum there is a flow of specific molecules that carry a certain amount of information [11]. Living organisms interact in many ways but for the majority of plants, animals, and microorganisms chemical signals are the main means of communication. The infinite number of these complex and heterogeneous chemical signals determines the complexity of the many different ecosystems found in nature [2—4, 13, 29, 50]. The versatility of chemical structures and their function as signals in ecosystems are very important for the diversity of life. In all communities chemical interactions are very important aspects of the adaptation of species to their environment and to one another. Moreover, species niche differentiation and community organization may be based mainly on chemical interactions [59]. In some cases secondary compounds, especially alkaloids, protect the organism against predators, competitors or pathogens. The ecological function of alkaloids as chemical signals in an ecosystem could account for the great variety in their structures, their selection in evolution, their specific location in the organism, and the rapidity of their translocation and turnover. Chemical interactions are dynamic processes and alkaloids therefore exist in a state of dynamic equilibrium and are not static end products of metabolism [38, 50]. By using "allelochemics" [9, 59] or "ecomons" [30, 43] one species may affect the growth, health, behaviour or population biology of organisms of another species, although these effects are in general due to a multiplicity of causes and compounds. Many chemical agents have two or three roles, while compounds that are toxic to one species may be ineffective against a close relative. Preservation, conservation and development of the species are the basic functions of these chemical interactions in a defined ecosystem (*cf.* [48]).

7.1 Alkaloids as Pigments

To attract animals to carry out pollination, plants depend quite naturally on their sexual system and floral structure (*e.g.* large and brightly coloured petals). In addition to the most important groups of flower pigments formed by flavonoids and carotenoids, some alkaloids also play a role as pigments. For instance, pteridines (*cf.* chapter 17.2), yellow, brown or red pigments, are found in the wing scales of butterflies, and betalains (*cf.* chapter 14.14), red or yellow pigments, occur in the flowers of certain families of *Centrospermae* and in the skin of pilei of *Amanita* and *Hygrocybe* species [61].

7.2 Alkaloids as Chemical Defence Agents (Repellents) and as Attractants

Bell [4] has suggested that plants synthesize a greater variety of secondary natural compounds than animals because plants cannot rely on physical mobility to escape from their predators.

The role of alkaloids in defence is related to their irritant, toxic and/or unpalatable properties and has been widely studied (*cf.* [33]). Two types are involved in affecting the feeding preferences of insects: chemical repellents (or feeding toxins) and chemical attractants (or feeding stimulants). Some insects and animals have evolved detoxification mechanisms, thereby making new ecological niches accessible to them.

Nearly every type of toxin can be utilized by a particular insect as a feeding stimulant. The alkaloid sparteine (*cf.* chapter 12.2) acts as a stimulant to the broom aphid *Acyrthrosiphon spartii* when feeding on *Sarothamnus scoparius*. The insect changes its feeding site according to where the highest concentration of alkaloid occurs within the host plant [53]. Sparteine is clearly important for the whole life cycle of the aphid.

The relationship between the cactus and fruit-fly in desert regions in western America involves both feeding attractants and repellents [22]. The attractant is common to all cacti, it is a sterol (*e.g.* in Senita cactus schottenol) and the fruit-fly uses this compound to synthesize its moulting hormone. On the other hand, alkaloids act as deterrents to particular *Drosophila* species. *Lophocereus schottii* (Senita cactus) contains the two alkaloids lophocereine and pilocereine (*cf.* chapter 14.2), which act as repellents to *Drosophila* feeders, only *D. pachea* being impervious to their toxic effects. In another cactus, *Carnegiea gigantea* (Saguaro) the alkaloid carnegeine acts as a repellent of *D. pachea* but not of *D. nigrospiracula*. The complex relationship is shown in Fig. 7.1.

Defence substances may provide their producers with an evolutionary advantage. In certain *Solanum* species alkaloids (*cf.* chapter 19.5) deter the Colorado beetle *Leptinotarsa decemlineata* [24, 54]. Resistance of potatoes to the Colorado beetle was first demonstrated in the wild form *Solanum demissum* and was found to be caused by the major steroid alkaloid of the leaves, the so-called demissine (*cf.* [49]).

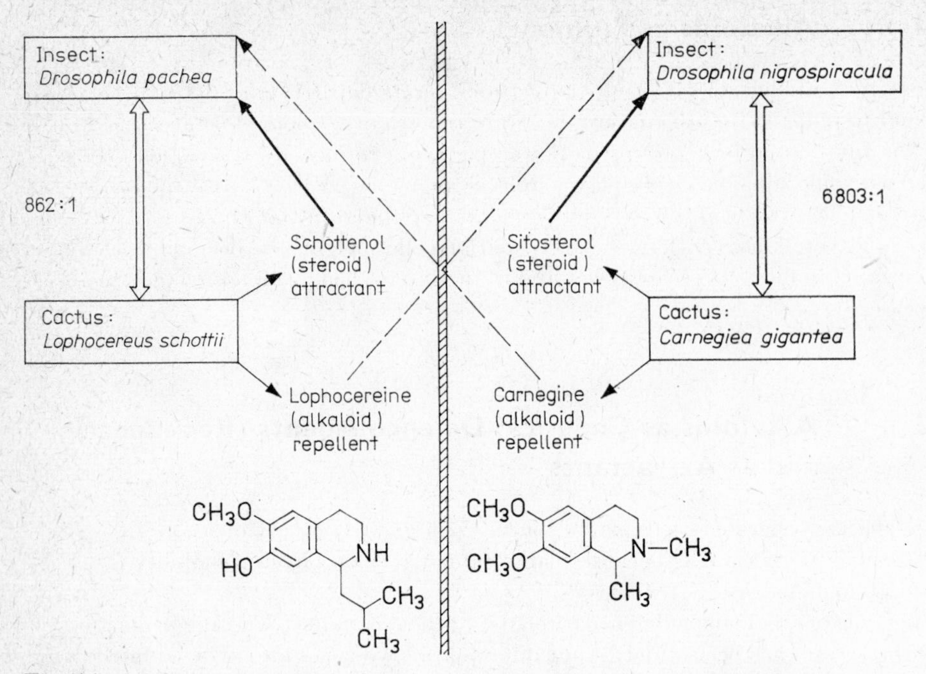

Fig. 7.1
Biochemical relationships between *Drosophila* species and *Cactaceae* in Western-American deserts (from [13])
Host-plant-specifity: for every 862 *D. pachea* resp. 6803 *D. nigrospiracula* exemplars one other *Drosophila* species was found.

This alkaloid has a structure closely related to that of solanine, the main alkaloid of the cultivated potato *S. tuberosum*. Tomatin, the major alkaloid in the tomato, *Lycopersicon esculentum*, is structurally similar to demissine and also acts as repellent. It is possible that these steroid alkaloids directly affect the steroid biosynthesis (*e.g.* ecdysone) of the beetle. In all 26 species of *Amaryllidoideae* (*e.g. Hymenocallis litteralis*) the alkaloid lycorine (*cf.* chapter 14.12) deters the locust *Schistocera gregaria* [52].

About two thousand plants are known to contain effective insecticides [5] and some plants combine insecticidal and repellent effects. Nicotine (*cf.* chapter 13.4) and anabasine (*cf.* chapter 13.4) act as powerful repellents of many insect groups.

In addition, plant alkaloids are also significant repellents of grazing animals [1, 55]. The *Compositae Senecio jacobaea* is rich in pyrrolizidine alkaloids (*cf.* chapter 11.3) and is avoided by sheep and cattle. Legumes of the genera *Lupinus*, *Baptisia*, *Cytisus* and *Genista* contain quinolizidine alkaloids (chapter 12.2) and are avoided by most herbivores. Further deterrents to grazing animals are gramine (*cf.* chapter 15.1) and hordenine (*cf.* chapter 14.1) in the grasses *Phalaris tuberosa* and *P. arundinacea*, the quinoline alkaloid perloline in *Lolium*- and *Festuca* species, and the pyrrolizidine alkaloid monocrotaline (*cf.* chapter 11.3) in *Crotalaria spectrabilis*.

Several alkaloids are also used for defence in animals. The major principle of the toxic repellent of the ant *Solenopsis saevissima* has been identified as a series of 2,6-dialkylpiperidines (*e.g.* 2-methyl-6-nonylpiperidine), which are structurally related to the highly toxic plant alkaloid coniine (2-propylpiperidine) of the hemlock, *Conium maculatum* (*cf.* chapter 12.1). Another example is the ant alkaloid solenopsin which shows haemolytic, insecticidal, and antibiotic activity [7, 16, 31, 37].

7.3 Alkaloids as Toxins

A great variety of toxic substances have been identified and many chemical structures are involved in such interactions as a form of protection of organisms against predators. Toxins affect primarily in small concentrations the metabolic processes of antagonistic predators. The toxicity depends on the amount taken over a given period, the age and physiological state of the animal, the mechanism of absorption and detoxification, as well as the mode of excretion. It is probably true that plant alkaloids, accumulated in the diet, mainly contribute to animal protection mechanisms [21, 41, 42]. The best known case is that of *Lepidoptera*. Alkaloids play an important role in the relationship between the *Compositae Senecio vulgaris* and *S. jacobaea* and the tiger moth *Arctia caja* and the cinnabar moth *Tyria jacobaeae* in temperate climates [41]. Both plants produce pyrrolizidine-type alkaloids (*e.g.* senecionine (*cf.* chapter 11.3)) which are highly toxic to many organisms. Nevertheless, the caterpillars of *T. jacobaeae* and *A. caja* are adapted to these toxic substances and complete their life cycle on these plants, where there are very few competitors. The toxic pyrrolizidine alkaloids are safely stored in the body of the insect and protect the moths against birds and other predators. Other insects, *e.g.* the Danaid butterflies, can store *Senecio* alkaloids after modification. Such interactions are common: Rothschild [42] has listed 23 *Lepidoptera*, 1 *Neuroptera*, 7 *Hemiptera*, 5 *Coleoptera*, 1 *Diptera* and 6 *Orthoptera* species that are capable of isolating and storing plant toxins.

Alkaloids involved in this process include, in addition to the pyrrolizidine group, the nicotine group, the alkaloids of *Lupinus*, and piperidine derivatives such as coniine [21].

Further examples showing the accumulation of toxic alkaloids in predators are surugatoxin (1) and saxitoxin (2) synthesized in certain algae (Fig. 7.2). Surugatoxin is found in the carnivorous marine gastropod *Babylonia japonica*. Its structure consists of a bromo-indol bound to a pteridine derivative.

Phytoplankton algae of the type *Gonyaulax* (*e.g. G. catenella, G. tamarensis*), *Gymodium*, *Prorocentrum*, and *Prymnesium* (dinoflagellates) can sometimes, depending on seasonal conditions, transmit the so-called saxitoxin (trialkyl-tetrahydropurine derivative) to mussels (*e.g. Mytilus*) and the Alaskan clam *Saxidomus giganteus*. The paralysing toxin is responsible for the toxicity of these marine invertebrates [23, 27, 28].

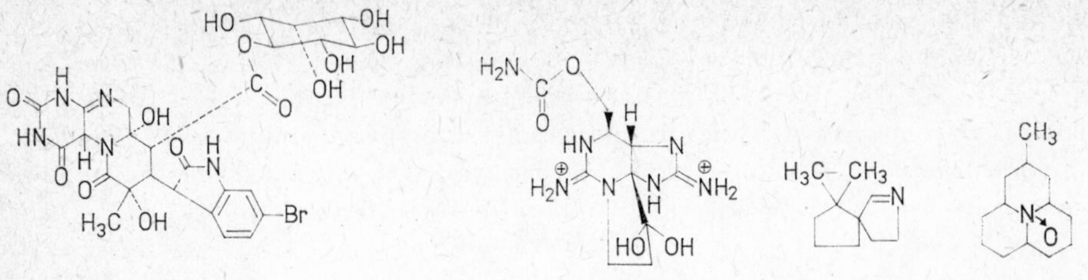

1 Surugatoxin 2 Saxitoxin 3 Polyzonimine 4 Coccinelline

Fig. 7.2
Some defence poisons

In addition, there is much evidence that poisonous alkaloids are synthesized *de novo* in animals. Alkaloids such as the quinazolones may be synthesized by the insects using biosynthetic paths similar to those found in higher plants. Two quinazolinones, glomerine (1,2-dimethyl-4-quinazolone) and homoglomerine (2-ethyl-1-methyl-4-quinazolone), have been identified in the European milliped *Glomeris marginata* [47]. These quinazolinones are closely related to the plant alkaloids arborine, glycorine, glycosmine, and peganine. Another milliped *Polyzonium rosalbum* forms polyzonimine (**3**), which acts as an irritant to predatory insects.

Furthermore, five compounds have been isolated from the secretion of the pygidial defensive glands of the staphylinid *Stenus comma* [45, 46]. The main component was an *N*-ethyl-3-(2-methylbutyl)piperidine alkaloid, the so-called stenusin. *Stenus comma* is a land beetle and its biotope is in the immediate neighbourhood of water. The beetle has to protect itself against predators and microorganisms, but also against the danger of drowning. Therefore, the alkaloid stenusin acts both as a toxic defence compound and as a spreading reagent.

The water-beetle *Ilybius fenestratus* produces methyl-8-hydroxy-quinoline 2-carboxylate in its defence secretion. The alkaloid is toxic to land animals and its antiseptic properties protect the beetle from penetration by microorganisms [44].

Coccinellidae (ladybirds) synthesize the *N*-oxid alkaloid coccinelline (**4**), which is one of a new class of alkaloids unknown in plants. After application of labelled 1-^{14}C- and 2-^{14}C-acetate, radioactive coccinelline was produced (*cf.* [2]).

7.4 Alkaloids as Pheromones

As reproduction of the species is one of the main purposes of living organisms, the attraction of members of the same species but of opposite sex is very important. Chemical signals are used to induce sexual attraction and "pheromone" is the accepted term used to describe them [6, 8, 18—20, 25, 51].

The existence of sexual pheromones has been proved in many insect species and other animals. Sex pheromones can have two types of effects: they attract members of the opposite sex, *e.g.* the long-distance attraction which the female butterflies exert on the male, and they can be aphrodisiacs.

There is good evidence that host plants actually provide certain insects with alkaloids as precursors of their sex pheromones *via* the diet [8, 14, 15].

Male danaid butterflies (*e.g. Danaus plexippus, D. gilippus, D. chrysippus, Lycorea ceres*) are attracted by plants of the families *Asteraceae, Boraginaceae, Fabaceae* and others containing pyrrolizidine alkaloids (*cf.* chapter 11.3) which they transform to male sex pheromones [10, 32]. The males store the transformed pyrrolizidines in their hairpencils at the tip of the abdomen. Here pyrrolizidone is released and its emanation inhibits the flight of females and promotes copulation [10, 26, 32].

Alkaloids are also used as pheromones by social insects (bees, ants, termites) to lay an odour trail from the nest to a food source and back. The chemical structure of such pheromones varies widely and they are highly specific. The leaf-cutting ants *Atta texana* and *A. cephalotes,* for instance secrete methyl-4-methylpyrrole-2-carboxylate as a trail pheromone. It is detected by *Atta* ants up to a concentration of 3.5×10^8(!) molecules per cm [40, 56]. The tropical ant *Monomorium pharaonis* uses 5-methyl-3-butyl-octahydroindolizine as a trail pheromone, which is derived from an alkaloid of plant origin [36, 58].

7.5 Alkaloids as Allelopathics

In addition to their involvement in plant-animal and animal-animal relations, certain alkaloids can be concerned in the relationships between higher plants and between higher plants and microorganisms.

Allelopathy is one aspect of the competition between plants. Phytotoxic substances may be released from one species by volatilization, washing of the surface, exudation from the plant roots and/or leaches from litter into the soil. They involve inhibition in the germination rate, growth, yield, and development of other species in the neighbourhood [34, 35]. Such allelopathic effects of chemicals may also be exerted between plants of the same type or between individuals of the same species ("auto-toxication").

The importance of alkaloids as seed germination inhibitors is reported, *e.g.* berberine (*cf.* chapter 14.10). The rhizomes of *Coptis japonica* that contain the alkaloid, grow slowly and usually take about 15 years to reach maturity. It seems that both endogenously and exogenously supplied berberine alkaloids influence plant cell growth [60].

Alkaloids may also play a role in the interaction between higher plants and microorganisms causing plant diseases [12, 17, 57].

Thus berberine, which occurs in the roots of *Mahonia trifoliata,* has a protective function in the plants attacked by the fungus *Phymatotrichum omnivorme.* The compound also performs bacteriostatic and bactericidal actions [60].

The hordatines A and B from Hordeum vulgare, which contain two *p*-coumaric acid residues linked to arginine, have a protective function against *Helminthosporium sativum.*

7.6 Conclusions

The examples given here suggest that at least some of the alkaloids produced in the different groups of organisms may be of ecological significance and that they have acquired functions which aid the development of various species. The main function is the protection of plants or animals against attack by their predators or microorganisms and/or an improvement of the ability of one species to compete with others in a particular habitat [55].

Most important in this respect is probably the use of alkaloids as repellents or toxins in defending against potential predators or microorganisms. The frequent location of alkaloids in exposed parts of the producer organism supports the idea of alkaloids acting as a defence mechanism: *e.g.* in the leaves, hairs, and cortex of plants and in the skin glands or defence glands of animals. The alkaloid content of these organs is often maintained in a state of equilibrium by synthesis and degradation and may be raised and lowered according to requirements [38, 39, 50].

In comparison with defence by products that, like polyphenols, reduce the parability of proteins, defence by alkaloids is less costly relative to precursors, energy etc. necessary for biosynthesis, because lower concentrations are sufficient. But the presence of alkaloids is not only determined by the ability to synthesize these compounds from specific precursors (*e.g.* tryptophane, tyrosine) and/or inability to achieve their degradation. As these products are toxic, their synthesis has necessarily required the co-evolution (sequential rather than simultaneous) of additional biochemical and/or structural systems for their excretion or safe accumulation. Alkaloids are stored in protected cell compartments surrounded by semi-permeable membranes with unidirectional transport, which isolate the toxic compounds. Moreover, detoxication of alkaloids, for instance by methylation or conjugation with carbohydrates, amino acids or proteins, seems to be relatively easy.

Similar aspects are observed in animals. Alkaloids are occasionally evident as defence substances in animals, but only in close relation to the development of mechanisms for the prevention of auto-intoxication. Alkaloids in plant cells do not, in general, protect plants from predators. On the contrary, some specialized herbivores use specific plant alkaloid producers avoided by other animals as an ecological niche.

But there are few data on the biochemical basis for the many mutualistic associations that exist between organisms. It is necessary to know more about the seasonal and diurnal (sometimes hourly) variations in concentration of alkaloids in organisms in both space and time. These changes could be of prime significance when considering the functions of alkaloids as chemical signals in various ecosystems.

The interpretation of the accumulation of certain alkaloids with irritant, toxic and/or unpalatable properties in plants as a defence mechanism against herbivores represents, however, only one aspect. It is possible that such compounds can directly influence the viability of the accumulating plants. Particularly under conditions of environmental stress the plant compartmentation may break down leading to autotoxic effects [9a]. In certain barley genotypes, the accumulation of the toxic indole alkaloid gramine is particularly marked under high temperature [12a]. Experiments indicate that such high-gramine barley genotypes with heat-inducible gramine accumulation show more severe symptoms of heat injury, e.g. chlorosis and necrosis of exposed leaves, than low-alkaloid types. This suggests that gramine accumulation may be a deleterious stress response in natural environments in which high temperature stress is probable.

As yet it is difficult to estimate the percentage of the several thousands of alkaloids known that are significant in ecology and to ascertain their importance. It is necessary to know more about the dependence and inter-relationships of plants, animals, and microorganisms which determine the habitat complexity in an ecosystem. The study of these complex interactions is still in its infancy and the difficulties in finding suitable experimental systems hinder the examination of the role played by alkaloids.

7.7 References

[1] Arnold, G. W.; Hill, J. L.: In: Harborne, J. B. (Ed.): Phytochemical Ecology. Academic Press, London 1972, p. 72.
[2] Barbier, M.: Introduction to Chemical Ecology. Longman, New York (1979).
[3] Beck, S. D.; Reese, J. C.: Recent Adv. Phytochem. **10** (1976) 41.
[4] Bell, E. A.: In: Bell, E. A.; Charlwood, B. V. (Eds.): Encyclopedia Plant Physiology. Vol. 8. Springer-Verlag, Berlin 1980, p. 11.
[5] Beye, F.: Plant Research and Development **7** (1978) 13.
[6] Birch, M. C. (Ed.): Pheromones. North-Holland Publishing Company, Amsterdam/London 1974.
[7] Blum, M. S.; Jones, T. H.; Hölldobler, B.; Fales, H. M.; Jaouni, T.: Naturwissenschaften **67** (1980) 144.
[8] Boppré, M.: Biologie in unserer Zeit **7** (1977) 161.
[9] Brown, W. L.; Eisner, T.; Whittaker, R. H.: Biosciences **20** (1970) 21.
[9a] Daday, H.: Heredity **20** (1965) 355.
[10] Edgar, J. A.; Culvenor, C. C. J.: Experientia **31** (1975) 393.
[11] Florkin, M.: Aspects Moléculaires de l'Adaptation et de la Phylogénie. Collection GPB Masson, Paris 1966.
[12] Friend, J.; Threlfall, D. R.: Biochemical Aspects of Plant Parasite Relationship. Academic Press, London 1976.
[12a] Hanson, A. D.; Traynor, P. L.; Ditz, K. M.; Reicosky, D. A.: Crop Sci. **21** (1981) 726.
[13] Harborne, J. B.: Introduction to Ecological Biochemistry. Academic Press, London/New York/San Francisco 1977.
[14] Hendry, L. B.; Wichmann, J. K.; Kurdenlang, D. M.; Mumma, R. O.: Science **188** (1975) 59.
[15] Hendry, L. B.; Wichmann, J. K.; Kurdenlang, D. M.; Mumma, R. O.: Science **192** (1976) 143.
[16] Hölldobler, B.: Oecologia **11** (1973) 371.
[17] Ingham, J. L.: Phytopath. Z. **78** (1973) 314.
[18] Karlson, P.; Butenandt, A.: Ann. Rev. Entomol. **4** (1959) 39.
[19] Karlson, P.; Lüscher, M.: Nature **183** (1959) 55.
[20] Karlson, P.; Schneider, D.: Naturwissenschaften **60** (1973) 113.
[21] Keeler, R. F.: Lloydia **38** (1975) 56.
[22] Kircher, H. W.; Heed, W. B.: Recent Adv. Phytochem. **3** (1970) 191.
[23] Kremer, B. P.: Naturwissenschaften **68** (1981) 101.

[24] Kuhn, R.; Low, I.: Chem. Ber. **94** (1961) 1088, 1096.
[25] Matsumoto, Y.: In: Wood, D. L.; Silverstein, R. M.; Nakajima, M. (Eds.): Control of Insect Behaviour by Natural Products. Academic Press, New York 1970, p. 189.
[26] Meinwald, J.; Meinwald, I. C.; Mazzacchi, P. H.: Science **164** (1969) 1174.
[27] Mebs, D.: Naturwiss. Rdsch. **30** (1977) 367.
[28] Mebs, D.: Naturwiss. Rdsch. **31** (1978) 508.
[29] Mothes, K.: Pharmazie **36** (1981) 199.
[30] Pasteels, J. M.: Ann. Soc. roy. Zool. Belg. **103** (1973) 103.
[31] Pedder, D. J.: Tetrahedron **32** (1976) 2275.
[32] Petty, R. L.; Boppré, M.; Schneider, D.; Meinwald, J.: Experientia **33** (1977) 1324.
[33] Rhoades, D. F.; Cates, R. G.: Recent Adv. Phytochem. **10** (1976) 168.
[34] Rice, E. L.: Allelopathy. Academic Press, New York/San Francisco/London 1974.
[35] Rice, E. L.: Bot. Rev. **45** (1979) 17.
[36] Ritter, F. J.; Rotgans, I. E. M.; Talman, E.; Verwiel, P. E. J.; Stein, F.: Experientia **29** (1973) 530.
[37] Ritter, F. J.; Persoons, C. J.: Neth. J. Zool. **25** (1975) 261.
[38] Robinson, T.: Phytochem. Bull. **6** (1973) 5.
[39] Robinson, T.: Science **184** (1974) 430.
[40] Robinson, S. W.; Moser, J. C.; Blum, M. S.; Amante: Insectes Soc. **21** (1974) 87.
[41] Rothschild, M.: In: Harborne, J. B. (Ed.): Phytochemical Ecology. Academic Press, London 1972, p. 1.
[42] Rothschild, M.: In: Van Emden, H. F.: Insect Plant Relationships. Blackwell Science Publishers, Oxford 1973, p. 59.
[43] Schaefer, M.: Naturwiss. Rdsch. **33** (1980) 128.
[44] Schildknecht, H.: Endeavour **30** (1971) 136.
[45] Schildknecht, H.: Angew. Chem. Int. Ed. Engl. **15** (1976) 214.
[46] Schildknecht, H.: In: Marini-Bettolo, G. M. (Ed.): Natural Products and the Protection of Plants. Elsevier, Amsterdam/Oxford/New York 1977, p. 59.
[47] Schildknecht, H.; Wenneis, W. F.; Weis, K. H.; Maschwitz, U. M.: Z. Naturforsch. **21** (1966) 121.
[48] Schlee, D.: Biol. Rdsch. **19** (1981) 189.
[49] Schreiber, K.: Kulturpflanze **16** (1968) 255.
[50] Seigler, D. S.: Biochem. Syst. Ecol. **5** (1977) 195.
[51] Shorey, H. H.: Animal Communication by Pheromones. Academic Press, New York 1976.
[52] Singh, R. P.; Pant, N. C.: Experientia **36** (1980) 552.
[53] Smith, P.: Naturwissenschaften **212** (1966) 213.
[54] Sturchow, B.; Low, I.: Entomol. Exp. Appl. **4** (1961) 133.
[55] Swain, T.: Ann. Rev. Plant Physiol. **28** (1977) 479.
[56] Tomlinson, J. H.; Silverstein, R. M.; Moser, J. C.; Brownlee, R. C.; Ruth, J. M.: Nature **234** (1971) 348.
[57] Wheeler, H.: Plant Pathogenesis. Springer-Verlag, Berlin 1975.
[58] Wheeler, J. W.: Lloydia **39** (1976) 53.
[59] Whittaker, R. H.; Feeny, P. P.: Science **171** (1979) 757.
[60] Yamamoto, H.: Bot. Mag. Tokyo **93** (1980) 307.
[61] Zahn, M.: Naturwiss. Rdsch. **31** (1978) 363.

8 Metabolism of Amino Acids Related to Alkaloid Biosynthesis

H. Reinbothe and J. Miersch

8.1 General Features of Amino Acid Metabolism and the Origin of the Amino-Nitrogen

Most alkaloids are derived from protein amino acids, such as the aromatic amino acids phenylalanine, tyrosine and tryptophan (*cf.* chapters 14 and 15), histidine (*cf.* chapter 18), lysine (*cf.* chapter 12), the closely related amino acids glycine and serine (*cf.* chapter 17), the members of the glutamate family, *e.g.* ornithine (*cf.* chapter 11), as well as from compounds closely related to true amino acids, such as anthranilic acid (*cf.* chapter 16) and nicotinic acid (*cf.* chapter 13). The idea that amino acids are the most important precursors of alkaloids was put foreward by Trier in 1912 [193] and was proved to be true in many experiments. In this chapter some general features of the amino acid metabolism, of the general nitrogen metabolism of plants and microorganisms and of the biosynthetic pathways of the alkaloid precursor amino acids and their regulation will be discussed. A number of excellent reviews exist in this field providing more detailed information [13, 25, 78, 79, 105, 117, 126, 127, 132, 134, 138, 196, 202].

Interest is focussed on plant amino acid metabolism because most of the known alkaloids are of plant origin. Unfortunately, however, information about the amino acid metabolism of alkaloid-producing plants is scarce. Most examples are therefore derived from plants that do not synthesize alkaloids. Nevertheless, the amino acid metabolism of different plants has many common features and generalization is possible.

8.1.1 The Origin of the Amino-Nitrogen

Green plants are exceptional when compared with other living beings because they are able to transform light energy into "chemical energy". "The reducing power" formed during photosynthesis is used for instance in the reductive assimilation of carbon dioxide, nitrate and sulfate. In addition several plant-microbe systems, such as the symbiotic system legume-Rhizobium, are capable of assimilating gaseous nitrogen. Most of the biologically fixed nitrogen available for agriculture is derived from rhizobia in symbiotic association with legumes. It may be stated that glutamine synthetase and ribulose-1,5-biphosphate carboxylase are the most important proteins in living organisms on which life on earth ultimately depends [78].

Nitrate as well as N_2 are reduced to ammonia by the use of electrons derived from the light-driven reactions of photosynthesis [67, 104]. The reduction of one molecule of carbon dioxide to the carbohydrate level requires 4 electrons, whereas the reduction of one molecule of NO_3^- to ammonia and the conversion of the latter into amino-N requires 10 electrons. On the basis of dry weight, the C/N ratio of the organic matter of plants is about 5:15, thus the amount of electrons necessary for nitrate assimilation is about one-fourth of that involved in the photosynthetic fixation of carbon dioxide.

Since 1970 evidence has been accumulated that the major route of the introduction of ammonia into amino acids is catalysed by the combined operation of a) glutamine synthetase (GS, *L*-glutamate : ammonia ligase (ADP-forming), EC 6.3.1.2) and b) glutamate synthase (*L*-glutamine : 2-oxoglutarate aminotransferase (GOGAT), *L*-glutamate : $NADP^+$ oxidoreductase (transaminating), EC 1.4.1.13, or *L*-glutamate : ferredoxin oxireductase (transaminating), EC 1.4.7.1 [47, 48]) according to the following equations (*e.g.* [95, 129, 210]):

a) *L*-Glutamate + NH_3 + ATP $\rightleftharpoons$ *L*-glutamine + ADP + P_i

b) *L*-glutamine + 2-oxoglutarate + Fd_{red} $\rightleftharpoons$ 2*L*-glutamate + Fd_{ox}

Net reaction:

NH_3 + ATP + 2-oxoglutarate + Fd_{red} $\rightarrow$ 2*L*-glutamate + ADP + P_i + Fd_{ox}

L-Glutamine as well as *L*-glutamate play a central role in the nitrogen metabolism of plants. Glutamine-N is readily transferred by transamination enabling the biosynthesis of many low-molecular compounds, such as NAD(P), carbamoylphosphate (see below), purines, pyrimidines as well as their nucleotides (*cf.* chapter 17), amino sugars, tryptophan (*cf.* chapter 8.2.1), histidine (*cf.* chapter 8.2.2), glutamate (*cf.* chapter 8.1.2), and *p*-aminobenzoate (being a precursor of the pteridines, *cf.* chapter 17). The general reaction of the glutamine-dependent transfer of amide-N by glutamine amidotransferase may be outlined as follows:

L-glutamine + acceptor compound $\rightleftharpoons$ aminated acceptor + *L*-glutamate.

In addition *L*-glutamate is an almost universal donor of the amino group in transamination reactions requiring pyridoxal phosphate:

L-glutamate + 2-oxoacid $\rightleftharpoons$ α-amino acid + 2-oxoglutarate.

Until 1970 NADPH-dependent glutamate dehydrogenase (GDH, *L*-glutamate : $NADP^+$ oxidoreductase (deaminating), EC 1.4.1.4) was considered significant in ammonia assimilation in all organisms including plants [139, 149, 186] and the root nodules of legumes [11, 52, 187]. The ability of GDH to utilize either NADPH or NADH was demonstrated in plant extracts, although the NADPH-dependent enzyme activity was usually much greater:

$$\text{2-oxoglutarate} + NH_3 + \text{NADPH} + H^+ \rightleftharpoons \textit{L}\text{-glutamate} + \text{NADP}^+ + H_2O.$$

It is suggested that in plant tissues GDH is not operative under "physiological conditions" of ammonia supply, because of the high K_m for ammonia [139, 152], which reflects a rather low substrate affinity compared with that of glutamine synthetase, which was estimated to be $1.9 \cdot 10^{-5}$ M [150]. There is evidence of the operation of a "glutamate synthase cycle" in plant tissues, the function of which depends upon the availability of ammonium and 2-oxoglutarate. If the supply of ammonium is limited, all the ammonium available is converted to glutamate, whereas if oxoglutarate is limited, all the available NH_4^+ and oxoglutarate are converted to glutamine, thus depriving the cell either of glutamine and free NH_4^+ or of oxoglutarate and glutamate. Several enzymes earlier supposed to be involved in ammonium assimilation in higher plants [163] are not significant, with the exception of the carbamoylphosphate synthesizing enzyme, which is of great importance both for UMP synthesis *via* the orotate pathway and for the formation of citrulline/arginine [88, 89].

The bacterial carbamoylphosphate synthetase (CPS) catalysing the irreversible formation of carbamoylphosphate (CP) uses the glutamine amide-*N* in marked preference to ammonia [158]. Carbamoylphosphate synthetase from various angiosperms has a pH optimum near 8.2 and is stimulated by Mg^{2+} and K^+. Activity is enhanced by *L*-ornithine, GMP, and GTP, but inhibited by UMP. Plant tissues probably contain only one kind of CPS providing CP for both UMP and arginine formation. In contrast in *Neurospora crassa* and in baking yeast as well as in animal tissues two different enzymes exist which synthesize carbamoylphosphate: an arginine pathway related CPS-A (CPS-I), which in animal tissues is mitochondrial, and a pyrimidine-specific CPS-P (CPS-II) [108, 200]. CPS-A requires *N*-acetyl-*L*-glutamate (NAG) as an allosteric effector [179]:

$$NH_4^+ + HCO_3^+ + 2\,\text{ATP} \xrightarrow{\text{NAG}} \text{carbamoylphosphate} + 2\,\text{ADP} + 2\,P_i$$

The synthesis of CPS-A of *Saccharomyces cerevisiae* [96] is regulated by two processes, one of which is specifically exerted by *L*-arginine and probably involves a repressor-operator type of interaction, the second is related by a more general control mechanism (*cf.* chapter 8.1.5) that regulates enzyme syntheses in the arginine pathway as well as in a number of other biosynthetic routes in amino acid metabolism [156, 157]. Control of CPS-A synthesis by *L*-arginine works through activation of the NAG synthase [179]. This control mechanism regulates the expression of gene cpa I that codes for the synthesis of the small subunit of CPS-A (mol.wt. 36 kdalton, showing glutaminase activity) but shows little influence on the production of the large subunit (mol.wt. 140 kdalton) encoded by the gene cpa II [158]. In *N. crassa* CPS-A and CPS-P yield two different subcellular pools of carbamoylphosphate, one of which is localized in the mitochondria and is related to arginine biosynthesis [15, 90]. As with *S. cerevisiae*, CP produced by each enzyme is not freely available to both arginine and nucleotide formation, because of feedback inhibition by the respective end products (see above).

8.1.2 The Glutamine Synthetase-Glutamate Synthase System

Glutamine synthetase (GS), an enzyme widely distributed in microorganisms, animals and plants, fixes ammonia in the amide-*N* of glutamine at the expense of the cleavage of ATP to ADP and P_i [95]. Glutamine synthetase isolated from various cells catalyses several partial reactions, *e.g.* γ-glutamyl transfer, arsenolysis, formation of 5-oxoproline as well as that of acyl phosphates, which in the presence of ADP results in the synthesis of ATP. The enzyme-bound intermediate, γ-glutamyl phosphate, is attacked by ammonia in such a way that *L*-glutamine is formed *via* a tetrahedral intermediate [65, 117]. *L*-Methionine sulfoximine is an inhibitory analogue of the transition state intermediate to the catalytic reaction [170]. *L*-Glutamine binds to the smaller subunit of GS ("glutaminase"), and ammonia released from the amide group is transferred to a binding site, which also accepts exogenously supplied ammonia. As with GS, almost every glutamine amidotransferase can use ammonia in place of glutamine [117].

Glutamine synthetase, glutamate synthase, and glutamate dehydrogenase are present in chloroplasts; GDH was also found in mitochondria [123, 132, 186]. By use of protoplasts, which are a convenient material for cell fractionation studies, it was shown that probably almost all of the Fd-dependent glutamate synthase of leaves is located in the chloroplasts [203], while only part of the total enzyme activity is cytosolic [85, 143]. Probably different isoenzymes of glutamine synthetase exist in chloroplasts and cytoplasma [109].

8.1.3 Nitrate Assimilation

Nitrate, the principal nitrogen source of most plants, has to be reduced to ammonium before it can be used in the biosynthesis of glutamine and glutamate [74, 77, 148]. Nitrate assimilation is performed by the combined operation of nitrate reductase, which reduces nitrate to nitrite, and nitrite reductase, which reduces nitrite to ammonium:

$$NO_3^- \xrightarrow{2e} NO_2^- \xrightarrow{6e} NH_4^+$$

Both enzymes are metalloproteins. Ferredoxin-dependent nitrate reductase is typical for cyanobacteria and some other photosynthetic prokaryotes, whereas pyridine nucleotide-dependent nitrate reductase was found in plants [67]:

$$NO_3^- + NAD(P)H + H^+ \xrightarrow{2e} NO_2^- + NAD(P)^+ + H_2O$$

According to the specificity shown either for NADH (EC 1.6.6.1, leaves of most angiosperms, *cf. e.g.* [81, 144]) or NADPH (EC 1.6.6.2), two classes of pyridine nucleotide-dependent enzymes may be distinguished. NAD(P)H-nitrate reductases possess two activities: first, a FAD-dependent diaphorase activity related to the NAD(P)H-activating moiety of the enzyme complex and second, a pyridine nucleotide-independent terminal activity containing molybdenum, which is able to reduce nitrate by reduced flavins or viologens. The flow of electrons from NAD(P)H to nitrate through the nitrate reductase system may be outlined as follows:

$$NAD(P)H \rightarrow [FAD \rightarrow cyt\ b\text{-}557 \rightarrow Mo] \rightarrow NO_3^-$$

Under "normal" conditions nitrate uptake and nitrate reductase activity are closely related [80]. Obviously, the rate of nitrate flux into the plants is correlated with the induction of nitrate reductase. However, nitrate may accumulate in plant cells, *e.g.* in the large central vacuole of leaf mesophyll cells [110], even in the presence of high nitrate reductase activity. A small "metabolic pool" of nitrate accessible to nitrate reductase exists besides a large "storage pool". Incoming nitrate does not fully mix with endogenous nitrate. A steady influx of NO_3^- has to be maintained to keep nitrate reductase induced [175]. In this context, it is noteworthy to keep in mind that nitrate can accumulate in large quantities in certain crop plants, *e.g.* in members of the *Chenopodiaceae* (spinach, beet), *Gramineae* (*Poaceae*), *Brassicaceae*, and *Asteraceae*. In extreme physiological conditions, concentrations of nitrate may exceed 2% fresh wt. corresponding to 17 to 24% dry wt. [79]. This fact might be alarming, because nitrate may be reduced to toxic nitrite in the human body which is known to react with amines to yield carcinogenic nitrosamines [207].

Ferredoxin-nitrite reductase is a characteristic of photosynthetic organisms, whereas NAD(P)H-nitrite reductase is present in such non-photosynthetic organisms as moulds, yeasts and bacteria. Nitrite reductase catalyses the reduction of nitrite to ammonium, which implies the transfer of 6 electrons. Ferredoxin (Fd)-dependent nitrite reductase is mainly located in the chloroplasts and is usually a soluble enzyme (EC 1.7.7.1). Its reaction may be given as follows:

$$NO_2^- + 6\,Fd_{red} + 8\,H^+ \xrightarrow{6e} NH_4^+ + 6\,Fd_{ox} + H_2O$$

In different nitrite reductases flavodoxin can substitute for Fd [198]. Purified Fd-nitrite reductase of higher plants contains a heme prosthetic group, which was identified as "siroheme" in spinach, previously shown to be a component of sulphite reductase. The iron-sulphur centre of the prosthetic group of the spinach enzyme is a tetranuclear cluster (4Fe – 4S) [97, 100]. The flow of electrons from Fd_{red} to nitrite through the nitrite reductase complex proceeds as follows:

$$Fd_{red} \rightarrow [4\,Fe - 4\,S] \rightarrow \text{siroheme} \rightarrow NO_2^-$$

Kinetic and steady-state turnover experiments suggest that nitrite bound as a ligand of the siroheme prosthetic group is reduced to ammonium without the occurrence of free intermediates.

The rate-controlling step in nitrate assimilation is the reduction of nitrate to nitrite. Nitrate reductase levels were shown to fluctuate in response to many environmental factors, such as light, temperature, pH, O_2 tension, water potential, nitrogen source etc. [75, 76, 104, 147, 199]. Furthermore, provision by the substrate to the nitrate assimilation system, and hence uptake, storage, and translocation of NO_3^-, may play important roles. In angiosperms, nitrate reductase is usually regarded as a substrate-inducible enzyme. However, fairly high enzyme levels were sometimes found even in the absence of nitrate [67]. It is well known that cytokinins as well as several organic nitro-compounds are able to induce nitrate reductase in place of nitrate [184]. Therefore, the proposed role of nitrate as an obligatory inducer is difficult to generalize. In addition an antagonistic effect of ammonium and amino acids [75] on the synthesis of the nitrate assimilatory enzymes has often been reported [91]. It appears that either ammonium itself or a product of ammonium assimilation plays a crucial role in the regulation of enzyme levels of the nitrate assimilation system.

8.1.4 N_2 Fixation by the Plant-Microbe Pair Legume-Rhizobium

Good deal has been learnt about the biology of the host plant, the rhizobia, and how the nitrogen-fixing symbiotic association between bacteria and plants is established [12]. In the course of the infection the rhizobia are released into the cytoplasm of recently divided cortical cells of the host plant, where they are enclosed by host membranes. The bacteria develop into the pleomorphic bacteroids capable of fixing N_2. Free-living rhizobia cannot subsist on N_2, although under special environmental conditions they produce nitrogenase.

The progress achieved in the enzymology and genetics of nitrogen fixation has recently been reviewed (*cf. e.g.* [71, 141]). The ultimate reductant for the nitrogenase of legume root nodules remains unclear, because the role of Fd_{red} in nitrogen-fixing soybean nodule bacteroids [14, 109] has not been clearly established.

Assimilation of the fixed nitrogen occurs *via* GS and glutamate synthase of plant origin [8, 36, 44]:

a) $1/2\,N_2 + 6\,ATP + 3e \rightarrow NH_3 + 6\,ADP + 6\,P_i$ (nitrogenase)
b) $NH_3 + ATP +$ glutamate $\rightarrow ADP + P_i +$ glutamine (GS)
c) glutamine + 2-oxoglutarate + $NAD(P)H \rightarrow 2$ glutamate + $NAD(P)^+$

Sum: $1/2\,N_2 + 7\,ATP + 3e + NAD(P)H +$ 2-oxoglutarate $\rightarrow 7\,ADP + 7\,P_i$
$+ NAD(P)^+ +$ glutamate

During nodule development the activities of GS and glutamate synthase showed a parallel increase, which was closely matched by a similar increase in nitrogenase as well as in leghemoglobin, which functions as an oxygen-binding hemeprotein of legumes. About 90% of the total glutamine synthetase was found in the plant fraction, whereas enzyme activity in the bacteroids remained low. The bulk of GS appears to be cytoplasmic, whereas the glutamate synthase is located in plastids. In nodulated legumes the bound nitrogen is translocated in the form of allantoin and allantoic acid [190]. Biosynthesis of these ureides proceeds by purine *de novo* synthesis and oxidative purine breakdown [164, 166, 174, 217].

8.1.5 Regulatory Principles in Amino Acid Metabolism: A Short Survey

The rate of metabolic fluxes through the pathways of amino acid biosynthesis is controlled by end product inhibition and repression [25, 134, 195, 196], while metabolite compartmentation is also important [54, 62, 86, 145, 146, 209, 221]. Amino acids are unequally distributed between the cytoplasmic portion of the cells and various organelles, such as the mitochondria, the chloroplasts and the vacuoles [40, 41, 49—51, 83, 111, 112, 214, 215]. Intercompartmental fluxes of amino acids are important in the regulation of metabolism and differential gene expression.

Intermediate compartmentation and substrate channelling take place by means of enzyme aggregation, such as multi-enzyme formation and multifunctionality of proteins [18]. In *N. crassa* for instance, 5 enzymes of the shikimate pathway catalysing the conversion of 3-deoxy-*D*-arabino heptulosonate-7-phosphate (DAHP) into 3-enoylshikimate-5-phosphate are localized at one polypeptide chain possessing 5 different active sites (domains) [201]. This "arom conjugate" represents a pentafunctional protein, the spatial arrangement of which is brought about by the peculiar tertiary structure of the polypeptide. In this way the substrate, DAHP, is channelled to the synthesis of chorismate, which is an important branching point in the metabolism of aromatic compounds (chapter 8.2.1). The "arom complex" is co-ordinately activated and protected against proteolysis by the first substrate. Co-ordinative protection against proteolytic attack is of special value, because it preserves the activity of the multifunctional protein. In times of energy deficit, for instance, a break in the metabolite channel converts the flow of carbon from the shikimate pathway into a competing catabolic system.

Multifunctionality of both anthranilate synthase and tryptophan synthase is discussed in chapter 8.2.1. Evidence for multifunctionality of enzymes requires careful genetic analysis [35], because the detection of the enzymes separately or as a multi-enzyme complex could be an artefact.

As mentioned above a general control mechanism exists in the amino acid metabolism of *S. cerevisiae* regulating the synthesis of the enzymes of the arginine pathway together with the enzymes of several other biosynthetic routes [118, 119, 120].

When all the amino acids used for protein biosynthesis are in excess, *i.e.* during growth in amino acid-rich medium, the various pathways are repressed. Upon growth in an unbalanced amino acid mixture or upon starvation for any short running amino acid whose synthesis is subject to this general control, the respective enzymes are derepressed. There is evidence that this mechanism is exerted by the amino-acyl-tRNA species and operates at the level of transcription. This general control mechanism might also be valid in plant cells with a need for certain protein amino acids, *e.g.* plant cell cultures.

8.2 Biosynthesis of Alkaloid Precursor Amino Acids

8.2.1 Biosynthesis of *L*-Phenylalanine, *L*-Tyrosine and *L*-Tryptophan

In the biosynthesis of the aromatic amino acids *L*-phenylalanine, *L*-tyrosine and *L*-tryptophan, a number of reactions result in the formation of chorismic acid, which gives prephenic acid by an enzymatic rearrangement (Fig. 8.1, reaction 1). Prephenic acid is the branching point in the formation of phenylalanine and tyrosine (reactions 2 and 4) [171, 195]. The last biosynthetic step is catalysed by aminotransferases (reactions 3 and 5), which have been studied in bacteria [84] and higher plants [56, 61, 211]. Chorismate mutase plays a central role in the biosynthesis of

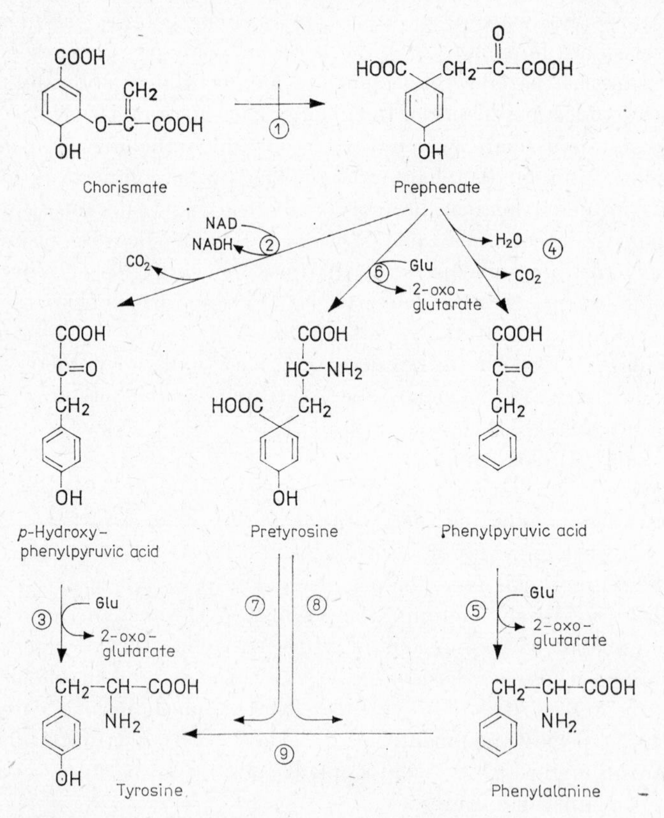

Fig. 8.1
Conversion of chorismate to tyrosine and phenylalanine
1 Chorismate mutase; 2 prephenate dehydrogenase; 3 aminotransferase; 4 prephenate dehydratase; 5 aminotransferase; 6 prephenate aminotransferase; 7 pretyrosine dehydrogenase; 8 pretyrosine dehydratase; 9 phenylalanine hydroxylase. The bold circle indicates the site of regulatory control.

phenylalanine and tyrosine [153, 211]. In numerous plants, there are three or two isoenzymes that are controlled by tryptophan (activator) as well as by phenylalanine and tyrosine (inhibitors) [56, 218].

In most bacteria, *e.g. E. coli*, tyrosine is formed *via* 4-hydroxyphenyl pyruvate (Fig. 8.1, reaction 1 to 3). In other bacteria including cyanobacteria, however, tyrosine and phenylalanine are synthesized *via* pretyrosine (reactions 1, 6 to 8) [87]. Both biosynthetic routes probably occur in plants because recently the enzymes of the pretyrosine pathway were found in cotyledons of *Phaseolus* [87] and the enzymes catalysing reactions 2 and 3 were isolated from tissues of other plant species [56, 213].

Hydroxylation of phenylalanine to tyrosine (Fig. 8.1, reaction 9) was reported in some microorganisms and higher plants [53, 70]. At least part of the phenylalanine and tyrosine biosynthetic pathway is located in chloroplasts [16, 27, 99].

The formation of *L*-tryptophan is common to all organisms [73, 132, 220]; in plants it is located in the chloroplasts [28, 66]. Tryptophan synthesis proceeds from chorismate via anthranilate (Fig. 8.2, reaction 1). Anthranilate is formed by anthranilate synthase, which catalyses the transfer of the amide-*N* of glutamine with the concomitant loss of the pyruvyl chain. The enzyme was found in several higher plants [213] and belongs to a family of glutamine-dependent amidotransferases (*cf.* chapter 8.1.2). It is monofunctional in some organisms and multifunctional in others. In the latter case anthranilate synthase activity is associated with other enzyme activities of the biosynthetic pathway leading to tryptophan [222].

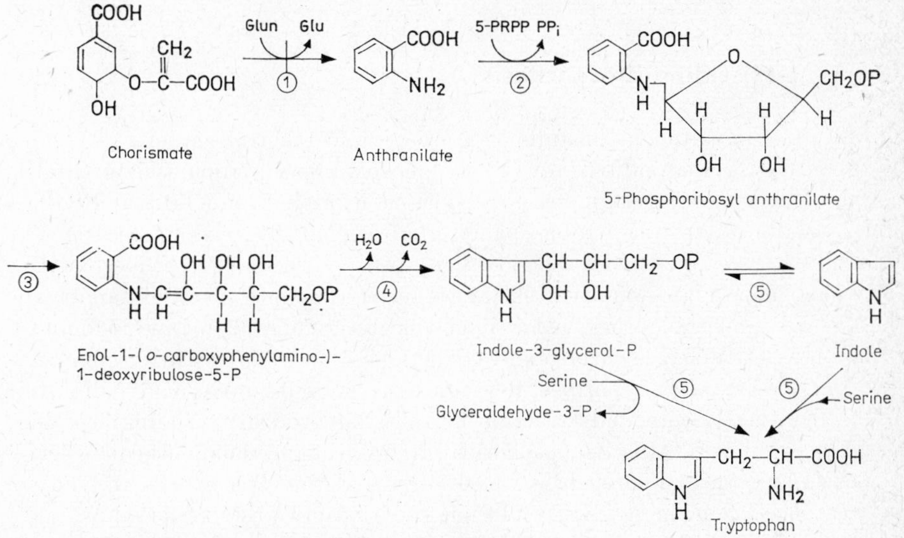

Fig. 8.2
Biosynthesis of tryptophan

1 Anthranilate synthase; 2 anthranilate-PRPP phosphoribosyl transferase; 3 *N*-5′-phosphoribosyl anthranilate isomerase; 4 indole-3-glycerol-P synthase; 5 tryptophan synthase. The bold circle indicates the site of regulatory control.

The mechanism of the action of anthranilate synthase was investigated most thoroughly in the case of the bifunctional enzyme. Anthranilate is converted into *N*-phosphoribosyl anthranilate by anthranilate phosphoribosyl transferase (Fig. 8.2, reaction 2), which in turn is isomerized by phosphoribosyl anthranilate isomerase (reaction 3) and then converted into indoleglycerol phosphate (reaction 4). Detailed information about these enzymes comes from experiments with maize and peas [68].

The flux of metabolites in tryptophan biosynthesis is controlled by tryptophan itself. Tryptophan exerts a feedback inhibition upon anthranilate synthase, which was investigated in detail with bacteria [64, 102, 183, 195, 196, 213] and higher plants [29, 213]. It is of interest that histidine limitation derepresses the enzymes of arginine and tryptophan biosynthesis in *N. crassa* [30].

Tryptophan gene distributions and tryptophan gene-enzyme realtionships were intensively studied in bacteria and fungi (*e.g. N. crassa, S. cerevisiae, Coprinus lagopus*) [35]. All higher plants studied so far have independent enzymes with phosphoribosyl transferase, phosphoribosyl anthranilate isomerase, indoleglycerol phosphate synthase and tryptophan synthase activity. Except for the usual association of the α- and β-chains of tryptophan synthase in a multi-enzyme complex [68] and some evidence for dimerization of indoleglycerol phosphate synthase in peas [35], there is no evidence of aggregate formation of any kind among the tryptophan synthesizing enzymes of angiosperms. In *Euglena gracilis*, however, all of the enzymes of tryptophan biosynthesis except anthranilate synthase are associated with a single polypeptide chain.

8.2.2 *L*-Histidine Biosynthesis

Imidazole ring formation of histidine is initiated by the transfer of the carbon-atom 2 and nitrogen-atom 1 from ATP to phosphoribosylpyrophosphate (PRPP) (reactions 1 to 6 in Fig. 8.3). First, a condensation product of ATP and PRPP is formed (reaction 1), then a pyrophosphate group is split off (reaction 2). The ATP ring is opened (reaction 3), and the ribose group from original PRPP is isomerized to a ribulose moiety (reaction 4). The Addition of the amido-*N* of glutamine and closure of the imidazole ring yields imidazoleglycerolphosphate and 5-amino-1-phosphoribosyl-4-imidazole carboxamide (reactions 5 and 6). The latter compound is used in the resynthesis of purines. Reactions 1 to 6 are followed by dehydratation forming an α-keto compound (reaction 7) and aminotransfer yielding histidinol phosphate (reaction 8). After dephosphorylation (reaction 9), the intermediate serves as a substrate for the dehydrogenase which forms histidine. Histidine is an inhibitor of the first enzyme in the pathway, ATP-phosphoribosyl transferase [155].

Histidine biosynthesis was thoroughly investigated in bacteria [155, 168]. Experiments with plants, however, are rare [130, 195]. ATP-phosphoribosyl transferase, imidazoleglycerol phosphate dehydratase and histidinol phosphate phosphatase were measured in plant extracts of barley, oat and pea shoots [212]. Histidinol dehydrogenase was shown to occur in cultivated cells of *Rosa* spec. [39].

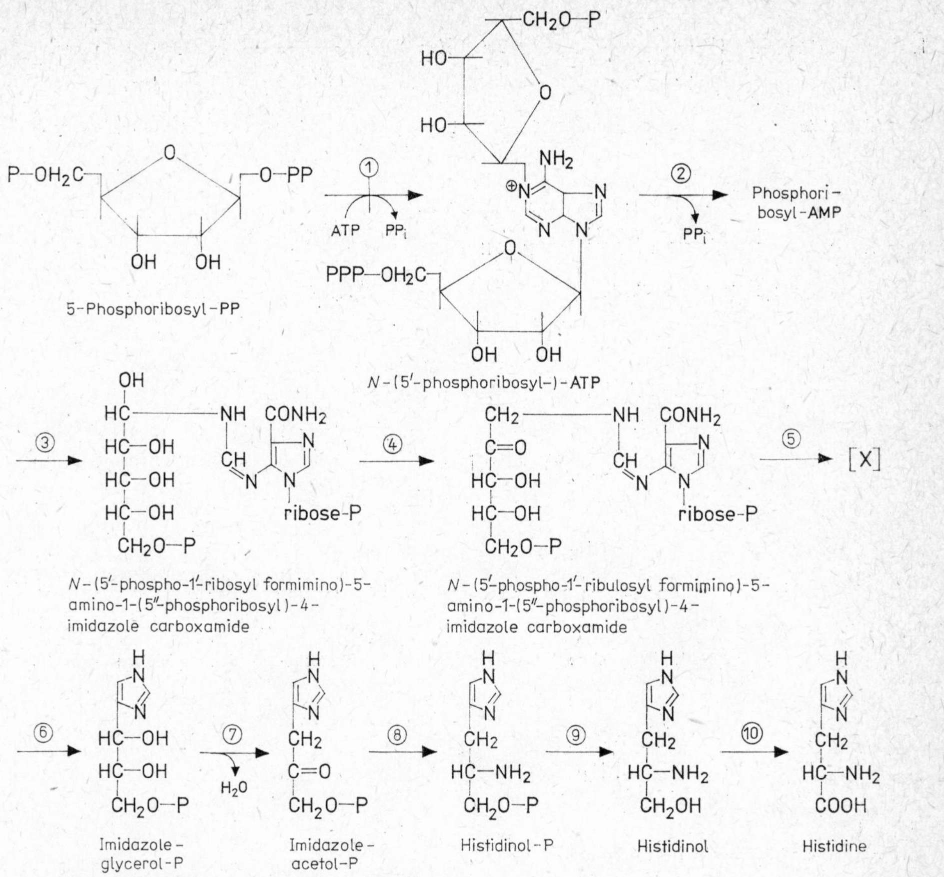

Fig. 8.3
Biosynthesis of histidine
1 ATP-phosphoribosyltransferase; 2 pyrophosphohydrolase; 3 phosphoribosyl-AMP cyclohydrolase; 4 ketol isomerase; 5 amidotransferase; 6 cyclase; 7 imidazole glycerol-P dehydratase; 8 aminotransferase; 9 histidinol-P phosphatase; 10 histidinol dehydrogenase. The bold circle indicates the site of regulatory control.

8.2.3 *L*-Lysine-Formation

Biosynthesis of *L*-lysine takes place *via* two ways, α,ε-diaminopimelic acid in plants and bacteria [31, 114, 136] and α-aminoadipic acid in fungi [219]. In higher plants the biosynthesis of lysine is located in the chloroplasts [135]. It begins with the condensation of aspartic semialdehyde with pyruvate (Fig. 8.4, reaction 3). The product formed cyclizes spontaneously and is then reduced by NADPH (reaction 4). Ring opening, succinylation (or acetylation) and transamination lead to succinyl (or acetyl)-diaminopimelate (reactions 5 and 6). Removal of the succinyl (or acetyl) group (reaction 7) results in the formation of *L*-diaminopimelate. The *meso*form of this substance as the final step is attacked by a carboxy-lyase (reactions 8 and 9).

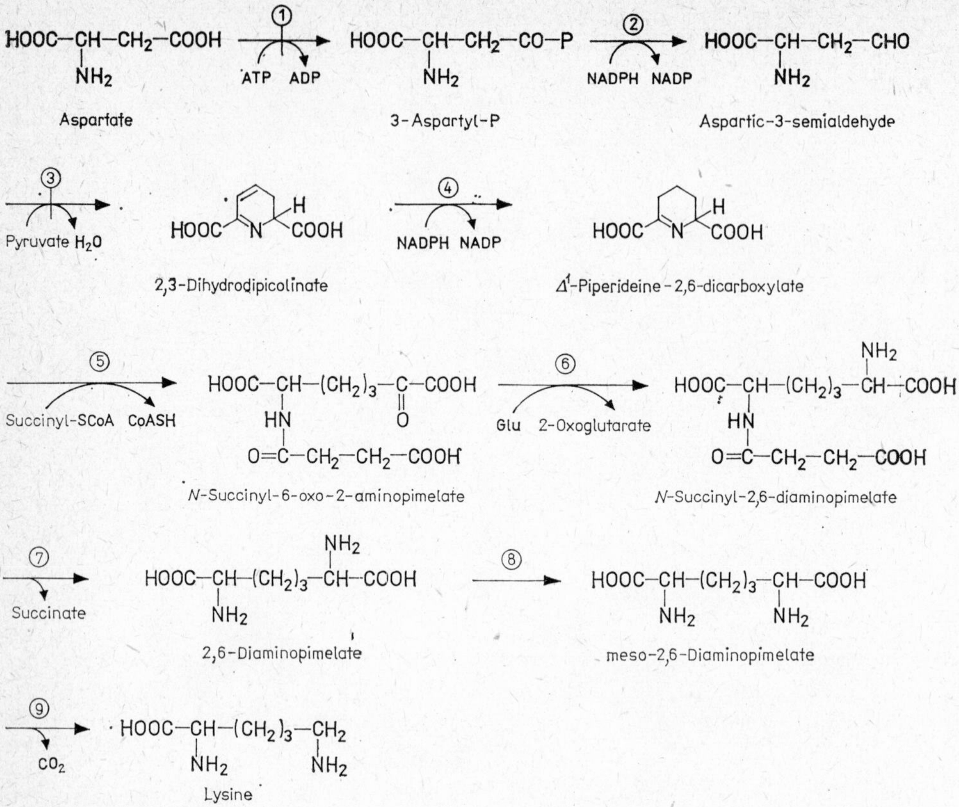

HOOC—CH(NH$_2$)—CH$_2$—COOH —①(ATP → ADP)→ HOOC—CH(NH$_2$)—CH$_2$—CO—P —②(NADPH → NADP)→ HOOC—CH(NH$_2$)—CH$_2$—CHO

Aspartate — 3-Aspartyl-P — Aspartic-3-semialdehyde

—③(Pyruvate, H_2O)→ 2,3-Dihydrodipicolinate —④(NADPH → NADP)→ Δ^1-Piperideine-2,6-dicarboxylate

—⑤(Succinyl-SCoA → CoASH)→ HOOC—CH(NH—CO—CH$_2$—CH$_2$—COOH)—(CH$_2$)$_3$—CO—COOH

N-Succinyl-6-oxo-2-aminopimelate

—⑥(Glu → 2-Oxoglutarate)→ HOOC—CH(NH—CO—CH$_2$—CH$_2$—COOH)—(CH$_2$)$_3$—CH(NH$_2$)—COOH

N-Succinyl-2,6-diaminopimelate

—⑦(Succinate)→ HOOC—CH(NH$_2$)—(CH$_2$)$_3$—CH(NH$_2$)—COOH

2,6-Diaminopimelate

—⑧→ HOOC—CH(NH$_2$)—(CH$_2$)$_3$—CH(NH$_2$)—COOH

meso-2,6-Diaminopimelate

—⑨(CO_2)→ HOOC—CH(NH$_2$)—(CH$_2$)$_3$—CH$_2$(NH$_2$)

Lysine

Fig. 8.4
Biosynthesis of lysine in higher plants

1 Aspartokinase; 2 aspartic semialdehyde dehydrogenase; 3 dihydrodipicolinate synthase; 4 dihydrodipicolinate dehydrogenase; 5 succinylase; 6 aminotransferase; 7 *N*-succinyl-α,ε-diaminopimelate deacylase; 8 α,ε-diaminopimelate epimerase; 9 diaminopimelate decarboxylase. The two bold circles indicate the sites of regulatory control.

In plants, the activity of aspartokinase is inhibited by either

a) *L*-lysine [23, 31, 58, 125, 128],

b) *L*-lysine together with *L*-threonine [1, 5] or

c) *L*-methionine or by *S*-adenosylmethionine [169, 178].

The presence of two isoenzymes of aspartokinase is described, each sensitive to only one amino acid [31, 38, 100, 130, 173]. Lysine also inhibits dehydrodipicolinate synthase (reaction 3) [31, 113, 125, 128].

Most of the enzymes of the overall reaction outlined in Fig. 8.4 were studied in bacteria. Knowledge of the corresponding enzymes of higher plants, however, is incomplete [114, 115, 128, 132, 181, 204, 205]. With *E. coli* mutants possessing a very low activity of lysyl-tRNA synthase [154] the genetics of the regulation of lysine biosynthetic enzymes were investigated [22, 125].

Several reports indicate that higher plants are also able to synthesize lysine *via* α-aminoadipic acid. This route was originally described for *Chytridio, Zygo-, Asco-, Basidiomycetes, Fungi imperfecti* and members of the *Euglenidae*. There is evidence that saccharopine as well as α-aminoadipate, *i.e.* intermediates of this pathway, occur in higher plants [114]. Presumably, these metabolites play a role in lysine degradation [136].

8.2.4 Biosynthesis of the Branched-chain Amino Acids *L*-Valine, *L*-Isoleucine and *L*-Leucine

In plants and bacteria the biosynthesis of leucine, valine, and the common intermediates of isoleucine biosynthesis are obviously catalysed by a similar set of enzymes (Fig. 8.5, reactions 2 to 5) [61, 132]. The carbon flow in this pathway [21] is controlled by

a) the inhibition of threonine deaminase (reaction 1) by isoleucine as well as by activation of this enzyme by valine [46],
b) feedback inhibition of acetolactate synthase (reaction 2) by leucine and valine [128], and
c) feedback inhibition of isopropylmalate synthase (reaction 6) by leucine [161].

Presumably, the control of the threonine deaminase in plants is a more complex phenomenon [26]. The biosynthesis of valine as well as that of isoleucine has recently been investigated in isolated chloroplasts [99, 135].

8.2.5 Selected Aspects of Glycine-Serine Metabolism in Plants

In plants, three different pathways lead to serine and glycine (Fig. 8.6). On the one hand, non-phosphorylated intermediates are involved (reactions 1 to 3) [92, 161, 162, 182]; on the other hand, phosphorylated intermediates have been found in germinating plant seedlings. The first enzyme of the phosphorylated pathway (reaction 4) was shown to be activated by methionine and inhibited by serine and various nucleotides [92, 180]. A third pathway is *via* glycolate (reactions 7 to 11) during photorespiration [32, 92, 192]. In mitochondria serine is formed from two molecules of glycine [72], including decarboxylation, and transfer of a hydroxymethyl group [34, 42, 57, 92, 208]. The mitochondrial interconversion of glycine and serine is related to the glycolate metabolism of higher plants and algae [7, 182, 192]. The following enzymes of the glycolate pathway are located in leaf peroxysomes [24, 92, 192]: glyoxylate-serine aminotransferase, glyoxylate-glutamate aminotransferase, and serine/glycolate decarboxylase [208]. But glycine may also be formed from serine by a hydroxymethyltransferase (reaction 12), which has been isolated from spinach leaves and chloroplasts [92, 93].

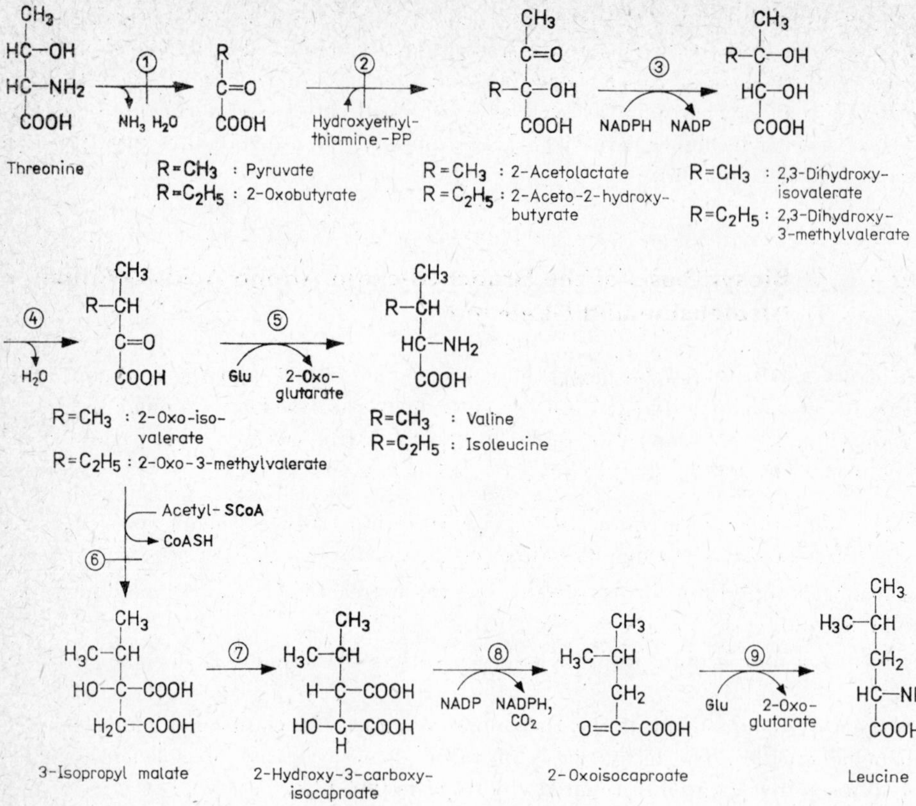

Fig. 8.5
Biosynthesis of branched-chain amino acids
1 Threonine deaminase; 2 acetolactate synthase; 3 acetolactate reducto-isomerase; 4 dihydroxy acid dehydratase; 5 aminotransferase; 6 isopropyl malate synthase; 7 isopropyl malate isomerase; 8 hydroxy-carboxyisocaproate decarboxylase; 9 leucine aminotransferase. The three bold circles indicate the sites of regulatory control.

8.2.6 Biosynthesis of the Sulphur Amino Acids *L*-Cysteine and *L*-Methionine

L-cysteine plays a central role in the sulphur metabolism of plants and algae. Sulphate reduction [9, 194] results in the formation of *L*-cysteine from which *L*-methionine and numerous other sulphur-containing plant constituents are derived. In plants, *L*-serine is converted into *L*-cysteine *via* *O*-acetyl-*L*-serine. The acyl derivative is sulphurated to form cysteine by cysteine synthase (Fig. 8.7, reaction 7), which has been purified from many plant species [9]. The biosynthesis of cysteine and homoserine was shown to occur in cytosol [10, 172] as well as in isolated chloroplasts [28, 101, 142]. In most organisms *L*-cystathionine is the main intermediate in the formation of methionine from cysteine (Fig. 8.7). Higher plants synthesize

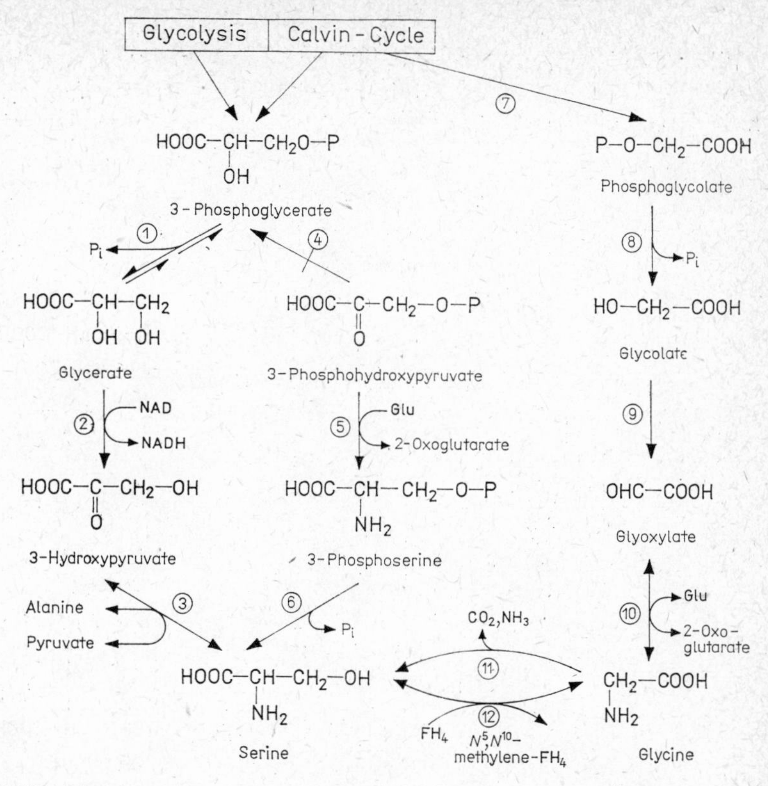

Fig. 8.6
Biosynthesis of serine and glycine
1 Phosphoglycerate phosphatase; 2 glycerate dehydrogenase; 3 aminotransferase; 4 phosphoglycerate dehydrogenase; 5 aminotransferase; 6 phosphoserine phosphatase; 7 ribulose-1,5-biphosphate oxygenase; 8 phosphoglycolate phosphatase; 9 glycolate oxidase; 10 aminotransferase; 11 glycine decarboxylase complex; 12 serine hydroxymethyltransferase. The bold circle indicates the site of regulatory control.

cystathionine *via* *O*-phospho-*L*-homoserine (reactions 1 to 3) as an activated form of homoserine [60, 128, 194], but not from *O*-acetyl-*L*-homoserine, which occurs in bacteria (reactions 3 and 6) [59]. The most important precursor in methionine synthesis, *O*-phospho-*L*-homoserine, is a product of homoserine phosphorylation (reaction 2) studied in homogenates of barley [2, 4] and peas [128, 189]. *Pisum sativum* and *Lathyrus sativus* form *O*-acetyl-*L*-homoserine esters, which accumulate in these species (reaction 6). Reduction of aspartic semialdehyde by homoserine dehydrogenase [43, 172, 206] (reaction 1) yields homoserine. The regulatory principles involved are the end-product control of the distinct isoenzymes of aspartokinase [23, 58, 169] and of homoserine dehydrogenase [43, 172, 206] as shown, for instance, with selected cell lines [23, 60, 82, 102, 213]. In the last step of cystathionine formation (reaction 3) [60] phosphohomoserine reacts with cysteine. Cystathionine is then

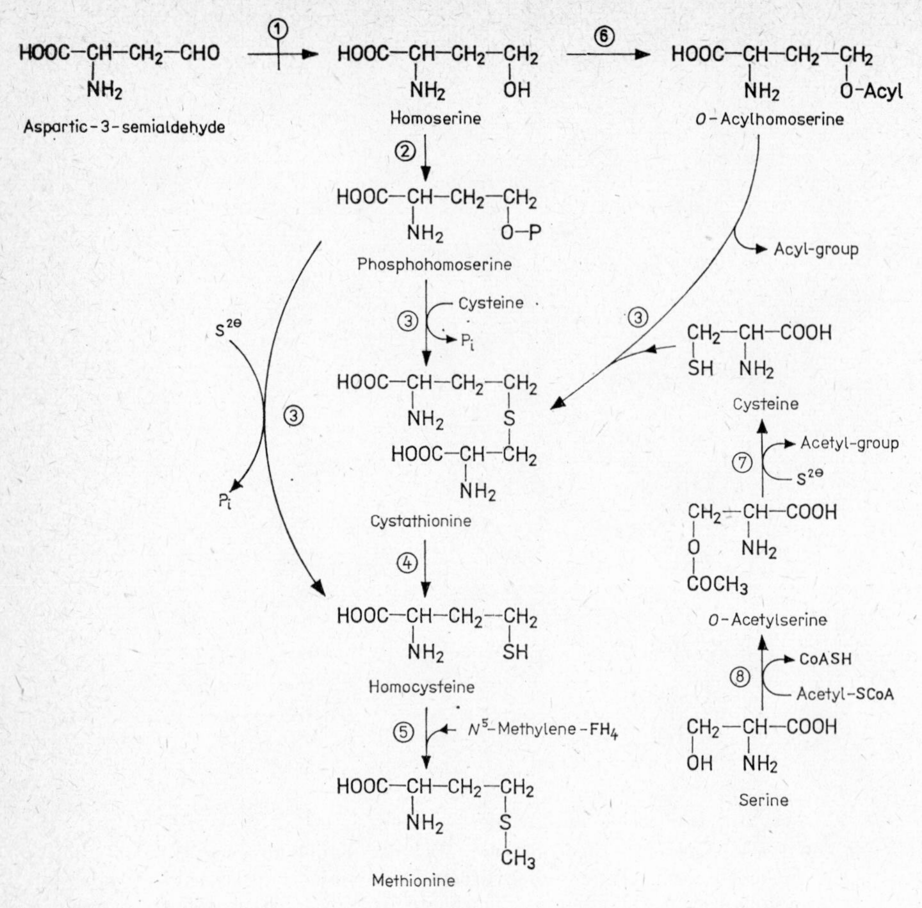

Fig. 8.7
Biosynthesis of sulfur-containing amino acids

1 Homoserine dehydrogenase; 2 homoserine kinase; 3 cystathionine-γ-synthase; 4 cystathionine-β-lyase; 5 tetrahydropteroyltriglutamate methyltransferase; 6 homoserine acyltransferase; 7 cysteine synthase; 8 acetyltransferase. The bold circle indicates the site of regulatory control.

transformed to homocysteine by cystathionine-β-lyase (reaction 4), which has been highly purified from *Spinacia* [61]. An alternative pathway to homocysteine including a direct sulphhydration catalysed by cystathionine synthase exists in *Chlorella* [60, 63].

The final reaction in *L*-methionine biosynthesis is the methylation of *L*-homocysteine (reaction 5) [23]. In microorganisms this step may require vitamin B_{12}. (*S*)-Adenosyl-*L*-methionine synthesized by the activation of methionine with ATP [3, 23, 60] is the donor of the methyl groups of many secondary products including the alkaloids. In a few special cases transmethylation proceeds with betaine, dimethylthetin or propiothetin as methyl donors [23, 60]. All methyl groups of these compounds are ultimately derived from *L*-serine by serine transhydroxymethylase catalysing the reversible cleavage of *L*-serine to glycine and N^5, N^{10}-methylene tetrahydrofolate [3, 4, 60].

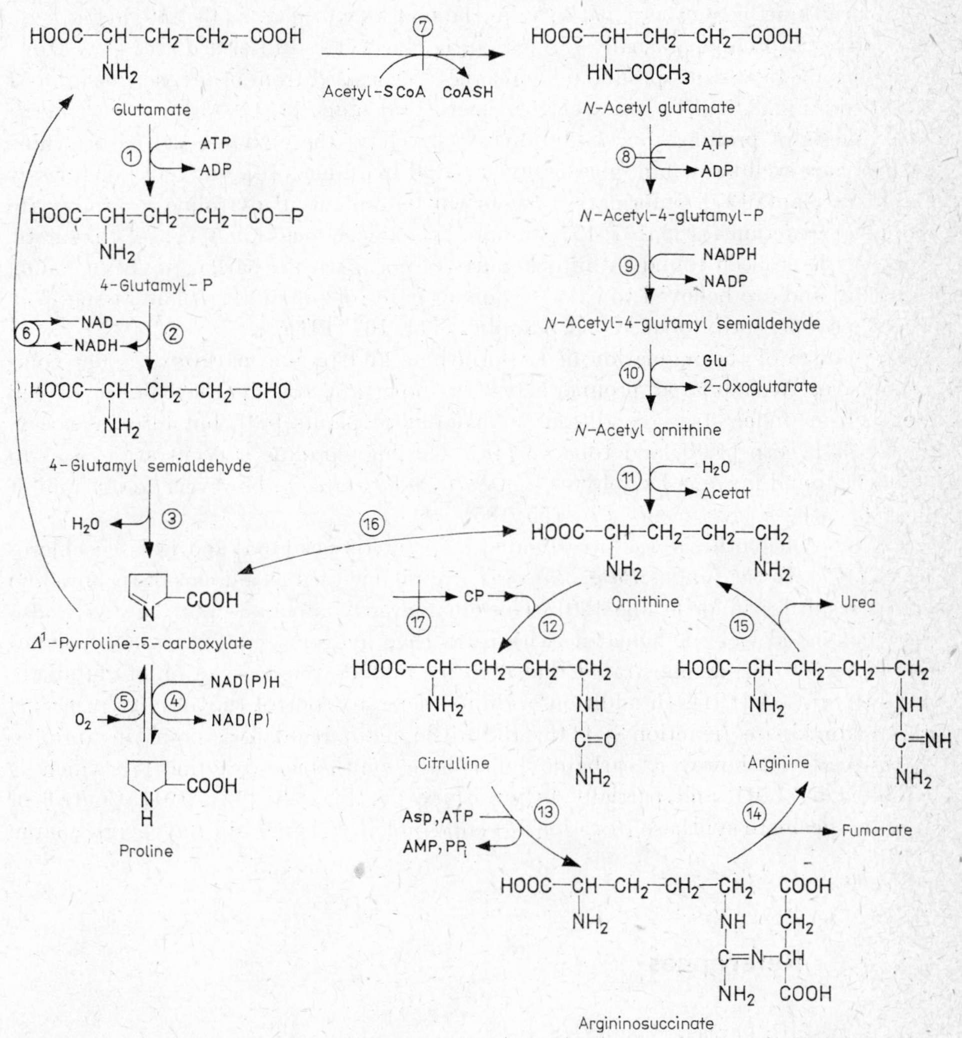

Fig. 8.8
Biosynthesis of proline, ornithine, and arginine

1 Glutamyl kinase; 2 glutamylsemialdehyde dehydrogenase; 3 spontaneous cyclization; 4 Δ^1-pyrroline-5-carboxylate reductase; 5 proline oxidase; 6 Δ^1-pyrroline-5-carboxylate dehydrogenase; 7 *N*-acetylglutamate synthase; 8 *N*-acetyl-γ-glutamokinase; 9 *N*-acetylglutamate semialdehyde dehydrogenase; 10 aminotransferase; 11 *N*-acetyl ornithinase; 12 ornithine transcarbamylase; 13 argininosuccinate synthase; 14 argininosuccinate lyase; 15 arginase; 16 ornithine aminotransferase; 17 carbamoylphosphate synthetase. The three bold circles indicate the sites of regulatory control.

8.2.7 Biosynthesis of the Glutamic Acid Family of Amino Acids: *L*-Ornithine, *L*-Arginine and *L*-Proline

Formation of *N*-acetyl-*L*-glutamate and *N*-acetyl-*L*-ornithine indicates the existence of the ornithine biosynthetic pathway in the majority of plants [137, 191] (Fig. 8.8). In contrast *L*-proline is usually formed *via* non-acetylated intermediates [17]. However, in some organisms proline may also be synthesized from ornithine (reactions 4 and 16 in Fig. 8.8) [61, 106, 211] *via* acetylated steps [121]. All enzymes of the biosynthesis of proline from *L*-glutamate as well as the enzymes of the ornithine pathway are well known in microorganisms and in animals [6], but to a much lesser extent in plants [37]. Information exists only for plant Δ^1-pyrroline-5-carboxylate reductase (reaction 4) [82, 167, 197] and proline oxidase (reaction 5). These enzymatic reactions have been found in mitochondria of corn, wheat, barley, soybean, mung bean [19] and are believed to exist in homogenates of some other plant tissues, *e.g.* wheat, peanut, corn, tobacco and pumpkin [114, 167, 191].

Knowledge of the regulation of the ornithine and proline pathways is not comprehensive. Accumulated proline plays an important role in the balance of cell metabolism under stress conditions in halophytic plants [63], but also in barley, wheat [33], corn [188], and tobacco [45]. The high proline concentrations (up to 600 mM) found in some halophytes, *e.g. Salicornia fruticosa*, however, do not inhibit the biosynthetic enzymes *in vitro* [55, 185].

Arginine biosynthesis was investigated in plant tissues [165] and isolated chloroplasts [177]. In the fungus *Panus tigrinus*, arginine is used as a donor of the amidino group resulting in ornithine [20]. This guanidino compound was, however, also decarboxylated yielding agmatine which was then hydrolysed [122]. *In vitro* studies (reactions 7 to 14) demonstrated that arginine inhibits the acetylation of glutamate with acetyl-CoA [191]. In addition arginine seems to control the activity of acetyl glutamate kinase (reaction 8) [116, 137]. The main regulatory point in arginine biosynthesis is, however, carbamoylphosphate synthetase (reaction 17) which is inhibited by UMP and partially by ornithine (*cf.* Fig. 8.8) [151, 191]. Control of argininosuccinate synthase (reaction 13) is possible by changes in the energy charge [176].

8.3 References

[1] Aarnes, H.: Physiol. Plant **32** (1974) 400.
[2] Aarnes, H.: Plant Sci. Lett. **7** (1976) 187.
[3] Aarnes, H.: Plant Sci. Lett. **10** (1977) 381.
[4] Aarnes, H.: Planta **140** (1978) 185.
[5] Aarnes, H.; Rognes, S. E.: Phytochemistry **13** (1974) 2717.
[6] Adams, E.; Frank, L.: Ann. Rev. Biochem. **49** (1980) 1005.
[7] Adrews, T. J.; Lorimer, G. H.; Tolbert, N. E.: Biochemistry **12** (1973) 11.
[8] Akkermans, A. D. L.; van Straten, J.; Roelofson, W.: In: Newton, W. E.; Postgate, J. R.; Rodriguez-Barrueco (Eds.): Proc. 2nd Int. Symp. on Nitrogen Fixation. London 1977, p. 591.

[9] Anderson, J. W.: In: Miflin, B. J. (Ed.): Amino Acids and their Derivatives. The Biochemistry of Plants, Vol. 5, p. 203. London/New York 1980.
[10] Ascano, A.; Nicholas, D. J. D.: Phytochemistry **16** (1976) 889.
[11] Awonaike, K. O.: Ph. D. Thesis, Univ. London 1980.
[12] Bauer, W. D.: Ann. Rev. Plant Physiol. **32** (1981) 407.
[13] Bender, D. A.: Amino Acid Metabolism. London 1975.
[14] Benemann, J. R.; Yoch, D. C.; Valentine, R. C.; Arnon, D. I.: Biochim. Biophys. Acta **226** (1971) 205.
[15] Bernhardt, S. A.; Davis, R. H.: Proc. Nat. Acad. Sci. USA **69** (1972) 1868.
[16] Bickel, H.; Schultz, G.: Phytochemistry **18** (1979) 498.
[17] Bidwell, R. G. S.; Durzan, D. J.: In: Davies, P. J. (Ed.): Historical and Current Aspects of Plant Physiology, p. 152. New York 1975.
[18] Bisswanger, H.; Schmincke-Ott, E. (Eds.): Multifunctional Proteins. New York 1980.
[19] Boggness, S. F.; Koeppe, D. E.: Plant Physiol. **62** (1978) 22.
[20] Boldt, A.; Miersch, J.; Reinbothe, H.: Phytochemistry **10** (1971) 731.
[21] Borstlap, A. C.: Acta Bot. Neerl. **24** (1975) 203.
[22] Boy, E.; Reinisch, F.; Richaud, D.; Patte, J.-C.: Biochemie **58** (1976) 213.
[23] Bright, S. W. J.; Lea, P. J.; Miflin, B. J.: In: Elliott, K.; Wheland, J. (Eds.): Ciba Foundation Symp. No. 72, p. 101. Sulphur in Biology. Amsterdam 1980.
[24] Brock, B. L.; Wilkinson, D. A.; King, J.: Canad. J. Biochem. **48** (1970) 486.
[25] Bryan, J. K.: In: Bonner, J.; Varner, J. E. (Eds.): Plant Biochemistry, 3rd ed., p. 525. New York 1976.
[26] Bryan, J. K.: In: Miflin, B. J. (Ed.): The Biochemistry of Plants, Vol. 5, p. 403. London/New York 1980.
[27] Buchholz, D.; Reupke, B.; Bickel, H.; Schultz, G.: Phytochemistry **18** (1979) 1109.
[28] Burdge, E. L.; Matthews, B. F.; Mills, W. R.; Widholm, J. M.; Wilson, K. G.; Carlson, L. R.; DeBonte, L. R.; Oaks, A.: Plant Physiol. **63** (1979) Suppl. 26.
[29] Carlson, J. E.; Widholm, J. M.: Physiol. Plant. **44** (1978) 251.
[30] Carsiotis, M.; Jones, R. F.; Wesseling, A. C.: J. Bacteriol. **119** (1974) 893.
[31] Cheshire, R. M.; Miflin, B. J.: Phytochemistry **14** (1975) 695.
[32] Chollet, R.; Ogren, W. L.: Bot. Rev. **41** (1975) 137.
[33] Chu, T. M.; Jussaitts, M.; Aspinall, D.; Paleg, L. G.: Physiol. Plant. **43** (1978) 254.
[34] Cladinin, M. T.; Cossins, E. A.: Phytochemistry **14** (1975) 387.
[35] Crawford, I. P.: In: Bisswanger, H.; Schmincke-Ott, E. (Eds.): Multifunctional Proteins. New York 1980, p. 151.
[36] Darrow, R. A.: In: Mora, J.; Ralacies, R. (Eds.): Glutamine: Metabolism, Enzymology, and Regulation, p. 139, New York 1980.
[37] Dashek, W. V.; Erickson, S. S.: Bot. Rev. **47** (1981) 349.
[38] Davies, H. M.; Miflin, B. J.: Plant Physiol. **62** (1978) 536.
[39] Davis, M. E.: Phytochemistry **10** (1971) 783.
[40] Davis, R. H.: Ann. Rev. Genetics **9** (1975) 39.
[41] Davis, R. H.: In: Nover, L.; Lynen, F.; Mothes, K. (Eds.): Cell Compartmentation and Metabolic Channelling, p. 239. Jena 1980.
[42] Day, D. A.; Wiskich, J. T.: Plant Physiol. **68** (1981) 425.
[43] DiCamelli, C. A.; Bryan, J. K.: Plant Physiol. **65** (1980) 176.
[44] Dilworth, M. J.: Ann. Rev. Plant Physiol. **25** (1974) 81.
[45] Dix, P. J.; Pearce, R. S.: Z. Pflanzenphysiol. **102** (1981) 243.
[46] Dougall, D. K.: Phytochemistry **9** (1970) 959.
[47] Dougall, D. K.: Biochem. Biophys. Res. Commun. **58** (1974) 639.
[48] Dougall, D. K.; Bloch, J.: Canad. J. Bot. **54** (1976) 2924.
[49] Dürr, M.: Ph. D. Thesis, ETH Zürich 1977.
[50] Dürr, M.; Boller, T.; Wiemken, A.: Arch. Microbiol. **105** (1975) 319.
[51] Dürr, M.; Krech, K.; Boller, T.; Wiemken, A.; Schwencke, J.; Nagy, M.: Arch. Microbiol. **121** (1979) 169.
[52] Duke, S. H.; Collins, M.; Soberalske, R. M.: Crop Sci. **20** (1980) 213.

[53] Endress, R.: Plant Physiol. **68** (1981) 272.
[54] Flint, H. J.; Porteous, D. J.; Kacser, H.: Biochem. J. **190** (1980) 1.
[55] Flowers, T. J.; Troke, P. F.; Yeo, A. R.: Ann. Rev. Plant Physiol. **28** (1977) 89.
[56] Gadal, P.; Boudet, A.; Robreau, G.: Plant Sci. Lett. **3** (1974) 55.
[57] Gardeström, P.; Bergman, A.; Ericson, I.: Plant Physiol. **65** (1980) 389.
[58] Gengenbach, B. G. F.; Walter, T. J.; Hibberd, K. A.; Green, C. E.: Crop Sci. **18** (1978) 472.
[59] Giovanelli, J.; Mudd, S. H.; Datko, A. H.: J. Biol. Chem. **253** (1978) 5665.
[60] Giovanelli, J.; Mudd, S. H.; Datko, A. H.: In: Miflin, B. J. (Ed.): The Biochemistry of Plants, Vol. 5, p. 454. London/New York 1980.
[61] Givan, C.: In: Miflin, B. J. (Ed.): The Biochemistry of Plants, Vol. 5, p. 329. London/New York 1980
[62] Glund, K.; Walther, R.: Biol. Rdsch. **29** (1982) 205.
[63] Göring, H.: In: Mineral Nutrition of Plants, Vol. 1, p. 103. Sofia 1979.
[64] Goto, Y.; Zalkin, H.; Keim, P.; Heinrikson, R.: J. Biol. Chem. **251** (1976) 941.
[65] Griffith, O. W.; Anderson, M. E.; Meister, A.: J. Biol. Chem. **254** (1979) 1205.
[66] Grosse, W.: Z. Pflanzenphysiol. **83** (1977) 249.
[67] Guerrero, M. G.; Vega, J. M.; Losada, M.: Ann. Rev. Plant Physiol. **32** (1981) 169.
[68] Hankins, C. N.; Largen, M. T.; Mills, S. F.: Plant Physiol. **57** (1976) 101.
[69] Hankins, C. N.; Mills, S. E.: J. Biol. Chem. **252** (1977) 235.
[70] Hanson, K. R.; Havir, E. A.: In: Swain, T.; Harborne, J. B.; Van Sumere, C. F. (Eds.): Recent Advances in Phytochemistry, Vol. 12, p. 91. New York 1978.
[71] Hardy, R. W. F.; Silver, W. S. (Eds.): A Treatise on Dinitrogen Fixation. New York 1977.
[72] Hartmann, T.; Ehmke, A.: Planta **149** (1980) 207.
[73] Haslam, E.: The Shikimate Pathway. Halstedt, New York/Toronto 1974.
[74] Haynes, R. J.; Goh, K. M.: Biol. Rev. **53** (1978) 465.
[75] Heimer, Y. M.; Riklis, E.: Plant Physiol. **64** (1979) 663.
[76] Herrero, A.; Flores, E.; Guerrero, M. G.: J. Bacteriol. **145** (1981) 175.
[77] Hewitt, E. J.: Ann. Rev. Plant Physiol. **26** (1975) 73.
[78] Hewitt, E. J.; Cutting, C. V. (Eds.): Nitrogen Assimilation of Plants. London 1979.
[79] Hewitt, E. J.; Hucklesby, D. P.; Mann, A. F.; Notton, B. A.; Rucklidge, G. J.: In [78], p. 255.
[80] Hewitt, E. J.; Hucklesby, D. P.; Notton, B. A.: In: Bonner, J.; Varner, J. E. (Eds.): Plant Biochemistry 3rd ed., p. 633. New York 1976.
[81] Hewitt, E. J.; Notton, B. A.: In: Coughlan, M. (Ed.): Molybdenum and Molybdenum-containing Enzymes, p. 273. Oxford 1980.
[82] Hibberd, K. A.; Walter, T.; Green, C. E.; Gengenbach, B. G. F.: Planta **148** (1980) 183.
[83] Huber-Wälchli, V.; Wiemken, A.: Arch. Microbiol. **120** (1979) 141.
[84] Jack, G. W.; McMahon, P. C.: Biochim. Biophys. Acta **523** (1978) 344.
[85] Jackson, C.; Dench, J. E.; Morris, P.; Lui, S. C.; Hall, D. O.; Moore, A. L.: Biochem. Trans. **7** (1979) 1122.
[86] Jauniaux, J.-C.; Urrestarazu, L. A.; Wiame, J.-M.: J. Bacteriol. **133** (1978) 1096.
[87] Jensen, R. A.; Pierson, D. L.: Nature **254** (1975) 667.
[88] Jones, M. E.: Ann. Rev. Biochem. **49** (1980) 253.
[89] Jones, R. W.; Sheard, R. W.: In [78], p. 521.
[90] Karlin, J. N.; Bowman, B. J.; Davis, R. H.: J. Biol. Chem. **251** (1976) 3948.
[91] Keßler, E.: Fortschr. Bot. **36** (1974) 99.
[92] Keys, A. J.: In: Miflin, B. J. (Ed.): The Biochemistry of Plants, Vol. 5, p. 359. London/New York 1980.
[93] Kisaki, T.; Yoshida, M.; Imai, A.: Plant Cell Physiol. **12** (1971) 275.
[94] Koch, B.; Wong, P.; Russell, S. A.; Howard, R.; Evans, H. J.: Biochem. J. **118** (1970) 773.
[95] Kondorosi, A.; Svab, Z.; Kiss, G. B.; Dixon, R. A.: Mol. gen. Genet. **151** (1977) 221.
[96] Lacroute, F.; Piérard, A.; Grenson, M.; Wiame, J. M.: J. Gen. Microbiol. **40** (1965) 127.

[97] Lancaster, J. R.; Vega, J. M.; Kamin, H.; Orme-Johnson, N. R.; Orme-Johnson, W. H.; Krueger, R. J.; Siegel, L. M.: J. Biol. Chem. **254** (1979) 1268.
[98] Lara, J. C.; Mills, S. E.: J. Bacteriol. **110** (1972) 1100.
[99] Larsen, P. O.; Cornwell, K. L.: Plant Physiol. **68** (1981) 292.
[100] Lea, P. J.; Miflin, B. J.: In: Gibbs, M.; Latzko, E. (Eds.): Encyclopedia of Plant Physiology, New. Ser. Vol. 6, p. 445. Berlin 1979.
[101] Lea, P. J.; Mills, W. R.; Wallsgrove, R. M.; Miflin, B. J.: In: Schiff, J. A.; Stanier, R. Y. (Eds.): The Origin of Chloroplasts. New York 1981.
[102] Lee, F.; Squires, C. L.; Yanofsky, C.: J. Mol. Biol. **103** (1976) 383.
[103] Legerton, T. L.; Weiss, R. L.: J. Bacteriol. **138** (1979) 909.
[104] Losada, M.; Guerrero, M. G.: In: Barber, J. (Ed.): Photosynthesis in Relation to Model Systems, p. 365. Amsterdam 1979.
[105] Losada, M.; Guerrero, M. G.; Vega, J. M.: In: Bothe, H.; Trebst, A. (Eds.): Biochemistry and Physiology of Nitrogen and Sulphur Metabolism. Berlin 1981.
[106] Lu, T. S.; Mazelis, M.: Plant Physiol. **55** (1975) 502.
[107] Ludwig, R. A.; Signer, E. R.: Nature **267** (1977) 245.
[108] Lusty, C. J.: Eur. J. Biochem. **85** (1978) 373.
[109] Mann, A. F.; Fentem, P.-A.; Stewart, G. R.: Biochem. Biophys. Res. Commun. **88** (1979) 515.
[110] Martinoia, E.; Heck, U.; Wiemken, A.: Nature **289** (1981) 292.
[111] Matile, P.: In: Bonner, J.; Warner, J. E. (Eds.): Plant Biochemistry, 3rd ed., p. 189. New York 1976.
[112] Matile, P.: Ann. Rev. Plant Physiol. **29** (1978) 193.
[113] Matthews, B. F.; Widholm, J. M.: Z. Naturforsch. **34c** (1979) 1177.
[114] Mazelis, M.: In: Miflin, B. J. (Ed.): The Biochemistry of Plants, Vol. 5, p. 541. London/New York 1980.
[115] Mazelis, M.; Creveling, R. K.; Food, J.: Biochemistry **2** (1978) 29.
[116] McKay, G.; Shargool, P. D.: Plant Sci. Lett. **9** (1977) 189.
[117] Meister, A.: In: Mora, J.; Ralacies, R. (Eds.): Glutamine: Metabolism, Enzymology, and Regulation, p. 1. New York 1980.
[118] Messenguy, F.: Mol. gen. Genet. **169** (1979) 85.
[119] Messenguy, F.; Colin, D.; Ten Have, J.-P.: Eur. J. Biochem. **108** (1980) 439.
[120] Messenguy, F.; Cooper, T. G.: J. Bacteriol. **130** (1977) 1253.
[121] Mestichelli, L. L. J.; Gupta, R. N.; Spenser, I. D.: J. Biol. Chem. **254** (1979) 640.
[122] Miersch, J.; Reinbothe, H.: Biochem. Physiol. Pfl. **162** (1971) 75.
[123] Miflin, B. J.: Planta **93** (1970) 160.
[124] Miflin, B. J.: In [134], p. 49.
[125] Miflin, B. J.: In: Smith, H. (Ed.): Regulation of Enzyme Synthesis and Activity in Higher Plants, p. 23. London/New York 1977.
[126] Miflin, B. J. (Ed.): In: The Biochemistry of Plants, Vol. 5: Amino Acids and their Derivatives. London 1980.
[127] Miflin, B. J.: In: Carlson, P. S. (Ed.): The Biology of Crop Productivity, p. 255. London 1980.
[128] Miflin, B. J.; Bright, S. W.; Davies, H. M.; Shewry, P. R.; Lea, P. J.: In: Hewitt, E. J.; Cutting, C. V. (Eds.): Nitrogen Assimilation of Plants, p. 335. London 1979.
[129] Miflin, B. J.; Lea, P. J.: Phytochemistry **15** (1976) 873.
[130] Miflin, B. J.; Lea, P. J.: Ann. Rev. Plant Physiol. **28** (1977) 299.
[131] Miflin, B. J.; Lea, P. J.: In [126], p. 169.
[132] Miflin, B. J.; Lea, P. J.: In: Encyclopedia of Plant Physiology, New. Ser., Vol. 14A: Boulter, D.; Parthier, B. (Eds.), p. 5. Berlin 1982.
[133] Miflin, B. J.; Lea, P. J.; Wallsgrove, R. M.: In: Mora, J.; Relacies, R. (Eds.): Glutamine: Metabolism, Enzymology, and Regulation. New York 1980, p. 213.
[134] Milborrow, B. V. (Ed.): Biosynthesis and its Control in Plants. London 1973.
[135] Mills, W. R.; Lea, P. J.; Miflin, B. J.: Plant Physiol. **65** (1980) 1166.
[136] Moller, B. L.: Plant Physiol. **54** (1974) 638.

[137] Morris, C. J.; Thompson, J. F.: Plant Physiol. **59** (1977) 684.
[138] Müntz, K.: Biol. Rdsch. **18** (1980) 261.
[139] Nagel, M.; Hartmann, T.: Z. Naturforsch. **35C** (1980) 406.
[140] Neuberger, A.: In: Neuberger, A. (Ed.): Amino Acid Metabolism and Sulphur Metabolism. Compreh. Biochem. **19A** (1981) 258.
[141] Newton, W. E.; Orme-Johnson, W. H. (Eds.): Nitrogen Fixation, Vol. I. Baltimore 1980.
[142] Ng, B. H.; Anderson, J. W.: Phytochemistry **18** (1979) 573.
[143] Nishimura, M.; Douce, R.; Akazawa, T.: Plant Physiol. **65**, suppl. (1980) 14.
[144] Notton, B. A.; Hewitt, E. J.: In [78], p. 227.
[145] Nover, L.; Lynen, F.; Mothes, K. (Eds.): Cell Compartmentation and Metabolic Channelling. Jena 1980.
[146] Nover, L.; Reinbothe, H.: In: Nover, L.; Luckner, M.; Parthier, B. (Eds.): Cell Differentiation. Molecular Basis and Problems. Jena/Heidelberg 1982, p. 23.
[147] Oaks, A.: In [78], p. 217.
[148] Oaks, A.; Gadal, P.: In [145], p. 245.
[149] Oaks, A.; Stulen, I.; Jones, K.; Winsear, M. J.; Misra, S.; Boesel, I. L.: Planta **148** (1980) 477.
[150] O'Neal, D.; Joy, K. W.: Plant Physiol. **54** (1974) 773; **55** (1975) 968.
[151] O'Neal, T. J.; Naylor, A. W.: Plant Physiol. **57** (1976) 23.
[152] Pahlich, E.; Gerlitz, C. H. R.: Phytochemistry **19** (1980) 11.
[153] Palmer, J. E.; Widholm, J. M.: Plant Physiol. **56** (1975) 233.
[154] Patte, J. C.; Boy, E.; Borne, F.: FEBS-Letters **43** (1974) 67.
[155] Pearson, S. M.; Koshland, D. E. jr.: J. Biol. Chem. **249** (1974) 4104.
[156] Piérard, A.; Messenguy, F.; Feller, A.; Hilger, F.: Mol. gen. Genet. **174** (1979) 163.
[157] Piérard, A.; Schröter, B.: J. Bacteriol. **134** (1978) 167.
[158] Powers, S. G.; Meister, A.: J. Biol. Chem. **253** (1978) 1258.
[159] Premakumar, R.; Sorger, G. J.; Gooden, D.: J. Bacteriol. **137** (1979) 1119.
[160] Quail, P. H.: Ann. Rev. Plant Physiol. **30** (1979) 425.
[161] Randall, D. D.; Tolbert, N. E.; Gremel, D.: Plant Physiol. **48** (1971) 480.
[162] Rehfeld, D. W.; Tolbert, N. E.: J. Biol. Chem. **247** (1972) 4803.
[163] Reinbothe, H.: In: Mothes, K.; Schütte, H. R. (Eds.): Biosynthese der Alkaloide, p. 40. Berlin 1969.
[164] Reinbothe, H.: Nova Acta Leopoldina **1976**, Suppl. Nr. 7, p. 152.
[165] Reinbothe, H.; Miersch, J.; Mothes, K.: In: Neuberger, A. (Ed.): Amino Acid Metabolism and Sulphur Metabolism. Compreh. Biochem. **19A** (1981) 51.
[166] Reinbothe, H.; Mothes, K.: Ann. Rev. Plant Physiol. **13** (1962) 129.
[167] Rena, A. B.; Splittstoesser, W. E.: Phytochemistry **14** (1975) 657.
[168] Rizzino, A. A.; Bresaliar, R. S.; Freundlich, M.: J. Bacteriol. **117** (1974) 449.
[169] Rognes, S. E.; Lea, P. J.; Miflin, B. J.: Nature **287** (1980) 357.
[170] Rowe, W. B.; Meister, A.: Biochemistry **15** (1973) 1578.
[171] Rubin, J. L.; Jensen, R. A.: Plant Physiol. **64** (1979) 727.
[172] Sainis, J.; Mayne, R. G.; Wallsgrove, R. M.; Lea, P. J.; Miflin, B. J.: Planta **152** (1981) 491.
[173] Sakano, K.; Komamine, A.: Plant Physiol. **61** (1978) 115.
[174] Schlee, D.; Reinbothe, H.: Phytochemistry **2** (1963) 231.
[175] Shaner, D. L.; Boyer, J. S.: Plant Physiol. **58** (1976) 499.
[176] Shargool, P. D.: FEBS-Letters **33** (1973) 348.
[177] Shargool, P. D.; Steeves, T.; Weaver, M. G.; Russel, M.: Canad. J. Biochem. **56** (1978) 273.
[178] Shewry, P. R.; Miflin, B. J.: Plant Physiol. **59** (1977) 69.
[179] Shigesda, K.; Aoyagi, A.; Tatibana, M.: Eur. J. Biochem. **85** (1978) 385.
[180] Slaughter, J. C.: Phytochemistry **12** (1973) 2627.
[181] Sodek, L.: Rev. Bras. Bot. **1** (1978) 65.
[182] Somerville, C. R.; Ogren, W. L.: Proc. Nat. Acad. Sci. USA **77** (1980) 2684.

[183] Squires, C. L.; Lee, F. D.; Yanofsky, C.: J. Mol. Biol. **92** (1975) 93.
[184] Srivastava, H. S.: Phytochemistry **19** (1980) 725.
[185] Stewart, G. R.; Lee, J. A.: Planta **120** (1974) 279.
[186] Stewart, G. R.; Mann, A. F.; Fentem, P. A.: In [60], p. 271.
[187] Stone, S. R.; Copeland, L.; Kennedy, I. R.: Phytochemistry **18** (1979) 1273.
[188] Thien, B. H.; Heinke, F.; Göring, H.: Wiss. Z. Päd. Hochschule Güstrow **1** (1980) 71.
[189] Thoen, A.; Rognes, S. E.; Aarnes, H.: Plant Sci. Lett. **13** (1978) 103.
[190] Thomas, R. J.; Schrader, L. E.: Phytochemistry **20** (1981) 361.
[191] Thompson, J. F.: In: Miflin, B. J. (Ed.): The Biochemistry of Plants, Vol. 5, p. 375. London/New York 1980.
[192] Tolbert, N. E.: In: Gibbs, M.; Latzko, E. (Eds.): Photosynthesis. Encyclopedia of Plant Physiology, New Ser., Vol. 6/2, p. 338. Berlin/Heidelberg/New York 1979.
[193] Trier, G.: Über einfache Pflanzenbasen und ihre Beziehungen zum Aufbau der Eiweißstoffe und Lecithin. Berlin 1912, S. 117.
[194] Trudinger, P. A.; Loughlin, R. E.: In: Neuberger, A. (Ed.): Amino Acid Metabolism and Sulphur Metabolism. Compreh. Biochem. **19A** (1981) 165.
[195] Umbarger, H. E.: Ann. Rev. Biochem. **47** (1978) 533.
[196] Umbarger, H. E.: In: Neuberger, A. (Ed.): Amino Acid Metabolism and Sulphur Metabolism. Compreh. Biochem. **19A** (1981) 1.
[197] Vansuyt, G.; Vallee, J. C.; Prevort, J.: Physiol. Veg. **17** (1979) 95.
[198] Vega, J. M.; Cardonas, J.; Losada, M.: Methods Enzymol. **69** (1980) 255.
[199] Vennesland, B.; Guerrero, M. G.: In: Gibbs, M.; Latzko, E. (Eds.): Encyclopedia of Plant Physiology, New Ser., Vol. 6, p. 425. Berlin 1979.
[200] Virden, R.: Biochem. J. **127** (1972) 503.
[201] Vitto, A.; Cole, K. W.; Gaertner, F. H.: In [145], p. 135
[202] Waller, G. R.; Nowacki, E. K.: Alkaloid Biology and Metabolism in Plants. New York 1978.
[203] Wallsgrove, R. M.; Keys, A. J.; Bird, I. F.; Cornelius, M. J.; Lea, P. J.; Miflin, B. J.: J. Exp. Bot. **31** (1980) 1005.
[204] Wallsgrove, R. M.; Lea, P. J.; Mills, W. R.; Miflin, B. J.: Plant Physiol. **63** (1979) Suppl. 26.
[205] Wallsgrove, R. M.; Mazelis, M.: FEBS-Letters **116** (1980) 189.
[206] Walter, T. J.; Connelly, J. A.; Gengenbach, B. G.; Wold, F.: J. Biol. Chem. **254** (1979) 1349.
[207] Walters, C. L.; Walker, R.: In [78], p. 637.
[208] Walton, W. J.; Butt, V. S.: Planta **153** (1981) 232.
[209] Wasternack, C.; Glund, K.; Walther, R.; Reinbothe, H.: Wiss. Z. Univ. Halle **31 M** (1982) 65.
[210] Weissman, G. S.: Plant Physiol. **57** (1976) 339.
[211] Whightman, F.; Forest, J. C.: Phytochemistry **17** (1978) 1455.
[212] Wiater, A.; Krajewska-Grynkiewicz, K.; Kloptowski, T.: Acta Biol. Polon. **18** (1971) 299.
[213] Widholm, J. M.: Canad. J. Bot. **54** (1976) 1523.
[214] Wiemken, A.: Methods Cell Biol. **12** (1975) 99.
[215] Wiemken, A.: In [145], p. 225.
[216] Wolf, E. C.; Weiss, R. L.: J. Biol. Chem. **255** (1980) 9189.
[217] Woo, K. C.; Atkins, C. A.; Pate, J. S.: Plant Physiol. **67** (1981) 1156.
[218] Woodin, T. S.; Nishioka, L.; Hsu, A.: Plant Physiol. **61** (1978) 949.
[219] Yinha, A. K.; Kurtz, M.; Bhattacharjee, J. K.: Appl. Microbiol. **26** (1973) 303.
[220] Yoshida, D.: Plant Cell Physiol. **10** (1969) 923.
[221] Zacharsky, C. A.; Cooper, T. G.: J. Bacteriol. **135** (1978) 490.
[222] Zalkin, H.: In: Bisswanger, H.; Schmincke-Ott, E. (Eds.): Multifunctional Proteins, p. 123. New York 1980.
[223] Zurawski, G.; Brown, K. D.: Biochim. Biophys. Acta **377** (1975) 473.

9 Basic Principles of Alkaloid Biosynthesis

H. W Liebisch

A few thousand *N*-heterocyclic bases are known to occur in higher plants, animals, and microorganisms. Surprisingly, there are only a limited number of precursors (some amino acids, mevalonate, acetate) and a few cyclization mechanisms that force the link between two reactive groups to generate a *N*-heterocyclic skeleton [19, 20]. Once formed, the ring system may undergo substitutions (hydroxylation, methoxylation, methylation, prenylation) or eliminations (demethylation, decarboxylation). These reactions can explain the numerous alkaloids derived from one basic skeleton but they cannot account for the enormous structural diversity of *N*-heterocyclic systems. The possible scope of biosynthesis, however, is greatly increased by secondary cyclization and by reactions involving ring fission at C—C or C—N bonds with subsequent recyclization of the rearranged fission products. Such series of reactions can lead to highly complex molecules that often bear little resemblance to the structures from which they are generated.

Because of the small number of primary precursors and cyclization mechanisms one has to consider the existence of key intermediates that may function as "relay" substances from which several metabolic pathways branch off. This may be true especially in a plant family or species where numerous structurally related alkaloids occur. By far the most impressive example for such a branched route is the huge number of iridoide indole alkaloids that accumulate in species of *Apocynaceae*.

In taxonomically distant plant families there seems to be a tendency to synthesize similar alkaloid structures along different pathways. This may be best illustrated by the different routes used to generate the quinolizidine system in *Asclepiadaceae* (*Cynanchum*), *Ericaceae* (*Vaccinium*), *Leguminosae* (*Lupinus*, *Ormosia*), *Lythraceae* (*Lythrum*), and *Nymphaceae* (*Nuphar*). Such behaviour often complicates biogenetic considerations.

With the elucidation of the basic principles of alkaloid biosynthesis an increasing number of *N*-heterocyclic bases has become available through biogenetic-type (biomimetic) synthesis [20, 34, 37].

9.1 Cyclization Mechanisms

Formation of *N*-heterocyclic ring systems usually takes place in the course of limited range of cyclization mechanisms [19, 20]: lactame formation, azomethine formation, Mannich-type condensation, and addition of amino groups to activated positions in aromatic or quinoide systems (Fig. 9.1). Lactame formation seems to be restricted mainly to the biosynthesis of dioxopiperazines (**3**) and peptide-containing alkaloids [15]. An enzyme-catalysed activation of the carboxyl group of an amino acid (**1**) with ATP or CoA is necessary prior to the reaction as for other non-ribosomal peptide bond formations.

Azomethines such as **7** are formed spontaneously as a result of the condensation of an amino group with a carbonyl function. In this case the action of the enzyme is confined to the formation of the reacting functional groups and to the reduction that withdraws the cyclic azomethine (**7**) from the equilibrium with the ring-open form **6**. In the example given in Fig. 9.1 the precursors (**4**, R = CH_2—CH_2—CH_3) and (**5**, R = COOH) used for the intramolecular azomethine condensation *via* routes

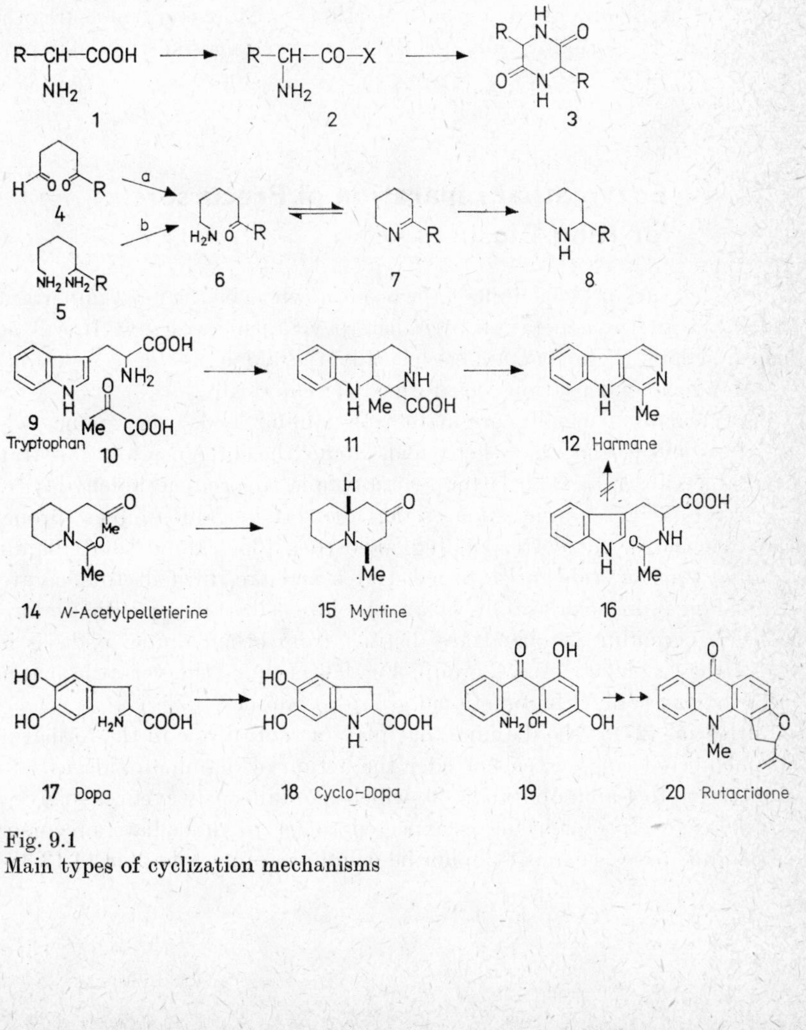

Fig. 9.1
Main types of cyclization mechanisms

a and b are generated in reactions catalysed by amino transferases. The final products are either coniine (**8**, R = $CH_2-CH_2-CH_3$) or pipecolic acid (**8**, R = COOH). If in **5** R = H then the aminoaldehyde (**6**) is yielded by the action of a diamino oxidase with Δ^1-piperideine (**7**, R = H) as the final product.

The vast majority of cyclizations follows a Mannich-type reaction characterized by the condensation of an electrophilic carbonyl group with two electrophilic reactants (amino group and aromatic C—H). An enzyme-catalysed Mannich-type reaction most frequently proceeds in a sterically controlled manner. As a reaction of strictosidine synthetase, strictosidine is produced while no trace of its epimer vincoside can be detected [23].

The intermediacy of an *N*-acylated compound such as **16** has been ruled out for several types of alkaloids in favour of the reaction between a suitable amine (**9**) and an α-keto acid (**10**) [12]. In some cases *N*-acylated intermediates were postulated for similar reactions as for the biosynthesis of myrtine (**15**) [35].

A nucleophilic amino group can easily react with an electrophilic centre of aromatic or quinoide systems. In this way cyclo-dopa (**18**) is derived from dopa (**17**) whereas phenazines, phenoxazines (*i.e.* **65**), and acridines (**20**) are formed from anthranilic acids *via* suitable intermediates such as (**19**) [38]. Moreover, there are other reactions resulting in the formation of *N*-heterocyclic systems like that of β-pyrazol-1-yl-*L*-alanine (**32**) [4].

9.2 Enzymatic Preparation of Precursors for Ring Closures

Amines and carbonyl reactants have been shown to be the most important immediate precursors for the generation of *N*-heterocyclic ring systems. It will now be demonstrated how these reactants, especially the aldehydes necessary for azomethine and Mannich condensations, are formed enzymatically.

Both reactants usually originate from amino acids, the amines by enzymatic decarboxylation and the α-keto acids under the influence of aminotransferases on a suitable substrate. The latter reaction fails to produce aldehydes from amines; this is the field of amine oxidases that convert biogenic amines through oxidative deamination to the corresponding aldehydes [36]. Mono- and diamine oxidases usually attack a wide range of substrates and are difficult to separate from each other. Polyamine oxidases have an absolute substrate specificity for polyamines such as spermidine (**26**) and are distinct from other amine oxidases in that they attack the secondary amino group. Fig. 9.2 outlines the versatile manner in which amine oxidases alter the functional groups of amines.

Putrescine (**27**) is the diamine that plays a central role in the formation of several five-membered ring systems. Under the action of diamine oxidases putrescine (**27**) is converted to 4-aminobutanal (**30**) which spontaneously cyclizes to give Δ^1-pyrroline (**33**). This reactive molecule is expected to be an immediate precursor of several indolizidine bases such as tylophorine or juliprosopine (chapter 14.13.3).

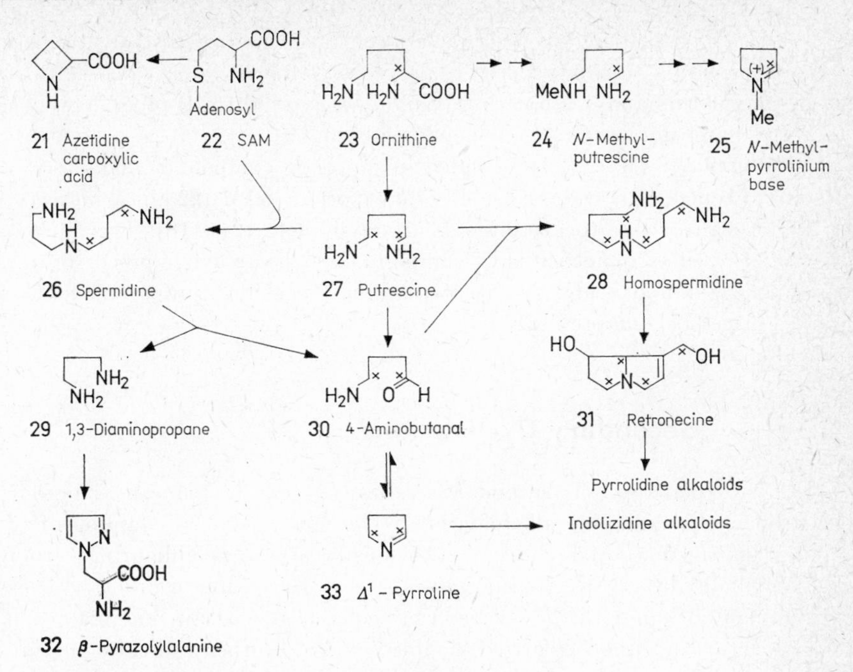

Fig. 9.2
Pathways to form immediate precursors for the intramolecular cyclization to five-membered ring systems (× labelled carbon atom)

One molecule of putrescine (**27**) and 4-aminobutanal (**30**) are combined by azomethine condensation and after reduction yield homospermidine (**28**), the established precursor of retronecine (**31**) [16]. Spermidine (**26**) is most probably synthesized in plants along the same routes as those observed in animals and bacteria [3]. According to this pathway an aminopropyl transfer from decarboxylated *S*-adenosyl-*L*-methionine (SAM, **22**) to putrescine (**27**) produces spermidine (**26**) in a reaction facilitated by spermidine synthetase (aminopropyl transferase). Spermidine (**26**) itself may condense with another reactant to give the macrocyclic spermidine alkaloids such as the oncinotines (**50**) and (**51**). Two molecules of spermidine (**26**) are also used for the generation of the symmetric retronecine precursor homospermidine (**28**) [26a, 27]. In the course of spermidine degradation another diamine is left, 1,3-diaminopropane (**29**) which in an unprecedented reaction yields the heterocyclic skeleton of *β*-pyrazol-1-yl-*L*-alanine (**32**) [4].

Some typical pyrrolidine alkaloids such as hygrine contain an *N*-methylpyrrolidine moiety. In this case *N*-methylation takes place at the level of the diamino acid ornithine (**23**), which on decarboxylation yields *N*-methylputrescine (**24**). Once more a specific diamine oxidase gives 5-*N*-methylaminobutanal and the *N*-methyl-Δ^1-pyrrolinium base (**25**) as immediate precursors. In this way [2-^{14}C] ornithine (**23**) is incorporated into the pyrrolidine and tropane alkaloids in a non-symmetric fashion.

[2-^{14}C] Lysine as the higher homologue of ornithine enters the heterocyclic skeleton of piperidine alkaloids such as pelletierine also in a non-symmetric manner but another reaction sequence is used to prevent symmetric intermediates. In a concerted decarboxylation and deamination lysine is converted to Δ^1-piperideine whereby the intermediate cadaverine remains enzyme-bound.

In Fig. 9.2 SAM (**22**) is included as another compound characterized by two reactive centres. Indeed, the methionine part of SAM (**22**) in a hitherto unknown reaction can cyclize to azetidine-2-carboxylic acid (**21**) [18]. This imino acid may be interpreted as an activated C_4 compound which can act as precursor for a number of unusual amino acids (*i.e.* nicotianamine), provided proper nucleophilic attack takes place [17] (chapter 12).

9.3 Secondary Cyclizations

The best explanation of "secondary cyclization" can be found in the group of tetrahydrobenzylisoquinoline alkaloids [19]. Through oxidative coupling of phenols [2] a key intermediate of high electronic density at the neighbouring carbon atom of a phenolic hydroxyl function permits the easy formation of radicals. These radicals are in equilibrium with the mesomeric forms (**34a** to **34c** in Fig. 9.3), each of which can react with either polarized double bonds or other radical structures to give C—C, C—O, or C—N bonds. By this oxidative coupling of phenols [2] simple tetrahydrobenzylisoquinolines such as **36** generate the polycyclic ring systems of the

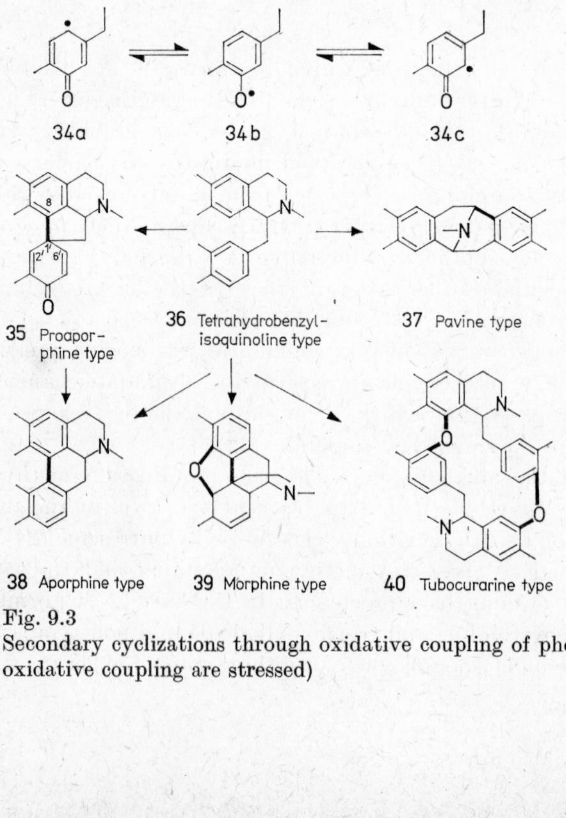

34a 34b 34c

35 Proaporphine type
36 Tetrahydrobenzylisoquinoline type
37 Pavine type

38 Aporphine type
39 Morphine type
40 Tubocurarine type

Fig. 9.3
Secondary cyclizations through oxidative coupling of phenols (bonds derived by oxidative coupling are stressed)

proaporphine (**35**), aporphine (**38**), protoberberine, cularine, pavine (**37**), and of the morphine (**39**) type [see 20]. By way of intramolecular oxidative coupling even dimeric alkaloids such as tubocurarine (**40**) are formed in this manner. Basing on corresponding couplings the multiplicity of *Colchicum*, *Erythrina*, and *Amaryllidaceae* bases is elaborated.

Another structural feature capable of creating secondary cyclizations is an azomethine bond. The conversion of the pelletierine-type structure into those of the *Lycopodium*, *Lythraceae*, or phenanthroquinolizidine alkaloids and into pseudopelletierine is apparently achieved through the intermediacy of an azomethine function, probably in a polarized form.

Even some aromatic amino acids containing dioxopiperazines like **41** show a tendency for secondary cyclizations [15] as illustrated in Fig. 9.4. One example is gliotoxin (**46**), a metabolite of various *Fungi imperfecti*. Aranotins such as **48** and bipolaramide (**47**) exhibit an interesting structural and biosynthetic resemblance to gliotoxin (**46**) [21, 22]. For aranotins (**48**) and gliotoxin (**46**) the precursor role of phenylalanine has been established [21] but the mechanism of the second cyclization has still not been proven. A most likely explanation for the formation of the dihydroaromatic system present in gliotoxin (**46**) is the intermediacy of benzene oxides such as **42**, probably at the dioxopiperazine level [25]. In this way the aromatic nucleus is susceptible to a nucleophilic attack from the dioxopiperazine nitrogen to produce the cyclohexadienol system of gliotoxin (**46**). The benzene oxide (**42**) is in equilibrium with the isomeric oxepine (**43**) and epoxidation of the latter should precede the cyclization to the aranotin-type dioxopiperazines.

9.4 Cleavage of C—N Bonds

Azomethines are in equilibrium with the ring-open amino-carbonyl structures. Also the lactame formation is easily reversible. In some cases secondary amines by oxidative reactions can be converted to lactames and thus provide a working point

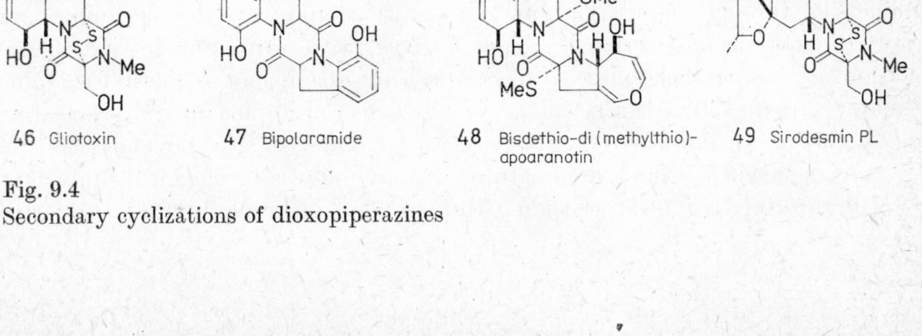

41 42 43 44 45

46 Gliotoxin 47 Bipolaramide 48 Bisdethio-di(methylthio)-apoaranotin 49 Sirodesmin PL

Fig. 9.4
Secondary cyclizations of dioxopiperazines

for ring opening. A special type of lactame ring fission with subsequent recyclization is the intramolecular transamination (ZIP reaction) discovered in the course of thermic or base-catalysed conversion (Fig. 9.5) of neooncinotine (**50**) to isooncinotine (**51**) [10]. As these spermidine alkaloids co-occur in *Oncinotis nitida* this reaction may also take place in nature.

One of the most important ring fission reactions used in the alkaloid field is the Hofmann elimination in which the heterocyclic nitrogen is rendered quarternary by exhaustive *N*-methylation and thus becomes susceptible to ring cleavage. An analogue of this reaction seems to operate *in vivo* and has been used to explain the origin of the seco-tropane alkaloids [33] such as physoperuvine, the free base of which is a mixture of the amino alcohol **52** and the amino ketone **53** [26b]. In particular quarternary aporphine bases such as magnoflorine (**54**) appear to be suitable precursors for a biochemical equivalent of the Hofmann elimination because the resulting double bond in the generated seco-aporphine alkaloids (as **55**) becomes part of a stable phenanthrene system (Fig. 9.5).

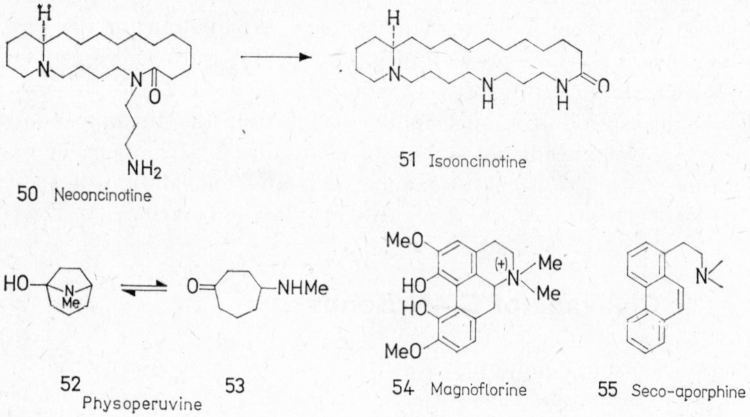

50 Neooncinotine

51 Isooncinotine

52 53 Physoperuvine

54 Magnoflorine

55 Seco-aporphine

Fig. 9.5
Ring fissions at lactame bonds and quarternary ammonium groups

Cyclic quarternary ammonium bases are shared in the conversion of several benzylisoquinoline alkaloids. Fig. 9.6 points to the protoberberines scoulerine (**57**) and stylopine (**56**) as versatile biogenetic intermediates because all three C—N bonds (indicated as *a*, *b*, and *c*) of the tertiary amine are susceptible to cleavage. Ring fission and recyclization are thus also bases for the expression of the structural diversity of benzylisoquinoline a basis in a single plant, *i.e. Chelidonium majus* [1]. Stylopine (**56**) through *N*-methylation is transformed into *N*-methylstylopine and decomposition of the quarternary base at *b* gives rise to an enamine that on recyclization, generates chelidonine (**59**). Simple ring opening of *N*-methylstylopine at *a* yields protopine (**60**). Also the further conversion of protopine (**60**) into rhoeadine (**61**) needs the intermediacy of a quarternary ammonium base. This novel mechanism of ring rearrangement is oxidative in nature and may follow a biochemical equivalent of a von Braun ring fission rather than a Hofmann type elimination [28]. Oxidative ring opening, probably of a ring C aromatized intermediate, followed by *N*-methylation will lead to narcotine (**58**).

In the course of oxidative ring openings a carbon atom is sometimes lost as demonstrated by the conversion of stephanine into aristolochic acid [34a].

Ring fission mechanisms are also known to alter the structure of primary metabolites as shown by the fate of tryptophan (**9**). This amino acid usually forms typical indole alkaloids. If the heterocyclic ring system of tryptophan (**9**) is cleaved at one of the C—N links, following rearrangement and recyclization a variety of other ring systems becomes accessible (Fig. 9.7). The biosynthesis of kynurenine (**62**) is known

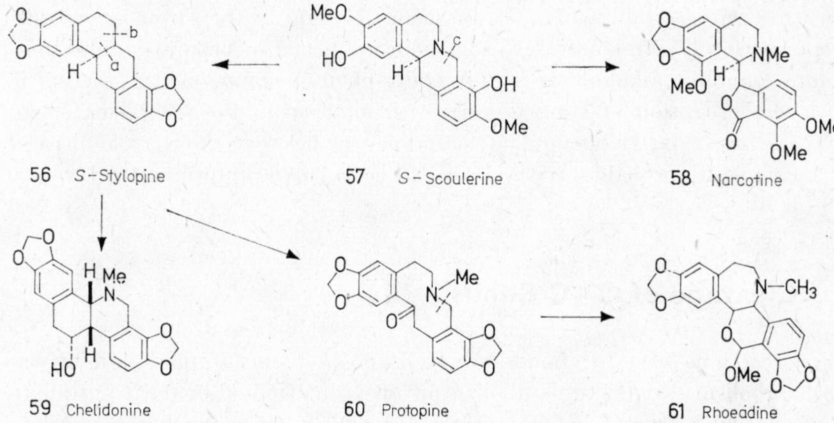

56 *S*-Stylopine 57 *S*-Scoulerine 58 Narcotine

59 Chelidonine 60 Protopine 61 Rhoeadine

Fig. 9.6
Cleavage of C—N bonds in the benzylisoquinoline group

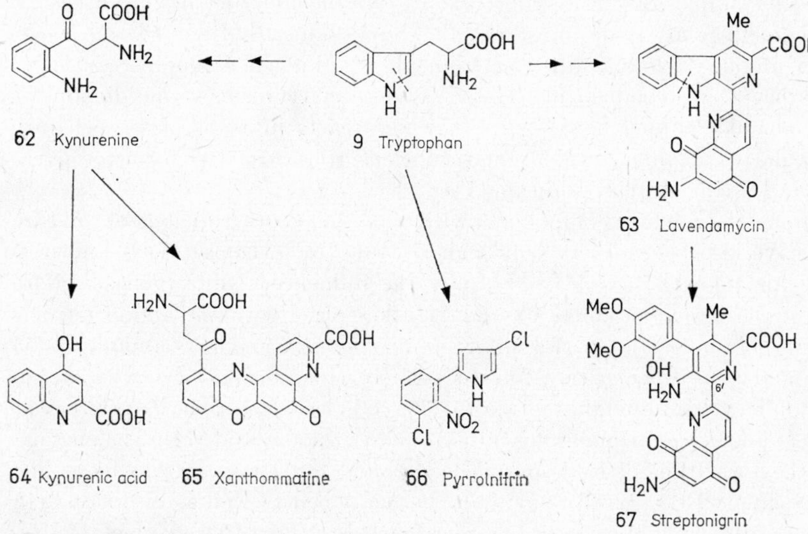

62 Kynurenine 9 Tryptophan 63 Lavendamycin

64 Kynurenic acid 65 Xanthommatine 66 Pyrrolnitrin 67 Streptonigrin

Fig. 9.7
Alkaloids derived by cleavage of tryptophan

to base on the elimination of C-2 of tryptophan. Cyclization of this intermediate **62** yields kynurenic acid (**64**) whilst dimerization gives phenoxazines such as xanthommatine (**65**). These examples show the indole nitrogen to become the hetero atom of other ring systems (Fig. 9.7 does not refer to the substantial number of quinoline, phenazine, phenoxazine, and acridine alkaloids formed from the tryptophan-derived anthranilic acid or an intermediate between chorismic acid and anthranilic acid [13]). Moreover, the indole nitrogen can become even a nitro or amino group whereas the former free amino group of tryptophan (**9**) becomes a hetero atom. These transformations have been recognized as the biogenetic basis of pyrrolnitrin (**66**) formation [8]. A recent example of a novel type of ring cleavage has been shown to occur in the biosynthetic sequence leading to the *Streptomyces* antibiotic streptonigrin (**67**). In contrast to the well-established incorporation of tryptophan (**9**) into quinoline alkaloids it is only the 4-phenylpicolinic acid moiety of **67** which, with the exception of C-6′, is derived from this amino acid. This fission takes place by non-oxidative opening of the indole ring between the pyrrole nitrogen and benzene rings [9], probably at the level of the harmane intermediate lavendamycin (**63**).

9.5 Cleavage of C—C Bonds

Ring fission reactions at C—C bonds are well-known to occur predominantly in microbial metabolism. Complete catabolism of an aromatic compound to aliphatic products ultimately requires cleavage of the benzene nucleus. Such reactions are initiated by oxidative attack and can also produce new groups of *N*-heterocyclic bases.

Two typical examples of ring fission (Fig. 9.4) are included in the biosynthesis of aranotines (*i.e.* **48**) and sirodesmins (*i.e.* **49**). Sirodesmin PL from *Phoma lingam* represents a unique skeleton, different from already known dioxopiperazines. Its biosynthesis has been explained in terms of a route *via* tyrosine-serine dioxopiperazine to phomamide (upper half as in **45**), which reacts in a complex mechanism involving oxidative opening of the aromatic nucleus, ring closure to form the pyrrolidine ring, and finally sulphur addition [7].

Ring fissions of aromatic compounds with subsequent rearrangement and recyclization have also been observed in higher plants [5]. *L*-Dopa (**17**) is known to undergo 2,3- or 4,5-extradiol cleavage under the influence of dioxygenases. If recyclization of the fission products **68** and **71** takes place with the amino nitrogen then the pigments muscaflavin (**72**) in *Amanita muscaria* and betalamic acid (**75**) in *Amanita* and *Centrospermae* are formed [24]. In *Amanita* and *Leguminosae* another type of recyclization takes place [6, 30, 32], leading to the *O*-heterocyclic non-protein amino acids stizolobinic acid (**73**) and stizolobic acid (**74**). The enzymes catalysing the formation of each pyrone compound are characterized by a high substrate specificity [31]. As all four products mentioned as well as cyclodopa (**18**) occur in *Amanita*, one might expect the rare case where at least five enzymes compete for one substrate and where the products of the enzymic reactions combine with each other to give the typical betalain pigments (Fig. 9.8). A 3,4-intradiol cleavage of similar diphenols has also been detected [14], but never a recyclization.

Cleavages of hydroaromatic C—C bonds are a typical feature of terpenoids. The monoterpenoic moieties of the iridoide indole and isoquinoline bases behave in the same manner and are thus a reason for structural deviations of some alkaloids from the basic skeleton.

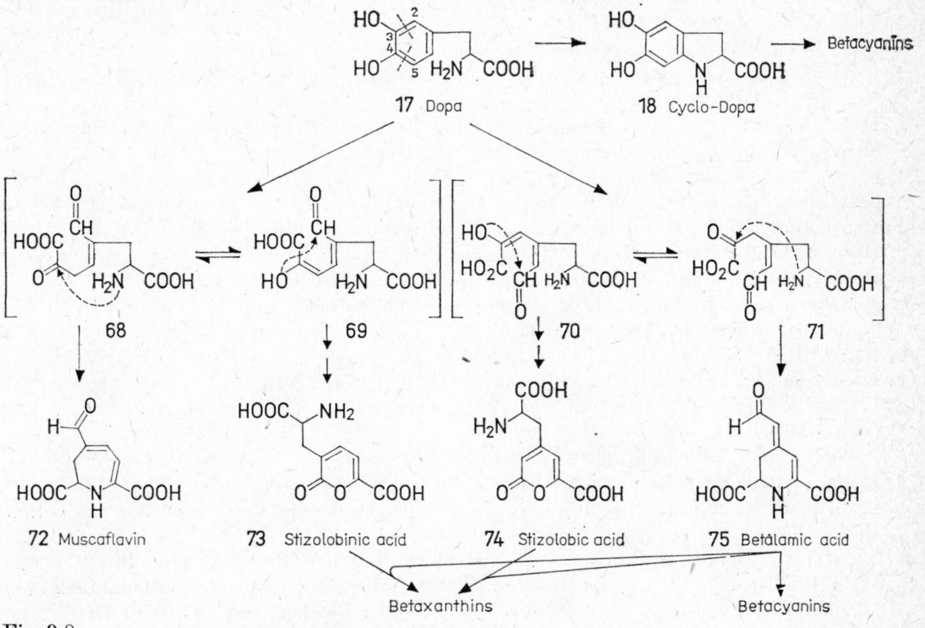

Fig. 9.8
Extradiol cleavage of dopa in the biosynthesis of the betalains

In this chapter some principles of the chemical basis of alkaloid formation are outlined. The occurrence of key intermediates as branching points is frequently found as, for example, with betalamic acid (75) as the structural basis of more than 50 betalain pigments (Fig. 9.8). Often such key intermediates, although already *N*-heterocyclic substances, can undergo secondary cyclizations or ring cleavages with subsequent rearrangements and recyclizations to increase the structural possibilities (Figs. 9.3 and 9.6). Other typical cases are the common intermediates for all types of *Lycopodium* alkaloids or the key compound from which the homoerythrina and the structurally quite different *Cephalotaxus* alkaloids [33a] branch. All examples are surpassed by the accumulation of several hundred iridoide indole alkaloids in *Catharanthus roseus*, all derived from tryptophan (9) and a common monoterpenoid C_9/C_{10} unit. Recent investigations have established that strictosidine is the sole precursor for the elaboration of at least the majority of iridoide indole bases. This is true not only for the *Apocynaceae* like *Catharanthus roseus* but also for alkaloids of taxonomically distant origin as in *Loganiaceae*, *Nyssaceae*, and *Rubiaceae* [11, 26, 29].

9.6 References

[1] Battersby, A. R.; Staunton, J.; Wiltshire, H. R.; Francis, R. J.; Southgate, R.: J. Chem. Soc. Perkin I **1975**, 1147.
[2] Battersby, A. R.; Taylor, W. I.: Oxidative Coupling of Phenols. Marcel Dekker, New York 1967.
[3] Billington, D. C.; Golding, B. T.; Narsereddin, I. K.: Chem. Commun. **1980**, 90.
[4] Brown, E. G.; Flayeh, K. A. M.; Gallon, J. R.: Phytochemistry **21** (1982) 863.
[5] Ellis, B. E.: Lloydia **37** (1974) 168.
[6] Ellis, B. E.: Phytochemistry **15** (1976) 489.
[7] Ferezou, J. P.; Quesneau-Thierry, A.; Servy, C.; Zissmann, E.; Barbier, M.: J. Chem. Soc. Perkin I **1980**, 1739.
[8] Floss, H. G.: Biosynthesis of some aromatic antibiotics. In: Corcoran, J. W. (Ed.): Antibiotics, Vol. IV. Springer-Verlag, Berlin 1981, p. 236.
[9] Gould, S. J.: J. Nat. Prod. **45** (1982) 38.
[10] Guggisberg, A.; Badawi, M. M.; Hesse, M.; Schmid, H.: Helv. chim. Acta **57** (1974) 414.
[11] Heckendorf, A. H.; Hutchinson, C. R.: Tetrahedron Lett. **1977**, 4153.
[12] Herbert, R. B.; Mann, J.: Chem. Commun. **1980**, 841.
[13] Herbert, R. B.; Mann, J.; Römer, A.: Z. Naturforsch. **37c** (1982) 159.
[14] Jaroszewski, J.; Ettlinger, M. G.: Phytochemistry **20** (1981) 819.
[15] Johne, S.; Gröger, D.: Pharmazie **32** (1977) 1.
[16] Khan, H. A.; Robins, D. J.: Chem. Commun. **1981**, 554.
[17] Kristensen, I.; Larsen, P. O.: Phytochemistry **13** (1974) 2791.
[18] Leete, E.; Davis, G. E.; Hutchinson, C. R.; Woo, K. W.; Chedekel, M. R.: Phytochemistry **13** (1974) 427.
[19] Liebisch, H. W.: Fortschr. chem. Forsch. **9** (1968) 534.
[20] Liebisch, H. W.: Cyclisierungsmechanismen bei der Alkaloidbiosynthese. In: Mothes, K.; Schütte, H. R. (Eds.): Biosynthese der Alkaloide. Deutscher Verlag der Wissenschaften, Berlin 1969, p. 101.
[21] Mabe, J. A.; Molloy, B. B.; Day, W. A.: Biochem. biophys. Res. Commun. **43** (1971) 588.
[22] Maes, C. M.; Steyn, P. S.; van Rooyen, P. H.; Rabie, C. J.: Chem. Commun. **1982**, 350.
[23] Mizukami, H.; Nordlöv, H.; Lee, S. L.; Scott, A. I.: Biochemistry **18** (1979) 3760.
[24] Musso, H.: Tetrahedron **35** (1979) 2843.
[25] Neuss, N.; Nagarajan, R.; Molloy, B. B.; Huckstep, L. L.: Tetrahedron Lett. **1968**, 4467.
[26] Phillipson, J. D.; Zenk, M. H. (Eds.): Indole and Biogenetically Related Alkaloids. Academic Press, London 1980.
[26a] Rana, J.; Robins, D. J.: J. Chem. Research **1983**, 146.
[26b] Ray, A. B.; Oshima, Y.; Hikino, H.; Kabuto, C.: Hetereocycles **19** (1982) 1233.
[27] Robins, D. J.; Sweeney, J. R.: J. Chem. Soc. Perkin I **1981**, 3083.
[28] Rönsch, H.: Eur. J. Biochem. **28** (1972) 123.
[29] Rueffer, M.; Nagakura, N.; Zenk, M. H.: Tetrahedron Lett. **1978**, 1593.
[30] Saito, K.; Komanine, A.: Eur. J. Biochem. **68** (1976) 237.
[31] Saito, K.; Komanine, A.: Eur. J. Biochem. **82** (1978) 385.
[32] Saito, K.; Komanine, A.; Hatanaka, S. I.: Z. Naturforsch. **33c** (1978) 793.
[33] Sahai, M.; Ray, A. B.: J. Org. Chem. **45** (1980) 3265.
[33a] Schwab, J. M.; Chang, M. N. T.; Parry, R. J.: J. Am. Chem. Soc. **99** (1977) 2368.
[34] Scott, A. I.: Biogenetic-type synthesis. In: Jones, B. J.; Sih, C. J.; Perlman, D. (Eds.): Application of Biochemical Systems in Organic Chemistry. Vol II, John Wiley & Sons, New York 1976, p. 555.
[34a] Sharma, V.; Jain, S.; Bhakuni, D. S.; Kapil, R. S.: J. Chem. Soc. Perkin I **1982**, 1153.
[35] Slosse, P.; Hootele, C.: Tetrahedron **37** (1981) 4287.
[36] Smith, T. A.: Plant Amines. In: Bell, E. A.; Charlwood, B. V. (Eds.): Encyclopedia of Plant Physiology, N. S. Vol. 8. Springer-Verlag, Berlin 1980, p. 433.
[37] Tramontini, M.: Synthesis **1973**, 703.
[38] Zschunke, H.; Baumert, A.; Gröger, D.: Chem. Commun. **1982**, 1263.

10 Simple Amines

H. R. Schütte

10.1 Aliphatic Monoamines

Lower aliphatic monoamines are widely distributed in the plant kingdom [49, 56, 57, 59] and are generally synthesized from the corresponding amino acids. In *Rhodophyceae*, a wide range of amines is formed by a non-specific particulate amino acid decarboxylase, leucine (1) being the best substrate (Fig. 10.1) [13–15]. In *Camellia sinensis* an alanine decarboxylase yields ethylamine [71]. However, in most flowering plants, aldehyde amination appears to be a more common route for amine biosynthesis [16, 17, 39]. The corresponding aminotransferase has been characterized from *Mercuriales perennis* [16]. There is no apparent relationship between the natural occurrence of amines and aldehyde amination activity [17]. Aldehyde amination probably occurs through the activity of a normal amino acid-keto acid transaminase which also reacts with aldehydes [72].

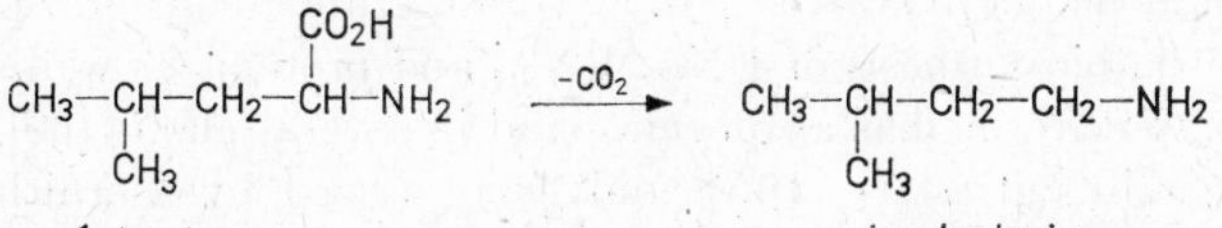

$$CH_3-CH(CH_3)-CH_2-CH(CO_2H)-NH_2 \xrightarrow{-CO_2} CH_3-CH(CH_3)-CH_2-CH_2-NH_2$$

1 Leucine   *iso*-Amylamine

Fig. 10.1
Formation of isoamylamine by decarboxylation of leucine

Galegine (5), dimethylallylguanidine, from *Galega officinalis* (*Fabaceae*) seems to be isoprenoidal derivative, but corresponding precursors such as mevalonate, dimethylallylamine, and leucine are not incorporated. On the other hand guanidoacetic acid (4) is incorporated intact into C-1 and C-2 and into the amidino group of galegine [46, 47, 65]. The amidino group comes from arginine (2) (Fig. 10.2) [43, 44]. Galegine (5) is formed in seedlings, leaves, flowers, and fruits, and especially in young leaves [43, 45]. The amidino group of hydroxygalegine (6) and sphaerophysine (3) is also derived from arginine (2) [2, 48]. The tetramethylene chain including the amino- and amidino group of sphaerophysine (in *Sphaerophysa salsula, Fabaceae*) is formed by arginine, probably *via* agmatine (7) (Fig. 10.2) [66].

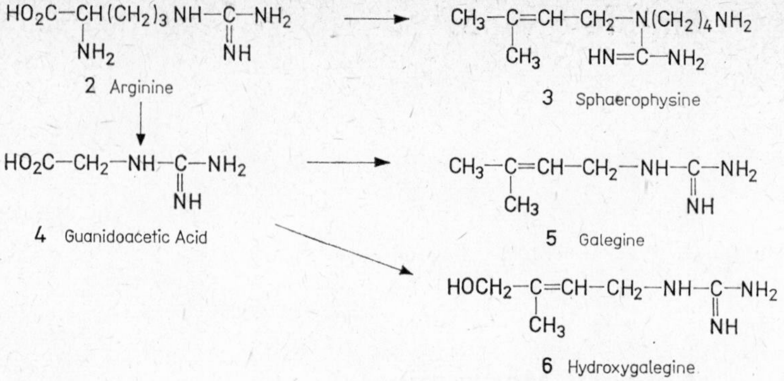

$HO_2C-CH(NH_2)(CH_2)_3NH-C(=NH)-NH_2$ → $CH_3-C(CH_3)=CH-CH_2-N(C(=NH)-NH_2)(CH_2)_4NH_2$

2 Arginine — 3 Sphaerophysine

$HO_2C-CH_2-NH-C(=NH)-NH_2$ → $CH_3-C(CH_3)=CH-CH_2-NH-C(=NH)-NH_2$

4 Guanidoacetic Acid — 5 Galegine

$HOCH_2-C(CH_3)=CH-CH_2-NH-C(=NH)-NH_2$

6 Hydroxygalegine

Fig. 10.2
Formation of galegine, hydroxygalegine, and sphaerophysine

10.2 Di- and Polyamines

The diamine putrescine (**10**) is closely related structurally to the two polyamines spermidine (**12**) and spermine (**13**). They are found in a wide variety of animals, bacteria, yeasts, and plants [55, 56, 69] and are important for several physiological processes such as the biosynthesis of DNA, RNA, and proteins as well as for stabilizing membranes. Cadaverine has also been found in several plants [60].

In mammalian cells, putrescine (**10**) is only synthesized by *L*-ornithine decarboxylase, which is also the rate-limiting enzyme for polyamine biosynthesis [30]. In higher plants, however, such as potassium-deficient barley, the halophyte *Limonium vulgare*, *Cucumis sativus*, *Glycine max*, *Daucus carota*, *Lathyrus sativus*, *Scorzonera hispanica*, putrescine is usually formed by decarboxylation of *L*-arginine (**2**) *via* the intermediate agmatine (**7**) (Fig. 10.3) [1a, 9, 23, 24, 28, 33, 35b, 40, 58, 61, 68]. *L*-Arginine decarboxylase and agmatine (**7**) are frequently found in plant extracts [12a, 53, 58]. The enzyme has been purified from seedlings of *Lathyrus sativus* [41]. In *Cucumis sativus* it is stimulated by cytokinins [68].

Agmatine iminohydrolase, which converts agmatine (**7**) to *N*-carbamylputrescine (**8**) and ammonia, has been detected in soybean seedlings [28], in extracts of tobacco plants [76], and in leaves and seeds of maize and sunflower seedlings [54]. A corresponding enzyme from groundnut cotyledons [51] was inhibited by polyamines such as spermidine (**12**) and spermine (**13**) [52]. Putrescine synthase from *Lathyrus sativus* contains multiple catalytic activities in the conversion of agmatine to putrescine [64a]. Potassium deficiency caused an increase of putrescine (**10**) concentration in different plants [3, 12, 34]. The arginine decarboxylase activity of *Glycine max* plants was 7 to 12 times greater with ammonium than with nitrate as nitrogen source and putrescine was accumulated [25]. It was suggested that the ammonium ion may be competing with potassium at specific sites within the cell [26].

However, in *Solanaceae* such as tobacco and *Lycopersicon esculentum* and also in *Euglena gracilis*, putrescine (**10**) is formed in the same way by arginine decarboxylase *via* agmatine (7) and also by decarboxylation of *L*-ornithine (9) [5, 18, 22, 32, 73, 75, 76]. The decarboxylation of *L*-arginine, catalysed by arginine decarboxylase and of ornithine, catalysed by ornithine decarboxylase, both take place with retention of configuration [35a, 48a, 73a]. But ornithine and not arginine (2), as in the case in most plants, is the main precursor. In tobacco cell cultures and in apical parts of tobacco plants, putrescine is often conjugated with cinnamic acids [10, 20, 20a, 31]. A *p*-fluorophenylalanine-resistant tobacco cell line has been described as accumulating 10 times more cinnamoyl putrescines (*e.g.* caffeoyl and feruloyl putrescine) than wild type cells [6, 7]. The increased synthesis of the cinnamic acids, which was evidently accompanied by increased synthesis of putrescine for cinnamoyl putrescine biosynthesis, was due to the higher activity of phenylalanine ammonia lyase [8]. Arginine (2) and ornithine (9) were equally well incorporated into the conjugates [5]. As in animal cells, α-difluoromethyl ornithine [29] was a very effective *in vitro* inhibitor of ornithine decarboxylase of tobacco cells. Putrescine was quickly transformed to γ-aminobutyric acid, succinic acid and malic acid in *Glycine max* [27].

Spermidine (12) and spermine (13) are probably biosynthesized in plants by the same route as in bacteria and animals, *i.e.* by decarboxylation of (*S*)-adenosyl-*L*-methionine. This gives (*S*)-methyladenosylhomocysteamine, which in turn donates its propylamine moiety in a stepwise manner to putrescine (10) and spermidine (12) respectively (Fig. 10.3) [4, 19, 69, 74]. Thus, labelled putrescine and arginine, a precursor of putrescine, have been incorporated into *L. sativus* and barley seedlings into spermidine (12) and spermine (13) [3, 40]. Two distinct and separable enzymes, namely spermidine synthase and spermine synthase, catalyse these transfer reactions [10a, 18a, 19, 50a]. The (*S*)-adenosyl-*L*-methionine decarboxylase from higher animals and yeast requires putrescine and spermidine for activation [11, 36, 37]. The corresponding decarboxylase from *Lathyrus sativus* seedlings is of prokaryotic type in being Mg^{2+}-dependent but putrescine-insensitive [67]. In some prokaryotes a novel route of spermidine synthesis prevails. Aspartic-β-semialdehyde forms a Schiff base with putrescine to be enzymatically reduced by an NADPH-dependent step to yield "carboxyspermidine" which in turn undergoes a pyridoxal phosphate-dependent enzymatic decarboxylation to give rise to spermidine (12) [70]. In *Lathyrus sativus* seedlings the classic pathway and the new route of spermidine biosynthesis co-exist. The latter is primarily restricted to spermidine synthesis [64]. Aging and senescence of pea leaves caused a progressive decrease in polyamine levels and diamine oxidase activity [62].

Both arginine (2) and ornithine (9) were incorporated efficiently into homospermidine (15) in sandalwood [21]. An enzyme catalysing the formation of homospermidine from putrescine (10) and NAD probably *via* the aldehyde (11) and the imine (14) was purified from *Lathyrus sativus* seedlings and its activity was demonstrated in sandalwood (*Santalium album*) leaves [63] (Fig. 10.3). Cell-free extracts of *Euglena gracilis* synthesized norspermidine and norspermine from 1,3-diaminopropane and labelled (*S*)-adenosylmethionine [73]. Cadaverine is formed in *Lupinus luteus* and *Sarothamnus scoparius* from lysine [50]. In *Lathyris sativus* lysine is a more efficient precursor of cadaverine than homoarginine [40, 42].

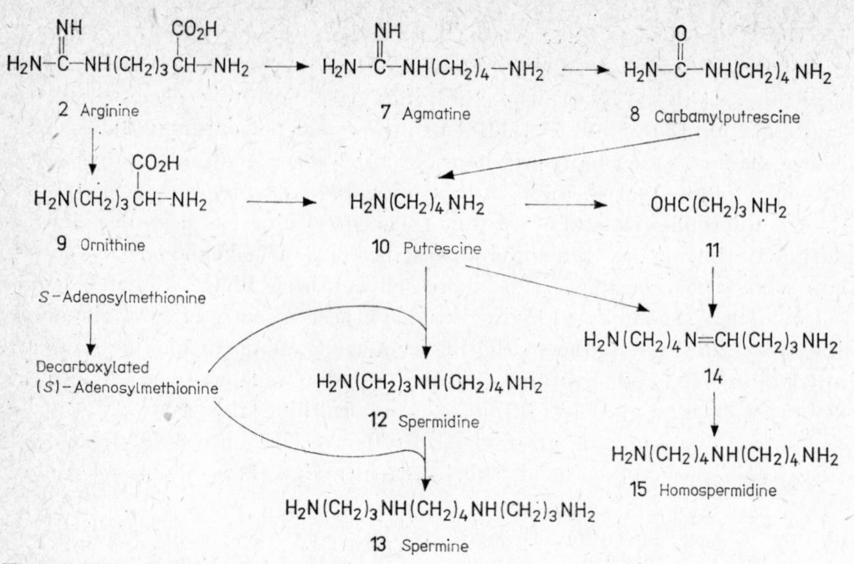

Fig. 10.3
Formation of putrescine, spermidine, spermine, and homospermidine

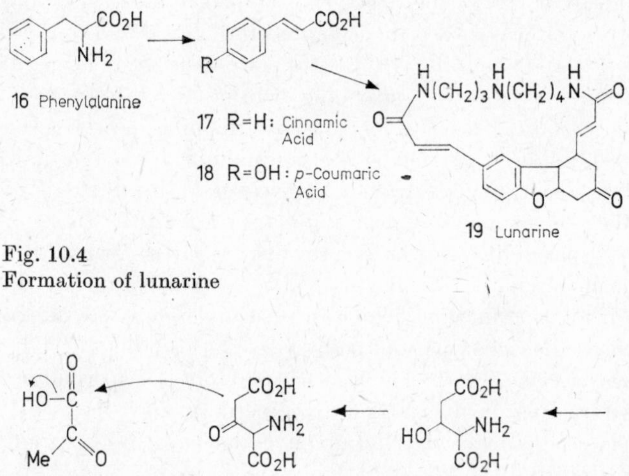

16 Phenylalanine

17 R=H: Cinnamic Acid

18 R=OH: p-Coumaric Acid

19 Lunarine

Fig. 10.4
Formation of lunarine

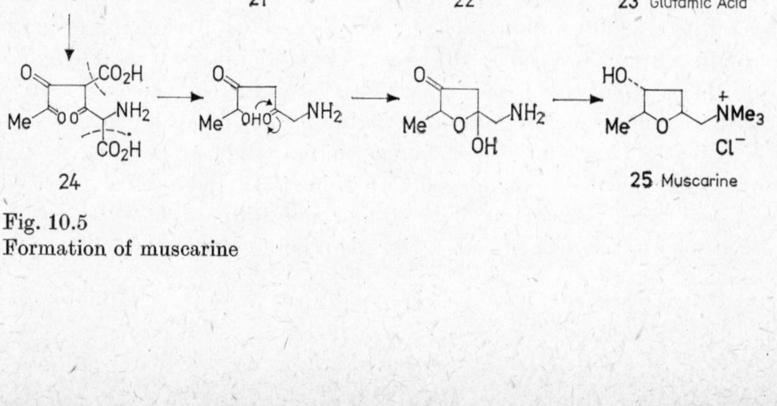

20 Pyruvic Acid

21

22

23 Glutamic Acid

24

25 Muscarine

Fig. 10.5
Formation of muscarine

Spermidine (**12**) and spermine (**13**) are constituents of several alkaloids [1]: lunarine (**19**) and lunaridine from *Lunaria biennis* contain a spermidine unit. Phenylalanine (**16**) but not tyrosine was incorporated *via* cinnamic (**17**) and *p*-coumaric acids (**18**) into lunarine (**19**). This most probably involves oxidative coupling to a dicarboxylic acid with prior or subsequent condensation with spermidine (Fig. 10.4) [38].

10.3 Muscarine

For muscarine (**25**), the main alkaloid of *Amanita muscaria*, a biosynthetic route has been discussed which starts from pyruvate (**20**) and glutamate (**23**) [35]. It has been suggested that glutamate is transformed to β-ketoglutamate (**21**) *via* β-hydroxyglutamate (**22**) followed by condensation with pyruvate. The hypothetical intermediate **24** then undergoes decarboxylations, reduction, cyclization, and methylation to yield muscarine as indicated (Fig. 10.5).

10.4 References

[1] Badawi, M. M.; Bernauer, K.; van den Broek, P.; Gröger, D.; Guggisberg, A.; Johne, S.; Kompis, I.; Schneider, F.; Veith, H. J.; Hesse, M.; Schmid, H.: Pure & Appl. Chem. **33** (1973) 81.
[1a] Bagni, N.; Torrigiani, P.; Barbieri, P.: Med. Biol. **59** (1981) 403.
[2] Barthel, A.; Reuter, G.: Pharmazie **23** (1968) 26.
[3] Basso, L. C.; Smith, T. A.: Phytochemistry **13** (1974) 875.
[4] Baxter, C.; Coscia, C. J.: Biochem. Biophys. Res. Commun. **54** (1973) 147.
[5] Berlin, J.: Phytochemistry **20** (1981) 53.
[6] Berlin, J.; Vollmer, B.: Z. Naturforsch. **34 C** (1979) 770.
[7] Berlin, J.; Widholm, J. M.: Plant Physiol. **59** (1977) 550.
[8] Berlin, J.; Widholm, J. M.: Phytochemistry **17** (1978) 65.
[9] Boldt, A.; Miersch, J.; Reinbothe, H.: Phytochemistry **10** (1971) 731.
[10] Cabanne, F.; Martin-Tanguy, J.; Martin, C.: Physiol. Veg. **15** (1977) 429.
[10a] Cohen, S.; Balint, R.; Sindhu, R. K.; Marcu, D.: Med. Biol. **59** (1981) 394.
[11] Coppoc, G. L.; Kallio, P.; Williams-Ashman, H. G.: Int. J. Biochem. **2** (1971) 673.
[12] Crocomo, O. J.; Basso, L. C.: Phytochemistry **13** (1974) 2659.
[12a] Dumortier, F. M.; Flores, H. E.; Shekhawat, N. S.; Galston, A. W.: Plant Physiol. **72** (1983) 915.
[13] Hartmann, T.: Phytochemistry **11** (1972) 1327.
[14] Hartmann, T.: Biochem. Physiol. Pflanzen **163** (1972) 1.
[15] Hartmann, T.: Biochem. Physiol. Pflanzen **163** (1972) 14.
[16] Hartmann, T.; Dönges, D.; Steiner, M.: Z. Pflanzenphysiol. **67** (1972) 404.
[17] Hartmann, T.; Ilert, H. I.; Steiner, M.: Z. Pflanzenphysiol. **68** (1972) 11.
[18] Heimer, Y. M.; Mizrahi, Y.; Bachrach, U.: FEBS-Letters **104** (1979) 146.
[18a] Hirasawa, E.; Suzuki, Y.: Phytochemistry **22** (1983), 103.
[19] Jänne, J.; Pösö, H.; Raina, A.: Biochim. Biophys. Acta **473** (1978) 241.
[20] Knobloch, K. H.; Berlin, J.: Planta Med. **42** (1981) 167.
[20a] Knobloch, K. H.; Beutnagel, G.; Berlin, J.: Planta **153** (1981) 582.
[21] Kuttan, R.; Rhadhakrishnan, A. N.: Biochem. J. **127** (1972) 61.

[22] Lafarge-Frayssinet, C.; Bertaux, O.; Valencia, R.; Frayssinet, C.: Biochem. Biophys. Acta **539** (1978) 435.
[23] Larher, F.: C. R. Acad. Sci. Ser. D **277** (1973) 1333.
[24] Larher, F.: C. R. Acad. Sci. Ser. D **279** (1974) 271.
[25] LeRudulier, D.; Goas, G.: Phytochemistry **14** (1975) 1723.
[26] LeRudulier, D.; Goas, G.: Physiol. Veg. **13** (1975) 125.
[27] LeRudulier, D.; Goas, G.: Physiol. Plant. **40** (1977) 87.
[28] LeRudulier, D.; Goas, G.: Physiol. Veg. **18** (1980) 609.
[29] Mamont, P. S.; Duchesne, M. C.; Grove, J.; Bey, P.: Biochem. Biophys. Res. Commun. **81** (1978) 58.
[30] Maudsley, D. V.: Biochem. Pharmacol. **28** (1979) 153.
[31] Mizusaki, S.; Tanabe, Y.; Noguchi, M.; Tamaki, E.: Phytochemistry **10** (1971) 1347.
[32] Mizusaki, S.; Tanabe, Y.; Noguchi, M.; Tamaki, E.: Plant Cell Physiol. **14** (1973) 103.
[33] Montague, M. J.; Armstrong, T. A.; Jaworski, E. G.: Plant Physiol. **63** (1979) 341.
[34] Murty, K. S.; Smith, T. A.; Bould, C.: Ann. Bot. **35** (1971) 687.
[35] Nitta, K.; Stadelmann, R. J.; Eugster, C. H.: Helv. Chim. Acta **60** (1977) 1747.
[35a] Orr, G. R.; Gould, S. J.: Tetrahedron Lett. **23** (1982) 3139.
[35b] Palavan, N.; Galston, A. W.: Physiol. Plant. **55** (1982) 438.
[36] Pösö, H.; Hannonen, P.; Himberg, J.-J.: Jänne, J.: Biochem. Biophys. Res. Commun. **68** (1976) 227.
[37] Pösö, H.; Sinervirta, R.; Jänne, J.: Biochem. J. **151** (1975) 67.
[38] Poupat, C.; Kunesch, G.: C. R. Acad. Sci. Ser. C **273** (1971) 433.
[39] Preusser, E.: Biol. Zentralbl. **94** (1975) 75.
[40] Ramakrishna, S.; Adiga, P. R.: Phytochemistry **13** (1974) 2161.
[41] Ramakrishna, S.; Adiga, P. R.: Eur. J. Biochem. **59** (1975) 377.
[42] Ramakrishna, S.; Adiga, P. R.: Phytochemistry **15** (1976) 83.
[43] Reuter, G.: Phytochemistry **1** (1962) 63.
[44] Reuter, G.: Arch. Pharmaz. **296** (1963) 516.
[45] Reuter, G.: Flora **154** (1964) 136.
[46] Reuter, G.; Barthel, A.: Pharmazie **22** (1967) 261.
[47] Reuter, G.; Barthel, A.; Steiniger, J.: Pharmazie **24** (1969) 358.
[48] Reuter, G.; Krone, I.: Pharmazie **24** (1969) 174.
[48a] Robins, D. J.: Phytochemistry **22** (1983) 1133.
[49] Schütte, H. R.: In: Mothes, K.; Schütte, H. R. (Eds.): Biosynthese der Alkaloide, p. 168. Deutscher Verlag der Wissenschaften, Berlin 1969.
[50] Schütte, H. R.; Knöfel, D.: Z. Pflanzenphysiol. **59** (1968) 80.
[50a] Sindhu, R. K.; Cohen, S.: Plant Physiol. **76** (1984) 219.
[51] Sindhu, R. K.; Desai, H. V.: Phytochemistry **18** (1979) 1937.
[52] Sindhu, R. K.; Desai, H. V.: Phytochemistry **19** (1980) 19.
[53] Smith, T. A.: Phytochemistry **2** (1963) 241.
[54] Smith, T. A.: Phytochemistry **8** (1969) 2111.
[55] Smith, T. A.: Phytochemistry **9** (1970) 1479.
[56] Smith, T. A.: Phytochemistry **14** (1975) 865.
[57] Smith, T. A.: Progress Phytochem. **4** (1977) 27.
[58] Smith, T. A.: Phytochemistry **18** (1979) 1447.
[59] Smith, T. A.: Encycl. Plant Physiol. **8** (1980) 433.
[60] Smith, T. A.; Wilshire, G.: Phytochemistry **14** (1975) 2341.
[61] Speranza, A.; Bagni, N.: Z. Pflanzenphysiol. **81** (1977) 226.
[62] Srivastava, S. K.; Ras, A. D. S.; Naik, B. I.: Ind. J. Exp. Biol. **19** (1981) 437.
[63] Srivenugopal, K. S.; Adiga, P. R.: Biochem. J. **190** (1980) 461.
[64] Srivenugopal, K. S.; Adiga, P. R.: FEBS-Letters **112** (1980) 260.
[64a] Srivenugopal, K. S.; Adiga, P. R.: J. Biol. Chem. **256** (1981) 9532.
[65] Steiniger, J.; Reuter, G.: Biochem. Physiol. Pflanzen **166** (1974) 275.
[66] Steiniger, S.; Reuter, G.: Pharmazie **29** (1974) 422.
[67] Suresh, M. R.; Adiga, P. R.: Eur. J. Biochem. **79** (1977) 511.

[68] Suresh, M. R.; Ramakrishna, S.; Adiga, P. R.: Phytochemistry **17** (1978) 57.
[69] Tabor, C. W.; Tabor, H.: Ann. Rev. Biochem. **45** (1976) 285.
[70] Tait, F. H.: Biochem. Soc. Trans. **4** (1976) 610.
[71] Takeo, T.: Phytochemistry **17** (1978) 313.
[72] Unger, W.; Hartmann, T.: Z. Pflanzenphysiol. **77** (1976) 255.
[73] Villanueva, V. R.; Adlakha, R. C.; Calvayrac, R.: Phytochemistry **19** (1980) 787.
[73a] Wigle, I. D.; Mestichelli, L. J. J.; Spenser, I. D.: Chem. Comm. **1982** 662.
[74] Williams-Ashman, H. G.; Canellakis, Z. N.: Perspect. Biol. Med. **22** (1979) 421.
[75] Yoshida, D.: Plant Cell Physiol. **10** (1969) 393.
[76] Yoshida, D.: Plant Cell Physiol. **10** (1969) 923.

11 Alkaloids Derived from Ornithine

H. W. Liebisch and H. R. Schütte

11.1 Tropane Alkaloids

11.1.1 Structure and Occurrence

All tropane alkaloids possess the bicyclic 1(*R*):5(*S*)-tropane (**1**) ring system (*N*-methyl-8-azabicyclo[3.2.1]octane) which is hydroxylated at least in position 3. 3α-Hydroxytropane (**2**, tropine) is the amino alcohol most frequently found in addition to its 3β-enantiomer (**3**, ψ-tropine), the di- and trihydroxylated tropanes, the 6,7-epoxide (scopoline) and the corresponding *N*-norcompounds. These bases can occur in free form but are usually esterified with aliphatic, aromatic and heterocyclic acids [30, 47, 87, 89, 101, 144]. Typical examples are hyoscyamine (**4**, 3α-[2′(*S*)-tropoyloxy]-tropane, in racemic form called atropine), hyoscine (**5**, the corresponding 6,7-epoxide, in racemic form called scopolamine), tigloidine (**6**, 3β-tigloyloxytropane), littorine (**7**, 3α-[2′(*R*)-hydroxy-3-phenylpropionyloxy]-tropane, cochlearine (**8**, 3α-*m*-hydroxybenzoyloxy tropane), meteloidine (**9**, 3α-tigloyloxytropane-6β,7β-diol), acetyltropine (**11**), and valtropine (**12**, 2′(*S*)-methylbutanoyl tropane) (Fig. 11.1).

With the exception of the tribe *Nicandrae*, tropane alkaloids occur in all tribes of *Solanaceae* [144], especially in species of *Atropa*, *Datura*, *Duboisia*, *Hyoscyamus* and *Scopolia*. Some other plant families containing tropane alkaloids are the *Convolvulaceae*, *Cruciferae* (cochlearine, **8**), or the *Rhizophoraceae* (brugine, **10**). In several species of *Erythroxylon* (*Erythroxylaceae*) coca alkaloids of similar structure have been detected, characterized by a 3β-hydroxy function and a carboxyl group at C-2 of the tropane nucleus, *i.e.* in ecgonine (**13**), cocaine (**14**), and catuabine A (**15**). Recently [126] some new sources for tropane alkaloids were found, all belonging to *Proteaceae*. From *Bellendena montana*, *Agastachys odorata* and two *Darlingia* species, all endemic in Australia, as well as from two *Knightia* species, endemic in New Caledonia, more than 30 quite unusual ester alkaloids have been isolated and structurally defined (Fig. 11.3). These alkaloids exhibit a number of unprecedented structural features, *i.e.* aliphatic or aromatic substituents at C-4, formation of γ-pyranes. In *Knightia* they are accompanied by ψ-tropine (**3**) congeners [115]. Ferruginine (**31**) and ferrugine (**38**) from *Darlingia* [21] are the first tropane alkaloids with no hydroxyl group. In *Darlingia darlingiana* eight additional biogenetically related pyrrolidine bases (Fig. 11.3) were observed [19]. In this review the upper bridge-head carbon atom of tropane alkaloids with its *R* configuration is denoted as C-1 (as in **1**) in all cases. Some bases are now termed C-4 substituted tropane

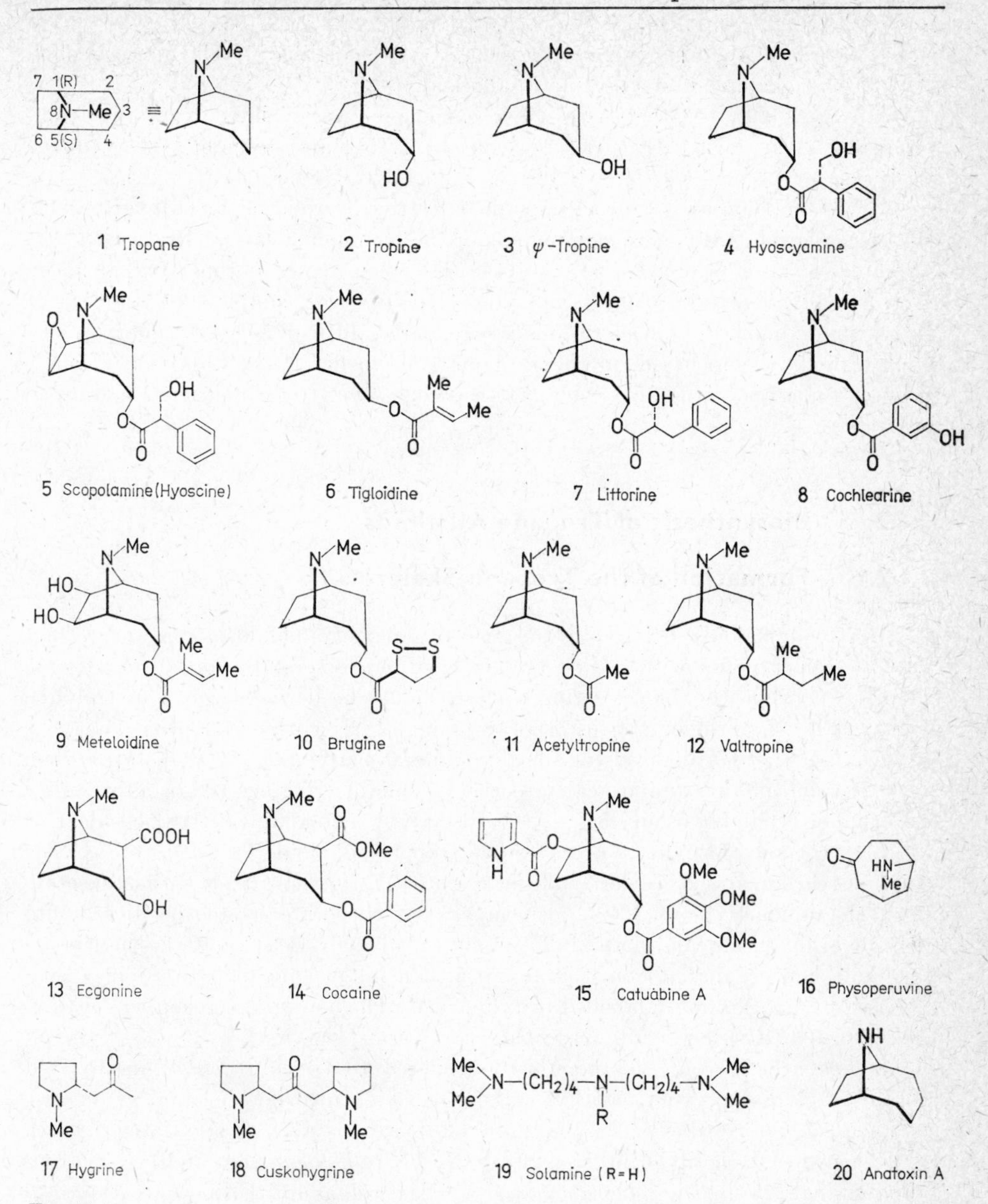

Fig. 11.1
Typical tropane alkaloids and pyrrolidine bases

alkaloids rather than 2-substituted as recommended earlier (*cf.* [14]), where the *S*-configurated bridge-head atom was C-1. In the same manner the numbers 6 and 7 are exchanged in some cases of dihydroxylated tropane bases.

Another tropane alkaloid with no hydroxyl group at C-3 is physoperuvine (**16**) from *Physalis peruviana* [136a, 137, 148]. As a free base physoperuvine represents an equilibrium of the aminoketone **16** and the corresponding bicyclic aminoalcohol. It is the first example of a secotropane alkaloid.

Anatoxin-A (**20**), together with other congeners, was isolated from the fresh-water blue-green algae *Anabaena flos-aquae* [27, 51] and resembles the tropane structure.

Tropane alkaloids are often accompanied by the pyrrolidine bases hygrine (**17**) and cuscohygrine (**18**) and by compounds of the common structure **19** with R = H (solamine) or an acyl moiety [41, 42, 144]. The occurrence of tropane alkaloids in systematically divergent plant families [144] points to a simple biogenetic route and chemotaxonomical considerations are rendered difficult. The chemotaxonomic interest should be focussed on the acid moieties rather than the amino alcohols because tropic and tiglic acid esters of tropanols seem to be restricted mainly to *Solanaceae*.

11.1.2 Biosynthesis of Tropane Alkaloids

11.1.2.1 Formation of the Tropane Skeleton

The tropane alkaloids belong to the first groups of alkaloids subjected to biosynthetic experiments with radioactive labelled precursors. Basing on the classic Robinson concept the bicyclic ring system should be derived from a pyrrolidine moiety with a C_3 bridge. The incorporation of [2-^{14}C] ornithine (**21**) (Fig. 11.2) into hyoscyamine (**4**) was reported for the first time as early as 1954 [95]. Since then the conversion of this amino acid into the pyrrolidine ring has been confirmed for hyoscyamine (**4**) and hyoscine (**5**) [84, 106—108, 110], meteloidine (**9**) [98], cochlearine (**8**) [103], cocaine (**14**) [56, 88], and other tropane alkaloids [87, 101, 115a]. The remaining carbon atoms are derived from acetate [75, 76], most likely *via* acetoacetate [106]. The radioactivity of [2-^{14}C] ornithine was incorporated stereospecifically into only one bridge-head carbon atom of tropine (**2**), namely C-1 [84, 107]. The heterocyclic nitrogen originates from the amino acid as shown by incorporation experiments with ^{14}C, ^{15}N doubly labelled ornithine (**21**). The δ-amino nitrogen was mainly used [100, 106, 108].

The biosynthetic route is therefore considered to continue *via* *N*-methylated intermediates that prevent ornithine (**21**) from being incorporated in a symmetric fashion. The intermediacy of *N*-methylated products was substantiated by the occurrence of δ-*N*-methylornithine (**22**) in *Atropa belladonna* [63] and δ-*N*-methylaminobutanal (**25**) in *Datura stramonium* [124] as well as by the incorporation into hyoscyamine (**4**) of labelled δ-*N*-methylornithine (**22**) [3, 8] and *N*-methylputrescine (**24**) [94, 104, 106]. A valuable result gave the specific incorporation of the carbon atom contiguous to the *N*-methylamino group of *N*-methylputrescine (**24**) into only one bridge-head carbon atom (most probably C-5) of the tropane (**1**) nucleus [96].

In equilibrium with δ-*N*-methylaminobutanal (**25**) is the *N*-methylpyrrolinium base (**26**), which on reaction with acetoacetate may give the hypothetic hygrine-

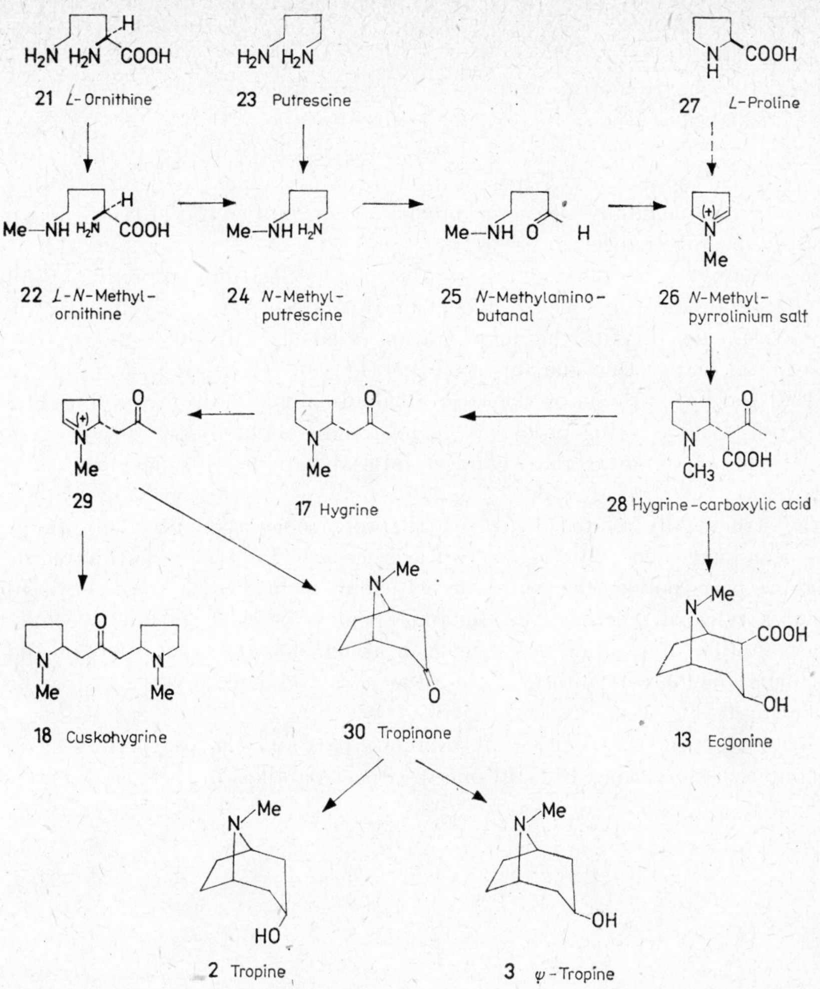

Fig. 11.2
Biosynthesis of tropane and coca alkaloids

carboxylic acid (**28**) [87, 105] and on spontaneous decarboxylation hygrine (**17**). This alkaloid seems to be a precursor of both the bicyclic tropane alkaloids and cuscohygrine (**18**), as indicated by feeding experiments [105, 132]. Application of resolved hygrines demonstrated that the 2(*R*)-enantiomer (**17**) was the favourite precursor of hyoscyamine (**4**) and hyoscine (**5**) in *D. innoxia* [120]. On the contrary, tropane alkaloids of *A. belladonna*, *Hyoscyamus niger*, and *Physalis alkekengi* became labelled with both enantiomers [121]. Hygroline (**72**) as possible reduction product of hygrine (**17**) neither in free nor in esterified form served as an immediate precursor for the biosynthesis of tropane alkaloids [122a, 122b]. The intermediacy of tropinone (**30**) is suggested where reduction may take place. An NADP(H)-specific tropine dehydrogenase (= tropine:$NADP^+$ oxidoreductase), isolated from sterile root cultures of *Datura stramonium*, catalyses the interconversion tropine (2) ⇌ tropinone (**30**) [81].

Putrescine (**23**) as a symmetric compound is not on the route between ornithine (**21**) and the tropane alkaloids. Nevertheless, this diamine (**23**) was incorporated into hyoscyamine (4) by three independent teams [74, 92, 99, 106, 107, 109]. Putrescine (**23**) may reach the path by *N*-methylation to *N*-methylputrescine (**24**). Other members of the ornithine family also act as precursors, *i.e.* proline (**27**) [108], although they were of minor importance.

If hygrine-carboxylic acid (**28**) undergoes a second cyclization and reduction of the carbonyl group under controlled stereochemical conditions, the bicyclic skeleton of ecgonine (**13**) is reached. With the incorporation of labelled ornithine (**21**), putrescine (**23**), acetate, and methionine into cocaine (**14**) in *Erythroxylon coca* [56, 88, 89b, 105, 162] the biosynthesis of the coca alkaloids along the route in Fig. 11.2 seems reasonable. A convincing proof has been brought about by the discovery of 2-carbomethoxy-3-tropinone as an advanced intermediate in the biosynthesis of cocaine (**14**) [89a].

For all the structurally related 4-substituted proteaceous alkaloids a biogenetic scheme has been suggested [20]. Instead of acetoacetate for the formation of the non-pyrrolidone part, polyketides with three to four acetate units were proposed as well as polyketides with benzoic (or cinnamic) acid moieties as starter molecules. With the assumption of pyrrolidine bases such as **32**, dehydrodarlinine (**35**), and dehydrodarlingianine (**36**) as intermediates, the expected pathways [20] may be altered according to Fig. 11.3.

The striking structural similarities of tropinone (**30**) and the pyrrolidine bases hygrine (**17**) and cuscohygrine (**18**) lead one to expect parallels in their endogenous

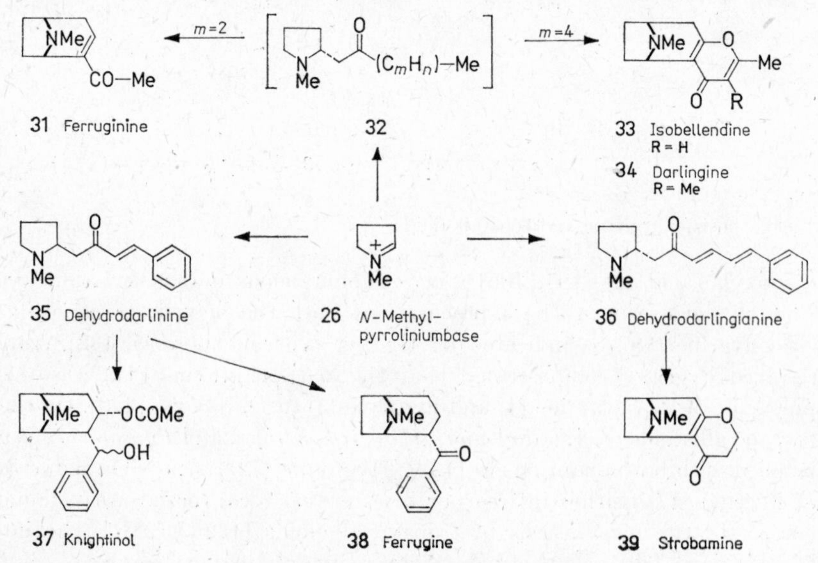

Fig. 11.3
Hypothetical plan for the formation of the *Proteaceae* alkaloids

formation. Indeed, hygrine (**17**) in *Nicandra physaloides* became specifically labelled following application of [2-^{14}C] ornithine (**21**) and [1-^{14}C] acetate·[132] with the radioactivity of the amino acid being confined to C-2 of the alkaloid. Further proof was the specific incorporation into cuscohygrine (**18**) of labelled ornithine (**21**) [88, 106], δ-*N*-methylornithine (**22**) [8], putrescine (**23**) [106], *N*-methylputrescine (**24**) [106], hygrine (**17**) [120, 121, 132], acetate, and aceto-acetate [7, 106] in *A. belladonna*, *E. coca*, or *Scopolia lurida*. In contrast to the tropane alkaloids the pathway to cuscohygrine (**18**) showed little stereochemical preference for either enantiomer of resolved hygrines (**17**) [120, 121] and at least in *E. coca* ornithine (**21**) was incorporated symmetrically [88] into both cuskohygrine (**18**) and cocaine (**14**), probably through putrescine (**23**).

Ornithine (**21**) has been shown to enter the pyrrolidine ring of nicotine through a symmetric intermediate. Further experimental work should clarify whether two independent pathways exist for the pyrrolidine nucleus in plants containing both tropane and tobacco alkaloids, *i.e.* in *Duboisia myoporoides*. The same is true for graftings of *Hyoscyamus* with *Nicotiana* where labelled ornithine (**21**) was incorporated into hyoscyamine (**4**) and nicotine to a comparable extent [110].

11.1.2.2 Esterification and Oxidation of Tropine

The further metabolism of tropine (**2**) involves esterification and oxidation steps. It seems likely that activation of the acid moieties takes place prior to the enzymatic esterification. Indeed, *in vitro* synthesis of hyoscyamine (**4**) from tropine (**2**) and tropic acid (**40**) in the presence of ATP and CoA has been reported [72]. Repetition with tropoyl coenzyme A ester [57] should be more successful. So-called atropine esterases, cleaving hyoscyamine (**4**), have been observed in several animal tissues [47, 101].

In vivo experiments established the incorporation of labelled tropine (**2**) into hyoscyamine (**4**), hyoscine (**5**), and several tropane tigloyl esters [17, 85, 94]. Further labelling and physiological experiments established the entire route leading to hyoscine (**5**) to proceed *via* tropine (**2**) $\rightarrow$ hyoscyamine (**4**) $\rightarrow$ 6β-hydroxyhyoscyamine (**43**) $\rightarrow$ 6,7-dehydrohyoscyamine (**46**) $\rightarrow$ hyoscine (**5**) (Fig. 11.4) [47, 101]. The tritium of [6β,7β-$^{3}H_2$] tropine (**2**) was completely lost [94], indicating an unusual mechanism of *cis*-dehydration, which was interpreted [47] as the nucleophilic attack of an enzyme at the 6α-site of 6β-hydroxyhyoscyamine (**43**).

Teloidine, the amino alcohol of meteloidine (**9**), originated from ornithine (**21**) *via* tropine (**2**) [85, 94, 98]. The further hydroxylation of tropine apparently occurred after esterification to 3α-tigloyloxytropane (**42**) [16, 17, 93] with retention of the configuration at C-6 and C-7 [94] in a cytochrome P 450 catalysed reaction [117]. This situation is similar to the formation of hyoscine (**5**) in that esterification of tropine (**2**) precedes oxidation at the heterocyclic nucleus (*cf.* Fig. 11.4).

The interconversion of the mono-, di-, and tri-tigloyloxytropanes has still not been fully explained. On the basis of investigations with labelled alkaloids and as a result of competitive feeding experiments of ^{14}C-tropine (**2**) with cold alkaloids a

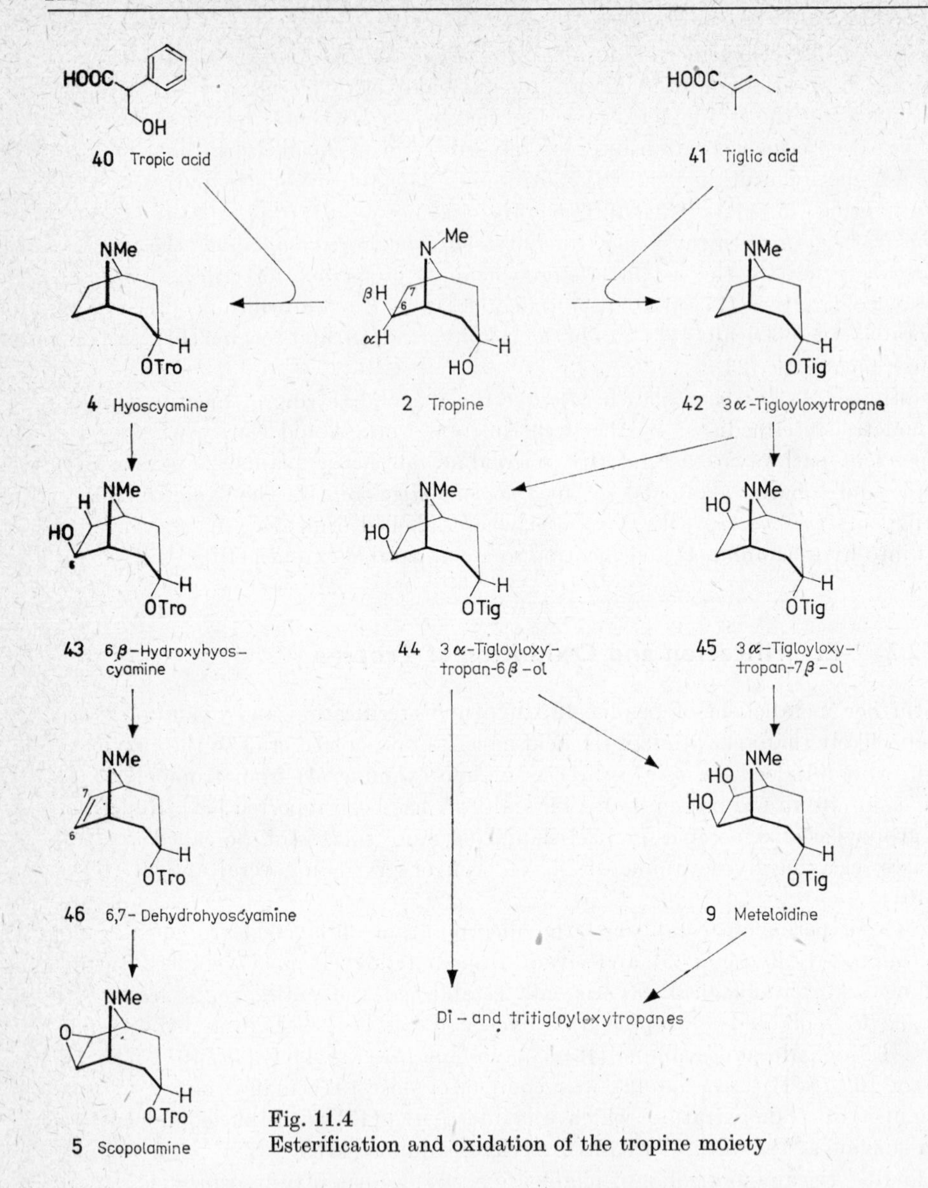

Fig. 11.4
Esterification and oxidation of the tropine moiety

tentative scheme (Fig. 11.4) has been outlined [16, 17]. A high rate of hydrolysis of the ester alkaloids [1, 11, 90, 93] further complicates the situation.

11.1.2.3 Biosynthesis of the Acid Moieties

In 1960 it was discovered [83] that tropic acid (**40**) was derived from phenylalanine (**50**). This experiment has been successfully repeated with several plants containing tropane alkaloids [56, 103, 113, 153], *cf.* also [87, 101]. Further investigations demonstrated the conversion of [1,3-^{14}C] phenylalanine (**50**) into tropic acid (**40**) with the same ratio of radioactivity in the carboxyl group and in C-2 as the starting material [113]. These results point to an intramolecular shift of the carboxyl group from C-2 to C-3, which has been further established by ^{13}C NMR measurements of tropic acid (**40**) generated from [1,3-^{13}C] phenylalanine (**50**) in *D. innoxia* [91]. Migration of the carboxyl group is facilitated by a hydride shift [89c] and is, thus, obviously different from the pathway whereby tropic acid (**40**) is formed chemically from phenylalanine (**50**) by the migration of the aromatic nucleus [103]. As most of the reactions at C-2 of an amino acid are facilitated by pyridoxalphosphate, the participation of this co-enzyme in the re-arrangement (Fig. 11.5) was postulated [87, 103]. After a series of electronic shifts the initially formed azomethine (**49**) should enable the carboxyl group to migrate from C-2 to C-3. The azomethine double bond shifts towards C-2 of the amino acid and hydrolytic cleavage forms α-formylphenylacetic acid (**54**), which is subsequently reduced to tropic acid (**40**). *In vitro* the latter reaction failed [58] because of the spontaneous decarboxylation

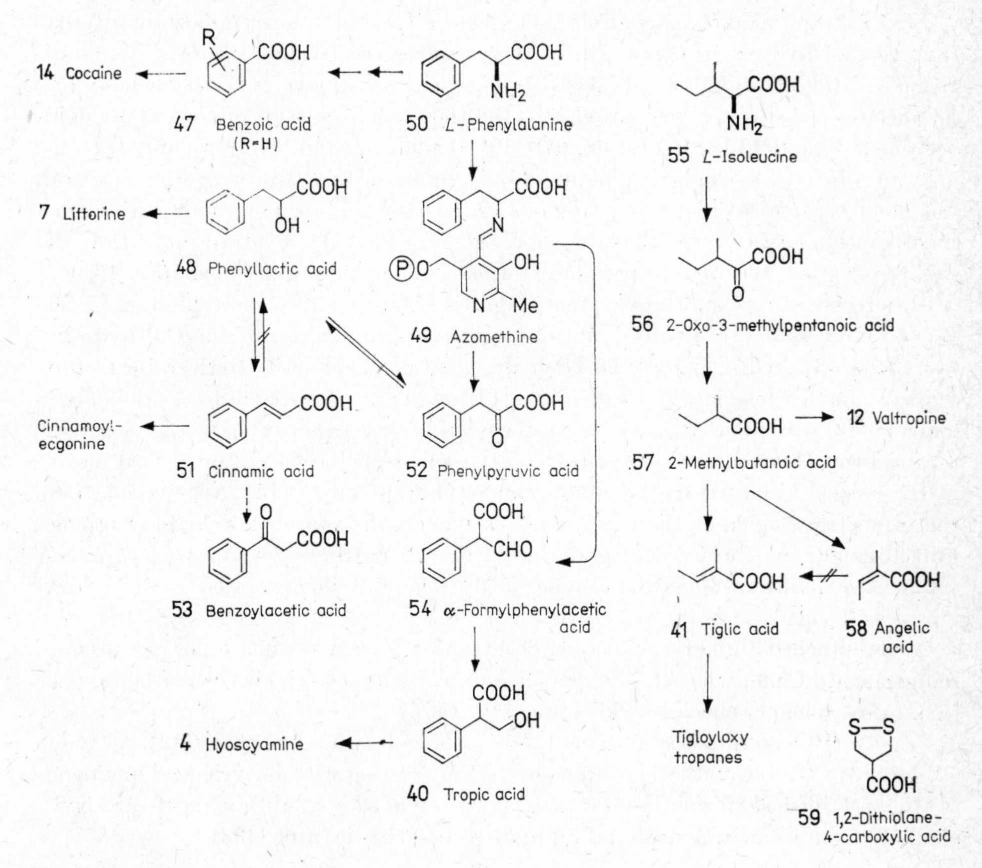

Fig. 11.5
Biosynthesis of the acid moieties of the tropane alkaloids

of the unstable acid (**54**). The re-arrangement does not occur at the ester level since neither [3β-^{3}H, ^{14}CO] littorine (**7**) nor [2-^{14}C] cinnamoyl-[*N*-methyl-^{14}C] tropine were incorporated intact into hyoscyamine (**4**) [90]. Precursors other than phenylalanine (**50**), provided they were incorporated specifically, seem to play a minor role in the biosynthesis of tropic acid (**40**) [4, 44, 61, 89b, 103].

Labelled phenylalanine (**50**) has also been incorporated into littorine (**7**) with the radioactivity confined to the expected sites in the phenyllactic acid (**48**) moiety of the alkaloid [43, 44]. [3-^{14}C] Phenylalanine (**50**) was also an efficient precursor of benzoic acid (**47**) in cocaine (**14**) [56, 89a], and of *m*-hydroxybenzoic acid in cochlearine (**8**) [102].

Cinnamic acid (**51**) does not seem to be on the way to either tropic (**40**) or phenyllactic acid (**48**) but may play a role as precursor of cinnamoylecgonine and of the truxillic acid moiety of truxillin (*cf.* [101]). Quite unusual acids such as trimethoxybenzoic and pyrrole-2-carboxylic acid have been detected as esterifying agents in the alkaloids (*i.e.* catuabine A, **15**) of the Brazilian *Catuaba* shrub, *E. vaccinifolium* [52].

Next to tropic acid (**40**) the aliphatic tiglic acid (**41**) is the most important esterifying agent of hydroxytropanes. Up to three molecules of tiglic acid (**41**) can be attached to the alkamine, *i.e.* in $3\alpha,6\beta,7\beta$-tritigloyloxytropane. *L*(+)-Isoleucine (**55**) has been successfully incorporated into the tiglic acid moieties of several alkaloids such as **6**, **9**, and **42** [9, 10, 14, 66, 97, 119]. Unlike 3α-tigloyloxytropane (**42**) the amount of the 3β isomer tigloidine (**6**) is limited. In *Physalis peruviana* both alkaloids became labelled after application of [U-^{14}C] isoleucine (**55**) with higher incorporation rates for the 3α-tigloyloxy isomer (**42**) [14]. By transamination and decarboxylation the amino acid (**55**) forms (*S*)-methylbutanoic acid (**57**) [9, 86]. The stereospecificity of the enzymatic dehydrogenation of 2-methylbutanoic acid (**57**) to tiglic acid (**41**) involves an antiperiplanar elimination of the C-2 hydrogen and the pro-3*R* hydrogen [66]. This dehydrogenation occurs prior to the esterification with tropine because labelled valtropine (**12**) was not converted into 3α-tigloyloxytropane (**42**) without cleavage of the ester bond [11]. Further proof of the biogenetic arrangement (Fig. 11.5) was the incorporation of isoleucine (**55**) into the acid moiety of the recently isolated 6β-[2(*S*)-methylbutyryloxy]-tropine [15]. Angelic acid (**58**), the geometric isomer of tiglic acid (**41**), is not on the route to tigloyloxytropanes [10] although the former acid is derived from the latter in *Cynoglossum officinale* [122]. Also hydroxy-2-methylbutanoic acids were not incorporated in tigloyloxytropanes [9, 119].

Other branched aliphatic acids were observed in the solanaceous genus *Schizanthus*, indigenous to Chile, where the tropane alkamines were esterified with either angelic (**58**), mesaconic, or senecioic acids [48a, 139, 150].

Brugine (**10**) contains the unusual 1,2-dithiolane-3-carboxylic acid [112]. Although still unproven, a biosynthetic route is probable, starting with butyric acid analogous to the established formation of asparagusic acid (= 1,2-dithiolane-4-carboxylic acid, **59**) from isobutyric acid and an unidentified SH-donator [133].

11.1.3 Physiology of Alkaloid Production in Intact Plants and Cell Cultures

Physiological experiments in the field of alkaloids are often carried out on solanaceous plants and were successful far earlier than the first biosynthetic investigations [127]. Because of the pharmacological interest of some tropane bases the genetic and biochemical aspects of alkaloid production have also been carefully investigated [31].

By grafting experiments the roots have been shown to be the major site of tropane alkaloid synthesis [128], *cf.* also [31, 101]. Obviously the alkaloids are transported throughout the whole plant, although green parts can biosynthesize alkaloids to some extent [110].

The alkaloid pattern varies considerably (*cf.* [31, 101]) with respect to the developmental stage [60, 147, 165, 166], environmental conditions [32—34], genetic varieties [64, 80], and chemical families [55]. A recently developed radioimmunoassay for scopolamine [169] will further improve present analytical procedures.

Some of the physiological problems could perhaps be solved using alkaloid-producing tissue or cell cultures. A growing number of publications deal with the occurrence and pattern of tropane alkaloids in cell cultures. Usually the total alkaloid content in cultures is far below that detected in the leaves of the same plant [38, 39, 68, 80, 82, 138, 146, 158—160, 163]. Initiation of calluses from different parts of a seedling showed a similar alkaloid pattern [38, 160]. Alkaloid contents increased when unorganized callus cultures were transferred to an auxine-free medium where root formation was initiated [38, 62a, 138, 160, 163]. Even alkaloid-producing plants could be regenerated from cultured cells of *D. innoxia* [67].

Biosynthetic experiments with cell cultures, however, seem to be more difficult in comparison with higher plants. The esterification of tropine (**2**) with a suitable acid has been detected [68, 82, 143, 145, 146, 159]. Acetyltropine (**11**) was formed, even in the presence of tropic acid [143].

Exogenously applied tropane alkaloids seem to be catabolized in both tissue cultures and intact plants very rapidly [62, 145].

11.2 Pyrrolidine Alkaloids

11.2.1 Structure and Occurrence

The group of pyrrolidine alkaloids comprises compounds of very different structures, *i.e.* proline (**27**) and its congeners (Fig. 11.6), the "typical" 2-substituted pyrrolidine bases of the hygrine (**17**) type (Fig. 11.7), a number of indolizidine alkaloids (Fig. 11.8), and some unusual structures (Fig. 11.9) produced by microorganisms. Pyrrolidine moieties are often attached to other heterocyclic bases as in nicotine (Fig. 13.11) or brevicolline (**76**). Amide-like bound pyrrolidine units occur in *Piper* species (Fig. 12.4 and Table 12.1). It is therefore difficult to distin-

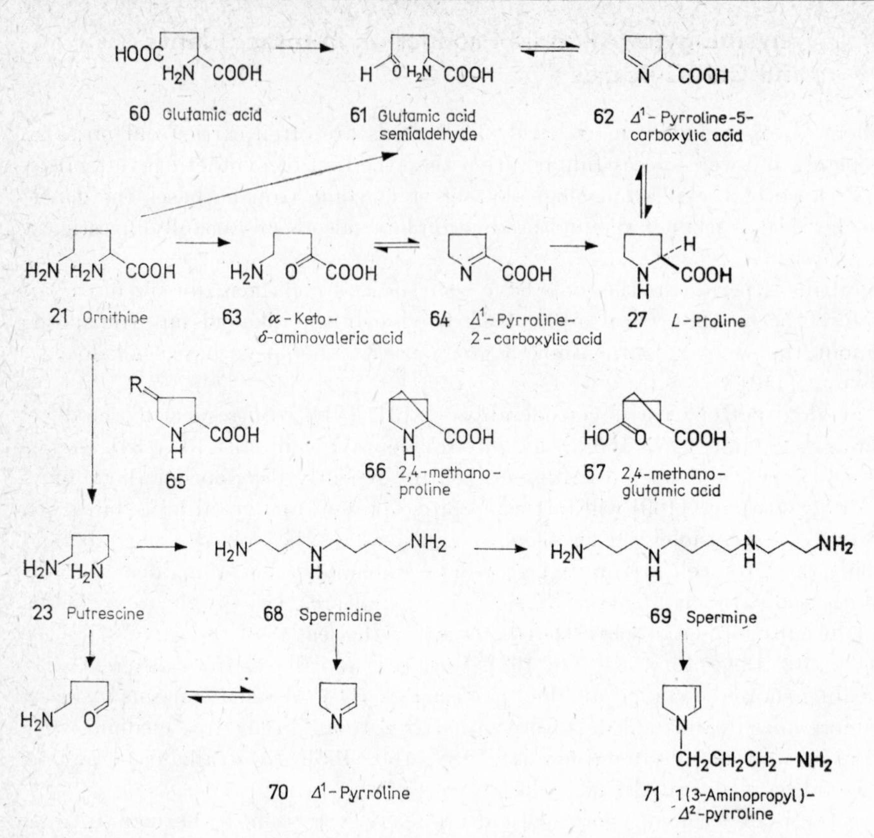

Fig. 11.6
Biosynthesis of some pyrrolidine compounds originating from ornithine or glutamine

guish the pyrrolidine alkaloids from other alkaloid groups, the more so because they occur in divergent plant families, animals and microorganisms [101, 144]. Plant families distinguished by numerous pyrrolidine alkaloids are *Orchidaceae*, *Erythroxylaceae*, *Eleagnaceae*, and *Solanaceae*.

The structure of some pyrrolidine bases resembles that of piperidine alkaloids, *i.e.* the pairs hygrine (**17**) and pelletierine (Fig. 12.1), nicotine and anabasine (Fig. 13.11), securinine and norsecurinine (Fig. 12.9), tylophorine (**75**) and tylocrebine. As these pairs of alkaloids can occur together, one may expect similarities in their biosynthesis. The most interesting example, although biosynthetically untested, is anahygrine (Fig. 12.6), where a pyrrolidine and a piperidine unit are on either side of a C_3 bridge.

It should be emphasized that not all naturally occurring pyrrolidine and indolizidine bases originate from ornithine. This holds true for shihunine (Fig. 14.57), slaframine and some indolizidine alkaloids (Fig. 12.19), and the compounds of the mesembrine type (Fig. 14.51).

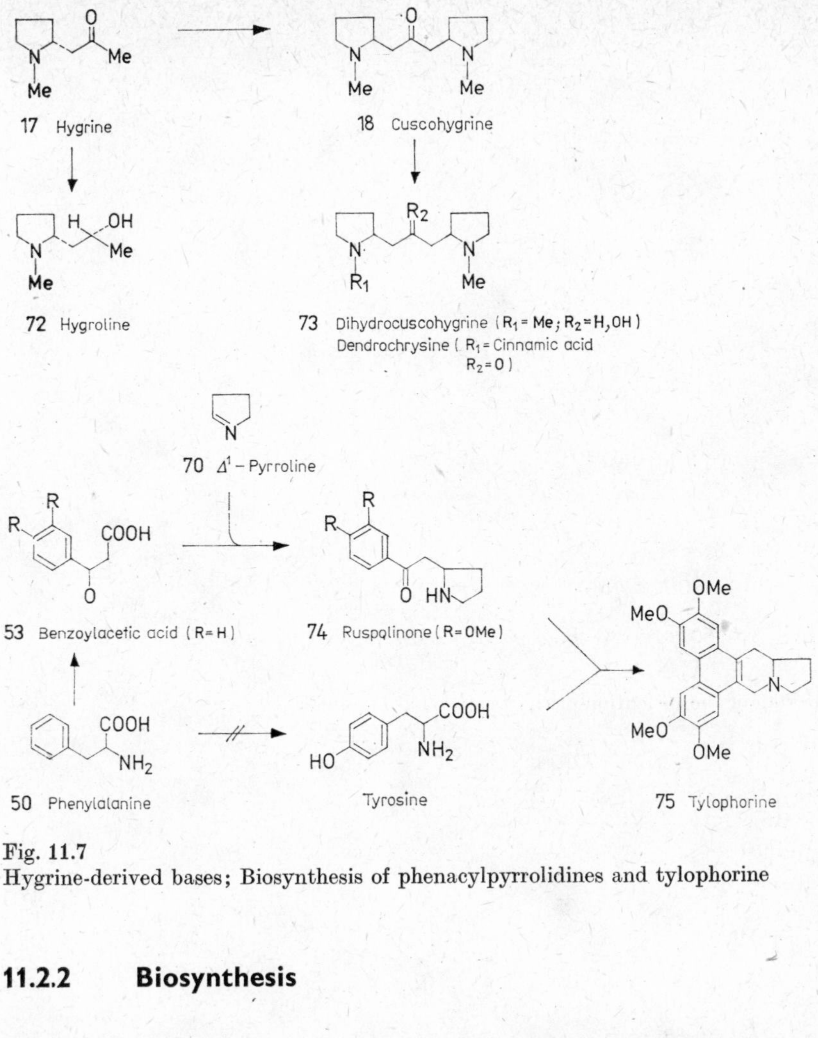

17 Hygrine

18 Cuscohygrine

72 Hygroline

73 Dihydrocuscohygrine (R_1 = Me; R_2 = H,OH)
Dendrochrysine (R_1 = Cinnamic acid
R_2 = O)

70 Δ^1 – Pyrroline

53 Benzoylacetic acid (R = H)

74 Ruspolinone (R = OMe)

50 Phenylalanine

Tyrosine

75 Tylophorine

Fig. 11.7
Hygrine-derived bases; Biosynthesis of phenacylpyrrolidines and tylophorine

11.2.2 Biosynthesis

Proline (**27**) as a protein constituent is universally distributed and its biosynthesis is well understood (Fig. 11.6). The diamino acid ornithine (**21**) in plants yields α-keto-δ-aminovaleric acid (**63**) after deamination, which is in equilibrium with its cyclized form Δ^1-pyrroline-2-carboxylic acid (**64**). A NADPH-dependent reduction of the latter gives rise to proline (**27**). The corresponding enzymes *L*-ornithine: 2-oxoacid aminotransferase and Δ^1-pyrrolinecarboxylate reductase, have been isolated from peanut (*Arachis hypogea*) [118]. Intact incorporation of [2-^{14}C, δ-^{15}N] ornithine (**21**) confirmed the heterocyclic nitrogen of proline (**27**) in *Datura stramonium* to be derived from the δ-amino group of the diamino acid **21** [108]. Similar experiments with [2-^{3}H, 5-^{14}C]- and [5-^{14}C ^{3}H$_2$] ornithine (**21**) resulted in the maintenance of the tritium at C-5 during proline (**27**) formation in *Datura stramonium, Nicotiana tabacum,* and *Lupinus angustifolius* [123]. Loss of tritium from the α-

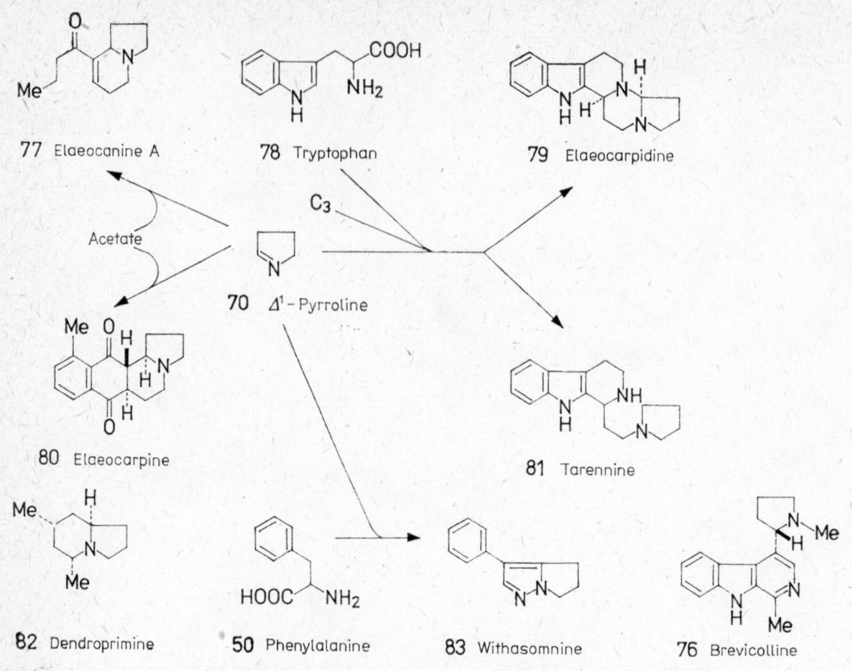

Fig. 11.8
Hypothetical plan of the formation of the *Elaeocarpus* alkaloids

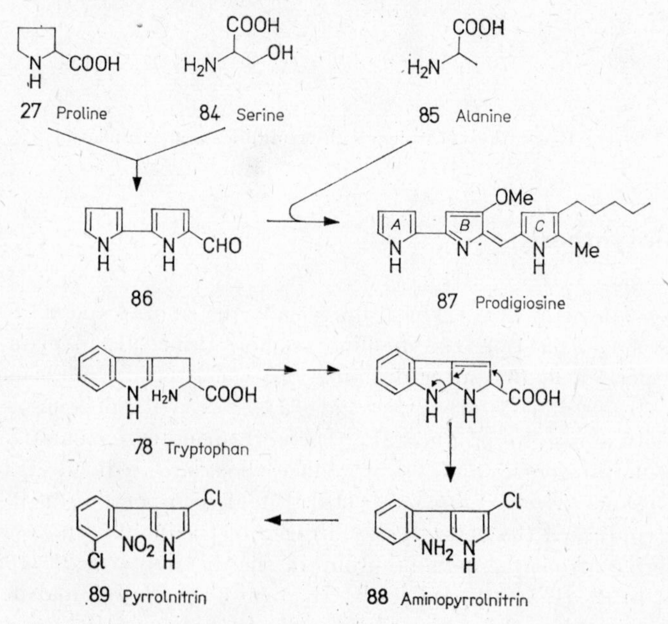

Fig. 11.9
Biosynthesis of the microbial pyrrole compounds prodigiosine and pyrrolnitrin

position of ornithine (**21**) disproved the alternative pathway *via* glutamic semialdehyde (**61**). In contrast, animals and microorganisms form proline (**27**) *via* glutamic semialdehyde (**61**) and Δ^1-pyrroline-5-carboxylic acid (**62**) (*cf.* [2]).

Numerous substituted prolines have been isolated, mainly from legumes [2, 12, 48]. Some of the recently discovered prolines are listed in Fig. 11.6. 4-Methylene-*D,L*-proline (**65**, R = CH_2) [54] and both the *D*- and *L*-enantiomers of 4-hydroxymethylproline (**65**, R = H, CH_2OH) were isolated from the loquat (*Eribotyra japonica, Rosaceae*) [53] while the closely related 2,4-methanoproline (**66**) was isolated from *Ateleia herbert smithii* (*Leguminosae*) [13]. Biosynthetic experiments have still not been carried out but the simultaneous occurrence of 4-methyleneproline (**65**, R = CH_2) with 4-methyleneglutamic acid [170] and of 2,4-methanoproline (**66**) with 2,4-methanoglutamic acid (**67**) [13] suggests mutual interconversion [12, 48]. *N*-methylation of proline (**27**) leads to stachhydrine, the methyl betaine of proline (*cf.* [101]).

Quite another type of substitution pattern is represented by *N*-ethoxycarbonyl-*L*-prolineamide isolated from *Arnica montana* (*Compositae*) [70].

In contrast to the deamination of the diamino acid ornithine (21) by an aminotransferase, the di- and polyamines were deaminated by diamine oxidases, present predominantly in legumes [155]. Pea seedling diamine oxidases efficiently converted putrescine (**23**) into Δ^1-pyrroline (**70**) [156, 157]. *N*-Methylputrescineoxidase has been studied in tobacco [125]. This enzyme produces *N*-methyl-Δ^1-pyrrolinium base (**26**) from *N*-methylputrescine (**24**) and is thus concerned in the biosynthesis of nicotine (*cf.* chapter 13), tropane alkaloids (*cf.* chapter 11.1) and several pyrrolidine bases. Enzymes are found in *Gramineae* that convert spermidine (**68**), originating from putrescine (**23**) and decarboxylated *S*-adenosylmethionine in a reaction catalysed by aminopropyl transferase (*cf.* [161]), and spermine (**69**) into pyrrolines. The most active of these enzymes are in barley and oat seedlings [154]. Δ^1-Pyrroline (**70**) and *N*-(3-aminopropyl)-Δ^2-pyrroline (**71**), both together with 1,3-diaminopropane, were obtained as reaction products (Fig. 11.6).

The 2-substituted pyrrolidine bases hygrine (**17**) and cuscohygrine (**18**) often accompany the tropane alkaloids [101, 144]. Due to their apparent structural similarity the common biosynthetic pathway outlined in chapter 11.1 (Fig. 11.2) is reasonable. Obviously, if untested, both alkaloids can be further metabolized. Reduction of the carbonyl group under sterically controlled conditions may give both enantiomers of hygroline (**72**). *L*-Hygroline has been discovered in *Erythroxylon* and *Cochlearia* while the *D*-enantiomer was found in *Carallia* and *Gynotroches* (*cf.* [144]) and both stereoisomers together in *Schizanthus hookeri* [48a, 117a]. In *Solanaceae* cuscohygrine (**18**) is present in higher amounts especially in *Scopolia lurida* [106]. It is one of the major alkaloids in Peruvian and Bolivian *Erythroxylon coca* leaves, in the former of which (—)-dihydrocuscohygrine (**73**) has also been detected [164]. In the course of aging of the tissues tropane alkaloids and probably also the pyrrolidine bases become demethylated. Following demethylation of cuscohygrine (**18**) a reaction with cis or trans cinnamic acid will give cis or trans dendrochrysine (**73**) in *Dendrobium chrysanthemum* [40]. A number of structurally related pyrrolidine alkaloids of the general type **74** have been isolated from the *Acanthaceae*

Ruspolia hypercrateriformis [142], for example ruspolinone (**74**, R = OCH_3). These alkaloids are of considerable importance in the biosynthesis of the phenanthrohexahydroindolizidine bases such as tylophorine (**75**). Alkaloids of this type have so far been isolated from only 5 of the 300 genera of the family *Asclepiadaceae* and from the *Moraceae Ficus septica* (*cf.* [49]). Biosynthetic evidence [18, 65, 129] established that C-9, C-7′ and ring *B* of tylophorine (**75**) arose from tyrosine whereas ring *A* together with C-10 and C-6′ were derived from phenylalanine (**50**) *via* cinnamic acid. The 5-membered ring *E* originated from ornithine (**21**). The biosynthetic scheme outlined in Fig. 11.7 was further supported by the incorporation of several substituted phenacylpyrrolidines of the ruspolinone (**74**) type [65]. Formation of the higher homologues, the phenacylpiperidines present in *Lobelia* species (*cf.* Chapter 12.1.3.2, Fig. 12.7), should take place by condensation of Δ^1-piperideine with benzoylacetic acid (**53**).

Within the family *Elaeocarpaceae* in only 7 species of the genus *Elaeocarpus* alkaloids were found which contain a hydrogenated indolizidine unit [73]. The *Elaeocarpus* species showed distinctive differences in alkaloid pattern. In general, two types of alkaloids were found, the C_{12} alkaloid elaeocanine A (**77**) from *E. kaniensis* and the C_{16} alkaloids of the elaeocarpine (**80**) type from *E. sphaericus*. A single indole alkaloid, elaeocarpidine (**79**), was isolated as the major alkaloid of *E. densiflorus*. Although tracer experiments have not been carried out, a reasonable hypothesis [73] is that Δ^1-pyrroline (**70**) reacts with a polyketide of the appropriate chain length to give the C_{12} and C_{16} bases. In a similar way elaeocarpidine (**79**) would be formed by the reaction of Δ^1-pyrroline (**70**) with tryptophan (**78**) and a C_3 unit (Fig. 11.8). It is interesting to note that dihydroelaeocarpidine as tarennine (**81**) was isolated from the *Rubiaceae Tarenna bipindensis* [25]. Several hexahydroindolizidine alkaloids (*cf.* [49]) such as dendroprimine (**82**) occur in *Orchidaceae* [22] and *Leguminosae* [69]; their biogenetic origin, however, is still uncertain.

Withania somnifera (*Solanaceae*), along with pyrrolidine, tropane and steroid alkaloids, contains a unique pyrazole base, withasomnine (**83**) [151]. Incorporation of labelled ornithine (**21**) and phenylalanine (**50**) confirmed that the sequence of withasomnine (**83**) formation is as presented in Fig. 11.8 [131].

Δ^1-Pyrroline (**70**) or its *N*-methylated congener (**26**) is also a likely intermediate on the way to brevicolline (**76**) because in *Carex brevicollis* (*Cyperaceae*) [1,4-^{14}C] putrescine (**23**) has been specifically incorporated into the alkaloid [114]. Simple pyrrolidine bases also occur in animals. 2,5-*trans*-Substituted pyrrolidines have been detected in the venom of the European thief-ant *Solenopsis fugax* [23] and the South African fire ant *Solenopsis punctaticeps* [134]. American fire-ant species contain the corresponding 2,6-substituted piperidines (Fig. 12.8). Another example of the spurious and divergent occurrence of pyrrolidine substances is the series of 2,3-substituted pyrrole compounds isolated from the marine sponge *Oscarella lobularis* [29].

Pyrrole-2-carboxylic acid is common as an esterifying agent and is present in several alkaloid families such as quinolizidine [116] or tropane alkaloids (**15** in Fig. 11.1). There have so far been no biosynthetic experiments but the route would probably start with ornithine (**21**).

A different pathway leads to the formation of the pyrrole units in porphyrines but as they are present in essentially all plants (as chlorophyll) and mammals (as haemoglobin) their biosynthesis should not be discussed in connection with alkaloids (*cf.* [101]). It is only possible to discuss briefly two recently determined pathways leading to unusual pyrrole structures in microorganisms.

a) Prodigiosine (**87**), a red pigment produced by *Serratia marcescens*, consists of three linked pyrrole units. Related members of the prodigine family have been isolated from *Streptomyces longiporus rubrum* [167] and *Actinomadura* species [50] too. The pigments differ mainly in the nature of ring *C* and its side-chain. Numerous biosynthetic experiments showed the pyrrole rings to originate in different ways: ring *A* from proline (**27**), ring *B* from serine (**84**), and ring *C* plus side-chain in each of the prodigines are formed by condensation of alanine (**85**) or glycine with a polyketide or fatty acid of appropriate chain length [50, 111, 167, 168]. The aldehyde **86** is a likely intermediate for all prodigine-type pigments (Fig. 11.9).

b) Pyrrolnitrin (**89**) is an antifungal antibiotic that, together with some analogues, is produced by various species of *Pseudomonas*. Among several potential precursors tested only tryptophan (**78**) was efficiently incorporated into pyrrolnitrin (**89**) [45]. Further experiments with ^{15}N-tryptophan (**78**) pointed to the conversion of this amino acid to pyrrolnitrin (**89**) *via* the postulated intermediates [28, 45, 46, 149] (Fig. 11.9). Thereby, the amino group of tryptophan became the pyrrole nitrogen while the indole nitrogen gave rise to the nitro group of pyrrolnitrin (**89**), probably *via* aminopyrrolnitrin (**88**).

These two examples may illustrate that, especially in microorganisms, pyrrolidine and pyrrole compounds are formed by routes other than the ornithine (**21**) pathway (*cf.* [100, 101]).

11.3 Pyrrolizidine Alkaloids

Pyrrolizidine alkaloids are most common in *Senecio* and *Crotolaria* species [139a]. They contain a characteristic base fragment, an amino alcohol, which is usually esterified with the so-called necic acids. These necic acids are branched-chain aliphatic mono- or dicarboxylic acids with 5, 6, 7, 8 or 10 C-atoms. Retronecine (**94**) (Fig. 11.10) is the most common base fragment of pyrrolizidine alkaloids. In senecionine (*Senecio magnificus*) it is esterified with senecic acid (**96**), in retrorsine (**95**) (*S. isatideus*) with isatinecic acid (**98**), and in seneciphylline (*S. douglasii*) with seneciphyllic acid (**97**) (Fig. 11.11).

In experiments with *Crotolaria spectabilis*, *C. retusa*, *Senecio douglasii*, *S. isatideus*, and *S. magnificus*, ornithine (**21**), arginine, putrescine (**23**), spermidine, and spermine have all been demonstrated to be specific precursors of retronecine (**94**) [6, 14a, 26, 71, 130, 135c, 136, 139b, 141, 141a]. Ornithine is a more efficient precursor than arginine. Arginine can be hydrolysed to ornithine by arginase, an enzyme that has

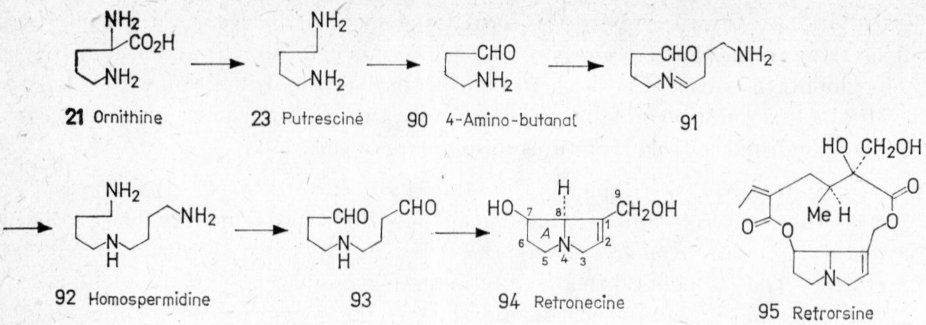

21 Ornithine → 23 Putrescine → 90 4-Amino-butanal → 91 → 92 Homospermidine → 93 → 94 Retronecine; 95 Retrorsine

Fig. 11.10
Biosynthesis of retronecine

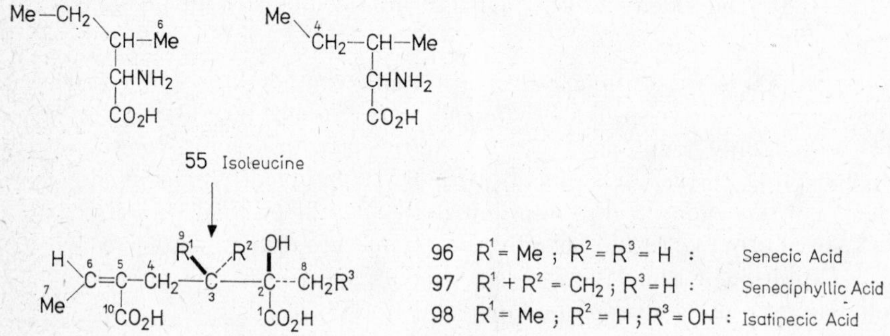

55 Isoleucine

96 R^1 = Me ; R^2 = R^3 = H : Senecic Acid
97 R^1 + R^2 = CH_2 ; R^3 = H : Seneciphyllic Acid
98 R^1 = Me ; R^2 = H ; R^3 = OH : Isatinecic Acid

Fig. 11.11
Biosynthesis of different necic acids

been detected in a number of higher plants. Degradations of retronecine (**94**) derived from [2-^{14}C]- and [5-^{14}C] ornithine and [1,4-^{14}C] putrescine have shown that in each case ca. 25% of the total radioactivity is located at C-9 and 25% at C-5 to C-7. This suggests the intermediacy of a symmetrical C_4 fragment, such as putrescine (**23**). Further degradation of retronecine has proved difficult and conflicting results have been obtained for the distribution of radioactivity in ring *A*. In experiments with *Senecio isatideus* the labelling pattern of [1,4-$^{13}C_2$]-, [2,3-$^{13}C_2$]- and [^{13}C, ^{15}*N*]-labelled putrescine incorporation into retronecine was consistent with the formation of a symmetrical C_4—N—C_4 intermediate, which was shown to be homospermidine (**92**) [59, 78, 79, 135a, 135b]. Therefore the pathway in Fig. 11.10 is likely for the formation of retronecine, in which ornithine (**21**) is decarboxylated to putrescine (**23**). This yields 4-aminobutanal (**90**) by amine oxidase action, which reacts to the intermediates **91** and **93**. These are probably the immediate precursors of the pyrrolizidine skeleton. Whether homospermidine (**92**) is on the normal pathway or can be transformed to **93** is not clear.

In young plants of *Heliotropium spathulatum* leaves were the main organs in which necine biosynthesis takes place [21a].

Most of the biosynthetic work on necic acids has been carried out with the senecic (**96**), seneciphyllic (**97**), and isatinecic acid (**98**) group. Isoleucine (**55**) and its biological precursor *L*-threonine [152] have been shown to be incorporated specifically into seneciphyllic acid (**97**), the necic acid component of seneciphylline, into senecic acid (**96**), the necic acid component of senecionine [35, 36], into monocrotalic acid, the necic acid component of monocrotaline [140], and into trichodesmic acid [37a]. Of the four stereo-isomers, only *L*-isoleucine was efficiently incorporated into senecic acid [37]. It has been established that it is formed from two molecules of isoleucine (Fig. 11.11), with the loss of the carboxygroups from both participating isoleucine molecules and with the direct incorporation of the C_5-units derived from this precursor without re-arrangement or degradation [36].

C-2 of isoleucine labels C-1 and C-10 of senecic acid, C-5 of the amino acid was located at C-9 and C-7, and C-6 of the amino acid was located at C-8 and C-4 of the necic acid.

In elucidating the pathway by which isoleucine is converted into the necic acids, five-carbon intermediates of the isoleucine metabolism, 2-methylbutanoic acid, angelic acid, and 2-methyl-3-oxobutanoic acid can be excluded as precursors for senecic acid (**96**) and seneciphyllic acid (**97**) in *Senecio douglasii* [5, 35]. One or probably two of the hydrogens at C-6 in isoleucine are retained during the conversion into C-4 of senecic acid (**96**).

11.4 References

[1] Achari, R.; Evans, W. C.; Newcombe, F.: Naturwissenschaften **56** (1969) 88.
[2] Adams, E.; Frank. L.: Annu. Rev. Biochem. **49** (1980) 1023.
[3] Ahmad, A.; Leete, E.: Phytochemistry **9** (1970) 2345.
[4] Ansanin, M.; Woolley, J. G.: J. Pharm. Pharmacol. **30** (1978) Suppl. 82 P.
[5] Bale, N. M.; Cahill, R.; Davies, N. M.; Mitchell, M. B.; Smith, E. H.; Crout, D. H. G.: J. Chem. Soc. Perkin I **1978**, 101.
[6] Bale, N. M.; Crout, D. H. G.: Phytochemistry **14** (1975) 2617.
[7] Baralle, F. E.; Gros, E. G.: Phytochemistry **8** (1969) 849; 853.
[8] Baralle, F. E.; Gros, E. G.: Chem. Commun. **1969**, 721.
[9] Basey, K.; Woolley, J. G.: Phytochemistry **12** (1973) 2197.
[10] Basey, K.; Woolley, J. G.: Phytochemistry **12** (1973) 2883.
[11] Basey, K.; Woolley, J. G.: Phytochemistry **14** (1975) 2201.
[12] Bell, E. A.: Progr. Phytochem. **7** (1981) 171.
[13] Bell, E. A.; Qureshi, M. Y.; Pryce, R. J.; Janzen, D. H.; Lemke, P.; Clardy, J.: J. Am. Chem. Soc. **102** (1980) 1409.
[14] Beresford, P. J.; Woolley, J. G.: Phytochemistry **13** (1974) 2143.
[15] Beresford, P. J.; Woolley, J. G.: Phytochemistry **13** (1974) 2511.
[16] Beresford, P. J.; Woolley, J. G.: Phytochemistry **14** (1975) 2205.
[17] Beresford, P. J.; Woolley, J. G.: Phytochemistry **14** (1975) 2209.
[18] Bhakuni, D. S.; Mangla, V. K.: Tetrahedron **37** (1981) 401.
[19] Bick, I. R. C.; Gillard, J. W.; Leow, H. M.: Austral. J. Chem. **32** (1979) 2523.
[20] Bick, I. R. C.; Gillard, J. W.; Leow, H. M.; Lounasmaa, M.; Pusset, J.; Sévenet, T.: Planta Med. **41** (1981) 379.
[21] Bick, I. R. C.; Gillard, J. W.; Woodruff, M.: Chem. & Ind. **1975**, 794.

[21a] Birecka, H.; Catalfamo, J. L.: Phytochemistry **21** (1982) 2645.
[22] Blomquist, L.; Leander, K.; Lüning, B.; Rosenblom, J.: Acta. Chem. Scand. **26** (1972) 3203.
[23] Blum, M. S.; Jones, T. H.; Hölldobler, B.; Fales, H. M.; Jaouni, T.: Naturwissenschaften **67** (1980) 144.
[24] Blum, M. S.; Rivier, L.; Plowman, T.: Phytochemistry **20** (1981) 2499.
[25] Boissier, J. R.; Combes, G.; Effler, A. H.; Klinga, K.; Schlittler, E.: Experientia **27** (1971) 677.
[26] Bottomley, W.; Geissman, T. A.: Phytochemistry **3** (1964) 357.
[27] Carmichael, W. W.; Biggs, D. F.; Peterson, M. A.: Toxicon **17** (1979) 229.
[28] Chang, C.; Floss, H. G.; Hurley, L. H.; Zmijewski, M.: J. Org. Chem. **41** (1976) 2932.
[29] Cimino, G.; Stefano, S. de; Minale, L.: Experientia **31** (1975) 1387.
[30] Clarke, R. L.: The Tropane Alkaloids. In: Manske, R. H. F. (Ed.): The Alkaloids, Vol. XVI. Academic Press, New York 1977, p. 83.
[31] Conklin, M. E.: Genetic and Biochemical Aspects of the Development of Datura. Karger, Basel 1976.
[32] Cosson, L.: Phytochemistry **8** (1969) 2227.
[33] Cosson, L.; Escudero-Morales, A.; Cougoul, N.: Plant Med. Phytother. **12** (1978) 319.
[34] Cosson, L.; Kuntzmann-Cougoul, N.: Herba Hung. **18** (1979) 135.
[35] Crout, D. H. G.; Benn, M. H.; Imaseki, H.; Geissman, T. A.: Phytochemistry **5** (1966) 1.
[36] Crout, D. H. G.; Davies, N. M.; Smith, E. H.; Whitehouse, D.: J. Chem. Soc. Perkin I **1972**, 671.
[37] Davies, N. M.; Crout, D. H. G.: J. Chem. Soc. Perkin I, **1974**, 2079.
[37a] Devlin, J. A.; Robins, D. J.: J. Chem. Soc. Perkin I, **1984**, 1329.
[38] Dhoot, G. K.; Henshaw, G. G.: Ann. Bot. **41** (1977) 943.
[39] Eapen, S.; Rangan, T. S.; Chadha, M. S.; Heble, M. R.: Plant Sci. Lett. **13** (1978) 83.
[40] Ekevag, U.; Elander, M.; Gawell, L.; Leander, K.; Lüning, B.: Acta Chem. Scand. **27** (1972) 1982.
[41] Evans, W. C.; Ghani, A.; Woolley, J. G.: J. Chem. Soc. Perkin I **1972**, 2017.
[42] Evans, W. C.; Somanabaudhu, A.: Phytochemistry **19** (1980) 2351.
[43] Evans, W. C.; Woolley, J. G.: Phytochemistry **8** (1969) 2183.
[44] Evans, W. C.; Woolley, J. G.: Phytochemistry **15** (1976) 287.
[45] Floss, H. G.: Biosynthesis of Some Aromatic Antibiotics. In: Corcoran, J. W. (Ed.): Antibiotics, Vol. IV. Springer-Verlag, Berlin 1981, p. 236.
[46] Floss, H. G.; Manni, P. E.; Hamill, R. L.; Mabe, J. A.: Biochem. biophys. Res. Commun. **45** (1971) 781.
[47] Fodor, G.: The Tropane Alkaloids. In: Manske, R. H. F. (Ed.): The Alkaloids, Vol. XVII. Academic Press, New York 1977, p. 352.
[48] Fowden, L.: Progr. Phytochem. **2** (1970) 203.
[48a] Gambaro, V.; Labbé, C.; Castillo, M.: Phytochemistry **22** (1983) 1838.
[49] Gellert, E.: J. Nat. Prod. **45** (1982) 50.
[50] Gerber, N. N.; McInnes, A. G.; Smith, D. G.; Walter, J. A.; Wright, J. L. C.; Vining, L. C.: Canad. J. Chem. **56** (1978) 1155.
[51] Gorham, P. R.; Carmichael, W. W.: Pure appl. Chem. **52** (1979) 165.
[52] Graf, E.; Lude, W.: Arch. Pharmaz. **311** (1978) 139.
[53] Gray, D. O.: Phytochemistry **11** (1972) 751.
[54] Gray, D. O.; Fowden, L.: Phytochemistry **11** (1972) 745.
[55] Gröger, D.: Planta Med. **28** (1975) 269.
[56] Groß, D.; Schütte, H. R.: Arch. Pharmaz. **296** (1963) 1.
[57] Gross, G. G.; Koelen, K. J.: Z. Naturforsch. **35c** (1980) 363.
[58] Gross, G. G.; Koelen, K. J.: Müller, A.; Schmidtberg, G.: Z. Naturforsch. **36c** (1981) 611.
[59] Grue-Sörensen, G.; Spenser, I. D.: J. Am. Chem. Soc. **103** (1981) 3208; Canad. J. Chem. **60** (1982) 643; J. Am. Chem. Soc. **105** (1983) 7401.

[60] Gupta, S.; Prabhakar, V. S.; Madan, C. L.: Planta Med. **23** (1973) 370.
[61] Hamon, N. W.; Eyolfson, J. L.: J. Pharmac. Sci. **61** (1972) 2006.
[62] Hamon, N. W.; Youngken, H. W.: Lloydia **34** (1971) 199.
[62a] Hashimoto, T.; Yamada, Y.: Planta med. **47** (1983) 195.
[63] Hedges, S. H.; Herbert, R. B.: Phytochemistry **20** (1981) 2064.
[64] Heltmann, H.: Herba Hung. **18** (1979) 101.
[65] Herbert, R. B.; Jackson, F. B.; Nicolson, I. T.: Chem. Commun. **1976**, 865; **1977**, 955.
[66] Hill, R. K.; Rhee, S. W.; Leete, E.; McGaw, B. A.: J. Am. Chem. Soc. **102** (1980) 7344.
[67] Hiroaka, N.; Tabata, M.: Phytochemistry **13** (1974) 1671.
[68] Hiroaka, N.; Tabata, M.; Konoshima, M.: Phytochemistry **12** (1973) 795.
[69] Hohenschutz, L.; Bell, E. A.; Jewers, P. J.; Leworthy, D. P.; Pryce, R. J.; Arnold, E.; Clardy, J.: Phytochemistry **20** (1981) 811.
[70] Holub, M.; Poplanski, J.; Sedmera, P.; Herout, V.: Coll. Czech. Chem. Commun. **42** (1977) 151.
[71] Hughes, C. A.; Letcher, R.; Warren, F. L.: J. Chem. Soc. **1964**, 4974.
[72] Jindra, A.; Staba, E. J.: Phytochemistry **7** (1968) 79.
[73] Johns, S. R.; Lamberton, J. A.: Elaeocarpus Alkaloids. In: Manske, R. H. F. (Ed.): The Alkaloids, Vol. XIV. Academic Press, New York 1973, p. 326.
[74] Kaczkowski, J.; Marion, L.: Canad. J. Chem. **41** (1963) 2651.
[75] Kaczkowski, J.; Schütte, H. R.; Mothes, K.: Naturwissenschaften **47** (1960) 304.
[76] Kaczkowski, J.; Schütte, H. R.; Mothes, K.: Biochim. Biophys. Acta **46** (1961) 588.
[77] Kalyanaraman, V. S.; Mahadevan, S.; Kumar, S. A.: Biochem. J. **149** (1975) 565; 577.
[78] Khan, H. A.; Robins, D. J.: Chem. Commun. **1981**, 146.
[79] Khan, H. A.; Robins, D. T.: Chem. Commun. **1981**, 554.
[80] Kibler, R.; Neumann, K. H.: Planta Med. **35** (1979) 354.
[81] Koelen, K. H.; Gross, G. G.: Planta Med. **44** (1982) 227.
[82] Kouassi, B.; Peaud-Lenoel, C.: C. R. Soc. Biol. **164** (1970) 46.
[83] Leete, E.: J. Am. Chem. Soc. **82** (1960) 612.
[84] Leete, E.: J. Am. Chem. Soc. **84** (1962) 55; Tetrahedron Lett. **1964**, 1619.
[85] Leete, E.: Phytochemistry **11** (1972) 1713.
[86] Leete, E.: Phytochemistry **12** (1973) 2203.
[87] Leete, E.: Planta Med. **36** (1979) 97.
[88] Leete, E.: Chem. Commun. **1980**, 1170; J. Am. Chem. Soc. **104** (1982) 1403.
[89] Leete, E.: Alkaloids Derived from Ornithine, Lysine, and Nicotinic Acid. In: Bell, E. A.; Charlwood, B. V. (Eds.): Encyclopedia of Plant Physiology, Vol. 8. Springer-Verlag, Berlin 1980, p. 65.
[89a] Leete, E.: J. Am. Chem. Soc. **105** (1983) 6727; phytochemistry **22** (1983) 699.
[89b] Leete, E.: Phytochemistry **22** (1983) 933.
[89c] Leete, E.: J. Am. Chem. Soc. **106** (1984) 7271.
[90] Leete, E.; Kirven, E. P.: Phytochemistry **13** (1974) 1501.
[91] Leete, E.; Kowanko, N.; Newmark, R. A.: J. Am. Chem. Soc. **97** (1975) 6826.
[92] Leete, E.; Louden, M. C. L.: Chem. & Ind. **3** (1963) 1725.
[93] Leete, E.; Lucast, D. H.: Phytochemistry **14** (1975) 2199.
[94] Leete, E.; Lucast, D. H.: Tetrahedron Lett. **1976**, 3401.
[95] Leete, E.; Marion, L.; Spenser, I. D.: Canad. J. Chem. **32** (1954) 1116.
[96] Leete, E.; McDonell, J. A.: J. Am. Chem. Soc. **103** (1981) 658.
[97] Leete, E.; Murrill, S. J. B.: Tetrahedron Lett. **1967**, 1727.
[98] Leete, E.; Nelson, S. J.: Phytochemistry **8** (1969) 413.
[99] Liebisch, H. W.: Abh. Dtsch. Akad. Wiss. Berlin; Klasse Chem., Geol., Biol., **1966**, No. 3, p. 525.
[100] Liebisch, H. W.: Fortschr. chem. Forsch. **9** (1968) 534.
[101] Liebisch, H. W.: Tropanalkaloide und Pyrrolidinbasen. In: Mothes, K.; Schütte, H. R. (Eds.): Biosynthese der Alkaloide. Deutscher Verlag der Wissenschaften, Berlin 1969, p. 183.
[102] Liebisch, H. W.; Bernasch, H.; Schütte, H. R.: Z. Chem. **13** (1973) 372.

[103] Liebisch, H. W.; Bhavsar, G. C.; Schaller H. J.: Beiträge zur Bildung der Tropasäure in vivo und in vitro. In: Mothes, K.; Schreiber, K.; Schütte, H. R. (Eds.): Biochemie und Physiologie der Alkaloide. Akademie Verlag, Berlin 1972, p. 233.
[104] Liebisch, H. W.; Maier, W.; Schütte, H. R.: Tetrahedron Lett. **1966**, 4079.
[105] Liebisch, H. W.; Peisker, K.; Radwan, A. S.; Schütte, H. R.: Z. Pflanzenphysiol. **67** (1972) 1.
[106] Liebisch, H. W.; Radwan, A. S.; Schütte, H. R.: Liebigs Ann. Chem. **721** (1969) 163.
[107] Liebisch, H. W.; Ramin, H.; Schöffinius, I.; Schütte, H. R.: Z. Naturforsch. **206** (1965) 1183.
[108] Liebisch, H. W.; Schütte, H. R.: Z. Pflanzenphysiol. **57** (1967) 434.
[109] Liebisch, H. W.; Schütte, H. R.; Mothes, K.: Liebigs Ann. Chem. **668** (1963) 139.
[110] Liebisch, H. W.; Shalaby, A. F.; Schütte, H. R.: Naturwissenschaften **53** (1966) 434.
[111] Lim, D. V.; Quadri, S. M. H.; Nichols, C.; Williams, R. P.: J. Bacteriol. **129**, (1977) 124.
[112] Loder, J. W.; Russell, G. B.: Austral. J. Chem. **22** (1969) 1271.
[113] Louden, M. L.; Leete, E.: J. Am. Chem. Soc. **84** (1962) 4507.
[114] Lovkova, M. J.; Klimenteva, N. I.; Lazurevskij, G. V.: Prikl. Biochim. Mikrobiol. (russ.) **15** (1979) 775.
[115] Lounasmaa, M.; Pusset, J.; Sevenet, T.: Phytochemistry **19** (1980) 949, 953.
[115a] Luanratana, O.; Griffin, W. J.: J. Nat. Prod. **45** (1982) 551.
[116] Machanda, A. H.; Nabney, J.; Young, D. W.: J. Chem. Soc. (C) **1968**, 615.
[117] Major, E. W. T.; Davies, J. J.; Woolley, J. G.: J. Pharm. Pharmacol. **30** (1979) Suppl. 81 P.
[117a] Martin, A. S.; Rovirosa, J.; Gambaro, V.; Castillo, M.: Phytochemistry **19** (1980) 2007.
[118] Mazelis, M.; Fowden, L.: Phytochemistry **8** (1969) 801.
[119] McGaw, B. A.; Woolley, J. G.: Phytochemistry **16** (1977) 1711.
[120] McGaw, B. A.; Woolley, J. G.: Phytochemistry **17** (1978) 257.
[121] McGaw, B. A.; Woolley, J. G.: Phytochemistry **18** (1979) 189.
[122] McGaw, B. A.; Woolley, J. G.: Phytochemistry **18** (1979) 1647.
[122a] McGaw, B. A.; Woolley, J. G.: Phytochemistry **21** (1982) 2653.
[122b] McGaw, B. A.; Woolley, J. G.: Phytochemistry **22** (1983) 1407.
[123] Mestichelli, L. J. J.; Gupta, R. N.; Spenser, I. D.: J. Biol. Chemistry **254** (1979) 640.
[124] Mizusaki, S.; Kisaki, T.; Tamaki, E.: Plant Physiol. **43** (1968) 93.
[125] Mizusaki, S.; Tanabe, Y.; Noguchi, M.; Tamaki, E.: Phytochemistry **11** (1972) 2757; Plant Cell Physiol. **14** (1973) 103.
[126] Motherwell, W. D. S.; Isaacs, N. W.; Kennard, O.; Bick, I. R. C.; Bremner, J. B.; Gillard, J.: Chem. Commun. **1971**, 133.
[127] Mothes, K.; Hieke, K.: Naturwissenschaften **31** (1943) 17.
[128] Mothes, K.; Trefftz, G.; Reuter, G.; Romeike, A.: Naturwissenschaften **41** (1954) 530.
[129] Mulchandani, N. B.; Iyer, S. S.; Badheka, L. P.: Phytochemistry **8** (1969) 1931; **10** (1971) 1047; **15** (1976) 1697.
[130] Nowacki, E.; Byerrum, R. J.: Life Sci. **1** (1962) 157.
[131] O'Donovan, D. G.; Forde, T. J.: Tetrahedron Lett. **1970**, 3637.
[132] O'Donovan, D. G.; Keogh, M. F.: J. Chem. Soc. C **1969**, 223.
[133] Parry, R. J.; Mizusawa, A. E.; Ricciardone, M.: J. Am. Chem. Soc. **104** (1982) 1442.
[134] Pedder, D. J.; Fales, H. M.; Jaouni, T.; Blum, M.; MacConnell, J.; Crewe, R. M.: Tetrahedron **32** (1976) 2275.
[135a] Rana, J.; Robins, D. J.: J. Chem. Res., Synop. **1983**, 146; C. A. **99** (1983) 191 760.
[135b] Rana, J.; Robins, D. J.: Chem. Comm. **1983**, 1222.
[135c] Rana, J.; Robins, D. J.: Chem. Commun. **1984**, 517.
[136] Rao, P. G.; Zutshi, U.; Soni, A.; Atal, C. K.: Planta med. **35** (1979) 279.
[136a] Ray, A. B.; Oshima, Y.; Hkino, H.; Kabuto, C.: Heterocycles **19** (1982) 1233.
[137] Ray, A. B.; Sahai, M.; Sethi, P. G.: Chem. Ind. **1976**, 454.
[138] Ray Bhandary, S. B.; Collin, H. A.; Thomas, E.; Street, H. E.: Ann. Bot. **33** (1969) 647.

[139] Ripperger, H.: Phytochemistry **18** (1979) 171.
[139a] Robins, D. J.: Fortschr. Chem. Org. Naturst. **41** (1982) 115.
[139b] Robins, D. J.: J. Chem. Res., Synop. **1983**, 326.
[140] Robins, D. J.; Bale, N. M.; Crout, D. H. G.: J. Chem. Soc. Perkin I **1974** 2082.
[141] Robins, D. J.; Sweeney, J. R.: Chem. Commun. **1979** 120; J. Chem. Soc. Perkin I **1981** 3083.
[141a] Robins, D. J.; Sweeney, J. R.: Phytochemistry **22** (1983) 457.
[142] Roessler, F.; Ganzinger, D.; Johne, S.; Schöpp, E.; Hesse, M.: Helv. Chim. Acta **61** (1978) 1200.
[143] Romeike, A.: Biochem. Physiol. Pflanzen **168** (1975) 87.
[144] Romeike, A.: Bot. Notiser. **131** (1978) 85.
[145] Romeike, A.; Aurich, O.: Phytochemistry **7** (1968) 1547.
[146] Romeike, A.; Koblitz, H.: Kulturpflanze **18** (1970) 169.
[147] Saint-Firmin, A.; Paris, R. R.: C. R. Acad. Sci. Ser. D **267** (1968) 1448.
[148] Sahai, M.; Ray, A. B.: J. Org. Chem. **45** (1980) 3265.
[149] Salcher, O.; Lingens, F.; Fischer, P.: Tetrahedron Lett. **1978**, 3097.
[150] San Martin, A.; Rovirosa, J.; Gambaro, V.; Castillo, M.: Phytochemistry **19** (1980) 2007.
[151] Schröter, H. B.; Neumann, D.; Katritzki, A. R.; Swinbourne, F. J.: Tetrahedron **22** (1966) 2895.
[152] Schütte, H. R.: Biosynthese niedermolekularer Naturstoffe, Gustav Fischer Verlag, Jena 1982.
[153] Schütte, H. R.; Liebisch, H. W.: Z. Pflanzenphysiol. **57** (1967) 440.
[154] Smith, T. A.: Phytochemistry **11** (1972) 899; **15** (1976) 633.
[155] Smith, T. A.: Phytochemistry **14** (1975) 865.
[156] Srivasta, S. K.; Prakash, V.: Phytochemistry **16** (1977) 189.
[157] Srivasta, S. K.; Prakash, V.; Naik, B. I.: Phytochemistry **16** (1977) 185.
[158] Staba, E. J.; Jindra, A.: J. pharmac. Sci. **57** (1968) 701.
[159] Stohs, S. J.: J. pharmac. Sci. **58** (1969) 703.
[160] Tabata, M.; Hayamoto, H.; Hiroaka, N.; Konoshima, M.: Phytochemistry **11** (1972) 949.
[161] Tabor, C. W.; Tabor, H.: Annu. Rev. Biochem. **45** (1976) 285.
[162] Tamprateep, P.; Tayler, E. H.; Ramstad, E.: Lloydia **26** (1963) 203.
[163] Thomas, E.; Street, H. E.: Ann. Bot. **34** (1970) 657.
[164] Turner, C. E.; Elsohly, M. A.; Hanus, L.; Elsohly, H. N.: Phytochemistry **20** (1981) 1403.
[165] Verzar-Petri, G.: Pharmazie **28** (1973) 603.
[166] Verzar-Petri, G.; Soti, F.; Horvath, L.: Herba Hung. **13** (1974) 77.
[167] Wasserman, H. H.; Shaw, C. K.; Sykes, R. J.: Tetrahedron Lett. **1974**, 2787.
[168] Wasserman, H. H.; Sykes, R. J.; Peverada, P.; Shaw, C. K.; Cushley, R. J.; Lipsky, S. R.: J. Am. Chem. Soc. **95** (1973) 6874.
[169] Weiler, E. W.; Stöckigt, J.; Zenk, M. H.: Phytochemistry **20** (1981) 2009.
[170] Welter, A.; Marlier, M.; Dardenne, G. A.: Phytochemistry **17** (1978) 131.

12 Lysine-derived Alkaloids

H. W. Liebisch and H. R. Schütte

12.1 Piperidine Alkaloids

Unlike other classes of alkaloids, *i.e.* tropane, isoquinoline, or indole alkaloids, the group of piperidine alkaloids includes bases common in the plant kingdom, in animals, and in microorganisms. Furthermore, more complex structures are possible from secondary reactions as indicated, for example, by the *Lycopodium* and *Securinega* alkaloids.

Increasing experimental evidence clearly demonstrates piperidine alkaloids to be derived biogenetically from either lysine (**1**), acetate, or mevalonate [44, 50, 94, 107]. The mevalonate pathway is restricted to a few structures only (*cf.* chapter 19). However, some piperidine bases originate from the acetate pathway and their structure is closely related to alkaloids built up from lysine (**1**). This may be illustrated best in the case of the lysine-derived piperidine nucleus of the *Punica* alkaloids such as pelletierine (**2**) and the *Conium* bases such as coniine (**3**) where both the side chain and the heterocyclic nucleus are formed from acetate (Fig. 12.1).

Some microorganisms contain 1-pyrindine bases such as **4** and **5** (Fig. 12.1), one of which is derived from a polyketide but the other from lysine (*cf.* chapter 12.1.8).

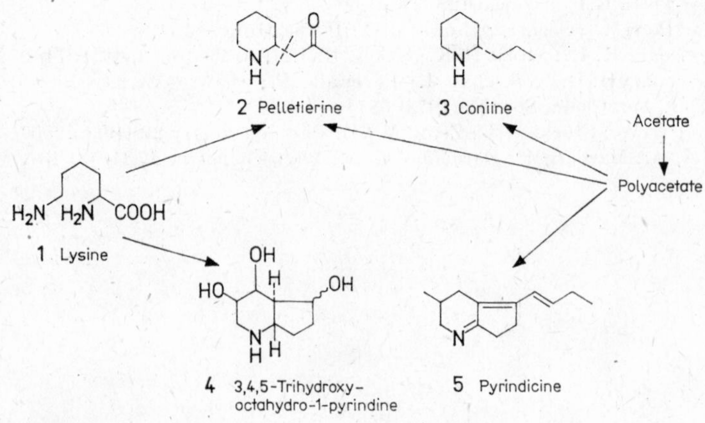

Fig. 12.1
Lysine and acetate pathway to piperidine alkaloids

This shows that structural relationships can be biogenetically misleading. Given the above evidence, it is necessary to include in this chapter not only the piperidine alkaloids originating from lysine (**1**) but also the acetate-derived piperidine bases.

If lysine (**1**) is the precursor of the C_5N skeleton of piperidine alkaloids then one nitrogen and, with the exception of pipecolic acid, the carboxyl group are to be eliminated. Both α-deamination and decarboxylation were mediated by common pyridoxal phosphate-dependent enzymes. Decarboxylation will give the symmetrical diamine cadaverine (**6**). In a number of piperidine alkaloids, however, [2-^{14}C]lysine (**1**) is incorporated asymmetrically and thus a symmetrical intermediate is ruled out. To account for this discrepancy a hypothesis has been made that the protection of one amino group in lysine (**1**) by a methyl (or acyl) function as that in **7** yields the asymmetrical *N*-methylcadaverine (**8**) on decarboxylation (Fig. 12.2). Deamination will now lead *via* **9** to the *N*-methylpiperideinium base (**10**). Reaction of the latter with the immediate precursors of the side-chain would result in the *N*-methylated piperidine alkaloids. This hypothesis was disproved by the observation that ε-*N*-methyllysine (**7**), although a naturally occurring amino acid, was not on the way to the piperidine alkaloids in *Sedum* species [82, 103].

Another hypothesis claims a "bound" cadaverine (**6**) which may be selectively deaminated. Such an enzyme-bound cadaverine is present as azomethine (**13**) when lysine (**1**), having reacted with pyridoxal phosphate (**11**) to **12**, is decarboxylated

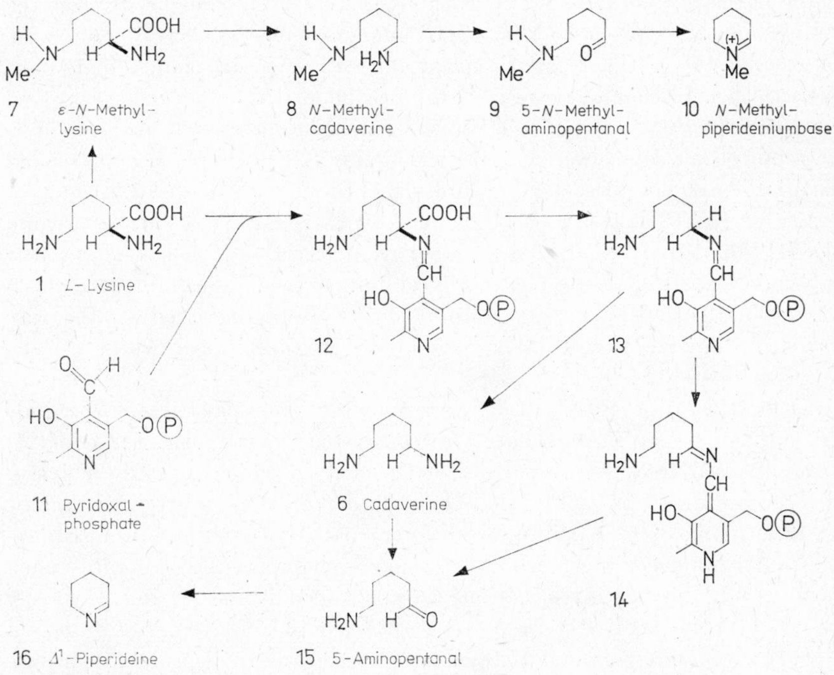

7 ε-N-Methyllysine
8 N-Methylcadaverine
9 5-N-Methylaminopentanal
10 N-Methylpiperideiniumbase
1 L-Lysine
12
13
11 Pyridoxalphosphate
6 Cadaverine
14
16 Δ¹-Piperideine
15 5-Aminopentanal

Fig. 12.2
Alternative routes for the asymmetrical incorporation of lysine

with retention of configuration [8, 38, 102, 103]. Only in this manner (Fig. 12.2) is the α-amino group of lysine (1) lost on the way *via* **14** to 5-aminopentanal (**15**) and its cyclization product Δ^1-piperideine (**16**). As an alternative to this stepwise mechanism a concerted oxidative decarboxylation has been postulated [97] that excludes the intermediate **14**.

Δ^1-Piperideine (**16**) is most likely the universal intermediate on the biosynthetic route to lysine-derived piperidine alkaloids. As the 1,2-double bond is not capable of tautomeric shift to the 1,6-position [92] this intermediate preserves the asymmetry of lysine (**1**) incorporation (*cf.* [93, 106]).

Symmetrical incorporation of lysine (**1**) may be realized by the shortened sequence **13** $\rightarrow$ cadaverine (**6**) $\rightarrow$ 5-aminopentanal (**15**). In the same manner, externally applied cadaverine (**6**) will be introduced into the pathway by deamination to 5-aminopentanal (**15**) under the influence of diamine oxidases.

12.1.1 Pipecolic Acids

Pipecolic acid (**21**) is a common imino acid [9, 36]. Chirality has been established in only a few instances and in almost all such cases $L(-)$-pipecolic acid (**21**) was found. $D(+)$-Pipecolic acid has so far been reported only as a constituent of some polypeptide antibiotics such as amphomycin [11].

So far, we know the piperideine carboxylic acids (**19** and **20**) (Fig. 12.3) are not intermediates in the biosynthesis of typical piperidine alkaloids but may act as the immediate precursors of pipecolic acid (**21**). Numerous feeding experiments were performed to investigate the biosynthesis of pipecolic acid in plants, mammals, and microorganisms. In each of these tissues pipecolic acid (**21**) was indeed derived from lysine (**1**). In order to define further the pathway *via* the alternative intermediates α-aminoadipic semialdehyde (**17**) and 5-amino-2-ketocaproic acid (**18**) and their cyclization products **19** and **20**, doubly-labelled precursors were introduced. Incorporation of [6-^{14}C^{3}H$_2$]lysine (**1**) took place with retention of all the tritium relative to ^{14}C [56, 57]. Loss of tritium label from [2-^{3}H, 6-^{14}C]lysine (**1**) on the way to pipecolic acid (**21**) was explained by assuming the presence of 5-amino-2-ketocaproic acid (**18**) [58]. Thus, the α-amino group was eliminated on the way to pipecolic acid (**21**) when isolated from *Sedum acre*, *Phaseolus vulgaris*, intact rats, and *Neurospora crassa* [32].

In contrast to this, on the basis of ^{15}N experiments, α-aminoadipic semialdehyde (**17**) was believed to be an intermediate between lysine (**1**) and pipecolic acid (**21**) in *Phaseolus vulgaris* [187]. A comparison of the way in which [2-^{3}H, 6-^{14}C]lysine (**1**) was incorporated into pipecolic acid (**21**) and the piperidine alkaloid sedamine (**54** in Fig. 12.7) in *Sedum acre* pointed to different pathways and thus the assumption of a common intermediate for both lysine-derived piperidine structures must be abandoned [58]. Further investigations with resolved lysine (**1**) unequivocally demonstrated pipecolic acid (**21**) in *Nicotiana glauca* [39, 102], *Lycopodium tristachyum* [117], *Decodon vertiallatus* [52], *Lolium* species, and *Zea mays* [7], in rats [45] as well as in *Pseudomonas putica* [122] to be derived from D-lysine (**1**). In contrast, the alkaloids in *Nicotiana*, *Sedum*, *Lycopodium* and *Decodon* originate from L-lysine (**1**).

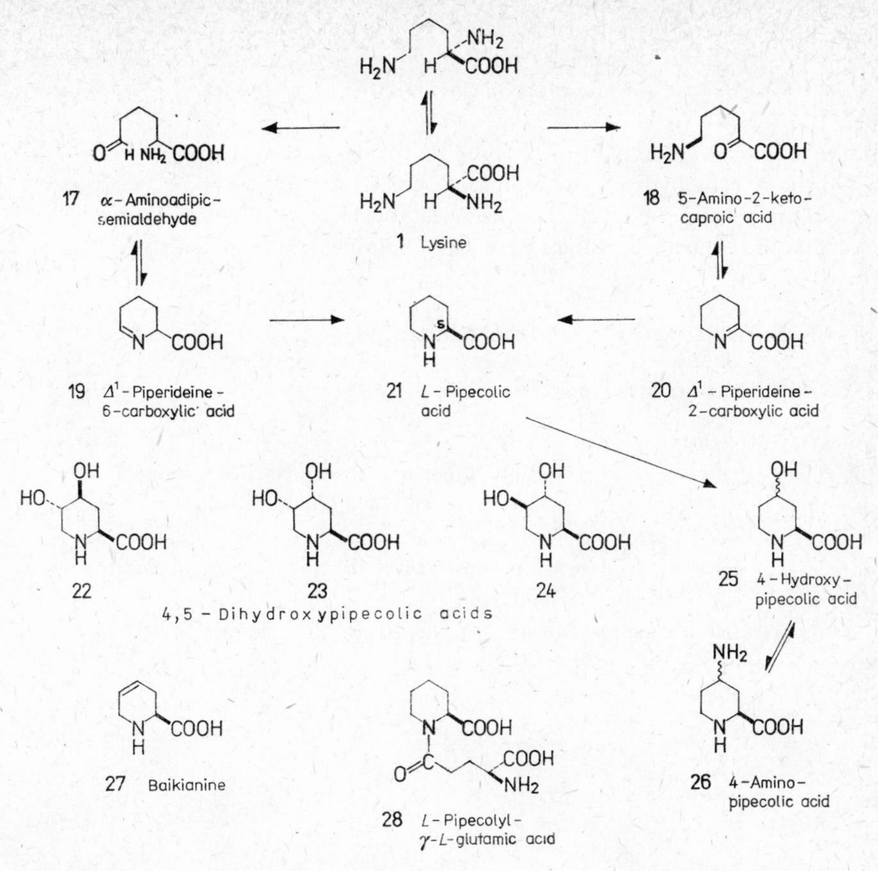

Fig. 12.3
Biosynthesis of pipecolic acids

An exception is the fungus *Rhizoctonia leguminicola* [46] where *L*-lysine (**1**) *via* **18** and **20** forms pipecolic acid (**21**), which is completely incorporated into piperidine alkaloids (*cf*. Fig. 12.19).

The last step in the formation of pipecolic acid (**21**) is catalysed by piperideine carboxylic acid reductase. It is reversible and initiates a path by which the imino acid is catabolized. Many substituted pipecolic acid metabolites are known that can be formed by oxidative attack at the heterocyclic nucleus rather than from hydroxylated lysine derivatives. Some of these substances are listed in Fig. 12.3, among them the 4- and 5-hydroxypipecolic acids (i.e. **25**, both cis and trans), common to legume seeds [9, 36, 159a]. All four possible stereoisomeric 4,5-dihydroxypipecolic acids (**22** to **24** and the 2,4-cis-4,5-trans-4,5-dihydroxypipecolic acid) have been isolated from various legume seeds [10a, 159a, 191]. The distribution data of the four monohydroxypipecolic acids as well as the four dihydroxypipecolic acids in *Calliandra* species appear to be of taxonomic significance [159a].

In *Strophanthus scandens* 4-hydroxypipecolic acid (25) originates from labelled pipecolic acid (21). Moreover, tracer evidence indicated not only that the hydroxy acid (25) is converted to 4-aminopipecolic acid (26) but also that the reaction is reversible [166].

Baikianin (27) is also common (even present in algae [38, 115]) and may arise from dehydration of a hydroxylated pipecolic acid. The existence of two further metabolites of pipecolic acid (21) has been established: *L*-pipecolyl-γ-*L*-glutamic acid (28) [26] and homostachydrine (the methyl betaine of pipecolic acid).

12.1.2 *N*-Substituted Piperidine Alkaloids

12.1.2.1 Alkaloids from *Piper* Species

For a long time the known alkaloids in leaves, fruits, stem, and roots of several species of *Piper* (*Piperaceae*) were mainly piperine (29) and a few other substances of similar structure. Medicinal interest and the insecticidal properties of *Piper* extracts [110, 124, 198] stimulated intensive investigations as a result of which more than 50 different alkaloids are known today. All are neutral compounds that contain an amide-like nitrogen derived from either piperidine, pyrrolidine, or isobutylamine. The basic structural features of the *Piper* alkaloids are given in Table 12.1, and some typical examples are shown in Fig. 12.4.

Acids	Bases
C_mH_x—COOH Aliphatic acids	H_2N-isobutyl Isobutylamine
RO-C_6H_4-C_nH_y—COOH Ring-substituted aromatic acids	HN (ring) Pyrrolidine
	HN (ring) Piperidine (and derivatives)

Table 12.1
Basic structural features of *Piper* alkaloids

Up till now no biosynthetic experiments on the origin of the *Piper* alkaloids have been reported. The possible ways in which they are formed, however, seem to be simple. The aliphatic acids should be formed along the polyketide pathway and have an even number of carbon atoms, ranging from 8 to 22. Within the group of substituted aromatic acids the odd number of C-atoms ranges from 3 to 11. Using a cinnamic acid as starter molecule the side-chain may increase *via* the polyketide route. With few exceptions, all chains are unsaturated. After suitable activation the acids can react with heterocyclic imines or isobutylamine to form the

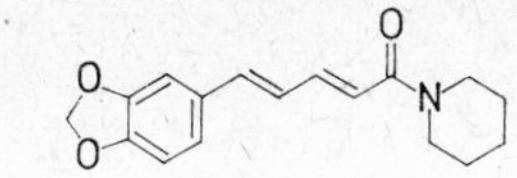

29 Piperine

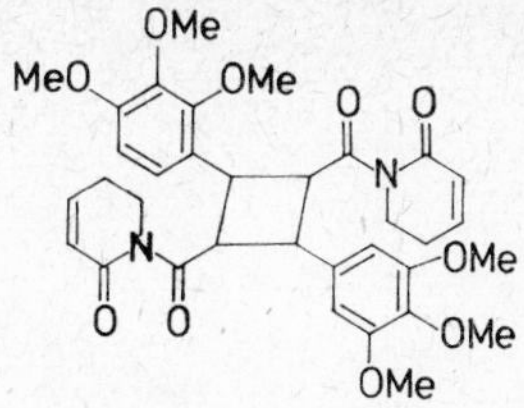

31 Piplartine dimer A

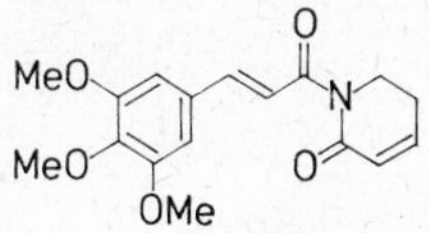

30 Piplartine (Piperlonguminine)

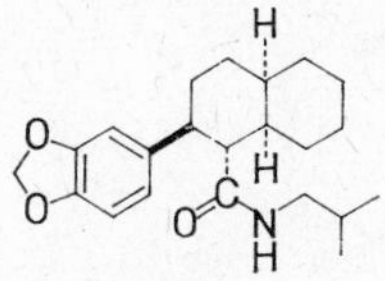

33 Cyclopiperstachine

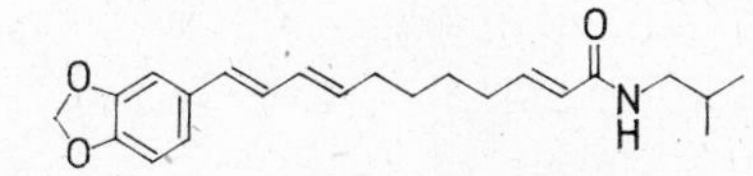

32 Piperstachine

Fig. 12.4
Alkaloids from *Piper* species

amide structures of the *Piper* alkaloids. The occurrence of free piperidine base in *Piper* species has often been reported. The enzymes for the formation of the amides seem to be of low specificity as outlined by the broad spectrum of acids used. Even *N*-formylpiperidine has been isolated [27]. Deviating from the above pattern (Table 12.1) a few amides with other amines have been discovered. The tendency of unsaturated compounds to cyclize is best demonstrated by the occurrence of piplartine (**30**) and piplartine-dimer A (**31**) in *P. tuberculatum* [34] and of piperstachine (**32**) together with cyclopiperstachine (**33**) in *P. trichostachyon* [72, 73].

Up till now about 12 species of *Piper* have been assayed for their alkaloid content. As a consequence of differing alkaloid pattern with regard to chemical types, the source of the plant, and considerable seasonal variations [5, 60] no clear chemotaxonomic classification has emerged. In *P. auritum* even aporphine-type alkaloids have been detected [61].

Piperidides and pyrrolidides of unsaturated aliphatic acids with 10 to 16 carbon atoms have recently been discovered in *Compositae* such as *Leucocyclus formosus* and species of *Achillea* [12, 43]. In these plants they occur with the previously known isobutylamides.

12.1.2.2 *N*-Substituted Bi-Piperidyl Alkaloids

Several genera of legumes such as *Adenocarpus*, *Ammodendron*, *Genista*, *Lupinus*, or *Retama* are known to contain a number of 1,4,5,6-tetrahydroanabasine alkaloids of the type shown in Fig. 12.5. Furthermore, hexahydroanabasine and 1,4,5,6-tetrahydro-1′,2′-dehydroanabasine are known bases.

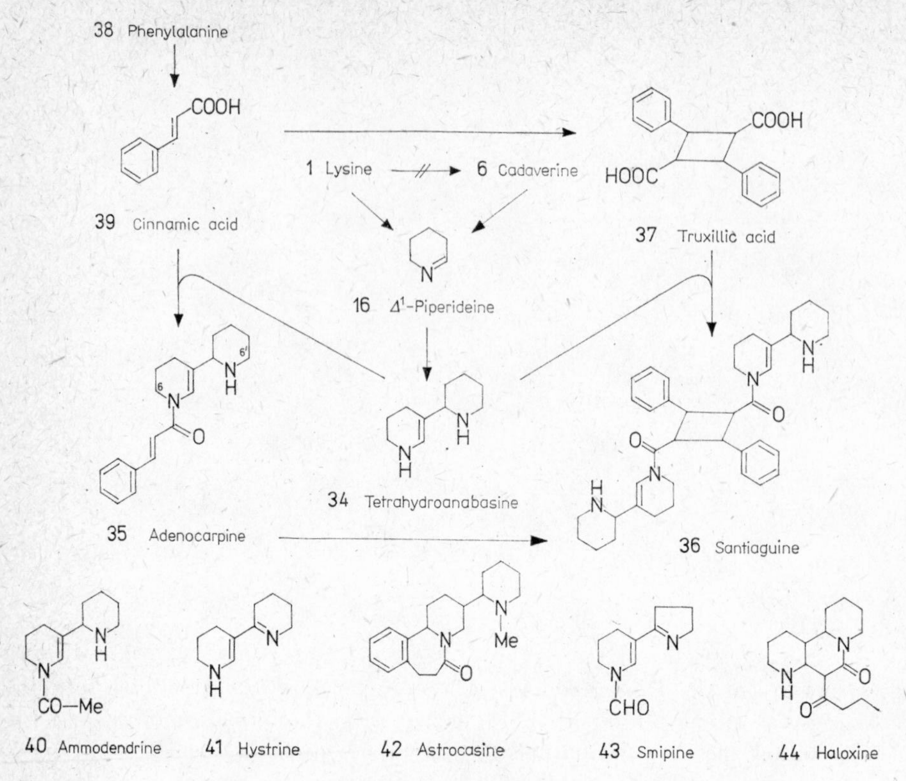

Fig. 12.5
Biosynthesis of *N*-substituted bi-piperidyl alkaloids

Tetrahydroanabasine (**34**) itself can be formed from cadaverine (**6**) by cell-free diamine oxidase preparations from seedlings of peas and lupins [64, 127]. Biosynthetic experiments in seedlings of *Adenocarpus viscosus* have shown the incorporation of [1,5-^{14}C]cadaverine (**6**) and [2,6,2′,6′-^{14}C]tetrahydroanabasine (**34**) into adenocarpine (**35**) [174]. [6,6′-^{3}H, 9-^{14}C]Adenocarpine (**35**) when applied to *A. foliosus* gave labelled santiaguine (**36**) with no change in its ^{3}H:^{14}C ratio [140]. This demonstrates the formation of the α-truxillic acid (**37**) moiety by condensation of the cinnamic acid residues of two adenocarpine (**35**) molecules.

Half of the radioactivity of [6-^{14}C]lysine (**1**) and [6-^{14}C]Δ^1-piperideine (**16**) was incorporated asymmetrically into C-6, and the remainder of the label probably into C-6′ of the bi-piperidyl moieties [141]. The complete route is: lysine (**1**) → Δ^1-piperideine (**16**) → tetrahydroanabasine (**34**) → adenocarpine (**35**) → santiaguine (**36**) (Fig. 12.5). Cadaverine (**6**) should be linked to this pathway by conversion into Δ^1-piperideine (**16**).

The α-truxillic acid (**37**) moiety of santiaguine (**36**) became labelled with [2-^{14}C]-phenylalanine (**38**) and [2-^{14}C]cinnamic acid (**39**) in *A. foliosus* [140, 141]. Incorporation probably proceeds *via* adenocarpine (**35**) since [2,4-^{14}C]α-truxillic acid (**37**) showed only a limited rate of incorporation. On application of labelled cinnamic acid (39) α-truxillic acid (37) has been isolated by isotopic dilution techniques, providing proof of this acid in a non-substituted form in plants. Usually α-truxillic acid is esterified or amide-linked to alkaloids (see tropane, pyrrolizidine, and *Piper* alkaloids).

Although experimental proof is still lacking, the biosynthetic route leading to other bi-piperidyl alkaloids may resemble that outlined in Fig. 12.5. This may be the case for ammodendrine (40), hystrine (41) [196], and a small group of alkaloids such as astrocasine (42), isolated from the *Euphorbiaceae Astrocasia phyllanthoides* [109].

Several bi-piperidyl alkaloids have been discovered in *Lupinus formosus* [35] along with the unprecedented alkaloid structure of smipine (43). An interesting hypothesis has been proposed for its biosynthesis, starting with an ammodendrine-like alkaloid.

Haloxine (44) was isolated together with halosaline (63 in Fig. 12.8) from the *Chenopodiaceae Haloxylon salicornicum* [120]. One should envisage its biogenesis in terms of a reaction in which the typical 2-substituted piperidine alkaloids are formed (Fig. 12.8) from Δ^1-piperideine (16) and a β-keto acid with subsequent elimination of the carboxyl group. Haloxine (44) may be the missing link where a lactam formation with the second piperidine nitrogen prevents the β-keto acid intermediate from being decarboxylated.

Occurrence of the above bases with lupin alkaloids is well known [33, 35, 197] and the formation of bi-piperidyl alkaloids from *N*-methylcytisine (138 in Fig. 12.23) has been suggested [128]. The hypothesis of mutual interconversion of several types of quinolizidine alkaloids and ammodendrine (40) in *Ammodendron karelinii* [87] is hardly substantiated by the experimental results available.

12.1.3 2- and 2,5-Substituted Piperidine Alkaloids

12.1.3.1 Alkaloids from *Punica* and *Withania*

The *Punica* alkaloids comprise a group of only three bases, pelletierine (45), *N*-methylpelletierine (48), and pseudopelletierine (50). These occur predominantly in the bark of the pomegranate, *Punica granatum* (*Puniceae*). *Withania somnifera* (*Solanaceae*) contains two unique piperidine bases, anaferine (46) and anahygrine (49) [190] along with pyrrolidine, tropane, and other types of alkaloids. In their chemical structures the alkaloids mentioned above (Fig. 12.6) may be considered as the higher homologues of tropinone, hygrine, and cuscohygrine (*cf.* Fig. 11.2). According to this structural similarity lysine (1) and acetate should be the precursors of the piperidine alkaloids.

Both [2-^{14}C]- and [6-^{14}C]lysine (1) were indeed specifically incorporated into the piperidine nucleus of pelletierine (45) [74] and *N*-methylpelletierine (48) [55, 56, 74, 105, 108, 143] in *Punica granatum* and *Sedum sarmentosum*. In both alkaloids the label of [6-^{14}C]lysine (1) was located exclusively in C-6 of the piperidine nucleus, thus excluding a symmetrical intermediate. The incorporation of [4,5-^{3}H, 6-^{14}C]-

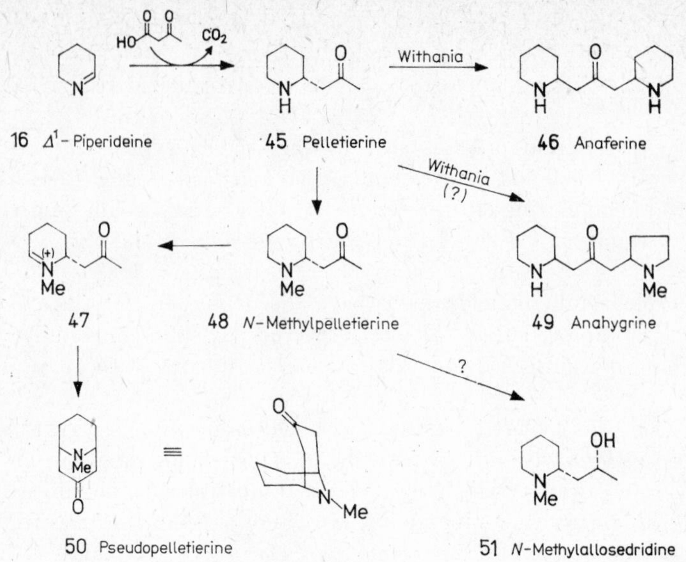

16 Δ^1-Piperideine

45 Pelletierine

46 Anaferine

47

48 *N*-Methylpelletierine

49 Anahygrine

50 Pseudopelletierine

51 *N*-Methylallosedridine

Fig. 12.6
Biosynthesis of 2,5-di-substituted piperidine alkaloids occurring in *Punica*, *Withania*, and *Sedum* species

lysine (**1**) took place with retention of the $^3H:^{14}C$ ratio [56]. This is entirely consistent with the assumption of consecutive decarboxylation and deamination of lysine (**1**) whereby the intermediate cadaverine (**6**) remains enzyme-linked (**13** in Fig. 12.2). More sophisticated biosynthetic experiments compared the incorporation values of doubly labelled lysine (**1**) with tritium in 2-, 4,5-, and 6-positions, and used optically pure lysine (**1**) in its radioactive form [102]. The results obtained permitted an unequivocal decision to be made regarding the chirality of the lysine (**1**) precursor. Only the *L*(−)-enantiomer was incorporated into *N*-methylpelletierine (**48**) and *N*-methylallosedridine (**51**) in *S. sarmentosum*.

The side-chain of both alkaloids became labelled after application of [^{14}C]acetate [74, 108, 143] with the carboxyl radioactivity in the carbonyl group. Acetoacetic acid has been proposed as an intermediate though β-hydroxybutyric acid, a more stable analogue of acetoacetate, was not incorporated intact [55, 56]. In connection with the biosynthesis of the side-chain of sedamine (**54** in Fig. 12.7) from phenylalanine (**38**), α-aminobutyric acid was unsuccessfully tested as a precursor [56].

Labelled lysine (**1**) and acetate were also specifically incorporated into pseudopelletierine (**50**) and anaferine (**46**) [74, 108, 143]. After establishing the precursor role of hygrine in the biosynthesis of cuscohygrine and tropine (Fig. 11.2), it should be mentioned that the piperidine analogues anaferine (**46**) and pseudopelletierine (**50**) originate in a similar way: [8-^{14}C]pelletierine (**45**) in *W. somnifera* gave labelled anaferine (**46**) with a high incorporation rate [74]. The same was true for the conversion of [8-^{14}C]*N*-methylpelletierine (**48**) into pseudopelletierine (**50**) in *P. granatum*, where the intermediacy of **47** would be expected.

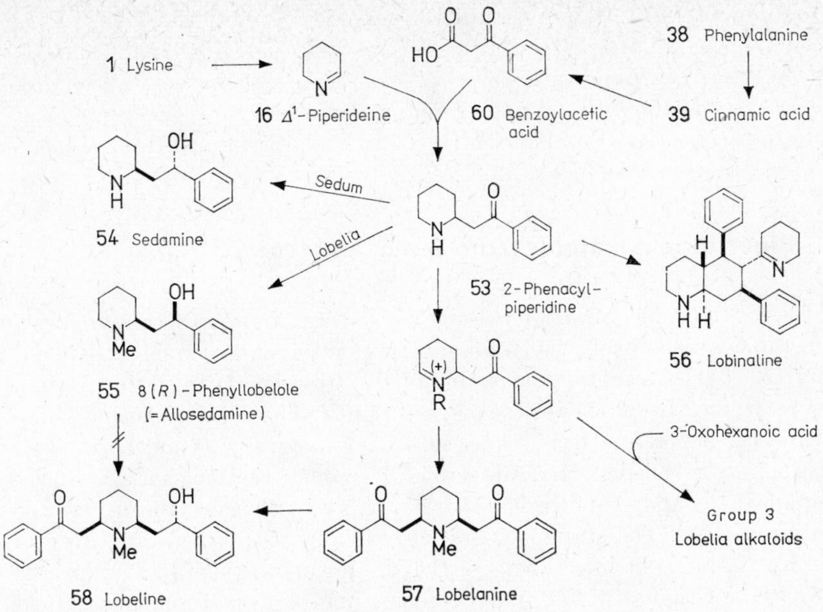

1 Lysine
16 Δ¹-Piperideine
60 Benzoylacetic acid
38 Phenylalanine
39 Cinnamic acid
54 Sedamine
Sedum
Lobelia
53 2-Phenacyl-piperidine
56 Lobinaline
55 8(R)-Phenyllobelole (=Allosedamine)
3-Oxohexanoic acid
Group 3 Lobelia alkaloids
58 Lobeline
57 Lobelanine

Fig. 12.7
Biosynthesis of *Lobelia* alkaloids with aromatic side-chains

The methyl groups of *N*-methylpelletierine (**48**) and pseudopelletierine (**50**) are derived from the *S*-methyl group of methionine [108]. The specific incorporation of [6-^{14}C]lysine (**1**) into the above-mentioned alkaloids (Fig. 12.6) means that a symmetrical intermediate must be ruled out. The efficient conversion of [1,5-^{14}C]-cadaverine (**6**) into *N*-methylpelletierine (**48**) and pseudopelletierine (**50**) must be interpreted in terms of a side-pathway (Fig. 12.2) in which the diamine **6** yields a normal intermediate, most probably 5-aminopentanal (**15**) [105, 108]. These results have been confirmed unequivocally by the selective incorporation of the tritium label from chirally labelled [1-$^{3}H_{1}$]cadaverine (**6**) into the piperidine alkaloids of *Sedum sarmentosum* [103]. This behaviour is similar to that of putrescine in the biosynthesis of tropane alkaloids (Fig. 11.2). Pipecolic acid (**21**) is not on the route to the *Punica* alkaloids [108].

Screening for minor alkaloids in *Punica granatum* gave no indication for bases with a reduced carbonyl group [108]. Obviously no enzyme capable of catalyzing the reduction step was present. On the contrary, *Sedum* species have been reported to contain *N*-methylpelletierine (**48**) together with the corresponding secondary alcohol *N*-methylallosedridine (**51**) and its 2(*S*),8(*S*)-analogue sedridine [37, 102]. The stereochemistry of *N*-methylpelletierine (**48**) from *Sedum* is still unknown. If the chirality of this alkaloid matches that of the 2(*R*)-*N*-methylpelletierine isolated from *Punica* then a stereospecifically controlled enzymic reduction would account for the origin of 2(*R*),8(*S*)-*N*-methylallosedridine in *S. sarmentosum*.

As a consequence of their simple biosynthetic formation pelletierine (**45**) and its

N-methyl compound **48** occur in a number of very different plant families, *e.g.* in the solanaceous genera *Datura, Duboisia, Salpiglossis,* and *Withania*; in *Sedum* species (*Crassulaceae*); in *Lupinus formosus* (*Leguminosae*) [35]; and in *Lycopodium labelliforme* (*Lycopodiaceae*). In the latter family pelletierine (**45**) is the intermediate on the route to the *Lycopodium* alkaloids (Fig. 12.10) and is formed from lysine (**1**) *via* a symmetrical intermediate.

12.1.3.2 Piperidine Alkaloids from *Sedum* and *Lobelia*

Several species of the genus *Sedum* (crassulaceaeous) are known to contain 2-substituted piperidine alkaloids of which sedamine (**54**) is the most common. Labelling experiments [53, 54] led to the conversion of [6-^{14}C]lysine (**1**) to [6-^{14}C]sedamine (**54**) in *S. acre*. Only *L*-lysine (**1**) was the precursor [102]. This alkaloid is thus shown to be yet another piperidine base whose derivation avoids a symmetrical intermediate. If chirally labelled [1-$^{3}H_1$]cadaverine (**6**) reaches the biogenetic route *via* 5-aminopentanal (**15**) and Δ^1-piperideine (**16**) (Fig. 12.2) then its label is incorporated specifically [103]. The entire skeleton of sedamine (**54**) is generated in combination of a C_5N fragment like **16** with a C_6C_2-fragment in which [2,3-^{14}C]phenylalanine (**38**) was incorporated [53, 54]. Phenylalanine (**38**) on deamination yields cinnamic acid (**39**). The next intermediate is most likely to be 3-hydroxy-3-phenylpropionic acid, oxidation of which leads to benzoylacetic acid (**60**) (Fig. 12.7).

Much more members of structurally related piperidine alkaloids have been found in species of *Lobelia*. From the biogenetic point of view one may divide *Lobelia* alkaloids according to substitution pattern into the following 5 subgroups:

a) 2-substituted bases such as 8-phenyllobelole (**55**) and 8-ethyllobelone (**64**),
b) those with aromatic side-chains at both C-2 and C-6 (*i.e.* lobeline (**58**)),
c) those with aliphatic substituents at C-2 and aromatic side-chains at C-6,
d) those with aliphatic substituents at both C-2 and C-6 such as 8,10-diethyllobelidione (**67**), and
e) alkaloids apparently formed by dimerization of group a bases (*i.e.* lobinaline (**56**)).

Labelled lysine (**1**) was incorporated into all alkaloids so far investigated, for example into lobeline (**58**) [75, 143, 144], 8,10-diethyllobelidione (**67**) [144], and lobinaline (**56**) [53, 59]. Lobeline (**58**) was shown to be derived through a symmetrical intermediate [143]. If the intermediacy of cadaverine (**6**) is taken into consideration, this route is inconsistent with the well-established asymmetrical incorporation of lysine (**1**) into *Sedum* and *Punica* alkaloids. It was therefore concluded that the symmetric intermediate lays further along the biosynthetic pathway. Lobelanine (**57**) may be the compound concerned because it was converted with the high incorporation rate of 12% into lobeline (**58**) in *Lobelia inflata* [142]. Lobinaline (**56**) may be formally dissected into two 8-phenyllobelones (or the corresponding secondary alcohols). Incorporation of [6-$^{14}C^{3}H_2$]lysine (**1**) was not symmetrical and with retention of the $^{3}H:^{14}C$ ratio incorporated into both halves of lobinaline (**56**) in *L. cardinalis* [59] (Fig. 12.7).

The aromatic side-chains became non-randomly labelled following application of radioactive phenylalanine (**38**), cinnamic acid (**39**) [53, 59, 75, 144], and 3-hydroxy-3-phenylpropionic acid [144]. An efficient conversion of [6-^{14}C]2-phenacylpiperidine (**53**) into lobinaline (**56**) established this compound as a likely intermediate [59].

Aliphatic side-chains, as in 8,10-diethyllobelidione (**67**), become radioactive on feeding labelled acetate. The intermediacy of 3-oxohexanoic acid (**61**) is expected to give compounds such as **65** and **66** (Fig. 12.8), which in turn should lose their terminal methyl groups [144].

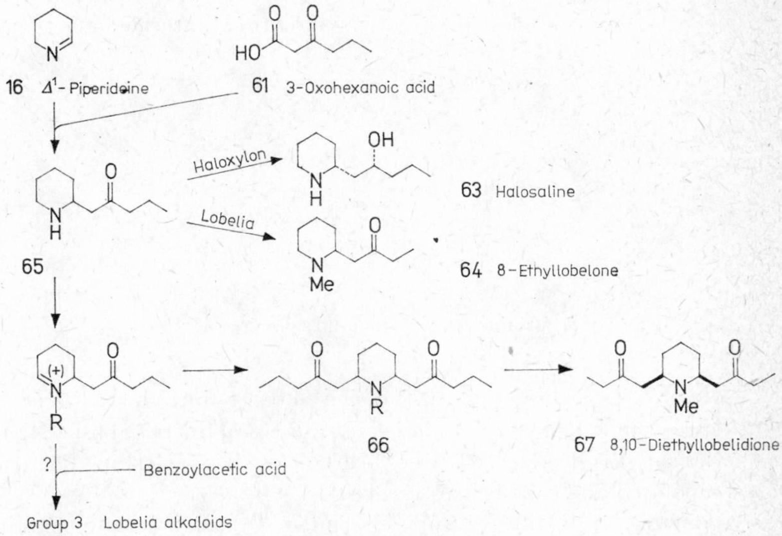

Fig. 12.8
Biosynthesis of *Lobelia* alkaloids with aliphatic side-chains

Haloxylon salicornicum and other *Chenopodeaceae* have become known as further sources of related piperidine alkaloids [120]. Halosaline (**63**) is a lysine-derived [139] 8-propyllobelole and assumed to be generated from the reaction of Δ^1-piperideine (**16**) with 3-oxohexanoic acid (**61**). It still contains the original C_5-side-chain in contrast to the typical *Lobelia* alkaloids with C_4-side-chains (i. e. **64** and **67**). Halosaline (**63**) in *H. salicornicum* occurs with haloxine (**44** in Fig. 12.5), a bipiperidyl alkaloid in which the 3-oxohexanoic acid (**61**) is preserved as an intact structure unit. Both alkaloids are thus missing links that support the biogenetic hypothesis outlined in Fig. 12.8.

12.1.4 Alkaloids with Condensed Piperidine Ring Systems

12.1.4.1 *Securinega* Alkaloids

Securinine (**68**) is the most abundant alkaloid of a group of unprecedented lysine-derived bases occurring in the *Euphorbiaceae* species *Securinega suffrutticosa* and *Phyllanthus discoides* [194]. The biosynthesis has been analysed in terms of a pathway that involves the condensation of Δ^1-piperideine (**16**) onto the aromatic nucleus of tyrosine (**69**). The incorporation of labelled lysine (**1**), cadaverine (**6**), and Δ^1-piperideine (**16**) is consistent with this hypothesis [40, 147, 164, 165]. Securinine (**68**) derived from [6-$^{14}C^{3}H_2$]lysine (**1**) was shown to retain all tritium relative to radiocarbon, indicating the α-amino group of the amino acid to be the source of the heterocyclic nitrogen [40]. Labels of [2-^{14}C]lysine (**1**) and [2-^{14}C]Δ^1-piperideine (**16**) were specifically converted into C-2 of the alkaloid (Fig. 12.9). A symmetrical intermediate can therefore be ruled out. All other lysine-derived alkaloids whose nitrogen lies at a ring junction (*i.e.* alkaloids of *Lupinus*, *Lycopodium*, and *Lythraceae*) incorporate this amino acid by way of a symmetrical intermediate.

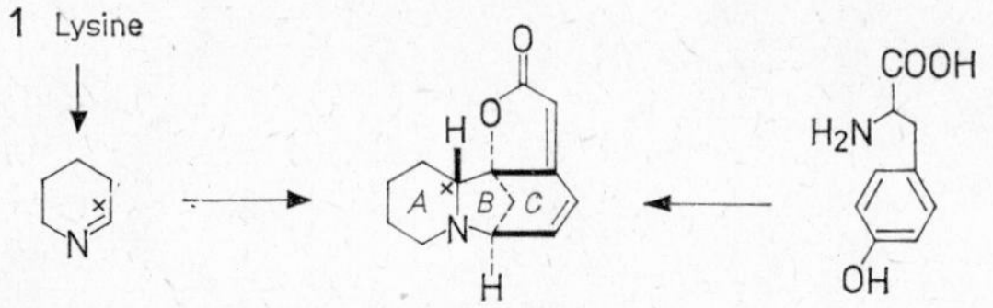

Fig. 12.9
Origin of securinine from lysine and tyrosine

After loss of its carboxyl group tyrosine (**69**) was introduced intact into rings *C* and *D* of securinine (**68**) in *Securinega suffrutticosa* [40, 147, 148, 164, 165] and the lactone carbonyl is derived from the β-carbon of the side-chain. Phenylalanine (**38**) was incorporated about 100 times less [164].

Some minor alkaloids differ from securinine (**68**) in that they contain additional hydroxyl groups [194]. Alkaloids of the nor-securinine type, however, represent the lower ring *A* homologues of securinine (**68**) and may thus be formed in a similar way using Δ^1-pyrroline instead of Δ^1-piperideine (**16**).

12.1.4.2 *Lycopodium* Alkaloids

Probably all *Lycopodium* species (club moss, *Lycopodiaceae*) contain alkaloids of either a common $C_{16}N_2$ type or precursors and conversion products thereof [114]. The incorporation of labelled lysine (**1**), cadaverine (**6**), and Δ^1-piperideine (**16**) into cernuine (**70**) [66] and lycopodine (**71**, a $C_{16}N$-type base) [18, 19, 51] established these alkaloids to be lysine-derived (Fig. 12.10). Chemical degradation of the lycopodine (**71**) molecule from *Lycopodium flabelliforme* showed the radioactivity of both [2-^{14}C]- and [6-^{14}C]lysine (**1**) to be equally distributed amongst 4 carbon atoms of the alkaloid, pointing to the symmetrical cadaverine (**6**) as an intermediate [51]. The remaining carbon atoms in the *Lycopodium* alkaloids under consideration should be derived from C_2 units, but feeding experiments with labelled acetate gave only small incorporation rates. This evidence led to a biogenetic hypothesis according to which the alkaloids may originate from two C_8N-units such as pelletierine (**45**).

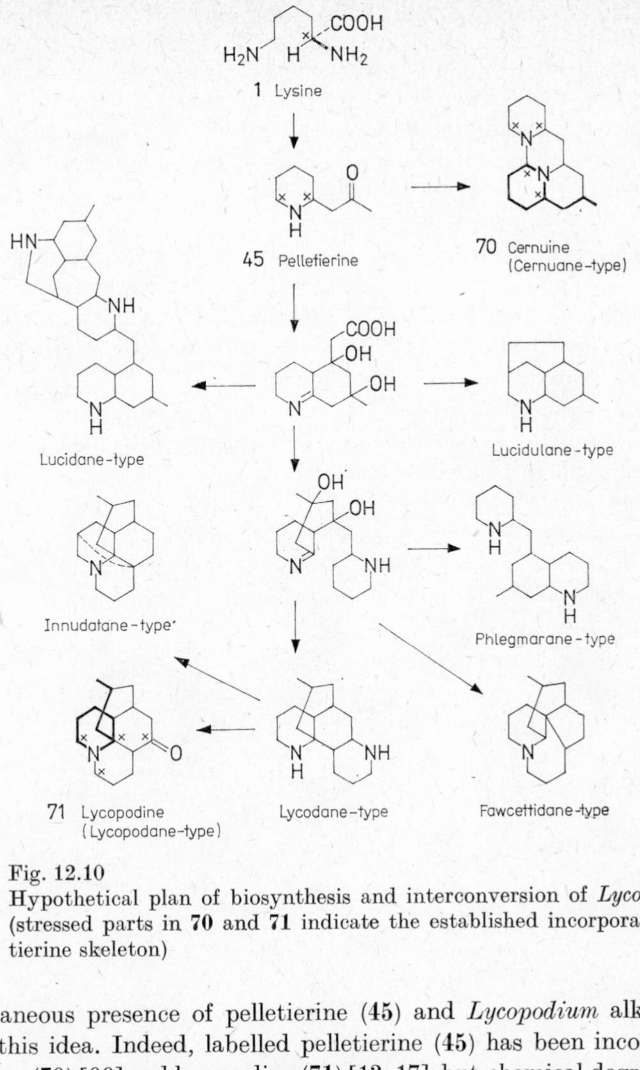

Fig. 12.10
Hypothetical plan of biosynthesis and interconversion of *Lycopodium* alkaloids (stressed parts in **70** and **71** indicate the established incorporation of the pelletierine skeleton)

The simultaneous presence of pelletierine (**45**) and *Lycopodium* alkaloids strongly supported this idea. Indeed, labelled pelletierine (**45**) has been incorporated intact into cernuine (**70**) [66] and lycopodine (**71**) [13, 17], but chemical degradation showed that only one unit became radioactive rather than two as predicted (Fig. 12.10). The apparent discrepancy between the incorporation of pelletierine (**45**) into only one half but its immediate precursor Δ^1-piperideine (**16**) into both halves of the molecule led to the proposal of a condensation between pelletierine (**45**) and a pelletierine-like C_8N unit [13, 14, 117]. In contrast to *Lycopodium*, in *Punica* and *Sedum* the *N*-methylpelletierine (**48**) originates from lysine (**1**) *via* an asymmetrical intermediate (*cf.* chapter 12.1.3.1).

The huge number of *Lycopodium* alkaloids have been arranged into 8 main structural types (Fig. 12.10), enabling a chemotaxonomic classification [14] of the *Lycopodium* alkaloids that closely resembles modern taxonomic divisions.

12.1.5 Miscellaneous Lysine-derived Piperidine Alkaloids

More complex substituted piperidine alkaloids have been detected as minor bases together with *Ormosia* and *Lythraceae* alkaloids (*cf.* chapter 12.2). Several alkaloids from different groups contain an additional piperidine ring as demonstrated in the case of the harmane alkaloid homobrevicolline (**72**). Preliminary biosynthetic experiments resulted in the incorporation of labelled lysine into this alkaloid in *Carex brevicollis* [88].

2-substituted piperidines structurally related to that of *Lobelia* (Fig. 12.7) have been detected in several other plants, *e.g.* *O*-Methylpleurospermine (**73**) as the first *Urticaceae* alkaloid in *Boehmeria platyphylla* [62] or febrifugine (**74**) in *Dichroa febrifuga*. Whereas the former is structurally related [65] to the cryptopleurine-type bases the latter is related to quinazoline alkaloids.

An unusual amino acid, mimosine (**77**), in *Leucaena* and *Mimosa* species (*Leguminosae*) occurs with pipecolic acids. Its γ-pyridone moiety is lysine-derived (*cf.* [44]). Disregarding the rapid incorporation [203] of serine (**76**, R = H), it has been recently established at the enzymic level that *O*-acetylserine (**76**, R = CH_3CO) is the immediate substrate in reactions yielding mimosine (**77**) in *Leucaena leucocephala* [129]. On glucosylation mimoside (**78**) is formed (Fig. 12.11); this has been accomplished in both intact seedlings and cell-free preparations [130].

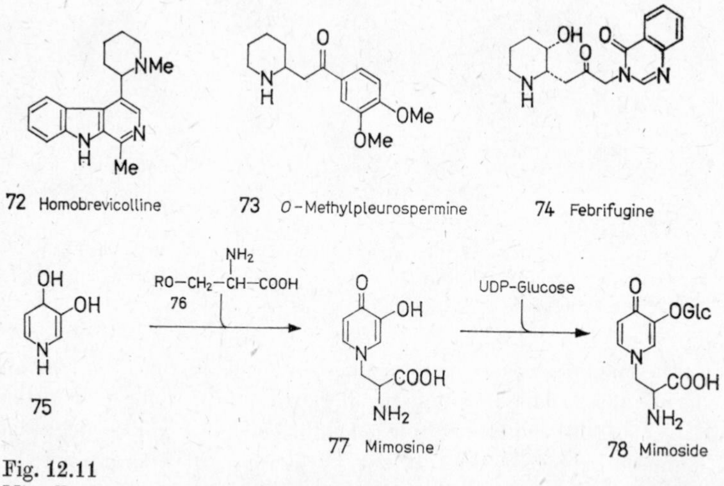

Fig. 12.11
Miscellaneous piperidine alkaloids. Biosynthesis of mimosine and mimoside

12.1.6 Acetate-derived Piperidine Alkaloids

12.1.6.1 *Conium* Alkaloids

A special biogenetic concept is realized in hemlock (*Conium maculatum*, *Umbelliferae*) because labelled acetate is incorporated into the piperidine-type alkaloids coniine (**3**) and conhydrine (**83**) [90, 91]. The postulated pathway in a linear arrangement of four acetate units has been substantiated by the incorporation of labelled octanoic acid, 5-oxooctanoic acid (**79**) and 5-oxooctanal (**80**) into *Conium* alkaloids [94, 100, 101]. The kinetics of $^{14}CO_2$ incorporation indicated a further route (Fig. 12.12): γ-coniceine (**82**) → coniine (**3**) → *N*-methylconiine (**86**) [28, 29].

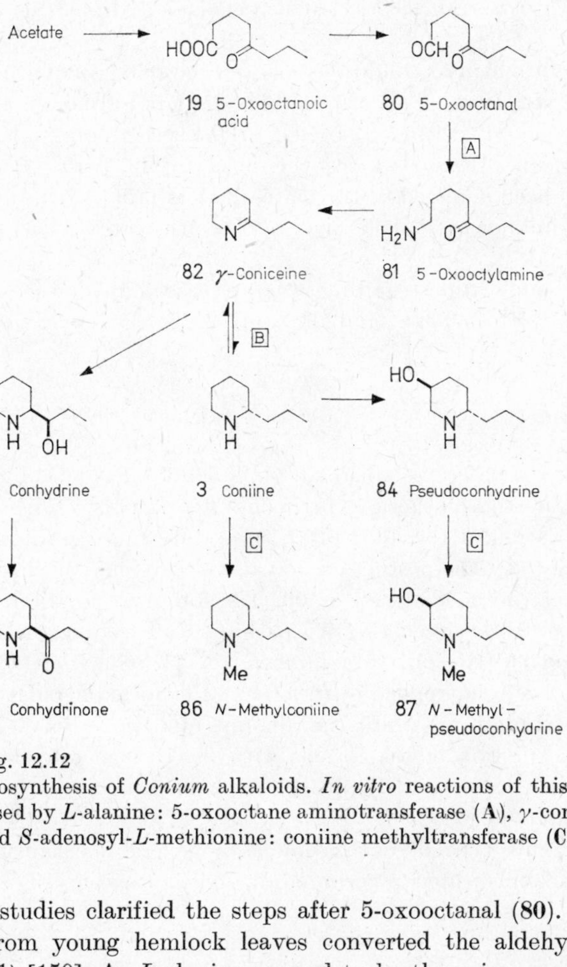

Acetate
19 5-Oxooctanoic acid
80 5-Oxooctanal
A
82 γ-Coniceine
81 5-Oxooctylamine
B
83 Conhydrine
3 Coniine
84 Pseudoconhydrine
C
85 Conhydrinone
86 *N*-Methylconiine
87 *N*-Methyl-pseudoconhydrine

Fig. 12.12
Biosynthesis of *Conium* alkaloids. *In vitro* reactions of this sequence were catalysed by *L*-alanine: 5-oxooctane aminotransferase (**A**), γ-coniceine reductase (**B**), and *S*-adenosyl-*L*-methionine: coniine methyltransferase (**C**)

Enzymatic studies clarified the steps after 5-oxooctanal (**80**). An active enzyme preparation from young hemlock leaves converted the aldehyde **80** into 5-oxooctylamine (**81**) [150]. As *L*-alanine seemed to be the prime source of nitrogen for this transamination, the enzyme may be considered to be *L*-alanine:5-oxooctanal aminotransferase (AAT) [154, 155]. On the basis of differing substrate requirements, two isoenzymes have been separated [155], one of which, AAT A, is located in the mitochondria whereas AAT B seems to be present in chloroplasts [157]. As the mitochondrial AAT A occurs in several alkaloid-free plants, it is assumed that AAT B is responsible for the formation of 5-oxooctylamine (**81**). This aminoketon cyclizes spontaneously to γ-coniceine (**82**). A NADPH-dependent γ-coniceine reductase from leaves and fruits converted γ-coniceine (**82**) to coniine (**3**) with the hydrogen originating from the B (*pro-S*) side of the pyridine coenzyme [153]. *In vivo* experiments demonstrated the reversibility of this reaction with (+)-coniine (**3**) as the preferred substrate [31, 95].

N-Methylation of coniine (**3**) to *N*-methylconiine (**86**) was catalysed by *S*-adenosyl-*L*-methionine:coniine methyltransferase from maturing *Conium* fruits [151, 152]. The same enzyme preparation methylated pseudoconhydrine (**84**) but not conhydrine (**83**) [156].

Under certain environmental conditions conhydrine (**83**), conhydrinone (**85**), and pseudoconhydrine (**84**), occurred in considerable amounts although away from the main biosynthetic route (Fig. 12.12) [95, 100, 101]. Furthermore, in a group of high altitude South African *Conium* species the hitherto unknown *N*-methylpseudoconhydrine (**87**) has been detected as the main alkaloid [156, 158]. *N*-methylconiine (**86**), together with unknown alkaloids, has also been isolated from cell cultures of *Conium maculatum* [116].

There is only one publication regarding *Conium* bases outside the genus *Conium*: the insectivorous *Sarracenia flava* (and probably *S. purpurea*, *Sarraceniaceae*) was found to contain coniine (**3**) [125].

12.1.6.2 Pinidine

Another acetate-derived piperidine alkaloid, pinidine (**91**), has been found in several *Pinus* species. The first biosynthetic experiments with *Pinus jeffreyi* demonstrated the specific incorporation of the label of [1-^{14}C]acetate into carbon atoms 2, 4, 6, and 9 of the alkaloid [98]. The postulated polyacetate (**88**) may cyclize with a nitrogen source *via* either intermediate **89** or **90**. The low incorporation rates made it impossible to distinguish between the two pathways by feeding the labelled dioxo intermediates **89** and **90** [96, 99]. In a further trial [1-^{14}C]malonate together with non-labelled acetate has been applied to *P. jeffreyi*. Chemical degradation of pinidine (**91**) proved C-2 to contain one-third the amount of radioactivity compared with C-9, because of the dilution of the former with inactive acetate [99]. As acetate is the starter unit in the formation of polyacetate chains, the biosynthesis of pinidine (**91**) should proceed *via* **89**.

One may favour this type of cyclization as a general mechanism to produce acetate-derived 2,6-disubstituted piperidine alkaloids in species of *Azima*, *Carica*, *Cassia* or *Prosopis* (see next chapter).

12.1.6.3 Alkaloids from *Azima* and *Carica*

A few macrocyclic alkaloids (**93**—**95**; Fig. 12.14) have been detected that consist of two halves of 2(*S*)-methyl-3(*S*)-hydroxypiperidine-6(*R*)-carboxylic acids [193]. As far as we know, these alkaloids are restricted to *Carica papaya* (= tropical melon tree, *Caricaceae*) and to *Azima tetracantha* (*Salvadoraceae*).

Preliminary biosynthetic experiments with *C. papaya* demonstrated labelled acetate to be incorporated into carpaine (**95**) in preference to lysine (**1**) or mevalonic acid [10]. Dehydrocarpaines were believed to act as late intermediates [200]. Assuming a polyacetate pathway (Fig. 12.14) it is interesting to note the identical carbon skeletons of the hypothetic acid **92** ($m = 7$) and myristic acid, $CH_3-(CH_2)_{12}-COOH$, occurring in the *Carica* seed oil.

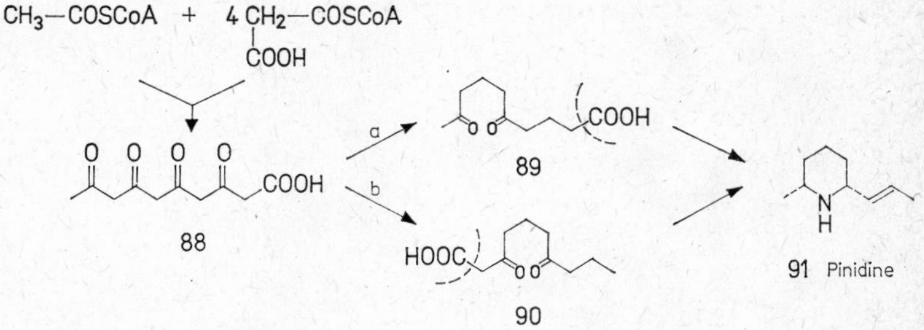

$CH_3-COSCoA + 4\,CH_2(COOH)-COSCoA$ → **88** → (a) **89** / (b) **90** → **91** Pinidine

Fig. 12.13
Biosynthesis of pinidine

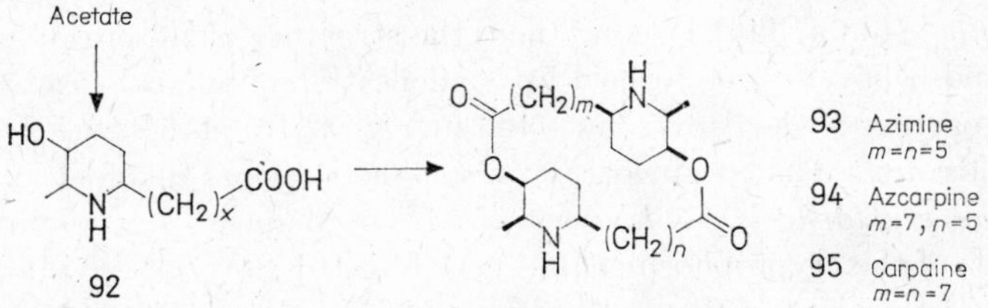

Acetate → **92** → **93** Azimine $m=n=5$; **94** Azcarpine $m=7, n=5$; **95** Carpaine $m=n=7$

Fig. 12.14
Proposed route for alkaloid formation in *Azima* and *Carica*

12.1.6.4 Alkaloids from *Cassia* and *Prosopis*

From several species of the genera *Cassia* (*Leguminosae*) and *Prosopis* (*Mimosaceae*) an increasing number of structurally related piperidine alkaloids has been isolated. Common features of these bases are:

a) methyl or hydroxymethyl groups at C-2 of the nucleus,
b) hydroxy or methoxy groups at C-3, and
c) a non-branched side-chain with 12 or 14 carbon atoms.

Four typical alkaloids (**96**—**99**) are given in Fig. 12.15, showing the differences in the stereochemical arrangement of the substituents.

Due to the structural similarity with the alkaloids in Fig. 12.14 one may propose a biogenetic route from acetate *via* a polyacetate. An interesting alternative [126] is that a phytosphingosine (**104**), generated from palmitoyl CoA (**102**) and serine (**76**), may be the immediate precursor of these types of alkaloids (Fig. 12.16). Because both routes use acetate/malonate to form a long chain intermediate, more sophisticated labelling experiments should permit a distinction between the alternative pathways.

A terminal carboxy group is usually missing, which prevents the alkaloids from dimerization in a way common for *Carica* bases. Along with the simple piperidine

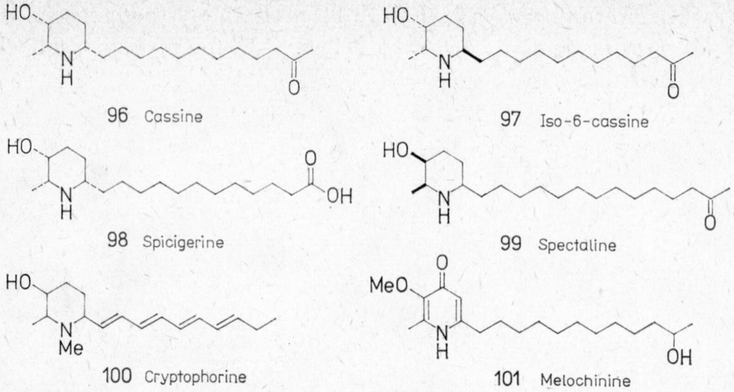

96 Cassine
97 Iso-6-cassine
98 Spicigerine
99 Spectaline
100 Cryptophorine
101 Melochinine

Fig. 12.15
Basic structures of *Cassia* and *Prosopis* alkaloids

bases, however, some more complex alkaloids have been isolated, in particular from *Prosopis juliflora* [6, 6a, 146]. Among them the structure of juliprosopine (**105**) was determined and believed to be formed by condensation of typical *Prosopis* alkaloids such as **107** together with Δ^1-pyrroline (**106** in Fig. 12.16) [146]. In Fig. 12.15 additional alkaloids are included since they resemble *Prosopis* alkaloids in structure. *Bathiorhamnus cryptophoris*, a *Rhamnaceae* from Madagascar, contains at least three alkaloids of the cryptophorine (**100**) type [15, 16]. Several γ-pyridone alkaloids were shown to be present in the *Sterculiaceae Melochia pyramidata* with melochinine (**101**) as the main alkaloid [119]. Although occurring in very different plant families, the alkaloids in Fig. 12.15 should arise along closely related pathways.

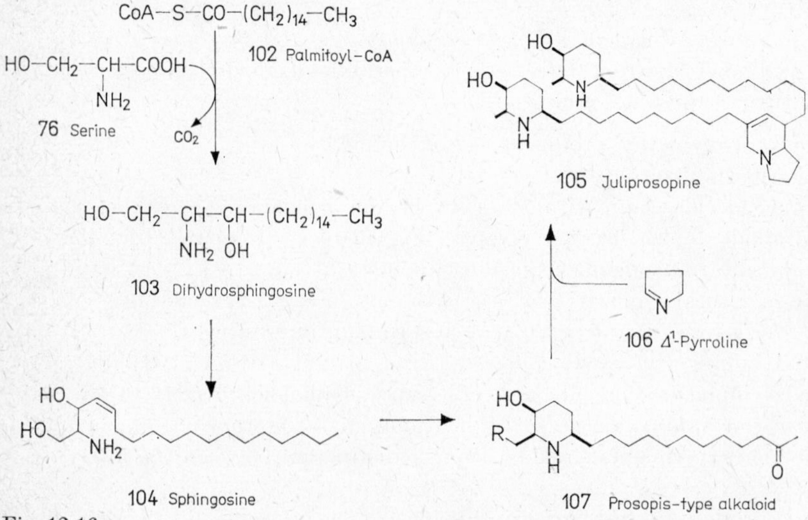

$CoA-S-CO-(CH_2)_{14}-CH_3$
102 Palmitoyl-CoA
$HO-CH_2-CH(NH_2)-COOH$
76 Serine
CO_2
$HO-CH_2-CH(NH_2)-CH(OH)-(CH_2)_{14}-CH_3$
103 Dihydrosphingosine
104 Sphingosine
105 Juliprosopine
106 Δ^1-Pyrroline
107 Prosopis-type alkaloid

Fig. 12.16
Hypothetical pathways for the formation of simple and condensed *Prosopis* alkaloids

12.1.7 Piperidine Alkaloids of Animal Origin

For a long time it seemed doubtful whether animals could produce alkaloids. At least for arthropods it is now well established that there are real alkaloids among the chemicals for defence and communication purposes [205], even piperidine bases. In *Aphaenogaster* ants anabaseine (**64** in Fig. 13.11) has been identified as a poison gland product, acting as attractant [210]. Chemical investigations on the venom from the fire ant (*Solenopsis saevissima*) led to the isolation of more than 12 piperidine alkaloids with the structures **108** and **109** and dehydrated forms thereof [71a, 113, 205]. These solenopsines exhibit a pronounced haemolytic, insecticidal, and antibiotic activity and are the only known non-proteinaceous venoms excreted when biting. Within the genus *Solenopsis* the different varieties of venom have a distinct alkaloid pattern that might enable their chemotaxonomical use (for pyrrolidine alkaloids in ants *cf.* Chapter 11.2). From the European ladybug (*Coccinella septempunctata*) a tricyclic defence substance, coccinelline (**111**), has been isolated [208] whose biogenetic intermediates resemble in some way the piperidine bases. With regard to the ring junctions three stereo-isomers exist, all of which occur in varieties of the genus *Coccinellidae* [205]. The distribution and pattern of the alkaloids indicate a good correlation with the aposematic colours and thus with the taxonomy of the *Coccinellidae* [149].

The unique structure of these bases makes it difficult to deduce their biosynthesis by analogy with known pathways, but the incorporation of labelled acetate favours the route outlined in Fig. 12.17 [207]. Adaline (**112**), a novel alkaloid from *Adalia*

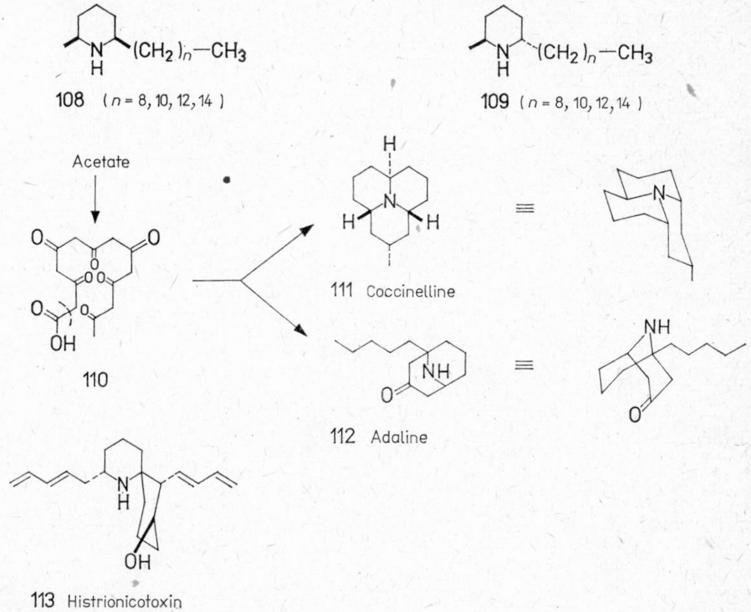

108 (n = 8, 10, 12, 14)

109 (n = 8, 10, 12, 14)

Acetate

110

111 Coccinelline

112 Adaline

113 Histrionicotoxin

Fig. 12.17
Piperidine alkaloids of animal origin

bipunctata (*Coccinellidae*), may branch off from a common intermediate such as **110** [206]. A 1-methyl analogue of adaline (**112**) together with a 6-methylisopelletierine were detected as defense alkaloids of *Cryptolaemus montronzieri* (*Coleoptera*: *Coccinellidae*) [14a].

Recently it became obvious that the neotropical frogs of the *Dendrobatidae* family produce alkaloids of very high pharmacological activity [25, 25a]. More than 100 different alkaloids have been isolated, none of them which have been found elsewhere in nature. With the exception of the steroidal batrachotoxine, all classes so far elucidated contain a piperidine nucleus: for example the allenic and acetylenic spiropiperidine alkaloids of the histrionicotoxine (**113**) type from *Dendrobates histrionicus* [204].

12.1.8 Piperidine Alkaloids of Microbial Origin

In microbial cells the majority of piperidine alkaloids seem to be acetate-derived. This is the case for the antibiotics of the cycloheximide (**115**) type, where both rings in *Streptomyces noursei* originate from the acetate/malonate pool (Fig. 12.18) [42, 76, 209].

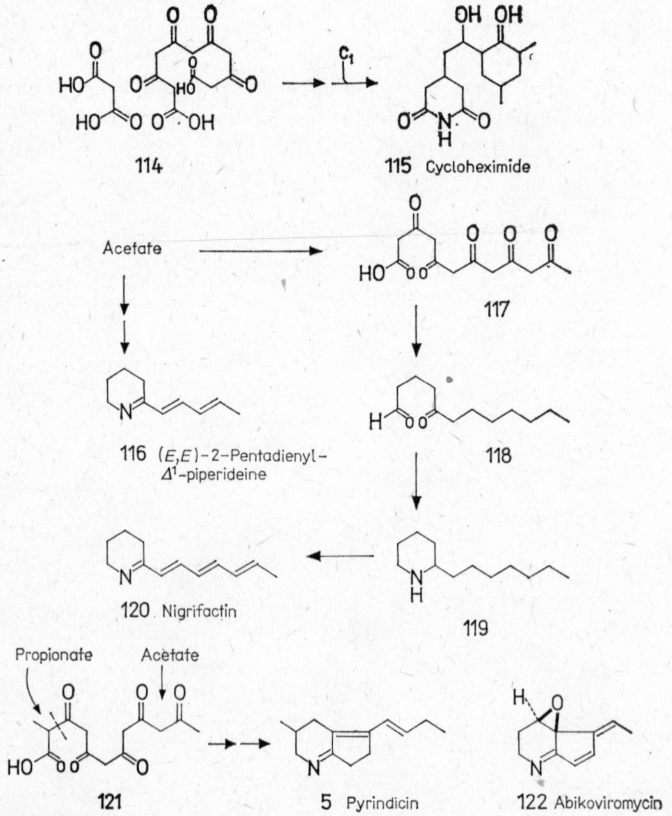

Fig. 12.18
Acetate-derived alkaloids of microbial origin

Several strains of *Streptomyces* and *Actinomyces* produce alkaloids such as nigrifactin (**120**) [202] and its lower homologue (*E,E*)-2-pentadienyl-Δ^1-piperideine (**116**) [83, 145]. The structural similarity between these alkaloids (Fig. 12.18) and the *Conium* bases (Fig. 12.12) suggests that a route exists for the formation of both types of compounds. This has been verified by the specific incorporation of [1-^{13}C]-acetate into nigrifactin (**120**) in *Streptomyces nigrifaciens* [201]. To confirm the conclusion of a linear arrangement of 6 acetate units the presumptive intermediates **117**–**119** were also administered. From the non-random incorporation the entire biosynthetic sequence leading to nigrifactin (**120**) has been postulated as outlined in Fig. 12.18, although the incorporation rates of the intermediates were lower in comparison with that of acetate.

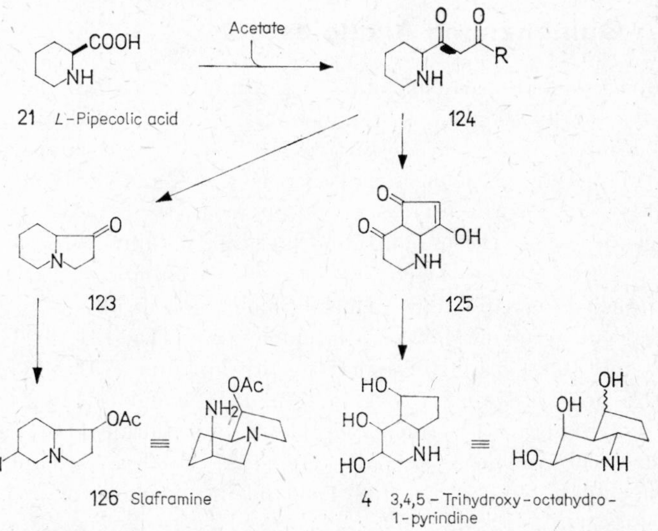

Fig. 12.19
Biosynthesis of slaframine and 3,4,5-trihydroxy-octahydro-1-pyrindine

The lower homologue **116** may be formed in a similar manner from 5 acetate units. Since the structure of pyrindicin (**5**) resembles that of nigrifactin (**120**), a biosynthetic route *via* a polyketide such as **121** might be expected. The ^{13}C-NMR signals of pyrindicin (**5**) labelled with either [1-^{13}C]acetate, [2-^{13}C]acetate, or [1-^{13}C]-propionate gave conclusive evidence for the biosynthetic pathway presented in Fig. 12.18 [71].

Several polypeptide antibiotics (*cf.* [42]) such as amphomycin [11] are known to contain pipecolic acid (**21**) moieties, which are present probably exclusively in the *D*-form. Other antibiotics possess an 4-oxo-pipecolic acid.

Structurally related piperidine alkaloids such as 3,4,5-trihydroxy-octahydro-1-pyrindine (**4**) and the parasympathomimetic slaframine (**126**, 1-acetoxy-6-amino-octahydroindolizine) have been isolated from the pathogenic fungus *Rhizoctonia leguminicola* [47–49]. Both alkaloids fall into the class of piperidine bases where biosynthesis from lysine (**1**) proceeds asymmetrically *via* Δ^1-piperideine carboxylic acid (**20**) and *L*-pipecolic acid (**21**) [23, 24, 46, 47, 49]. The remaining C_2 units are derived from malonate and are made available *via* a pipecolic acid-dependent decarboxylase [23, 24]. Results of labelling experiments show the participation of a C_8N intermediate such as **124**, common to both alkaloids (Fig. 12.19). The intermediacy of **123** and **125** was established by both feeding experiments and enzymatic investigations. These results show that the origins of the two 1-pyrindine alkaloids pyrindicin (**5** in Fig. 12.18) and 3,4,5-trihydroxy-octahydro-1-pyrindine (**4**) differ considerably and therefore a pathway cannot be suggested for bases of this type, *e.g.* abikoviromycin (**122**) [105], prior to biosynthetic investigations.

12.2 Quinolizidine Alkaloids

Quinolizidine alkaloids particularly occur in different families of *Leguminosae*. The simplest alkaloid of this group is lupinine (**127**), the main alkaloid of *Lupinus luteus*. Moreover different tetracyclic derivatives are known such as sparteine (**129**), and lupanine (**131**) as well as higher oxidized derivatives and skeleton variants such as matrine (**128**).

The suggestion that the quinolizidine skeleton is built up from two or three C_5 units derived from lysine (**1**) *via* its decarboxylation product cadaverine (**6**) has been confirmed [159, 167, 168]. Thus lysine is specifically incorporated *via* the symmetrical cadaverine as shown for lupinine (**127**) [41a, 131, 149b, 171, 181, 182, 195], sparteine (**129**) [41a, 81, 149a, 170, 173], lupanine (**131**) [132, 133, 137, 172, 184, 186], hydroxylupanine (**134**) [172, 175, 184], matrine (**128**) [1a, 2–4, 85, 89, 169, 180, 192, 199], angustifoline (**137**) [189], multiflorine [188], anagyrine (**133**) [87], cytisine (**139**), and methylcytisine (**138**) [179] as well as into aphylline (**141**) and aphyllidine (**140**) [1, 76a, 77, 78, 138, 181] and into sophocarpine (**142**) and pachycarpine [86, 87] (Fig. 12.20).

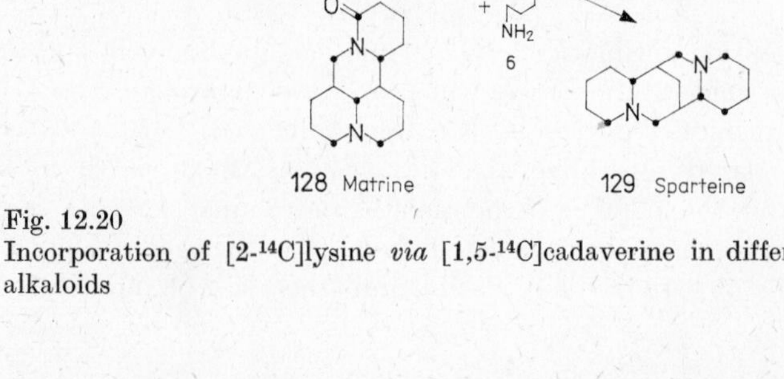

Fig. 12.20
Incorporation of [2-^{14}C]lysine *via* [1,5-^{14}C]cadaverine in different quinolizidine alkaloids

More detailed results for the biosynthetic pathway of quinolizidine alkaloids have been obtained from experiments with cell suspension cultures. Such cultures of *L. polyphyllus*, *Sarothamnus scoparius*, and *Baptisia australis* accumulate alkaloids about 2 to 3 orders less than those from differentiated plants [213, 218–221]. The alkaloid level can be raised, however, when cadaverine is fed. In all the cultures lupanine (**131**) figures as the only main alkaloid. This is independent of the alkaloid composition of the respective plants, which is significantly different in *S. scoparius* (sparteine as main alkaloid) and *B. australis* (*N*-methylcytisine and anagyrine as main alkaloids). It is concluded that the lupanine pathway is basic for the quinolizidine alkaloids and that the other alkaloids are derived from it. In cell suspension cultures only the basic pathway is evident while the more advanced sequences are missing.

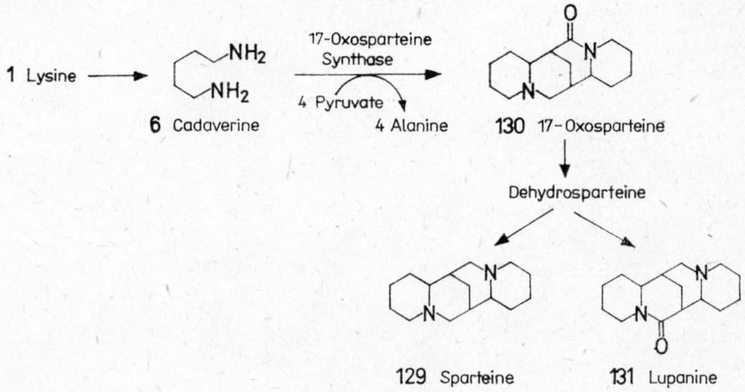

Fig. 12.21
Enzymatic synthesis of lupin alkaloids

The first step in the conversion of lysine (**1**) into quinolizidine alkaloids is its decarboxylation to yield cadaverine (**6**), which was detected by trapping methods [176]. A lysine decarboxylase has been isolated from *L. polyphyllus* leaf chloroplasts [63, 166a]. The existence of a cadaverine-pyruvate transaminating enzyme system (17-oxosparteine synthase) was also established in *L. polyphyllus* cell cultures. It catalyses the biosynthesis of lupin alkaloids with 17-oxosparteine (**130**) as the main product of the enzymatic process according to the reaction 3 cadaverine + 4 pyruvate → 17-oxosparteine + 4 alanine (Fig. 12.21) [212, 216]. The participation of diamine oxidase [137, 177] has been ruled out; the deamination product of cadaverine is not released from the enzyme. The enzyme system most probably catalyses the formation of tetracyclic alkaloids in a channelled manner without releasing free intermediates.

A new model for the bisoynthesis of the lupin alkaloids has therefore been suggested involving bound intermediates (Fig. 12.22) [216]. A separate amino group is proposed as a carrier to which the intermediates remain bound until the end-product is released. The synthesis is initiated by binding the first cadaverine unit to pyridoxal phosphate (step 1/2) and transfer of the resulting semi-aldehyde to

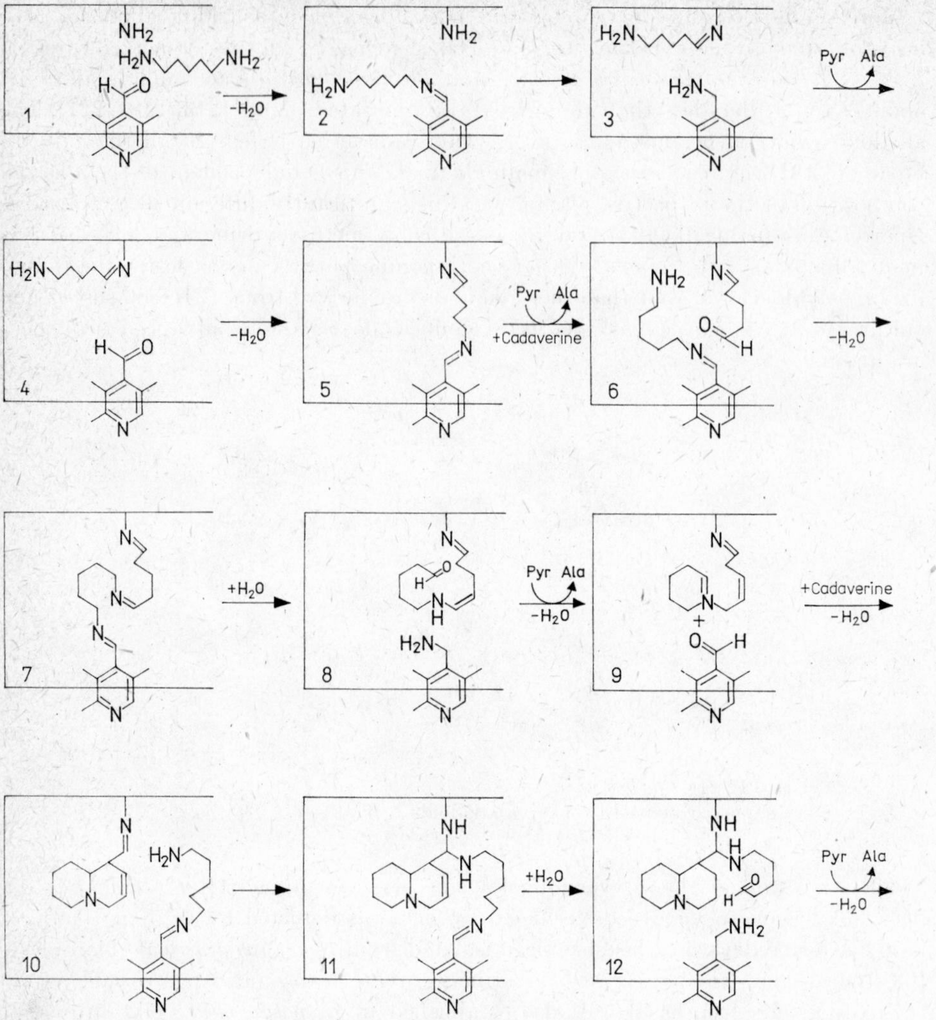

Fig. 12.22
Model mechanism for the synthesis of 17-oxosparteine catalysed by 17-oxosparteine synthase [216]

the carrier amino group. The following steps include transamination of pyridoxamine phosphate, transamination of the second amino group of the first cadaverine unit, binding the second cadaverine unit, and condensation of the cadaverine units by Schiff base formation. Then cyclization yields the quinolizidine ring system involving tautomerization, followed by Mannich reaction with the formation of a reactive carbonium ion at C-7. After condensation of the third cadaverine unit, cyclization to the tetracyclic system occurs, involving a mechanism similar to that described above. After double-bond shift, 17-oxosparteine (**130**) is released by hydrolysis as the key product. Lupinine (**127**) might be obtained by terminating the reaction after step **10**. The bound nature of the reaction intermediates explains the failure of the *in vivo* tracer studies to isolate intermediates between cadaverine

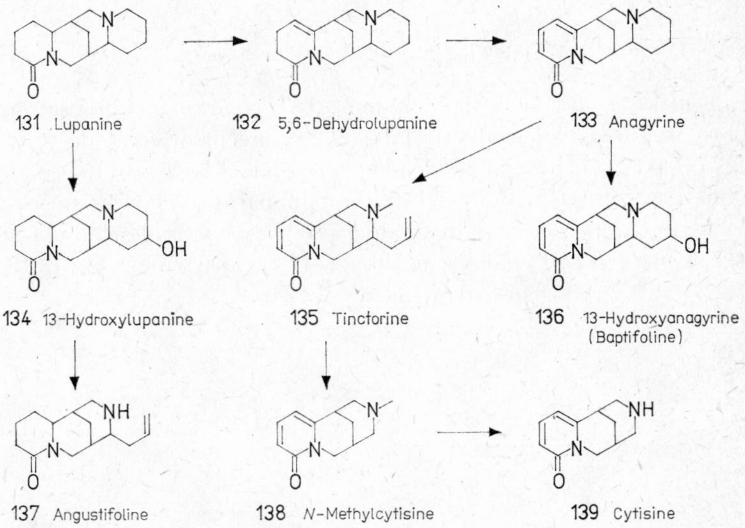

131 Lupanine
132 5,6-Dehydrolupanine
133 Anagyrine
134 13-Hydroxylupanine
135 Tinctorine
136 13-Hydroxyanagyrine (Baptifoline)
137 Angustifoline
138 *N*-Methylcytisine
139 Cytisine

Fig. 12.23
Transformation of lupanine to other quinolizidine alkaloids

and tetracyclic alkaloids [137, 168, 172, 175, 176, 185]. Moreover, with *L. polyphyllus* and *L. albus* it has been shown that the 17-oxosparteine synthase is located in chloroplasts, where it seems to be membrane-bound or membrane-associated [168, 215b, 217].

The lupin alkaloids are associated with chloroplasts and are restricted to the epidermis and subepidermis cell layer [211, 218a]. Since chloroplasts are also the site of lysine formation [118, 121, 123] the enzymes of lysine biosynthesis and those of the biosynthesis of lysine-derived lupin alkaloids are localized in the same subcellular compartment. Consequently alkaloid production is up to 4 to 10 times higher in photomixotrophic cell suspension cultures of *L. polyphyllus* with developed chloroplasts than in chlorophyll-free heterotrophic cultures [213]. Whereas lysine decarboxylase activity is present in all parts of a flowering *L. polyphyllus* plant, 17-oxosparteine synthase is only active in leaf extracts [215]. Lysine decarboxylase and 17-oxosparteine synthase were activated by reduced thioredoxin [214]. A diurnal fluctuation of quinolizidine alkaloid accumulation has been observed in legume plants and photomixotrophic cell suspension cultures [215a].

Sparteine (**129**) was converted into lupanine (**131**) and other lupin alkaloids in *in vivo* and *in vitro* experiments with different species of *Leguminosae* [79, 134—137, 168, 175, 178, 183, 188, 189]. This suggests that sparteine and lupanine are precursors of the other more oxidized lupin alkaloids. However, short time experiments

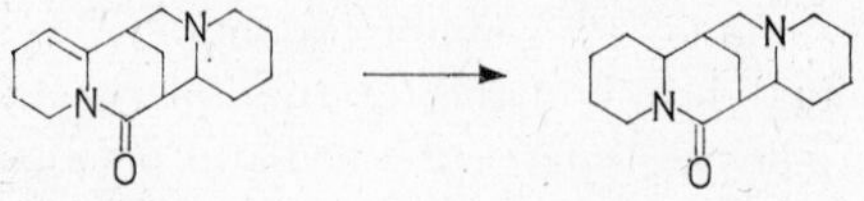

140 Aphyllidine 141 Aphylline

Fig. 12.24
Conversion of aphyllidine into aphylline

with labelled CO_2 indicated that lupinine (**127**), sparteine (**129**), and lupanine (**131**) can be synthesized independently [20—22]. A precursor common to both sparteine and lupanine [22, 41] might be dehydrosparteine, which could derive from 17-oxosparteine (**130**) by reduction and water elimination. This is consistent with the result that lupanine is the main alkaloid produced by crude chloroplasts of *L. polyphyllus* incubated with cadaverine (**6**) or 17-oxosparteine (**130**) [217] (Fig. 12.21) and produced by cell suspension cultures.

128 Matrine 142 Sophocarpine

143 Sophoridine 144 3-Hydroxysophoridine

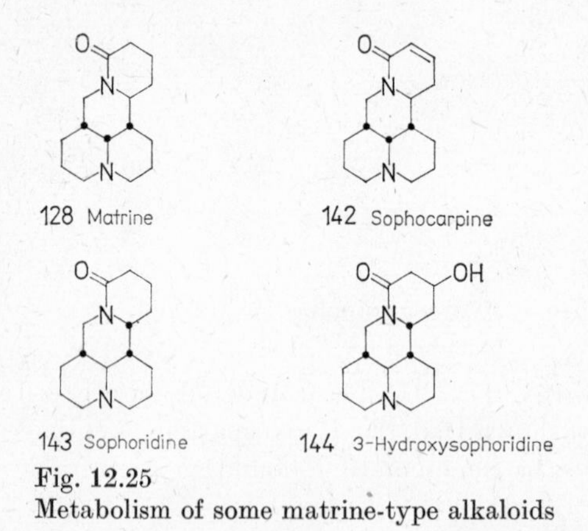

Fig. 12.25
Metabolism of some matrine-type alkaloids

Lupanine has been converted into 13-hydroxylupanine (**134**) and its ester in *L. angustifolius* and *L. albus* [175, 183], into angustifoline (**137**) in *L. angustifolius*, and into anagyrine (**133**), cytisine (**139**), *N*-methylcytisine (**138**), and 13-hydroxyanagyrine (**136**) in *Baptisia leucopheya* and *Ammodendron karelinii* (Fig. 12.23) [87, 135]. A tigloyl-CoA:13-hydroxylupanine O-tigloyl-transferase could be demonstrated in *L. albus* seedlings [215c]. The pathway of lupanine (**131**) to the pyridone bases probably proceeds *via* 5,6-dehydrolupanine (**132**) to anagyrine (133), which on the one hand can be hydroxylated and esterified to 13-hydroxyanagyrine (**136**) and its esters and on the other hand can be degraded to tinctorine (**135**), *N*-methylcytisine (**138**), and cytisine (**139**). A similar pathway with lupanine as the key alkaloid has been found by pulse feeding $^{14}CO_2$ to *Thermopsis rhombifolia* [20].

Anabasis aphylla is capable of transforming aphyllidine (**140**) into aphylline (**141**) [112] (Fig. 12.24) and of metabolising the alkaloids to CO_2 [68, 69, 111]. Aphylline is not converted into other alkaloids [78]. Similar transformations are also found

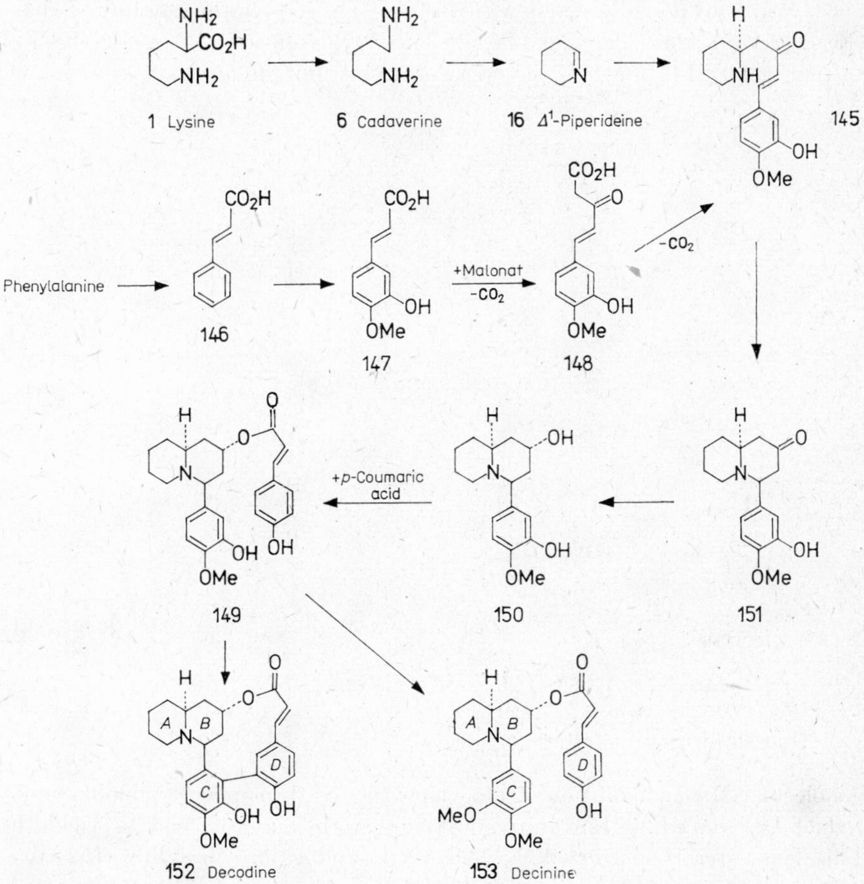

Fig. 12.26
Biosynthesis of decodine and decinine

for the matrine skeleton. Thus matrine can be converted into sophocarpine and sophoramine in *Sophora alopecuroides* and *Ammothamnus lehmanni*, and sophoridine, having a different configuration from matrine, can be transformed into 3-hydroxysophoridine [1a, 2, 84—86].

The quinolizidine system is also a constituent of some *Lythraceae* alkaloids such as cryogenine (**158**), lythrumine (**160**), decodine (**152**), decinine (**153**), vertine, and lythrine (enentiomers or isomers of **153**) [162]. But in experiments with *Heimia* *salicifolia* for cryogenine, vertine, and lythrine and with *Decodon verticillatus* for decodine and decinine it was found that lysine (**1**) is incorporated only into ring *A* *via* cadaverine (**6**) and Δ^1-piperideine (**16**) [52, 64a, 81]. The labelling pattern after feeding phenylalanine showed incorporation of two intact C_6-C_3 units yielding rings *B*, *C* and *D* [67, 80, 160, 161]. Pelletierine, earlier suggested as precursor, is not incorporated. In *Heimia salicifolia* a metabolic interconversion of different cryogenine derivatives has been found [30]. Δ^1-Piperideine (**16**) derived from lysine (**1**) *via* cadaverine (**6**) possibly reacts with the acid **148** to give the key intermediate (**145**) (Figs. 12.25, 12.26). The acid **148** is formed by extension of the side chain of the cinnamic acids (**146** and **147**) by a two-carbon unit donor such as malonyl-CoA.

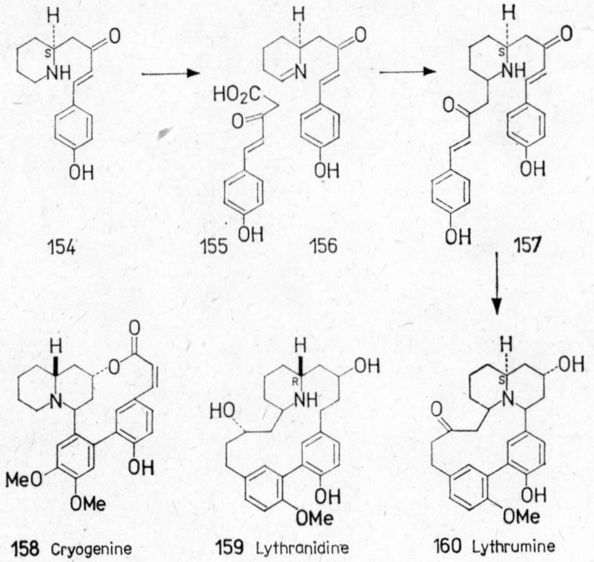

154 155 156 157

158 Cryogenine **159** Lythranidine **160** Lythrumine

Fig. 12.27
Biosynthesis of lythrumine

Intramolecular Michael addition of **145** yields the (*S*)-trans-phenylquinolizidinone (**151**), which is reduced to **150** and esterified with *p*-coumaric acid to yield the ester **149**. This ester is converted by phenolic coupling into decodine (**152**) (*o,o*-coupling) and decinine (**153**) (*o,p*-coupling). Compounds corresponding in structure to the phenylquinolizidinone (**151**), the phenylquinolizidinol (**150**), and its phenylpropanoid ester such as **149** have been found in *Heimia salicifolia* [162, 163]. A similar pathway is assumed for the alkaloids of the *Lythrum* group involving the intermediacy of a 2-substituted piperidine (**154**) (Fig. 12.27), analogous to **145**, which reacts as **156** with a second acid unit (**155**) to form **157** [67]. Following Michael addition, reduction and phenolic coupling yield lythrumine (**160**). For the biosynthesis of lythranidine (**159**), a piperidine alkaloid, the corresponding 2-substituted piperidine (**154**) (*R*-isomer) is transformed in the same way but without Michael addition.

12.3 References

[1] Abdusalamov, B. A.: Chim. Prir. Soedin. **1980**, 56; C. A. **92** (1980) 177497.
[1a] Abdusalamov, B. A.: Chim. Prir. Soedin. **1984**, 3; C. A. **100** (1984) 188 864.
[2] Abdusalamov, B. A.; Aslanov, K. A.; Sadykov, A. S.: Chim. Prir. Soedin. **1977**, 549; C. A. **87** (1977) 164322.
[3] Abdusalamov, B. A.; Aslanov, K. A.; Sadykov, A. S.: Biochimija **36** (1971) 290; C. A. **75** (1971) 45685.
[4] Abdusalamov, B. A.; Takanaev, A. A.; Sadykov, A. S.; Aslanov, K. A.: Uzb. Biol. Ž. **13** (1969) 10; C. A. **71** (1969) 27925.h.
[5] Addae-Mensah, I.; Torto, F. G.; Dimonyekra, C. I.; Baxter, I.; Sanders, J. K. M.: Phytochemistry **16** (1977) 757.
[6] Ahmad, V. U.; Basha, A.; Haque, W.: Z. Naturforsch. **33b** (1978) 347.
[6a] Ahmad, V. U.; Qazi, S.: Z. Naturforsch. **38b** (1983) 660.
[7] Aldag, R. W.; Joung, J. L.: Planta **95** (1970) 187.
[8] Battersby, A. R.; Murphy, R.; Staunton, J.: J. Chem. Soc. Perkin I **1982**, 449.
[9] Bell, A. E.: in: Bell, A. E.; Charlwood, B. V. (Eds.): Encyclopedia of Plant Physiology, N. S. Vol. 8. Springer-Verlag, Berlin 1980, p. 403; Progr. Phytochemistry **7** (1981) 171.
[10] Bevan, C. W. L.; Ogan, A. V.: Phytochemistry **3** (1964) 591.
[10a] Bleecker, A. B.; Romeo, J. T.: Phytochemistry **22** (1983) 1025.
[11] Bodansky, M.; Singler, G. F.; Bodansky, A.: J. Am. Chem. Soc. **95** (1973) 2352.
[12] Bohlmann, F.; Zdero, C.: Chem. Ber. **106** (1973) 1328.
[13] Braekman, J. C.; Gupta, R. N.; MacLean, D. B.; Spenser, J. D.: Canad. J. Chem. **50** (1972) 2591.
[14] Braekman, J. C.; Nyembo, L.; Symoens, J. J.: Phytochemistry **19** (1980) 803.
[14a] Brown, W. V.; Moore, B. P.: Austral. J. Chem. **35** (1982) 1255.
[15] Bruneton, J.; Cavé, A.: Tetrahedron Lett. **1975**, 739.
[16] Bruneton, J.; Cavé, A.; Paris, R. R.: Plant. Med. Phytother. **9** (1975) 21; C. A. **83** (1975) 28399.
[17] Castillo, M.; Gupta, R. N.; Ho, Y. K.; MacLean, D. B.; Spenser, I. D.: J. Am. Chem. Soc. **92** (1970) 1074.
[18] Castillo, M.; Gupta, R. N.; Ho, Y. K.; MacLean, D. B.; Spenser, I. D.: Canad. J. Chem. **48** (1970) 2911.
[19] Castillo, M.; Gupta, R. N.; MacLean, D. B.; Spenser, I. D.: Canad. J. Chem. **48** (1970) 1893.
[20] Cho, Y. D.; Martin, R. O.: Canad. J. Biochem. **49** (1971) 971.
[21] Cho, Y. D.; Martin, R. O.: Hanguk Saenghwa HaKhoe Ch. **10** (1977) 147; C. A. **90** (1979) 51492.
[22] Cho, Y. D.; Martin, R. O.; Anderson, J. N.: J. Am. Chem. Soc. **93** (1971) 2087.
[23] Clevenstine, E. C.; Broquist, H. P.; Harris, T. M.: Biochemistry **18** (1979) 3658.
[24] Clevenstine, E. C.; Walter, P.; Harris, T. M.; Broquist, H. P.: Biochemistry **18** (1979) 3663.
[25] Daly, J. W.: Fortschr. Chem. Org. Naturst. **41** (1982) 205.
[25a] Daly, J. W.; Tokuyama, T.; Fujiwara, T.; Highet, R. J.; Karle, I. L.: J. Am. Chem. Soc. **102** (1980) 830.
[26] Dardenne, G.; Casimir, J.; Sørensen, H.: Phytochemistry **13** (1974) 1515.
[27] Debrauwere, J.; Verzele, M.: Bull. Soc. Chim. Belg. **84** (1975) 167.
[28] Dietrich, S. M. C.; Martin, R. O.: J. Am. Chem. Soc. **80** (1968) 1921.
[29] Dietrich, S. M. C.; Martin, R. O.: Biochemistry **8** (1969) 4163.
[30] Dobberstein, R. H.; Edwards, J. M.; Schwarting, A.: Phytochemistry **14** (1975) 1769.
[31] Fairbairn, J. W.; Ali, A. A. E. R.: Phytochemistry **7** (1968) 1599.
[32] Fangmeier, N.; Leistner, E.: J. Biol. Chemistry **255** (1980) 10205.
[33] Faugeras, G.: Ann. Pharm. France **28** (1971) 241.
[34] Filko, R. B.; De Souza, M. P.; Mattos, M. E. O.: Phytochemistry **20** (1981) 345.
[35] Fitch, W. L.; Dolinger, P. M.; Djerassi, C.: J. Org. Chem. **39** (1974) 2974.

[36] Fowden, L.: Progr. Phytochem. **2** (1970) 203.
[37] Franck, B.: Chem. Ber. **91** (1958) 2803.
[38] Gerdes, H. J.; Leistner, E.: Phytochemistry **18** (1979) 771.
[39] Gilbertson, T. J.: Phytochemistry **11** (1972) 1737.
[40] Golebiewski, W. M.; Horsewood, P.; Spenser, I. D.: Chem. Commun. **1976**, 217.
[41] Golebiewski, W. M.; Spenser, I. D.: J. Am. Chem. Soc. **98** (1976) 6726.
[41a] Golebiewski, W. M.; Spenser, I. D.: Chem. Commun. **1983**, 1509; J. Am. Chem. Soc. **106** (1984) 1441, 7925.
[42] Gottlieb, D.; Shaw, P. D. (Eds.): Antibiotics. Springer-Verlag Berlin/Heidelberg/New York 1967.
[43] Greger, H.; Grenz, M.; Bohlmann, F.: Phytochemistry **20** (1981) 2579.
[44] Groß, D.: Fortschr. Chem. Org. Naturst. **29** (1971) 1.
[45] Grove, J. A.; Gilbertson, T. J.; Hammerstedt, R. H.; Henderson, L. M.: Biochim. Biophys. Acta **184** (1969) 329.
[46] Guengerich, F. P.; Broquist, H. P.: Biochemistry **12** (1973) 4270.
[47] Guengerich, F. P.; Broquist, H. P.: Bioorg. Chem. **1978** (2) 97.
[48] Guengerich, F. P.; DiMari, S. J.; Broquist, H. P.: J. Am. Chem. Soc. **96** (1973) 2055.
[49] Guengerich, F. P.; Snyder, J. J.; Broquist, H. P.: Biochemistry **12** (1973) 4264.
[50] Gupta, R. N.: Lloydia **31** (1968) 318.
[51] Gupta, R. N.; Castillo, M.; MacLean, D. B.; Spenser, I. D.; Wrobel, J. T.: J. Am. Chem. Soc. **90** (1968) 1360.
[52] Gupta, R. N.; Horsewood, P.; Koo, S. H.; Spenser, I. D.: Canad. J. Chem. **57** (1979) 1606.
[53] Gupta, R. N.; Spenser, I. D.: Chem. Commun. **1966**, 893.
[54] Gupta, R. N.; Spenser, I. D.: Canad. J. Chem. **45** (1967) 1275.
[55] Gupta, R. N.; Spenser, I. D.: Chem. Commun. **1968**, 85.
[56] Gupta, R. N.; Spenser, I. D.: Phytochemistry **8** (1969) 1937.
[57] Gupta, R. N.; Spenser, I. D.: J. Biol. Chemistry **244** (1969) 88.
[58] Gupta, R. N.; Spenser, I. D.: Phytochemistry **9** (1970) 2329.
[59] Gupta, R. N.; Spenser, I. D.: Canad. J. Chem. **49** (1971) 384.
[60] Hänsel, R.; Leuckert, C.; Schulz, G.: Z. Naturforsch. **21b** (1966) 530.
[61] Hänsel, R.; Leuschke, A.; Gomez-Pompa, A.: Lloydia **38** (1975) 529.
[62] Hart, N. K.; Johns, S. R.; Lamberton, J. A.: Austral. J. Chem. **21** (1968) 1397, 2579.
[63] Hartmann, T.; Schoofs, G.; Wink, M.: FEBS Letters **115** (1980) 35.
[64] Hasse, K.; Berg, P.: Naturwissenschaften **44** (1957) 584.
[64a] Hedges, S. H.; Herbert, R. B.; Wormald, P. C.: Chem. Commun. **1983**, 144.
[65] Herbert, R. B.; Jackson, F. B.; Nicolson, I. T.: Chem. Commun. **1976**, 865.
[66] Ho, Y. K.; Gupta, R. N.; MacLean, D. B.; Spenser, I. D.: Canad. J. Chem. **49** (1971) 3352.
[67] Horsewood, P.; Golebiewski, W. M.; Wrobel, J. T.; Spenser, I. D.; Cohen, J. F.; Comer, F.: Canad. J. Chem. **57** (1979) 1615.
[68] Ibraeva, B. S.; Lovkova, M. Y.; Klyshev, L. K.: Izvest. Akad. Nauk SSSR, Ser. Biol. **1980**, 356; C. A. **93** (1980) 66209.
[69] Ibraeva, B. S.; Lovkova, M. Y.; Klyshev, L. K.: Izvest. Akad. Nauk SSSR, Ser. Biol. **1981**, 97; C. A. **94** (1981) 99864.
[70] Impellizzeri, G.; Mangiafico, S.; Oriente, G.; Piatelli, M.; Sciuto, S.; Fattorusso, E.; Magno, S.; Santacroce, C.; Sica, D.: Phytochemistry **14** (1975) 1549.
[71] Iwai, Y.; Kumano, K.; Omura, S.: Chem. Pharm. Bull. **26** (1978) 736.
[71a] Jones, T. H.; Blum, M. S.; Fales, H. M.: Tetrahedron **38** (1982) 1949.
[72] Joshi, B. S.; Viswanathan, N.; Gawad, D. H.; Balakrishnan, V.: Helv. Chim. Acta **58**; (1975) 2295.
[73] Joshi, B. S.; Viswanathan, N.; Gawad, D. H.; Balakrishnan, V.; Philipsborn, W. von; Quick, A.: Experientia **31** (1975) 880.
[74] Keogh, M. F.; O'Donovan, D. G.: J. Chem. Soc. C **1970**, 1792.
[75] Keogh, M. F.; O'Donovan, D. G.: J. Chem. Soc. C **1970**, 2470.
[76] Kharatyan, S.; Puza, M.; Spizek, J.; Dolezilova, L.; Vanek, Z.: Chem. & Ind. **1963**, 1038.

[76a] Klyshev, L. K.: Izv. Akad. Nauk Kaz. SSR, Ser. Khim. **1984**, 22.
[77] Klyshev, L. K.; Moissev, R. K.; Klimentjeva, N. J.; Ibraeva, B. S.: Izvest. Akad. Nauk Kaz. SSR, Ser. Biol. **15** (1977) 1; C. A. **87** (1977) 164318.
[78] Klyshev, L. K.; Moissev, R. K.; Klimentjeva, N. J.; Ibraeva, B. S.: Vestn. Akad. Nauk Kaz. SSR, Ser. Biol. **15** (1977) 67; C. A. **87** (1977) 130602.
[79] Knöfel, D.; Schütte, H. R.: Z. Pflanzenphysiol. **64** (1971) 387.
[80] Koo, S. H.; Comer, F.; Spenser, I. D.: Chem. Commun. **1970**, 897.
[81] Koo, S. H.; Gupta, R. N.; Spenser, I. D.; Wrobel, J. T.: Chem. Commun. **1970**, 396.
[82] Korzan, P.; Gilbertson, T. J.: Phytochemistry **13** (1974) 435.
[83] Kumada, Y.; Naganawa, H.; Hamada, M.; Takeuchi, T.; Umezawa, H.: J. Antibiotics **27** (1974) 726.
[84] Kushmuradov, Y. K.; Aslanov, K. A.; Schütte, H. R.; Kuchkarov, S.: Chim. Prir. Soedin. **1977**, 244; C. A. **87** (1977) 98901.
[85] Kushmuradov, Y. K.; Gross, D.; Schütte, H. R.: Phytochemistry **11** (1972) 3441.
[86] Kushmuradov, Y. K.; Schütte, H. R.; Aslanov, K. A.; Kuchkarov, S.: Chim. Prir. Soedin. **1976**, 776; C. A. **86** (1977) 86200.
[87] Kushmuradov, Y. K.; Schütte, H. R.; Aslanov, K. A.; Kuchkarov, S.: Chim. Prir. Soedin. **1977**, 247; C. A. **87** (1977) 98902.
[88] Lazurjewski, G.; Terentjewa, I.: Heterocycles **4** (1976) 1783.
[89] Leeper, F. J.; Grue-Soerensen, G.; Spenser, I. D.: Canad. J. Chem. **59** (1981) 106.
[90] Leete, E.: J. Am. Chem. Soc. **85** (1963) 3523.
[91] Leete, E.: J. Am. Chem. Soc. **86** (1964) 2509.
[92] Leete, E.: J. Am. Chem. Soc. **91** (1969) 1697.
[93] Leete, E.: In: Bell, E. A.; Charlwood, B. V. (Ed.): Encyclopedia of Plant Physiology, N. S. Vol. 8. Springer-Verlag, Berlin 1980, p. 65.
[94] Leete, E.: Accounts Chem. Res. **3** (1971) 100.
[95] Leete, E.; Adityachaudhury, N.: Phytochemistry **6** (1967) 219.
[96] Leete, E.; Carver, R. A.: J. Org. Chem. **40** (1975) 2151.
[97] Leete, E.; Chedekel, M. R.: Phytochemistry **11** (1972) 2751.
[98] Leete, E.; Juneau, K. N.: J. Am. Chem. Soc. **91** (1969) 5614.
[99] Leete, E.; Lechleiter, J. C.; Carver, R. A.: Tetrahedron Lett. **1975**, 3779.
[100] Leete, E.; Olson, J. O.: Chem. Commun. **1970**, 1651.
[101] Leete, E.; Olson, J. O.: J. Am. Chem. Soc. **94** (1972) 5472.
[102] Leistner, E.; Gupta, R. N.; Spenser, I. D.: J. Am. Chem. Soc. **95** (1973) 4040.
[103] Leistner, E.; Spenser, I. D.: J. Am. Chem. Soc. **95** (1973) 4715.
[104] Leistner, E.; Spenser, I. D.: Chem. Commun. **1975**, 378.
[105] Liebisch, H. W.: Abh. Dtsch. Akad. Wiss. Berlin, Klasse Chem., Geol., Biol., **1966**, No. 3, p. 525.
[106] Liebisch, H. W.: Fortschr. Chem. Forsch. **9** (1968) 534.
[107] Liebisch, H. W.: In: Mothes, K.; Schütte, H. R. (Eds.): Biosynthese der Alkaloide. Deutscher Verlag der Wissenschaften, Berlin 1969, p. 275.
[108] Liebisch, H. W.; Marekov, N.; Schütte, H. R.: Z. Naturforsch. **23b** (1968) 1116.
[109] Lloyd, H. A.: Tetrahedron Lett. **1965**, 1761, 4537.
[110] Loder, J. W.; Moorhouse, A.; Russel, G. B.: Austral. J. Chem. **22** (1969) 1531.
[111] Lovkova, M. Y.; Ibraeva, B. S.; Klyshev, L. K.: Prikl. Biochim. Mikrobiol. **14** (1978) 635; C. A. **89** (1978) 160230.
[112] Lovkova, M. Y.; Ibraeva, B. S.; Klyshev, L. K.: Prikl. Biochim. Mikrobiol. **14** (1978) 818; C. A. **90** (1979) 118123.
[113] MacConnell, J. G.; Blum, M. S.; Fales, H. M.: Tetrahedron **26** (1971) 1129.
[114] MacLean, D. B.: In: Manske, R. H. F. (Ed.): The Alkaloids. Vol. 14 p. 347. Academic Press, New York 1973.
[115] Maeda, M.; Hasegawa, Y.; Hashimoto, H.: Agric. Biol. Chem. **44** (1980) 2725.
[116] Mahrenholz, R. M.; Carew, D. P.: Lloydia **29** (1966) 376.
[117] Marshall, W. D.; Nguyen, T. T.; MacLean, D. B.; Spenser, I. D.: Canad. J. Chem. **53** (1975) 41.

[118] Mazelis, M.; Miflin, B. J.; Pratt, H. M.: FEBS-Letters **64** (1976) 197.
[119] Medina, E.; Spiteller, G.: Liebigs Ann. Chem. **1981**, 538.
[120] Michel, K. H.; Sandberg, F.; Haglid, F.; Norin, T.: Acta Pharm. Suecica **4** (1967) 97.
[121] Miflin, B. J.; Lea, P. J.: Ann. Rev. Plant Physiol. **28** (1977) 295.
[122] Miller, D. L.; Rodwell, V. W.: J. Biol. Chem. **246** (1971) 2758.
[123] Mills, W. R.; Wilson, U. G.: Planta **142** (1978) 153.
[124] Miyakado, M.; Nakayama, I.; Yoshioka, H.: Agric. Biol. Chem. **44** (1980) 1701.
[125] Mody, N. V.; Henson, R.; Hedin, P. A.; Kokpol, V.; Miles, D. H.: Experientia **32** (1976) 829.
[126] Moriyama, Y.; Duan-Huynh, D.; Monneret, C.; Khuong-Huu, Q.: Tetrahedron Lett. **1977**, 825.
[127] Mothes, K.; Schütte, H. R.; Simon, H.; Weygand, F.: Z. Naturforsch. **14 b** (1959) 49.
[128] Murakoski, I.; Kidoguchi, E.; Haginiwa, J.; Ohmiya, S.; Higashiyama, K.; Otomasu, H.: Phytochemistry **20** (1981) 1407.
[129] Murakoski, I.; Kuramoto, H.; Haginiwa, J.; Fowden, L.: Phytochemistry **11** (1972) 177.
[130] Murakoski, I.; Ohmiya, S.; Haginiwa, J.: Chem. Pharm. Bull. **19** (1971) 2655.
[131] Nowacki, E.: Genetica Polon. **5** (1964) 189.
[132] Nowacki, E.; Byerrum, R. U.: Biochem. biophysic. Res. Commun. **7** (1962) 58.
[133] Nowacki, E.; Byerrum, R. U.: Bull. Acad. Polon. Sci., Ser. Sci. biol. **12** (1964) 489.
[134] Nowacki, E. K.; Waller, G. R.: Phytochemistry **14** (1975) 155.
[135] Nowacki, E. K.; Waller, G. R.: Phytochemistry **14** (1975) 161.
[136] Nowacki, E. K.; Waller, G. R.: Phytochemistry **14** (1975) 165.
[137] Nowacki, E. K.; Waller, G. R.: Rev. Latinoamer. Quim. **8** (1977) 49.
[138] Nurimov, E.; Lovkova, M. Y.; Abdusalamov, B. A.: Prikl. Biochim. Mikrobiol. **13** (1977) 628; C. A. **87** (1977) 148764.
[139] O'Donovan, D. G.; Creedon, P. B.: Tetrahedron Lett. **1971**, 1341.
[140] O'Donovan, D. G.; Creedon, P. B.: J. Chem. Soc. C **1971**, 1604.
[141] O'Donovan, D. G.; Creedon, P. B.: J. Chem. Soc. Perkin I **1974**, 2524.
[142] O'Donovan, D. G.; Forde, T.: J. Chem. Soc. C **1971**, 2889.
[143] O'Donovan, D. G.; Keogh, M. F.: Tetrahedron Lett. **1968**, 265.
[144] O'Donovan, D. G.; Long, D. J.; Forde, E.; Guary, P.: J. Chem. Soc. Perkin I **1975**, 415.
[145] Onda, M.; Konda, Y.; Narimatsu, Y.; Tanaka, H.; Awaya, J.; Omura, S.: Chem. Pharm. Bull. **22** (1974) 2916.
[146] Ott-Longoni, R.; Viswanathan, N.; Hesse, M.: Helv. Chim. Acta **63** (1980) 2119.
[147] Parry, R. J.: Tetrahedron Lett. **1974**, 307.
[148] Parry, R. J.: Chem. Commun. **1975**, 144.
[149] Pasteels, J, M.; Deroe, C.; Tursch, B.; Braekman, J. C.; Daloze, D.; Hootele, C.: J. Insect Physiol. **19** (1973) 1771; C. A. **79** (1973) 144274n.
[149a] Rana, J.; Robins, D. J.: Chem. Commun. **1983**, 1335.
[149b] Rana, J.; Robins, D. J.: Chem. Commun. **1984**, 81; J. Chem. Res., Synop. **1984**, 164.
[150] Roberts, M. F.: Phytochemistry **10** (1971) 3057.
[151] Roberts, M. F.: Phytochemistry **13** (1974) 1841.
[152] Roberts, M. F.: Phytochemistry **13** (1974) 1847.
[153] Roberts, M. F.: Phytochemistry **14** (1975) 2393.
[154] Roberts, M. F.: Phytochemistry **16** (1977) 1381.
[155] Roberts, M. F.: Phytochemistry **17** (1978) 107.
[156] Roberts, M. F.: Planta Med. **39** (1980) 216.
[157] Roberts, M. F.: Plant Cell Reports **1** (1981) 10.
[158] Roberts, M. F.; Brown, R. T.: Phytochemistry **20** (1981) 447.
[159] Robinson, R.: The Structural Relations of Natural Products. Clarendon Press, Oxford 1955.
[159a] Romeo, J. T.; Swain, L. A.; Bleecker, A. B.: Phytochemistry **22** (1983) 1615.
[160] Rother, A.; Schwarting, A. E.: Chem. Commun. **1969**, 1411.
[161] Rother, A.; Schwarting, A. E.: Phytochemistry **11** (1972) 2475.

[162] Rother, A.; Schwarting, A. E.: Lloydia **38** (1975) 477.
[163] Rother, A.; Schwarting, A. E.: Phytochemistry **17** (1978) 305.
[164] Sankawa, V.; Ebizuka, Y.; Yamasaki, K.: Phytochemistry **16** (1977) 561.
[165] Şankawa, V.; Yamasaki, K.; Ebizuka, Y.: Tetrahedron Lett. **1974**, 1867.
[166] Schenk, W.; Schütte, H. R.; Mothes, K.: Flora **152** (1962) 590.
[166a] Schoofs, G.; Teichmann, S.; Hartmann, T.; Wink, M.: Phytochemistry **22** (1983) 65–69.
[167] Schütte, H. R.: In: Gröger, D.; Schröter, H. B.; Schütte, H. R. (Eds.): Festschrift K. Mothes. Gustav Fischer Verlag, Jena 1965, p. 435.
[168] Schütte, H. R.: In: Mothes, K.; Schütte, H. R. (Eds.): Biosynthese der Alkaloide. Deutscher Verlag der Wissenschaften, Berlin 1969, p. 324.
[169] Schütte, H. R.; Aslanow, H.; Schäfer, C.: Arch. Pharmaz. **295** (1962) 34.
[170] Schütte, H. R.; Bohlmann, F.; Reusche, W.: Arch. Pharmaz. **294** (1961) 610.
[171] Schütte, H. R.; Hindorf, H.: Z. Naturforsch. **19b** (1964) 855.
[172] Schütte, H. R.; Hindorf, H.: Liebigs Ann. Chem. **685** (1965) 187.
[173] Schütte, H. R.; Hindorf, H.; Mothes, K.; Hübner, G.: Liebigs Ann. Chem. **680** (1964) 93.
[174] Schütte, H. R.; Kelling, K. L.; Knöfel, D.; Mothes, K.: Phytochemistry **3** (1964) 249.
[175] Schütte, H. R.; Knöfel, D.: Z. Pflanzenphysiol. **57** (1967) 188.
[176] Schütte, H. R.; Knöfel, D.: Z. Pflanzenphysiol. **59** (1968) 80.
[177] Schütte, H. R.; Knöfel, D.; Heyer, O.: Z. Pflanzenphysiol. **55** (1966) 110.
[178] Schütte, H. R.; Lehfeldt, J.: Z. Naturforsch. **19b** (1964) 1085.
[179] Schütte, H. R.; Lehfeldt, J.: J. prakt. Chem. **24** (1964) 143.
[180] Schütte, H. R.; Lehfeldt, J.; Hindorf, H.: Liebigs Ann. Chem. **685** (1965) 194.
[181] Schütte, H. R.; Mothes, K.: Z. Chem. **3** (1963) 278.
[182] Schütte, H. R.; Nowacki, E.: Naturwissenschaften **46** (1959) 493.
[183] Schütte, H. R.; Nowacki, E.; Kovacs, H. P.; Liebisch, H. W.: Arch. Pharmaz. **296** (1963) 438.
[184] Schütte, H. R.; Nowacki, E.; Schäfer, C.: Arch. Pharmaz. **295** (1962) 20.
[185] Schütte, H. R.; Sandke, G.; Lehfeldt, J.: Arch. Pharmaz. **297** (1964) 118.
[186] Schütte, H. R.; Schäfer, C.: Naturwissenschaften **48** (1961) 669.
[187] Schütte, H. R.; Seelig, G.: Z. Naturforsch. **22b** (1967) 824.
[188] Schütte, H. R.; Seelig, G.: Liebigs Ann. Chem. **711** (1968) 221.
[189] Schütte, H. R.; Seelig, G.; Knöfel, D.: Z. Pflanzenphysiol. **63** (1970) 393.
[190] Schwarting, A. E.; Bobbitt, J. M.; Rother, A.; Atal, C. K.; Khanna, K. L.; Leary, J. D.; Walter, W. G.: Lloydia **26** (1963) 258.
[191] Shewry, P. R.; Fowden, L.: Phytochemistry **15** (1976) 1981.
[192] Shibata, S.; Sankawa, Ü.: Chem. & Ind. **1963**, 1161.
[193] Smalberger, T. M.; Rall, G. J. H.; De Waal, H. L.; Arndt, R. R.: Tetrahedron **24** (1968) 6417.
[194] Snieckus, V.: In: Manske, R. H. F. (Ed.): The Alkaloids. Vol. 14. Academic Press, New York 1973, p. 425.
[195] Soucek, M.; Schütte, H. R.: Angew. Chem. **74** (1962) 901.
[196] Steinegger, E.; Moser, C.; Weber, P.: Phytochemistry **7** (1968) 849.
[197] Steinegger, E.; Schlunegger, E.: Pharm. Acta Helv. **45** (1970) 369.
[198] Su, H. C. F.; Horvat, R.: J. Agric. Food Chem. **29** (1981) 115.
[199] Takanaev, A. A.; Abdusalamov, B. A.: Uzb. Biol. Ž. **15** (1971) 20; C. A. **75** (1971) 115946.
[200] Tang, C. S.: Phytochemistry **18** (1979) 651.
[201] Terashima, T.; Idaka, E.; Kishi, Y.; Goto, T.: Chem. Commun. **1973**, 75.
[202] Terashima, T.; Kuroda, Y.; Kaneko, Y.: Tetrahedron Lett. **1969**, 2535.
[203] Tiwari, H. P.; Penrose, W. R.; Spenser, I. D.: Phytochemistry **6** (1967) 1245.
[204] Tokuyama, T.; Uenoyama, K.; Brown, G.; Daly, J. W.; Witkop, B.: Helv. Chim. Acta **57** (1974) 2597.
[205] Tursch, B.; Braekman, J. C.; Daloze, D.: Experientia **32** (1976) 401.

[206] Tursch, B.; Braekman, J. C.; Daloze, D.; Hootele, C.; Losman, D.; Karlsson, R.; Pasteels, J. M.: Tetrahedron Lett. **1973**, 201.
[207] Tursch, B.; Daloze, D.; Braekman, J. C.; Hootele, C.; Pasteels, J. M.: Tetrahedron **31** (1975) 1541.
[208] Tursch, B.; Daloze, D.; Dupont, M.; Pasteels, J. M.; Tricot, M. C.: Experientia **27** (1971) 1380.
[209] Vanek, Z.; Puza, M.; Cudlin, J.; Dolezilova, L.; Vondracek, M.: Biochem. Biophys. Res. Commun. **17** (1964) 532.
[210] Wheeler, J. W.; Olubajo, O.; Storm, C. B.; Duffield, R. M.: Science **211** (1981) 1051.
[211] White, H. A.; Spencer, M.: Canad. J. Bot. **42** (1964) 1481.
[212] Wink, M.; Hartmann, T.: FEBS-Letters **101** (1979) 343.
[213] Wink, M.; Hartmann, T.: Planta Med. **40** (1980) 149.
[214] Wink, M.; Hartmann, T.: Plant Cell Reports **1** (1981) 6.
[215] Wink, M.; Hartmann, T.: Z. Pflanzenphysiol. **102** (1981) 337.
[215a] Wink, M.; Hartmann, T.: Z. Naturforsch. **37c** (1982) 369.
[215b] Wink, M.; Hartmann, T.: Plant Physiol. **70** (1982) 74.
[215c] Wink, M.; Hartmann, T.: Planta **156** (1982) 560.
[216] Wink, M.; Hartmann, T.; Schiebel, H. M.: Z. Naturforsch. **34c** (1979) 704.
[217] Wink, M.; Hartmann, T.; Witte, L.: Z. Naturforsch. **35c** (1980) 93.
[218] Wink, M.; Hartmann, T.; Witte, L.; Schiebel, H. M.: J. Nat. Products **44** (1981) 14.
[218a] Wink, M.; Heinen, H. J.; Vogt, H.; Schiebel, H. M.: Plant Cell Reports **3** (1984) 230.
[219] Wink, M.; Schiebel, H. M.; Witte, L.; Hartmann, T.: Planta med. **44** (1982) 15.
[220] Wink, M.; Witte, L.; Hartmann, T.; Theuring, C.; Volz, V.: Planta med. **48** (1983) 253.
[221] Wink, M.; Witte, L.; Schiebel, H. M.; Hartmann, T.: Planta med. **38** (1980) 238.

13 Alkaloids Derived from Nicotinic Acid

D. Gross

Nicotinic acid (**12**) is the common precursor of a large number of natural products possessing the pyridine nucleus. It is derived biogenetically by at least two different pathways which will be discussed here in relation to the biosynthesis of alkaloids derived from nicotinic acid and biogenetically related pyridine compounds. The pyridine ring of several natural products has been found to arise not from nicotinic acid but from acetate, glycerol, aspartate, lysine and other metabolically active intermediates.

13.1 Biosynthesis of Nicotinic Acid

13.1.1 The Tryptophan Pathway

Early extensive studies on the biosynthetic route to nicotinic acid (**12**) indicated that it is derived in animals and certain microorganisms from the protein amino acid, tryptophan (**1**). As illustrated in Fig. 13.1, the oxidative degradation of tryptophan leads to quinolinic acid (**7**), which has been found to serve as the universal intermediate of nicotinic acid (**12**) and many other naturally occurring pyridine compounds. Quinolinic acid (**7**) is inserted into the pyridine nucleotide cycle (Fig. 13.2), where it is converted to the pyridine nucleotides (**8**, **9**, **10**), nicotinamide (**11**) and nicotinic acid (**12**). These compounds represent the obligatory intermediates of the pyridine nucleotide cycle, which operates in much the same manner in a great variety of organisms and differs only in the biosynthesis of quinolinic acid (**7**) [31].

The biosynthesis of **7** and **12** by the tryptophan pathway takes place in animals and certain bacteria, fungi and algae, but not in *E. coli*, *B. subtilis*, *M. tuberculosis* and some other microorganisms. In higher plants also, nicotinic acid does not derive from tryptophan as shown by the non-incorporation of labelled tryptophan into trigonelline (**13**) or the alkaloids of *Nicotiana* and *Ricinus* (*cf.* reviews [4 a, 36, 37] and references therein).

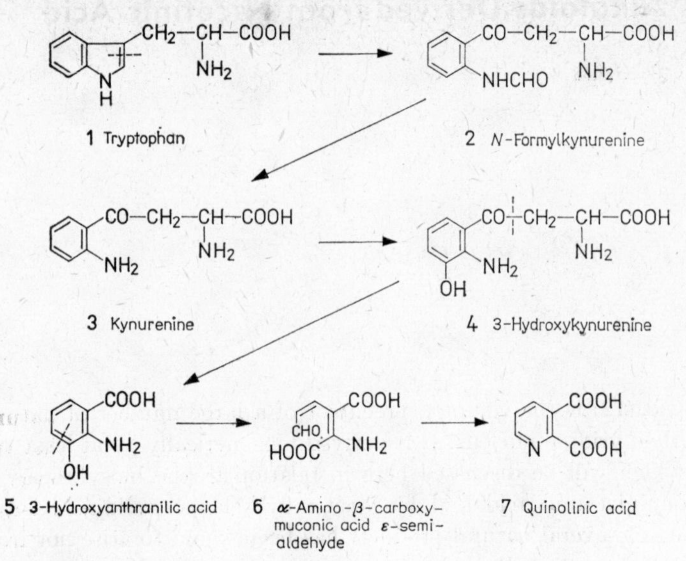

Fig. 13.1
Biosynthesis of quinolinic acid by the tryptophan pathway

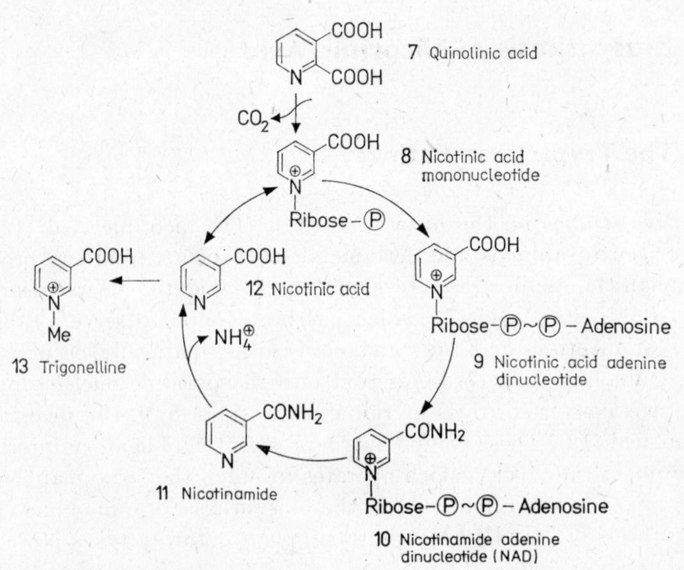

Fig. 13.2
The pyridine nucleotide cycle

13.1.2 Nicotinic Acid Derived by the $C_3 + C_4$ Pathway

The non-tryptophan pathway to nicotinic acid (**12**) has been studied predominantly with different bacteria (*M. tuberculosis* [2], *E. coli* [127, 128], *Serratia marcescens* [142], *Clostridium butylicum* [50] and others). Extensive experiments in which various [^{14}C]-labelled compounds have been supplied both to intact cells and cell extracts showed that C-2, C-3 and C-7 of **12** are derived from C-2, C-3 and C-4 of a four-carbon compound related to succinic acid or aspartic acid (**15**) (Fig. 13.3). For example, [4-^{14}C]aspartic acid applied to *M. bovis* strain BCG, was incorporated into **12** to give label only in the carboxyl group [119], whereas nicotinic acid formed from [3-^{14}C]aspartate in *M. tuberculosis* was found to contain all of the ^{14}C in the 3 position [2]. After administration of [1,4-^{14}C, ^{15}N]aspartate to *M. bovis* strain BCG the carboxyl group of **12** still possessed all the radioactivity; the ratio ^{14}C/^{15}N was approximately as expected, assuming that one molecule of aspartic acid, including its amino nitrogen, is incorporated into **12** and that the α-carboxyl group is lost [41].

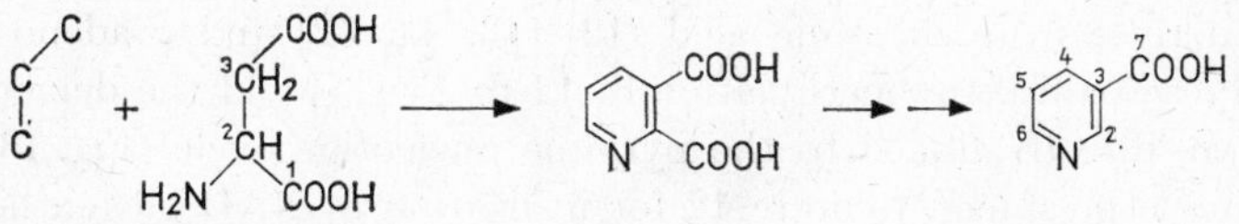

14 C_3 unit **15** Aspartic acid **7** Quinolinic acid **12** Nicotinic acid

Fig. 13.3
Biosynthesis of quinolinic acid and nicotinic acid by the $C_3 + C_4$ pathway

The carbons 4, 5 and 6 of **12** are derived from a three-carbon unit related to glycerol (**14**) as shown by the incorporation of [^{14}C]-labelled glycerols and metabolically related C_3 compounds **14** into **12** in *Mycobacterium* species [2, 39, 41], *E. coli* [50, 128] and anaerobically growing yeast [1, 45]. Similar results on the biosynthesis of quinolinic acid (**7**) were obtained by *in vitro* experiments with cell-free extracts of *E. coli* [12].

Quinolinic acid is the universal intermediate of both the tryptophan (Fig. 13.1) and the non-tryptophan pathways (Fig. 13.3). At present no metabolite in the conversion of aspartate (**15**) and the glycerol equivalent **14** into **12** has been identified, the mechanism of the $C_3 + C_4$ pathway(s) is therefore still unknown.

A different mechanism for the biosynthesis of **12** has been found in *Clostridium butylicum* [139, 143]. Its cell extracts are able to formylate aspartate (**15**) to *N*-formylaspartate (**15a**), which acts as an intermediate of **12**; the carbons 7, 3, 2 and 6 of nicotinic acid are therefore derived from C-2, 3, 4 and the formyl carbon of *N*-formylaspartate (Fig. 13.4).

There are few studies in which the biosynthesis of free nicotinic acid in higher plants has been investigated (*e.g.* [14 a, 52], for discussion *cf.* [4a]), but a large volume of biosynthetic work confirms that the pyridine ring of numerous derivatives of nicotinic acid and most pyridine alkaloids is derived from $C_3 + C_4$-formed quinolinic acid (**7**) *via* nicotinic acid (**12**) as discussed below.

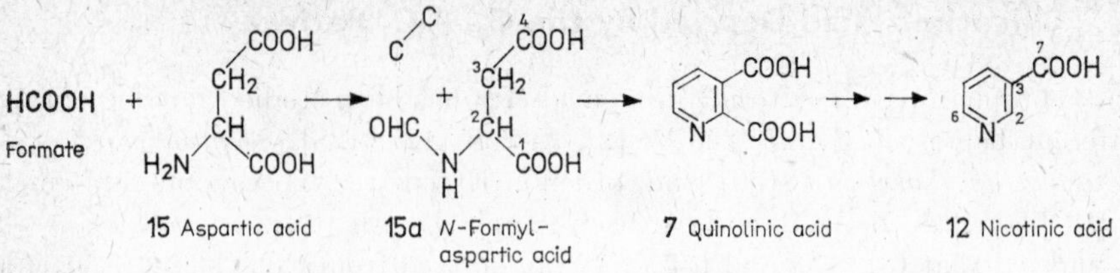

HCOOH + → + → → →

Formate

15 Aspartic acid 15a *N*-Formyl-aspartic acid 7 Quinolinic acid 12 Nicotinic acid

Fig. 13.4
Biosynthesis of nicotinic acid in cell-free extracts of *Clostridium butylicum*

13.2 Derivatives of Nicotinic Acid

13.2.1 Trigonelline

Trigonelline (**13**) found in various higher plants and in certain plant cell suspension cultures derives from nicotinic acid (**12**) [43, 44, 52] and *S*-adenosyl methionine [4, 52]. Presumably, the methylation of **12** to **13** [14] and the demethylation of **13** [44, 53] are directly linked to the pyridine nucleotide cycle (Fig. 13.2). Therefore, trigonelline is presumed to store **12** for utilization in NAD biosynthesis within the metabolic cycle of the plant.

13.2.2 Glycosides

The glucosyl esters of nicotinic acid, buchanine (**19**), and 1,3,6-O-trinicotinoyl-*α*-*D*-glucopyranose, have been isolated from *Cryptolepis buchanani* [24]; however, their biosyntheses have not yet been analysed. [6-^{14}C]Nicotinic acid applied to roots and leaves of tobacco plants is converted to its *N*-glucoside (**16**) and to 6-hydroxy-nicotinic acid, both representing major products of nicotinic acid metabolism in that plant, but not involved in the direct route of nicotine biosynthesis [115]. The specific incorporation of [4-^{14}C]aspartic acid (**15**) into **16** in *N. tabacum* confirms that the carbon atoms 2, 3 and 4 of aspartate are utilized in the biosynthesis of nicotinic acid [140] as shown in Fig. 13.3.

Plant tissue cultures have been used to study the metabolism of **12**. After the application of nicotinic acid to parsley cell suspension cultures, it was not only decarboxylated and degraded by ring fission but also converted in high yield to nicotinic acid-*N*-*α*-*L*-arabinoside (**17**), which has also been isolated from cell suspension cultures of *Daucus carota*, *Nicotiana tabacum* and *N. glauca* [66, 67, 68]. Surprisingly, in cell cultures of *Phaseolus aureus*, *Glycine max*, *Cicer arietinum* and *Chenopodium rubrum*, nicotinic acid (**12**) and nicotinamide (**11**) were mainly converted to trigonelline (**13**) [44, 67]. Therefore, it is postulated that trigonelline (**13**) and nicotinic acid-*N*-arabinoside (**17**) occur alternately in plant cell cultures, but both seem to act as a reserve form for nicotinic acid.

13.2.3 Nicotianine

Nicotianine (**18**), found in all organs of tobacco plants, is a pyridinium-containing *L*-amino acid [125] that is also derived from **12** as shown by the incorporation of [6-^{14}C]nicotinic acid administered to leaves and roots of *N. tabacum* [115].

13.2.4 4- and 5-Substituted Nicotinic Acids

The *Rubiaceae Nauclea diderrichii* contains four closely related pyridine alkaloids (**20**–**23**, Fig. 13.5) which were identified as 5-substituted derivatives of methyl nicotinate [108]. At present no information is available on their biosynthesis, but from the occurrence with dihydro-β-carboline alkaloids linked at C-1 to the 5 position of methyl nicotinate, a biogenetic relationship of the pyridine moieties of these alkaloids may exist. The biosynthetic pathway to methyl 4-aminonicotinate (**24**) occurring in *Fontanesia phillyreoides* and isolated with the artificial monoterpene alkaloid fontaphillin (**25**) [8] has not yet been suggested.

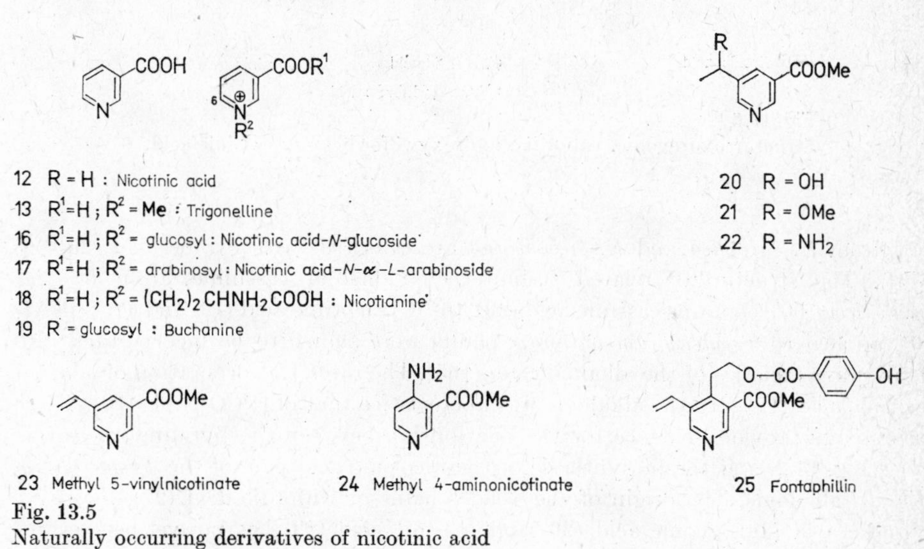

Fig. 13.5
Naturally occurring derivatives of nicotinic acid

13.3 Alkaloids Containing Nicotinic and Substituted Nicotinic Acids

13.3.1 *Celastraceae* Alkaloids

Besides the simple derivatives of nicotinic acid (**16**–**25**) shown in Fig. 13.5, a number of structurally related sesquiterpene alkaloids from the family *Celastraceae* are known which are all polyesters of hydroxylated agarofurans and contain either a nicotinic or substituted nicotinic acid nucleus (*cf.* [146]). For example, maytoline, maytine and maytolidine from *Maytenus ovatus* and *M. serrata* [62, 63], cathidine D and the cathedulins 3, 4, 5 and 6 from *Catha edulis* [6, 10], celapanine, celapanigine and celapagine from *Celastrus paniculatus* [151] are characterized by the presence of a nicotinoyl residue. There is no information available on the steps of biosynthesis.

The nicotinic acids isolated from the more complex *Celastraceae* alkaloids, found in *Euonymus europaea* and *E. alatus* and in *Tripterygium wilfordii*, all contain a C_5 carboxylic acid substituent at the 2 position (Fig. 13.6). The *Tripterygium wilfordii* alkaloids, wilforine, wilforgine, wilfordine, wilfortrine (*cf.* [146]) and isoevonine, alatamine and wilfordine [23, 160] from *E. europaea* and *E. alatus*, contain wilfordic acid (**27**) and hydroxywilfordic acid (**28**), while evonine and related alkaloids from *E. europaea* and *E. sieboldiana* are esterified with evonic acid (**26**) (*cf.* [146]). The structurally related sesquiterpene alkaloid cassinine, from *Cassine matabelica* [150], contains cassinic acid (**29**). [6-^{14}C]Nicotinic acid (**12**) and [^{14}CO]NAD (**10**), applied to *Tripterygium wilfordii* plants were shown to be incorporated into the pyridine moieties of the alkaloid esters [65]. The rate of incorporation of labelled nicotinic acid (**12**) into the alkaloids was about twice that of [^{14}CO]NAD (**10**). These experiments provide evidence for the relationship between the pyridine nucleotide cycle (Fig. 13.2) and the biosynthesis of the nicotinic acid part of the *Tripterygium wilfordii* alkaloids. The origin of the α-side-chains of wilfordic acid (**27**), hydroxywilfordic acid (**28**), evonic acid (**26**) and cassinic acid (**29**) has not yet been established.

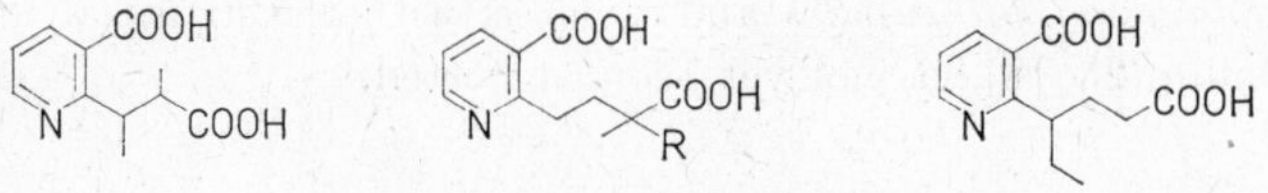

26 Evonic acid

27 R=H : Wilfordic acid
28 R=OH : Hydroxywilfordic acid

29 Cassinic acid

Fig. 13.6
Structures of evonic, wilfordic, hydroxywilfordic, and cassinic acid

13.3.2 Steroidal Alkaloids

Some steroidal alkaloids such as epipachysamine B from *Pachysandra terminalis* [55], rostratine and dihydrorostratine from *Marsdenia rostrata* [30, 148] or condurangamine A and B from the bark of *Cortex condurango* [129], also contain a nicotinoyl residue. The biosynthetic pathways of these alkaloids have also not yet been clarified, but it may be assumed that the esterification of nicotinic acid occurs at a late stage in their biosynthesis.

13.4 Alkaloids Derived from Nicotinic Acid

13.4.1 Pyridone Alkaloids

Several simple α-pyridone alkaloids are also derived biogenetically from nicotinic acid (**12**), *e.g.* ricinidine (**30**) from *Trewia nudiflora* [29], ricinine (**31**) from the castor bean plant *Ricinus communis*, 1-methyl-3-carbomethoxy-6-pyridone (**32**) from

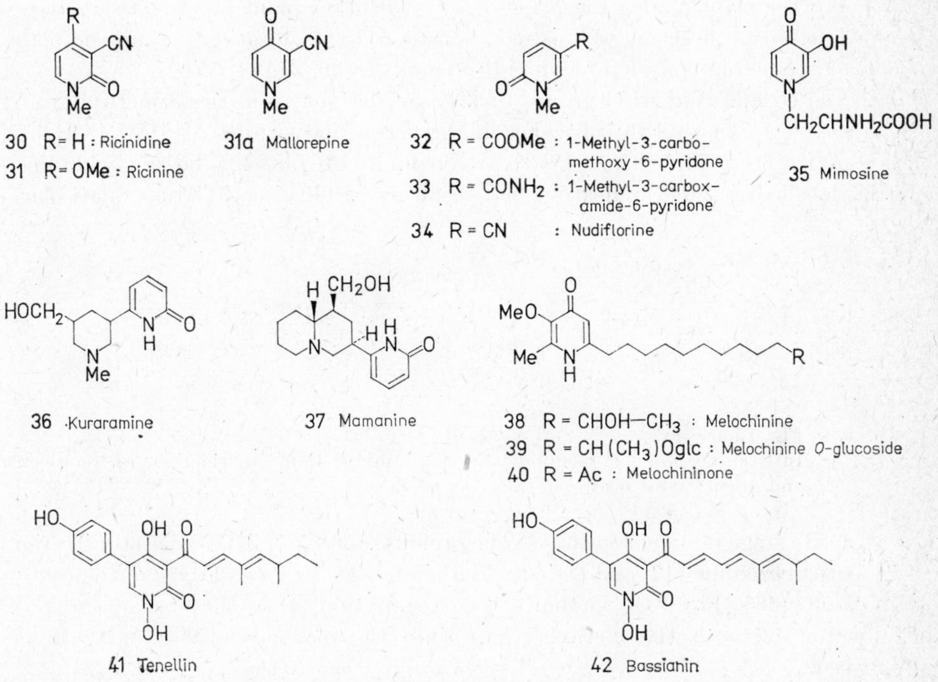

30 R = H : Ricinidine
31 R = OMe : Ricinine
31a Mallorepine
32 R = COOMe : 1-Methyl-3-carbomethoxy-6-pyridone
33 R = $CONH_2$: 1-Methyl-3-carboxamide-6-pyridone
34 R = CN : Nudiflorine
35 Mimosine
36 Kuraramine
37 Mamanine
38 R = $CHOH-CH_3$: Melochinine
39 R = $CH(CH_3)Oglc$: Melochinine *O*-glucoside
40 R = Ac : Melochininone
41 Tenellin
42 Bassianin

Fig. 13.7
Naturally occurring pyridones

Ampelocera ruizzii [9], 1-methyl-3-carboxamide-6-pyridone (**33**) [135], nudiflorine (**34**) [122, 135] from *Trewia nudiflora* (Fig. 13.7) and the γ-pyridone alkaloid mallorepine (**31a**) from *Mallotus repandus* [47]. The amino acid mimosine (**35**), however, is derived from lysine (*cf.* chapter 12.1.5). Kuraramine (**36**) from *Sophora flavescens* [123] and mamanine (**37**) from *S. chrysophylla* [54] may be metabolites of the biosynthetic pathway of the quinolizidine alkaloids (*cf.* chapter 12.2).

Melochinine (**38**), its *O*-glucoside (**39**) and melochininone (**40**) recently isolated from leaves of *Melochia pyramidata* are the first representatives of a so far unknown class of pyridone alkaloids [109]. There is no definite evidence regarding their biosynthesis, but it appears that these alkaloids arise from acetate units. This assumption is supported by the fact that structurally related piperidine alkaloids have also been found in higher plants, for example the *Cassia* and *Prosopis* alkaloids that are assumed to be acetate-derived (*cf*. chapter 12.1.6.4).

Tenellin (**41**) and bassianin (**42**), both produced as pigments by the fungus *Beauveria tenella* [25, 107], are also substituted α-pyridones. As they are not derived from nicotinic acid (**12**), their biosynthesis is described in chapter 14.13.5.

13.4.1.1 Ricinine

In *Ricinus communis* only one simple alkaloid, ricinine (**31**), occurs. As far as we know, no other plant among the *Euphorbiaceae* or other plant families accumulates ricinine. Its biosynthesis, as well as metabolism and translocation, within the plant have been studied in considerable detail (reviews *cf*. [36, 37, 84, 154]).

Both the *O*- and *N*-methyl groups of (**31**) are derived from the *S*-methyl group of methionine by methyl transfer. Feeding experiments with [3-^{14}C, $^{15}N-^{14}CH_3$]*N*-methyl-β-alanine or [4,6-^{14}C, $N-^{14}CH_3$]trigonelline (**13**) [38, 40] did not yield specifically labelled ricinine, indicating that in all probability methylation takes place

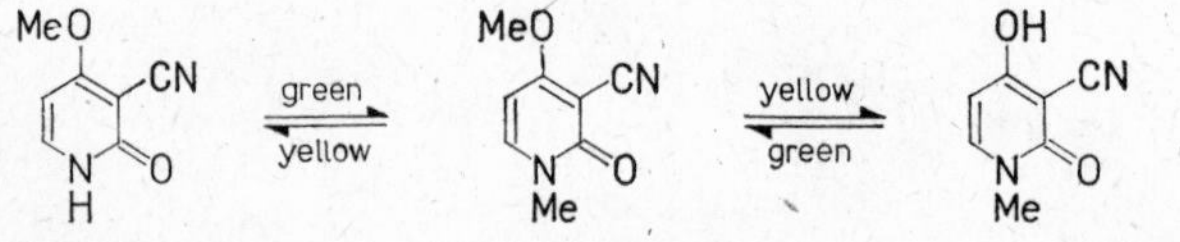

MeO CN N O H
green / yellow
MeO CN N O Me
yellow / green
OH CN N O Me

43 *N*-Demethyl-ricinine 31 Ricinine 44 *O*-Demethyl-ricinine

Fig. 13.8
Interconversion of ricinine and *N*- and *O*-Demethylricinine in senescent and green castor bean plant leaves

at a very late stage in its biosynthesis. The rapid formation of **31** after administration of *N*-demethylricinine (**43**) and *O*-demethylricinine (**44**) to green leaves of the castor bean plant [144] (Fig. 13.8) supports this assumption. But, the correct order of introduction of both the methyl groups into the molecule still remains to be established.

Quinolinic acid (**7**) is an efficient precursor of ricinine [43, 48, 51, 92, 152, 156, 162] and glycerol, succinic acid and related compounds are also good and specific precursors of ricinine in the castor bean plant. It has been demonstrated on many occasions that the carbons 2 and 3 as well as the cyano (C-8) arise from the carbon atoms 2, 3 and 4 of a related C_4 fragment. The question whether or not aspartate is incorporated intact with its amino nitrogen to give the α-pyridone nitrogen has been studied from the incorporation of [4-^{14}C, ^{15}N]aspartate (**15**) into ricinine (**31**) [162]. The ratio ^{14}C (C-8):^{15}N (α-pyridone ring) provides evidence for the direct incorporation of aspartate; however, a substantial amount of ^{15}N label was also found in the cyano group.

The carbon atoms 4, 5 and 6 of **31** derive from an intact C-3 unit (**14**) shown by early feeding experiments both with [1,3-^{14}C] and [2-^{14}C]glycerol. The tritium from C-2 of tritiated glycerol (**45**) was not incorporated into **31** while the tritium from

C-1 was retained (Fig. 13.9) [132]. Thus, a derivative of dihydroxyacetone, incorporated in a non-symmetrical fashion, is presumed to be the intermediate of the carbon atoms 4, 5 and 6 of ricinine.

During the conversion of quinolinic acid (**7**) to ricinine (**31**) the α-carboxy group is lost by decarboxylation, while the carboxy group at position 3 is converted to the carboxamide. The latter undergoes intramolecular dehydration to form the cyano group as shown by the intact incorporation of [^{3}H, $^{15}NH_2$]nicotinamide (**11**) into ricinine [152]. The γ-pyridone alkaloid mallorepine (**31a**) isolated from the *Euphorbiaceae Mallotus repandus* has been considered to be a missing link on the pathway from nicotinamide (**11**) to ricinine (**31**) [47].

In 1957 it was suggested that intermediates of the pyridine nucleotide cycle, such as NADH, might be involved in the biosynthesis of **31** [92]. Later it was found that the pyridine moieties of the pyridine nucleotides (**8**—**10**) were incorporated

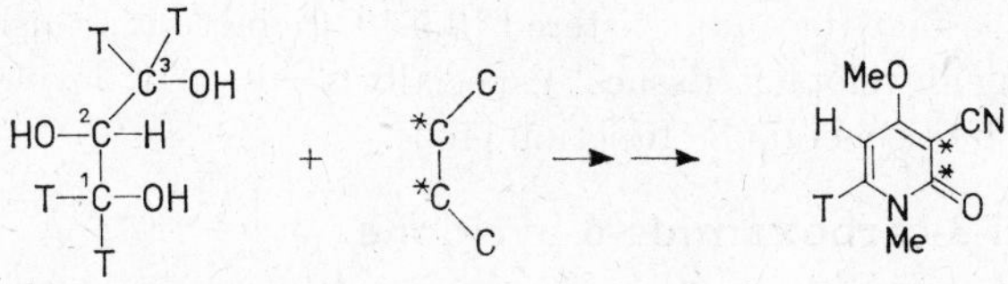

Fig. 13.9
Incorporation of [1,3-^{3}H] glycerol and C_4-compounds into ricinine (the labelling pattern of ^{14}C indicated with asterisks)

into **31** and therefore it was assumed that the pyridine nucleotide cycle is a necessary intermediate in the biosynthesis of ricinine from quinolinic acid [156]. This assumption is supported by the fact that inhibitors of the NAD synthetase such as azaserine and azaleucine decrease the incorporation rate of [^{14}C]-labelled quinolinic acid (**7**) into ricinine [51]. The conversion of **7** to **8** is catalysed by quinolinic acid phosphoribosyl transferase. The increase in specific activity of that enzyme just prior to an increased rate of synthesis of ricinine also supports the hypothesis that the pathway for pyridine nucleotide biosynthesis has been activated to supply precursors for ricinine biosynthesis [105]. The results of these experiments indicate an interdependency between the pyridine nucleotide cycle and ricinine biosynthesis, but the branch point for ricinine biosynthesis is still unknown. On the other hand, experiments in which exogenously supplied NAD (**10**) caused no decrease in the incorporation of quinolinic acid (**7**) into ricinine (**31**) [48] lead to the suggestion that ricinine is biosynthesized by a more or less direct route from quinolinic acid independent of the cycle. With regard to the biosynthetic pathway of ricinine two possibilities are being considered.

The decision regarding the operating pathway(s) can only be made by isolation and characterization of the enzymes involved in the biosynthetic pathway(s) of ricinine and by isolation and structural determination of the actual intermediates between quinolinic acid and ricinine.

Studies to determine the site of alkaloid bioynthesis have shown that all organs of the *Ricinus communis* plant are capable of synthesizing ricinine. The most active growing portions are more efficient in their ability to perform the biosynthesis and, moreover, the alkaloid biosynthesis is one of the earliest metabolic processes initiated by growth and differentiation [154].

Experiments on the metabolism of administered [3,5-^{14}C]ricinine have shown that the α-pyridone ring of ricinine is metabolized to respiratory carbon dioxide by the castor plant in both dark and light [153, 156]. With the exception of the *O*- and *N*-demethyl derivatives of ricinine (**43**, **44**), other metabolites have not yet been structurally elucidated.

Excised senescent yellow leaves of the castor bean plant, which do not normally contain ricinine and its demethylated products, rapidly demethylate administered ricinine to *N*-demethylricinine (**43**) [144, 155] and to *O*-demethylricinine (**44**) [64], both identified as metabolites of ricinine. The interconversion of *O*- and *N*-demethylricinine with ricinine in senescent and green castor plant leaves, showing the possibility of a link between ricinine catabolism and aging of the plant, is illustrated in Fig. 13.8. The fact that administered [3,5-^{14}C]ricinine is translocated from aging leaves to young developing tissues, especially seeds and developing fruits, suggests that ricinine has a metabolic function [155].

13.4.1.2 1-Methyl-3-carboxamide-6-pyridone

Trewia nudiflora is the closest relative of the genus *Ricinus* and contains some alkaloids (**30**, **33**, **34**) structurally related to ricinine (Fig. 13.7). Biosynthetic experiments with *T. nudiflora* plants have shown that [6-^{14}C] or [7-^{14}C]nicotinic acid (**12**) and [^{14}CO]nicotinamide (**11**) are incorporated into **33** and it was suggested that the pyridine nucleotide cycle also exists in *Trewia nudiflora* [135]. As outlined in Fig.

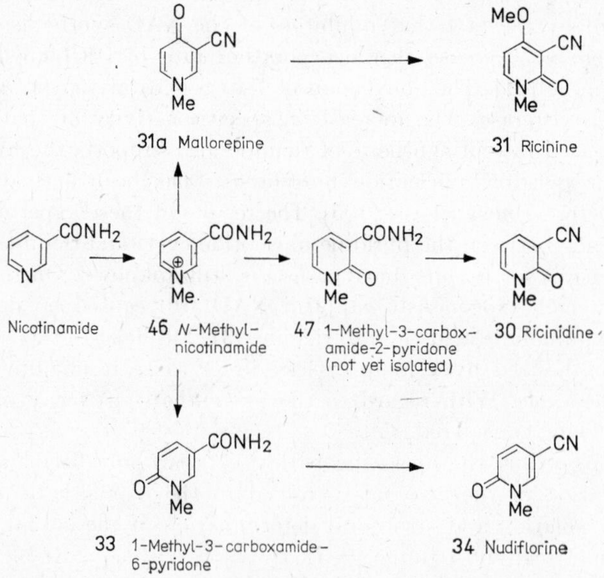

31a Mallorepine
31 Ricinine
11 Nicotinamide
46 *N*-Methyl-nicotinamide
47 1-Methyl-3-carboxamide-2-pyridone (not yet isolated)
30 Ricinidine
33 1-Methyl-3-carboxamide-6-pyridone
34 Nudiflorine

Fig. 13.10
Proposed biosynthesis of the α-pyridone alkaloids

13.10 *N*-methyl-nicotinamide (**46**) formed from nicotinamide (**11**) is postulated to be the immediate precursor both of 1-methyl-3-carboxamide-6-pyridone (**33**) [135] and of nudiflorine (**34**) and its isomer ricinidine (**30**) [122].

These α-pyridone alkaloids are formed from a pyridine compound by enzymatic hydroxylation of both the 2- and 6-positions. The conversion of a particular pyridine compound to a pyridone structure has been studied by pyridinium oxidizing enzymes that catalyse the oxidation of 3-cyano-1-methylpyridinium salts to form the corresponding 4- and 6-pyridones [27, 28]. Of 11 species of the family *Euphorbiaceae* at least 7 have been shown to possess enzymatic oxidizing activity on salts of pyridinium compounds. The enzyme system isolated from *Ricinus communis* and *Trewia nudiflora* consists of 3 active enzyme entities each oxidizing the substrate *in vitro* to 4- and 6-pyridones. It is postulated that this enzyme of the family *Euphorbiaceae* is phylogenetic specific.

13.4.2 Nicotine and Structurally Related Alkaloids

Nicotine (**49**), a 3-substituted pyridine connected to a *N*-methylpyrrolidine ring, is the most important member of the group of classic tobacco alkaloids. As Fig. **13.11** shows they differ in the substituent on the nitrogen of the pyrrolidine ring (**50**—**58**), or in the nature of the five- and six-membered *N*-heterocyclic nucleus linked to C-3 of the pyridine nucleus (**59**—**71**). Nicotine occurs not only in various *Nicotiana* species but also in many other plant families. In *Nicotiana* plants, (—)-nicotine as well as nornicotine (**48**) and anabasine (**65**) have 2′(*S*) configuration. In addition to these major alkaloids more than 20 minor alkaloids (generally accounting for about 2 to 5 per cent of the total) have been identified in tobacco leaves or in air-cured tobacco [7, 60, 106, 111—113, 136, 157].

A very large number of biosynthetic studies have been carried out to establish the pathway leading to nicotine (**49**) and the major tobacco alkaloids (*cf.* reviews [17, 19, 36, 37, 73, 75, 80, 84, 100] and references therein).

13.4.2.1 Nicotine, Nornicotine and *N*-Substituted Nornicotines

Origin of the Pyridine Ring. More than 15 years ago it was repeatedly demonstrated that nicotinic acid (**12**), labelled in different positions of the ring with ^{14}C, tritium or deuterium, is incorporated into the pyridine ring of nicotine with loss of its carboxyl group [18, 26, 141, 147, 161]. Later, these findings were confirmed by feeding experiments with [5,6-^{14}C, $^{13}C_2$]nicotinic acid to *Nicotiana tabacum* and *N. glauca* [79].

Studies on the biosynthesis of ricinine (**31**) in *Ricinus communis* (*cf.* 13.4.1.1) and nicotinic acid (**12**) in certain bacteria (*cf.* 13.1.2) have shown that the carbons 2 and 3 of the pyridine ring of nicotine (**49**) are also derived from a four-carbon dicarboxylic acid metabolically related to succinic acid or aspartic acid (**15**), while the carbons 4, 5 and 6 arise from an intact C_3-unit related to glycerol (**14**) or glyceraldehyde-3-phosphate, as indicated in Fig. 13.3. The experiments indicating a

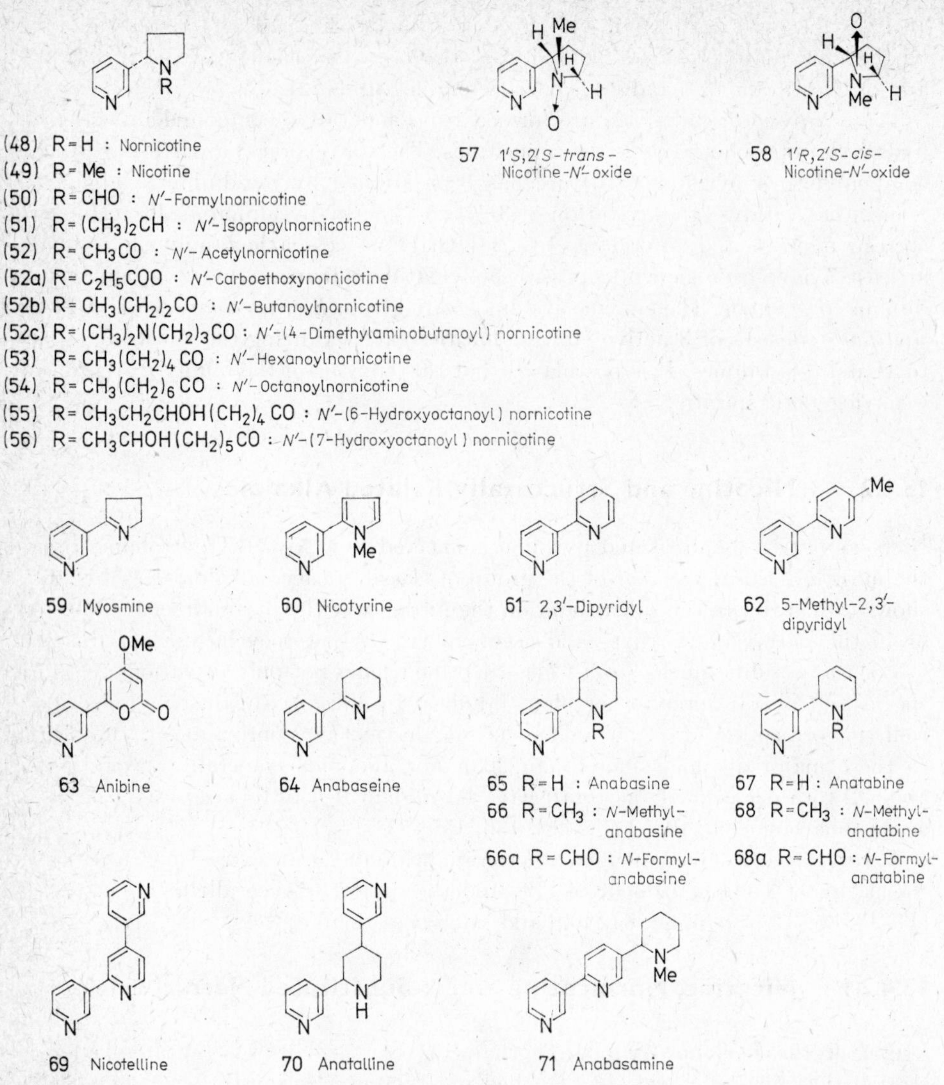

Fig. 13.11
Nicotine and structurally related alkaloids

specific incorporation of related C_3- and C_4-compounds into the pyridine ring of nicotine, have been reviewed elsewhere (*e.g.* [36, 37]) and will not be described here.

Quinolinic acid (**7**) has also been shown to be an efficient precursor of the pyridine ring moiety of nicotine in the tobacco plant [26, 161]. Its conversion to nicotinic acid (**12**) has been proven to occur in the tobacco plant, also *via* the pyridine nucleotide cycle (Fig. 13.2) [26]. Biosynthetic experiments using [2-^{3}H] and [2,3,7-^{14}C]-nicotinic acid indicated that the pyridine ring of nicotinic acid does not become symmetrical during its conversion into nicotine [18, 141, 161]. Thus, the point of attachment of the *N*-methylpyrrolidine ring of nicotine is at the C-3 position from which the carboxyl group is lost and replaced by the pyrrolidine ring. An enzyme system that catalyses the O_2 dependent release of $^{14}CO_2$ from [7-^{14}C]nicotinic acid was isolated from the tobacco root but not from leaves and stems [11].

The hydrogen at C-6 of nicotinic acid is lost during the biosynthesis of nicotine [18, 147]. This finding was confirmed by the administration of [6-^{14}C, ^{3}H]nicotinic

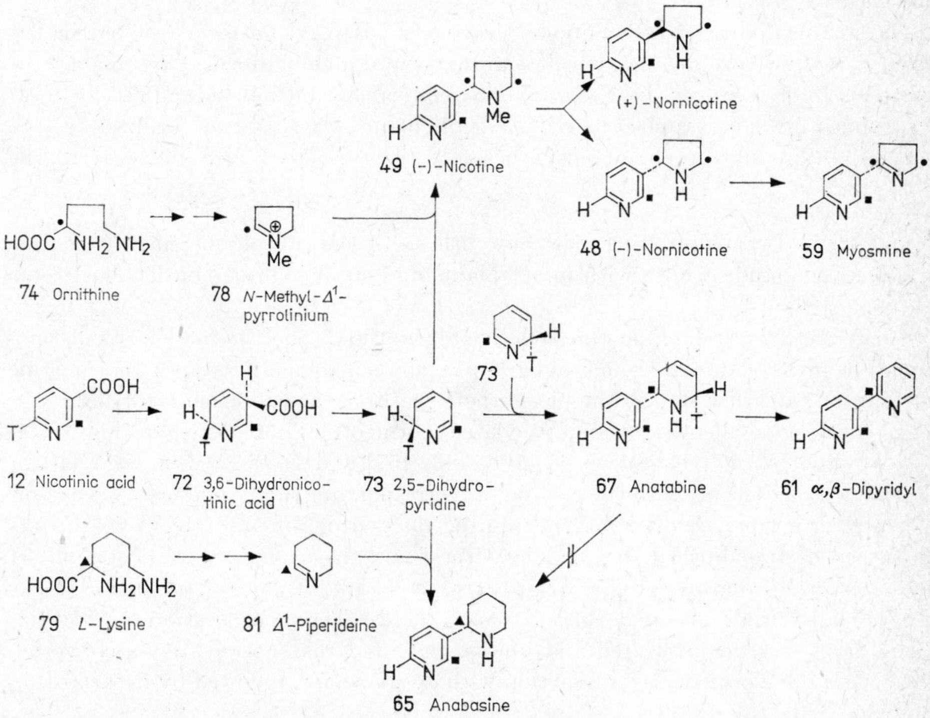

Fig. 13.12
Biosynthesis of the *Nicotiana* alkaloids.
The pattern of labelling in the different alkaloids derived from [2-^{14}C]-labelled ornithine, nicotinic acid and lysine indicated with the symbols ● ■ ▲. The fate of the hydrogen at C-6 of nicotinic acid is illustrated by means of a tritium label.

acid to intact tobacco plants. The acid was incorporated into nicotine with a 98% loss of tritium relative to ^{14}C [79, 93], while all the tritium in nicotine derived from [2-^{3}H]nicotinic acid was located at C-2 of the pyridine ring [18, 93]. Thus, the five carbons and the nitrogen found in the pyridine ring of nicotine are derived, intact, from nicotinic acid as suggested 50 years ago by Trier in his revision of Winterstein's "Die Alkaloide".

It has been suggested that 3,6-dihydronicotinic acid (72) in Fig. 13.12 is the activated form of nicotinic acid (12) (*cf.* [19, 79, 80]) and that the hydrogen added at C-6 during the stereospecific reduction of **12** to **72** is that which is retained in the final nicotine. In the case of anatabine (**67**) it was demonstrated that the initial reduction of **12** occurs by introduction of a hydrogen at C-6 in the *pro*-(*R*) position [81]. As indicated in Fig. 13.12, a hypothetical route leading from **12** to **49** includes the concerted decarboxylation of **72** and reaction with the *N*-methyl-Δ^1-pyrrolinium salt (**78**) with loss of tritium at C-6 [79, 80]. This assumption might be supported by the *in vitro* biomimetic synthesis of nicotine from glutardialdehyde, ammonia and *N*-methyl-Δ^1-pyrrolinium acetate (**78**) yielding nicotine *via* a dihydropyridinium intermediate (**73**) [74].

The application of [5,6-^{14}C]5-fluoronicotinic acid to *N. tabacum* resulted in the formation of 5-fluoronicotine, which is an unnatural alkaloid formed by an aberrant synthesis [86]. However, in *N. glauca* this conversion did not occur [83]. [4-^{14}C]4-Methylnicotinic acid applied to *N. tabacum* plants was also not utilized by the enzymes responsible for the biosynthesis of nicotine from nicotinic acid in the tobacco plant [91].

Origin of the Pyrrolidine Ring. The biosynthesis of the pyrrolidine ring of nicotine (**49**) has been studied both with intact plants and sterile cultures of detached roots of tobacco plants.

The *N*-methyl group of nicotine is derived from the C_1-pool *via S*-adenosyl methionine. The introduction of the methyl group occurs not by methylation of nornicotine (**48**) but at an earlier step in the biosynthetic pathway as will be shown later.

In contrast to nicotine, the *N'*-acylated nornicotines (**50**, **52**—**56**) (Fig. 13.11) isolated from *Nicotiana tabacum* [7, 106, 113, 157] and *Duboisia hopwoodii* [103a] may be formed from nornicotine and a corresponding activated acyl-precursor, partially in a non-enzymatic manner during flue-curing, but no tracer experiments have yet been published on the biosynthesis of these alkaloids. Compound **51** contains an *N*-isopropyl group that is extremely rare in natural products. It has been isolated from air-cured Burley tobacco [112]. The incorporation of [2-^{14}C]nicotine in **51** was demonstrated and it was suggested that nornicotine is converted to *N'*-isopropylnornicotine by reaction with acetoacetate, followed by decarboxylation and reduction [85].

Robinson and Trier suggested that the pyrrolidine ring of nicotine may be related biogenetically to the diamino acid ornithine (**74**). In early studies it was established that the administration of radioactive ornithine (labelled with ^{14}C in position 2 or 5) to tobacco plants results in the formation of nicotine (**49**) and nornicotine (**48**), which are labelled equally at C-2' and C-5' of their pyrrolidine rings (Fig. 13.12 and 13.13). Glutamic acid and proline, closely related to ornithine, also act as precursors of the pyrrolidine ring (*cf.* [36, 37, 80] and references therein). Recent biosynthetic experiments with [2-^{14}C]ornithine and [2,3-^{13}C, 5-^{14}C]ornithine fed to *N. glutinosa* [77, 98] as well as the exposure of *N. tabacum* plants to [^{14}C, ^{13}C]carbon dioxide [77, 124] validated the previous work on the symmetrical incorporation of **74** into the pyrrolidine ring. Moreover, feeding of α- and δ-^{15}N-labelled [2-^{14}C]-ornithines (**74**) to a sterile culture of *N. tabacum* roots demonstrated that the nitrogen of the δ-amino but not of the α-amino group is utilized in the formation of the pyrrolidine ring [90]. Thus, ornithine is an established precursor of nicotine biosynthesis. The finding that feeding of urea to tobacco tissue cultures increased the nicotine content [131] can be explained by the suppression of enzyme activity in the urea cycle, thereby increasing the availability of ornithine for nicotine biosynthesis.

All these results indicate that ornithine is incorporated *via* a symmetrical intermediate, logically putrescine (**75**), which has been shown to be also an efficient precursor of the pyrrolidine ring (*cf*. [36, 80]). The formation of **75** by decarboxylation of **74** is catalysed by the ornithine decarboxylase, which has been detected in tobacco roots [118]. The decarboxylation of *L*-ornithine to putrescine in intact *N*. tabacum takes place with retention of configuration [158a]. The origin of **75** from arginine *via* agmatine and *N*-carbamylputrescine has also been found in tobacco plants and these intermediates were therefore postulated to be possible precursors

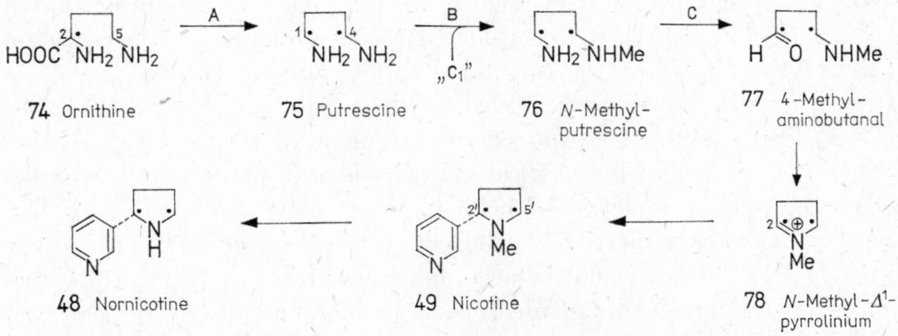

Fig. 13.13
Biosynthesis of nicotine and nornicotine from [2-^{14}C]ornithine.
The positions labelled with ^{14}C are indicated by heavy dots

of nicotine [163]. But, as mentioned before, the symmetrical labelling pattern found at C-2′ and C-5′ of the pyrrolidine ring of nicotine after feeding of [2-^{14}C]ornithine, can only be explained by a pathway *via* a symmetrical intermediate.

The introduction of the *N*-methyl group occurs at the step of putrescine (**75**) and produces *N*-methylputrescine (**76**). This reaction is catalysed by the putrescine *N*-methyltransferase, which has been obtained from a cell-free extract of tobacco roots [116, 118]. *N*-Methylputrescine (**76**) labelled in the methylamino group with ^{15}N and additionally in the methyl group with ^{14}C or in position 1 with ^{14}C or ^{13}C/^{14}C was totally incorporated into the pyrrolidine ring of nicotine [94, 137, 138]. These results indicate that there is no degradation or demethylation of **76** during its conversion to nicotine.

The subsequent step in the biosynthetic pathway leading to nicotine involves the oxidation of *N*-methylputrescine to 4-methylaminobutanal (**77**), which has been isolated in radioactive form from tobacco roots after feeding with [2-^{14}C]-ornithine (**74**), [1,4-^{14}C]putrescine (**75**) or [$^{14}CH_3$]methionine; the [^{14}C]-labelled butanal (**77**) was also incorporated into nicotine (**49**) [57, 114]. The oxidative deamination of *N*-methylputrescine proceeds with stereospecific loss of the *pro*-(*S*) hydrogen atom of C-4 [158a]. The *N*-methylputrescine oxidase, which catalyses the oxidative deamination of the primary amino group of **76**, was detected in tobacco roots and shows high specificity for the substrate [117, 118].

Cyclization of the butanal (**77**) affords the *N*-methyl-Δ^1-pyrrolinium salt (**78**), found to be the immediate precursor of the *N*-methylpyrrolidine ring of nicotine. The administration of [2-^{14}C]*N*-methyl-Δ^1-pyrrolinium chloride (**78**) to *N. tabacum* plants yielded radioactive nicotine, whose total radioactivity was located at C-2′ of the pyrrolidine ring [70]. In view of these results the biosynthetic route outlined in Fig. 13.13 can be accepted as the main pathway from ornithine to nicotine.

The intact incorporation of [2-^{14}C, ^{15}N^{14}CH$_3$]δ-*N*-methylornithine into the pyrrolidine ring (labelled only at C-2′ and at the *N*-methyl group) [33] as well as the formation of unnatural nicotine derivatives from supplied 1-, 2- and 3-methyl substituted pyrrolinium precursors [134] may be explained by not completely specific enzymes. Although [5-^{14}C]Δ^1-pyrrolidine-5-carboxylic acid was incorporated symmetrically into nicotine [61], it is considered to be probably not on the main pathway of pyrrolidine ring biosynthesis.

Nevertheless, there is controversy regarding the origin of the *N*-methyl-Δ^1-pyrrolinium salt (**78**), which is the immediate precursor of the pyrrolidine ring. Short-term feeding of [^{14}C] and [^{14}C, ^{13}C]carbon dioxide to tobacco plants afforded nicotines which showed a symmetrical labelling pattern at C-2′ and C-5′ [77, 124, 165], while others were asymmetrically labelled at these carbons [49, 133]. Any asymmetrical labelling pattern is not in accordance with the use of ornithine and putrescine as preformed precursors of nicotine. Therefore, the alternative pathway from CO_2 to **78** in which an active C_2-compound such as glycolaldehyde serves as the precursor will not be discussed here (for reviews dealing with that problem (*cf.* [17, 75, 80]).

Metabolism of Nicotine. From studies on the metabolism of nicotine (**49**) in different biological systems it has now been established that in *Nicotiana* plants and *Nicotiana* cell suspension cultures (—)-nicotine is not an inert end product of metabolism, but that it is *N*-demethylated in an irreversible reaction to racemic nornicotine (**48**) [5, 71, 88]. The racemization step involves loss of the hydrogen from the C-2′ position of nicotine [88]; however, the intermediates in the demethylation process are still unknown. The nicotine-*N*-oxides (**57**, **58**) isolated from *N. tabacum*, *N. affinis* and *N. sylvestris* [130] were only poorly incorporated into nicotine [3, 80].

In tobacco plants nornicotine (**48**) is further converted to myosmine (**59**), indicated by its isolation after feeding of (—)-nornicotine to *N. glutinosa* and *N. tabacum* plants [59] and by incorporation of [^{14}C]nicotine into myosmine [88]. Feeding [2′-^{14}C]myosmine to *N. glauca* did not produce labelled nicotine or nornicotine.

The stepwise degradation of nicotine not only by plants but also by animals and bacteria has been reviewed by several authors [34, 36, 56, 80, 100] and will not be described in detail. For the same reason the extensive work on the formation of nicotine and related alkaloids in cell suspension cultures of *Nicotiana* species is excluded.

13.4.2.2 Anatabine and α, β-Dipyridyl

The biosynthesis of (—)-anatabine (**67**) and α,β-dipyridyl (**61**), both minor alkaloids in *Nicotiana tabacum* and *N. glutinosa*, has been demonstrated to be quite different from the biogenetic route leading to anabasine (**65**), which differs from anatabine by only two hydrogens. The piperidine ring of **65**, but not the piperideine ring of **67**, is derived from lysine (**79**), shown by the non-incorporation of [2-^{14}C]lysine into anatabine [58, 76]. The administration of [2-^{14}C] and [6-^{14}C]nicotinic acids (**12**) to *N. glauca* and *N. tabacum* afforded labelled anatabine [76, 97], which was found to have equal labelling at C-2/C-2′ and C-6/C-6′, respectively (Fig. 13.12). These results indicated that both rings of anatabine are derived from nicotinic acid which is utilized as the common precursor. [5,6-^{14}C, $^{13}C_2$]Nicotinic acid applied to *N. glauca* and *N. tabacum* yielded anatabine with isotopic enrichment at C-5/C-6 and C-5′/ C-6′ [79] indicating that both the ring systems of anatabine originate from nicotinic acid. [5,6-^{14}C, ^{13}C, 6-^{3}H]Nicotinic acid administered to tobacco plants afforded anatabine labelled in both rings with ^{14}C and ^{13}C as before, but the tritium activity was localized only at C-2′ of the piperideine ring [81]. The observed elimination of tritium from C-6 of nicotinic acid is consistent with previous work on the biosynthesis of nicotine (*cf.* 13.4.2.1). Since the chirality at C-6′ of anatabine (**67**) was found to be (*S*), the reduction of **12** to **72** occurs by introduction of a hydrogen in the *pro*-(*R*) position [81]. **72** is suggested to be an intermediate in the course of anatabine biosynthesis. It was proposed that **67** is formed either by condensation of **72** and 2,5-dihydropyridine (**73**), formed by decarboxylation of **72**, or by a dimerization of two molecules of **73** which on dehydrogenation affords anatabine (Fig. 13.12) [75, 79, 80]. The intermediacy of **73** during the biosynthesis of anatabine in its formation from **12** in the tobacco plant was supported by a biomimetic synthesis of anatabine, in which baikiain (1,2,3,6-tetrahydropyridine-2-carboxylic acid) was converted *in situ* to anatabine [82, 94a].

Anabaseine (**64**), isolated very recently from the venom of *Aphaenogaster* ants [158], differs from anatabine (**67**) only in the position of the double bond in the piperideine ring. It would be interesting to establish whether its piperideine ring is derived from lysine (**79**) or nicotinic acid (**12**) but at present there is no information available concerning its biosynthesis.

α,β-Dipyridyl (**61**) is formed from nicotinic acid (**12**) *via* anatabine (**67**) as shown by the incorporation of [2-^{14}C]nicotinic acid and [2′-^{14}C, ^{13}C]anatabine administered to tobacco plants [96]. There was no conversion of anatabine into anabasine (**65**), indicating the different pathways of these alkaloids as outlined in Fig. 13.12.

The biosynthesis of 5′-methyl-α,β-dipyridyl (**62**) isolated from *N. tabacum* [157] has not yet been studied.

Anabasamine (**71**) has been isolated from *Anabasis aphylla* seeds [121], but in spite of the incorporation of [2-^{14}C]lysine (**79**) administered to *A. aphylla* [102], the origin of the two pyridine rings and the *N*-methylated piperidine ring has not yet been precisely established.

13.4.2.3 Anabasine

Origin of the Pyridine Ring. Anabasine (**65**) represents the major alkaloid of *Nicotiana glauca*. Its pyridine ring is known to originate not from lysine (**79**) but from nicotinic acid (**12**) as shown by the non-random incorporation of [^{3}H, ^{14}C] and [^{14}C, ^{13}C]ring-labelled nicotinic acids fed to tobacco plants [79, 97, 147]. Moreover it was confirmed by the incorporation of [2-^{14}C]nicotinic acid into both rings of anatabine (**67**), but into only the pyridine ring of anabasine (**65**), that the latter alkaloid is not formed by reduction of anatabine [97]. In accordance with the biosynthesis of nicotine (**49**) and anatabine (**67**), nicotinic acid (**12**) is incorporated into anabasine with elimination of the hydrogen at C-6 and loss of its carboxyl group (Fig. 13.12).

The administration of unnatural [5,6-^{14}C, ^{13}C]5-fluoronicotinic acid to *N. glauca* afforded labelled unnatural 5-fluoroanabasine, thus indicating that 5-fluoronicotinic acid can also be utilized as an unnatural substrate by the enzymes involved in the biosynthesis of anabasine [83].

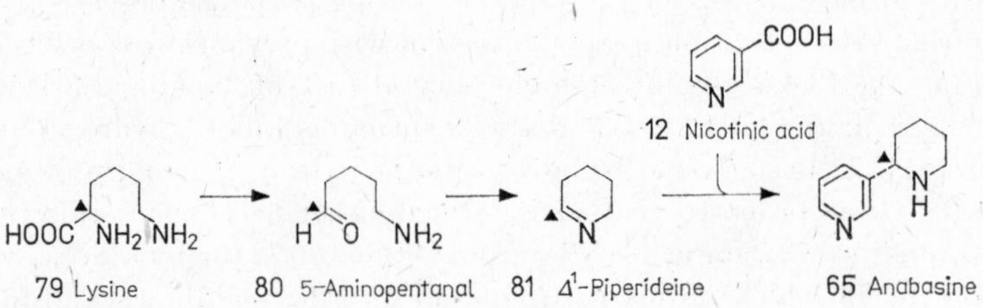

COOH
N
12 Nicotinic acid
HOOC NH2 NH2 → H O NH2 → N → N N H
79 Lysine 80 5-Aminopentanal 81 Δ^1-Piperideine 65 Anabasine

Fig. 13.14
Biosynthesis of anabasine from [2-^{14}C]lysine.
The positions labelled with ^{14}C are indicated by the symbol ▲.

Origin of the Piperidine Ring. The piperidine ring of anabasine corresponds to the pyrrolidine ring of nicotine (**49**). The transformation of [2′-^{14}C]nicotine to anabasine can be excluded [71, 88] although a relationship between these alkaloids has been observed [101, 110]. According to Robinson's theory lysine (**79**) acts as precursor of the piperidine nucleus of anabasine. This was confirmed by the fact that, after application of [2-^{14}C]lysine to *N. rustica*, *N. glauca* and *N. tabacum* as well as *Anabasis aphylla*, only the piperidine ring of anabasine was labelled [35, 69, 103]. Feeding of [2-^{14}C]lysine gave anabasine with label only at C-2′. On the basis of these experiments an asymmetrical incorporation of lysine must be true, in contrast to the mode of incorporation of ornithine (**74**) into nicotine (**49**) (Fig. 13.12).

The piperidine ring of **65** is derived from *L*-lysine, as shown by the better incorporation of the *L*- than the *D*-form [32], and by radioisotopic methods with a mixture of [4,5-$^{3}H_{2}$]*L*-lysine and [6-^{14}C]*DL*-lysine [99]. In tracer experiments using [2-^{14}C, α-^{15}N] and [2-^{14}C, ε-^{15}N]lysines, it has been demonstrated that the nitrogen of the piperidine ring originates from the ε-amino group of lysine [89]. The incorporation of *L*-lysine into anabasine is believed to occur by the pathway indicated in Fig. 13.14.

[1,5-^{14}C]Cadaverine [69, 72, 126] and [6-^{14}C]Δ^1-piperideine (**81**) [72] have also been shown to be effective precursors of anabasine both in *N. glauca* and *Anabasis aphylla*. As one would expect [1,5-^{14}C]cadaverine was incorporated symmetrically, but results from the asymmetrical incorporation of lysine mentioned above suggest that "free" cadaverine (*i.e.* non-enzyme bound), formed by decarboxylation of lysine, may not be an obligatory intermediate in the normal biosynthetic pathway leading from lysine (**79**) to Δ^1-piperideine (**81**) and anabasine (**65**). More detailed information concerning the conversion of lysine to piperidine alkaloids *via* Δ^1-piperideine can be found in the reference [99] and chapter 12.1 (Fig. 12.2).

Anabasine may also arise from a quite different pathway, as shown by the *in vitro* incorporation of [1,5-^{14}C]cadaverine into both the pyridine and piperidine rings of anabasine using a cell-free extract prepared from pea or lupin seedlings [120], but this route has not yet been found in intact higher plants.

The incorporation of [2-^{14}C]*N*-methyl-Δ^1-piperideinium chloride into *N*-methyl-anabasine (**66**) in *N. tabacum* and *N. glauca* plants [87] seems to occur by an aberrant synthesis which starts from an exogenously supplied unnatural precursor and is catalysed by more or less unspecific enzymes involved in the anabasine biosynthesis. [2-^{14}C]Lysine fed to *N. glauca* and *N. tabacum* was not incorporated into *N*-methyl-anabasine [87], which is therefore not an intermediate in the biosynthesis of anabasine from lysine.

13.4.3 Dioscorine

Dioscorine (**83**) is an alkaloid found in *Dioscorea hispida* and related species. For a long time the biosynthetic route leading to **83** with its unusual isoquinuclidine nucleus remained unknown although several routes were proposed. The unsaturated lactone ring of **83** was shown to be derived from acetate units, whereas lysine was

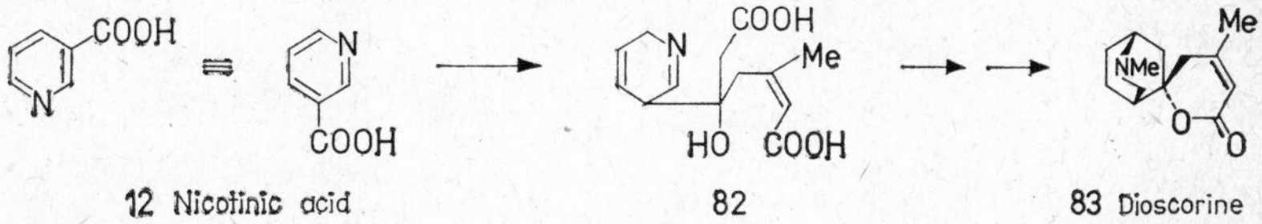

12 Nicotinic acid 82 83 Dioscorine

Fig. 13.15
Proposed biosynthetic pathway of dioscorine

not incorporated [95]. [6-^{14}C]Δ^1-Piperideine (**81**) gave scrambling of the label [95] and therefore it is excluded as a specific precursor of the isoquinuclidine ring; but after administration of [2-^{14}C]nicotinic acid (**12**) to *D. hispida* plants all the label was located at C-3 [78]. [5,6-^{14}C, $^{13}C_2$]Nicotinic acid was also incorporated in a specific manner. As outlined in Fig. 13.15 it is suggested that **12** is reduced to **72** which undergoes concerted decarboxylation and condensation with a branched acetate-derived C_8-unit to yield the intermediate **82**. The further conversion to **83** includes shift of a double bond in the dihydropyridine ring, reduction, aldol condensation, decarboxylation and *N*-methylation. The labelling pattern found in dioscorine after feeding of isotopically labelled nicotinic acids is in full agreement with that pathway.

13.5 Pyridine Compounds of Different Biogenetic Origin

Figure 13.16 shows some further natural products with the pyridine structure which have been isolated from microorganisms and higher plants: hernagine (**84**) from *Hernandia nymphaefolia* [159]; duckein (**85**) from *Aniba duckei* [15]; halfordinol (**86**) and related oxazole alkaloids from *Halfordia scleroxyla* [16], *Aeglopsis chevalieri* [22], *Aegle marmelos* [13, 104] and *Micromelium zeylanicum* [7a]; the 2,6-lutidine-3,5-dicarboxylic acid (**87**) occurring in esterified form in the *Amaryllidaceae* alkaloids clivonine, clivimine and cliviamartine [20, 21]; homarine (**88**) and the *Areca*-alkaloids (**89**—**92**). The biosynthesis of these alkaloids has not yet been studied.

The entire C_8 carbon skeleton of pyridoxol (**93**) is derived from the carbon atoms of glycerol. The carbon atoms 2′, 3, 4′, 5′ and 6 of **93** arise from C-1 and C-3 of gly-

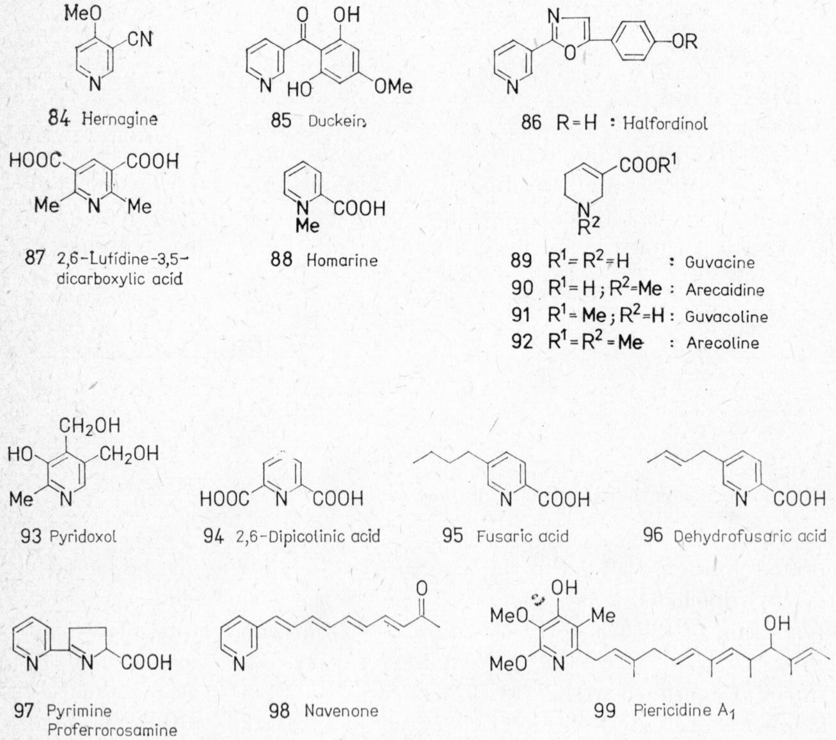

84 Hernagine

85 Duckein

86 R=H : Halfordinol

87 2,6-Lutidine-3,5-dicarboxylic acid

88 Homarine

89 $R^1=R^2=H$: Guvacine
90 $R^1=H$; $R^2=Me$: Arecaidine
91 $R^1=Me$; $R^2=H$: Guvacoline
92 $R^1=R^2=Me$: Arecoline

93 Pyridoxol

94 2,6-Dipicolinic acid

95 Fusaric acid

96 Dehydrofusaric acid

97 Pyrimine Proferrorosamine

98 Navenone

99 Piericidine A_1

Fig. 13.16
Naturally occurring pyridine compounds

cerol, while C-2, -4 and -5 originate from C-2 of glycerol. In a minor route C-5′ and C-5 are supplied by glycolaldehyde and not by glycerol (*cf.* [149] and references therein). 2,6-Dipicolinic acid (**94**) is formed from aspartate and pyruvate (*cf.* [36, 37]). The carbon atoms 2, 3 and 4 and the carboxyl group of fusaric acid (**95**) are derived from aspartate, while C-5 and C-6 and the n-butyl side-chain is formed from three acetate units (*cf.* [36, 37]). The piericidines (**99**) are constructed from four acetate and five propionate units [164]. The biosynthetic pathway leading to the alarm pheromone, navenone (**98**), isolated from *Navanax inermis* [145], is not known but acetate units may serve as precursors. The pyridine skeleton of pro-ferrorosamine A (**97**), isolated from *Pseudomonas roseus fluorescens* and identical to pyrimine, seems to be formed from *L*-asparagine [46].

13.6 References

[1] Ahmad, F.; Moat, A. G.: J. Biol. Chem. **241** (1966) 775.

[2] Albertson, J. N.; Moat, A. G.: J. Bacteriol. **89** (1965) 540.

[3] Alworth, W. L.; Liberman, L.; Ruckstahl, J. A.: Phytochemistry **8** (1969) 1427.

[4] Antony, A.; Gopinathan, K. P.; Vaidyanathan, C. S.: Indian J. Exp. Bot. **13** (1975) 39.

[4a] Arditti, J.; Tarr, J. B.: Am. J. Bot. **66** (1979) 1105.

[5] Barz, W.; Kettner, M.; Hüsemann, W.: Planta Med. **34** (1978) 73.

[6] Baxter, R. L.; Crombie, L.; Simmonds, D. J.; Whiting, D. A.: J. Chem. Soc., Chem. Commun. **1976**, 463, 465.

[7] Bolt, A. J. N.: Phytochemistry **11** (1972) 2341.

[7a] Bowen, I. H.; Perera, K. P. W. C.: Phytochemistry **21** (1982) 433.

[8] Budzikiewicz, H.; Horstmann, C.; Pufahl, K.; Schreiber, K.: Chem. Ber. **100** (1967) 2798.

[9] Burnell, R. H.; Benoin, P. R.; Khalil, M. F.: Lloydia **38** (1975) 444.

[10] Cais, M.; Ginsburg, D.; Mandelbaum, A.; Smith, R. M.: Tetrahedron **31** (1975) 2727.

[11] Chandler, J. L. R.; Gholson, R. K.: Phytochemistry **11** (1972) 239.

[12] Chandler, J. L. R.; Gholson, R. K.: Biochim. Biophys. Acta **264** (1972) 311.

[13] Chatterjee, A.; Majumder, R.: Indian J. Chem. **9** (1971) 763.

[14] Chen, S. C.; Godavari, H. R.; Waygood, E. R.: Canad. J. Bot. **52** (1974) 707.

[14a] Cooper, J. L.; Hilton, B. L.; Arditti, J.; Tarr, J. B.: New Phytol. **91** (1982) 621.

[15] Correa, D. de Barros; Gottlieb, O. R.: Phytochemistry **14** (1975) 271.

[16] Crow, W. D.; Hodgkin, J. H.: Tetrahedron Lett. (1963) 85; Austral. J. Chem. **17** (1964) 119; **21** (1968) 3075.

[17] Dalton, D. R.: In: Gassmann, P. G. (Ed.): The Alkaloids. Marcel Dekker, New York/Basel, 1979, p. 151.

[18] Dawson, R. F.; Christman, D. R.; D'Adamo, A.; Solt, M. L.; Wolf, A. P.: J. Am. Chem. Soc. **82** (1960) 2628.

[19] Dawson, R. F.; Osdene, T. S.: Recent Adv. Phytochem. **5** (1972) 317.

[20] Döpke, W.; Roshan, S. A.: Z. Chem. **20** (1980) 374.

[21] Döpke, W.; Bienert, M.; Burlingame, A. L.; Schnoes, H. K.; Jeffs, P. W.; Farrier, D. S.: Tetrahedron Lett. **1967**, 451.

[22] Dreyer, D. L.: J. Org. Chem. **33** (1968) 3658.

[23] Dubrakova, L.; Dolejs, L.; Tomko, J.: Collect. Czech. Chem. Commun. **38** (1973) 2132.

[24] Dutta, S. K.; Sharma, B. N.; Sharma, P. V.: Phytochemistry **17** (1978) 2047; **19** (1980) 1278.

[25] El Basyouni, S. H.; Brewer, D.; Vining, L. C.: Canad. J. Bot. **46** (1968) 441.

[26] Frost, G. M.; Yang, K. S.; Waller, G. R.: J. Biol. Chem. **242** (1967) 887.
[27] Fu, P.; Kobus, J.; Robinson, T.: Phytochemistry **11** (1972) 105.
[28] Fu, P.; Robinson, T.: Phytochemistry **9** (1970) 2443.
[29] Ganguly, S. N.: Phytochemistry **9** (1970) 1667.
[30] Gellert, E.; Summonds, R. E.: Austral. J. Chem. **27** (1974) 919.
[31] Gholson, R. K.: Nature **212** (1966) 933.
[32] Gilbertson, T. L.: Phytochemistry **11** (1972) 1737.
[33] Gilbertson, T. J.; Leete, E.: J. Am. Chem. Soc. **89** (1967) 7085.
[34] Gorrod, J. W.; Jenner, P.: Essays Toxicol. **6** (1975) 35.
[35] Griffith, T.; Griffith, G. D.: Phytochemistry **5** (1966) 1175.
[36] Gross, D.: Pyridinalkaloide. In: Mothes, K.; Schütte, H. R. (Eds.): Biosynthese der Alkaloide. Deutscher Verlag der Wissenschaften, Berlin 1969.
[37] Gross, D.: Progr. Chem. Org. Nat. Prod. **28** (1970) 109.
[38] Gross, D.; Banditt, P.; Kurbatov, J. W.; Schütte, H. R.: Z. Pflanzenphysiol. **58** (1968) 410.
[39] Gross, D.; Feige, A.; Stecher, R.; Zureck, A.; Schütte, H. R.: Z. Naturforsch. **20b** (1965) 1116.
[40] Gross, D.; Müller, D.; Schütte, H. R.: Z. Naturforsch. **24b** (1969) 705.
[41] Gross, D.; Schütte, H. R.; Hübner, G.; Mothes, K.: Tetrahedron Lett. **1963**, 541.
[42] Gupta, K. C.; Miller, R. L.: Lloydia **40** (1977) 303.
[43] Hadwiger, L. A.; Badiei, S. E.; Waller, G. R.; Gholson, R. K.: Biochem. Biophys. Res. Commun. **13** (1963) 466.
[44] Heeger, V.; Leienbach, K. W.; Barz, W.; Z. Physiol. Chem. **357** (1976) 1081.
[45] Heilmann, H. D.; Lingens, F.: Z. Physiol. Chem. **349** (1968) 231.
[46] Helbing, A. M.; Viscontini, M.: Helv. Chim. Acta **59** (1976) 2284.
[47] Hikino, H.; Tamada, M.; Yen, K. Y.: Planta Med. **33** (1978) 385.
[48] Hiles, R. A.; Byerrum, R. U.: Phytochemistry **8** (1969) 1927.
[49] Hutchinson, C. R.; Hsia, M.-T. S.; Carver, R. A.: J. Am. Chem. Soc. **98** (1976) 6006.
[50] Isquith, A. J.; Moat, A. G.: Biochem. Biophys. Res. Commun. **22** (1966) 565.
[51] Johnson, R. D.; Waller, G. R.: Phytochemistry **13** (1974) 1493.
[52] Joshi, J. G.; Handler, P.: J. Biol. Chem. **235** (1960) 2981.
[53] Joshi, J. G.; Handler, P.: J. Biol. Chem. **237** (1962) 3185.
[54] Kadooka, M. M.; Chang, M. Y.; Fukami, H.; Scheuer, P. J.; Clardy, J.; Solheim, B. A.; Springer, J. P.: Tetrahedron **32** (1976) 919.
[55] Kikuchi, T.; Uyeo, S.; Nishinaga, T.: Tetrahedron Lett. **1965**, 1993.
[56] Kisaki, T.; Maeda, S.; Koiwai, A.; Mikami, Y.; Sasaki, T.; Matsushita, H.: Beitr. Tabakforsch. **9** (1978) 308; C. A. **91** (1979) 189737.
[57] Kisaki, T.; Mizusaki, S.; Tamaki, E.: Arch. Biochem. Biophys. **117** (1966) 677.
[58] Kisaki, T.; Mizusaki, S.; Tamaki, E.: Phytochemistry **7** (1968) 323.
[59] Kisaki, T.; Tamaki, E.: Phytochemistry **5** (1966) 293.
[60] Koiwai, A.; Mikami, Y.; Matsushita, H.; Kisaki, T.: Agric. Biol. Chem. **43** (1979) 1421.
[61] Krampl, V.; Zielke, H. R.; Byerrum, R. U.: Phytochemistry **8** (1969) 843.
[62] Kupchan, S. M.; Smith, R. M.: J. Org. Chem. **42** (1977) 115.
[63] Kupchan, S. M.; Smith, R. M.; Bryan, R. F.: J. Am. Chem. Soc. **92** (1970) 6667.
[64] Lee, H. J.; Waller, G R.: Phytochemistry **11** (1972) 965.
[65] Lee, H. J.; Waller, G. R.: Phytochemistry **11** (1972) 2233.
[66] Leienbach, K. W.; Barz, W.: Z. Physiol. Chem. **357** (1976) 1069.
[67] Leienbach, K. W.; Heeger, V.; Barz, W.: Z. Physiol. Chem. **357** (1976) 1089.
[68] Leienbach, K. W.; Heeger, V.; Neuhann, H.; Barz, W.: Planta Med. Suppl. **1975**, 148.
[69] Leete, E.: J. Am. Chem. Soc. **78** (1956) 3520; **80** (1958) 4393.
[70] Leete, E.: J. Am. Chem. Soc. **89** (1967) 7081.
[71] Leete, E.: Tetrahedron Lett. **1968**, 4433.
[72] Leete, E.: J. Am. Chem. Soc. **91** (1969) 1697.
[73] Leete, E.: Adv. Enzymol. **32** (1969) 373.
[74] Leete, E.: J. Chem. Soc., Chem. Commun. **1972**, 1091.

[75] Leete, E.: Biosynthesis **1** (1972) 158; **2** (1973) 106; **3** (1975) 113; **4** (1976) 97; **5** (1977) 136.
[76] Leete, E.: J. Chem. Soc., Chem. Commun. 1975, 9.
[77] Leete, E.: J. Org. Chem. **41** (1976) 3438.
[78] Leete, E.: J. Am. Chem. Soc. **99** (1977) 648; Phytochemistry **16** (1977) 1705.
[79] Leete, E.: Bioorg. Chem. **6** (1977) 237.
[80] Leete, E.: Recent Adv. Chem. Comp. Tob. Tob. Smoke Symp. **1977**, 365.
[81] Leete, E.: J. Chem. Soc., Chem. Commun. **1978**, 610.
[82] Leete, E.: J. Chem. Soc., Chem. Commun. **1978**, 1055.
[83] Leete, E.: J. Org. Chem. **44** (1979) 165.
[84] Leete, E.: Encycl. Plant Physiol. **8** (1980) 65.
[85] Leete, E.: Phytochemistry **20** (1981) 1037.
[86] Leete, E.; Bodem, G. E.; Manuel, M. F.: Phytochemistry **10** (1971) 2687.
[87] Leete, E.; Chedekel, M. R.: Phytochemistry **11** (1972) 2751.
[88] Leete, E.; Chedekel, M. R.: Phytochemistry **13** (1974) 1853.
[89] Leete, E.; Gros, E. G.; Gilbertson, T. J.: J. Am. Chem. Soc. **86** (1964) 3907.
[90] Leete, E.; Gros, E. G.; Gilbertson, T. J.: Tetrahedron Lett. **1964**, 587.
[91] Leete, E.; Leete, S. A. S.: J. Org. Chem. **43** (1978) 2122.
[92] Leete, E.; Leitz, F. H. B.: Chem. & Ind. (1957) 1572.
[93] Leete, E.; Liu, Y.-Y.: Phytochemistry **12** (1973) 593.
[94] Leete, E.; McDonell, J. A.: J. Am. Chem. Soc. **103** (1981) 658.
[94a] Leete, E.; Mueller, M. E.: J. Am. Chem. Soc. **104** (1982) 6440.
[95] Leete, E.; Pinder, A. R.: Phytochemistry **11** (1972) 3219.
[96] Leete, E.; Ranbom, K. C.; Riddle, R. M.: Phytochemistry **18** (1979) 75.
[97] Leete, E.; Slattery, S. A.: J. Am. Chem. Soc. **98** (1976) 6326.
[98] Leete, E.; Yu, M. L.: Phytochemistry **19** (1980) 1093.
[99] Leistner, E.; Gupta, R. N.; Spenser, I. D.: J. Am. Chem. Soc. **95** (1973) 4040, 4715; J. Chem. Soc., Chem. Commun. **1975**, 378.
[100] Lovkova, M. Y.: Izvest. Akad. Nauk SSSR, Ser. Biol. (1974) 821.
[101] Lovkova, M. Y.; Iljin, G. S.; Minozhedinova, N. S.: Priklad. Biochim. Mikrobiol. **9** (1973) 595.
[102] Lovkova, M. Y.; Nurimov, E.: Priklad. Biochim. Mikrobiol. **13** (1977) 129.
[103] Lovkova, M. Y.; Nurimov, E.; Iljin, G. S.: Biochimija **39** (1974) 388.
[103a] Luanratana, O.; Griffin, W. J.: Phytochemistry **21** (1982) 449.
[104] Manandhar, M. D.; Shoeb, A.; Kapil, R. S.; Popli, S. P.: Phytochemistry **17** (1978) 1814.
[105] Mann, D. F.; Byerrum, R. U.: Plant Physiol. **53** (1974) 603.
[106] Matsushita, H.; Tsujino, Y.; Yoshida, D.; Saito, A.; Kisaki, T.; Kato, K.; Noguchi, M.: Agric. Biol. Chem. **43** (1979) 193.
[107] McInnes, A. G.; Smith, D. G.; Wat, C. K.; Vining, L. C.; Wright, J. L. C.: J. Chem. Soc., Chem. Commun. **1974**, 281.
[108] McLean, S.; Murray, D. G.: Canad. J. Chem. **48** (1970) 867.
[109] Medina, E.; Spiteller, G.: Chem. Ber. **112** (1979) 376; **114** (1981) 814; Liebigs Ann. Chem. **1981**, 538, 2096.
[110] Minozhedinova, N. S.; Lovkova, M. Y.; Ibrayeva, B. S.: Izvest. Akad. Nauk SSSR, Ser. Biol. (1978) 390.
[111] Miyano, M.; Matsushita, H.; Yasumatsu, N.; Nishida, K.: Agric. Biol. Chem. **43** (1979) 1607.
[112] Miyano, M.; Matsushita, H.; Yasamatsu, N.; Nishida, K.: Agric. Biol. Chem. **43** (1979) 2205.
[113] Miyano, M.; Yasumatsu, N.; Matsushita, H.; Nishida, K.: Agric. Biol. Chem. **45** (1981) 1029.
[114] Mizusaki, S.; Kisaki, T.; Tamaki, E.: Plant Physiol. **43** (1968) 93.
[115] Mizusaki, S.; Tanabe, Y.; Kisaki, T.; Tamaki, E.: Phytochemistry **9** (1970) 549.
[116] Mizusaki, S.; Tanabe, Y.; Noguchi, M.; Tamaki, E.: Plant Cell Physiol. **12** (1971) 633.

[117] Mizusaki, S.; Tanabe, Y.; Noguchi, M.; Tamaki, E.: Phytochemistry **11** (1972) 2757.
[118] Mizusaki, S.; Tanabe, Y.; Noguchi, M.; Tamaki, E.: Plant Cell Physiol. **14** (1973) 103.
[119] Mothes, E.; Gross, D.; Schütte, H. R.; Mothes, K.: Naturwissenschaften **48** (1961) 623.
[120] Mothes, K.; Schütte, H. R.; Simon, H.; Weygand, F.: Z. Naturforsch. **14b** (1959) 49.
[121] Mukhamedzhanov, Z.; Aslanov, K. A.; Sadykov, A. S.; Leontev, V. B.; Kiryukhin, V. K.: Chim. Prir. Soedin. **4** (1968) 158.
[122] Mukherjee, R.; Chatterjee, A.: Tetrahedron **22** (1966) 1461.
[123] Murakoshi, I.; Kidoguchi, E.; Haginiwa, J.; Ohmiya, S.; Higashiyama, K.; Otomasu, H.: Phytochemistry **20** (1981) 1407.
[124] Nakane, M.; Hutchinson, C. R.: J. Org. Chem. **43** (1978) 3922.
[125] Noguchi, M.; Sakuma, H.; Tamaki, E.: Arch. Biochem. Biophys. **125** (1968) 1017; Phytochemistry **7** (1968) 1861.
[126] Nurimov, E.; Lovkova, M. Y.; Abdusamalov, B. A.: Prikl. Biochim. Mikrobiol. **10** (1974) 785.
[127] Ogasawara, N.; Chandler, J. L. R.; Gholson, R. K.; Rosser, R. J.; Andreoli, A. J.: Biochim. Biophys. Acta **141** (1967) 199.
[128] Ortega, M. V.; Brown, G. M.: J. Am. Chem. Soc. **81** (1959) 4437; J. Biol. Chem. **235** (1960) 2939.
[129] Pailer, M.; Ganzinger, P.: Monatsh. Chem. **106** (1975) 37.
[130] Phillipson, J. D.; Handa, S. S.: Phytochemistry **14** (1975) 2683.
[131] Ravishankav, G. A.; Mehta, A. R.: Experientia **37** (1981) 1143.
[132] Robinson, T.: Phytochemistry **17** (1978) 1903.
[133] Rueppel, M. L.; Mundy, B. P.; Rapoport, H.: Phytochemistry **13** (1974) 141.
[134] Rueppel, M. L.; Rapoport, H.: J. Am. Chem. Soc. **92** (1970) 5528; **93** (1971) 7021.
[135] Sastry, S. D.; Waller, G. R.: Phytochemistry **11** (1972) 2241.
[136] Schmelz, I.; Hoffmann, D.: Chem. Rev. **77** (1977) 295.
[137] Schütte, H. R.; Maier, W.; Mothes, K.: Acta Biochim. Polon. **13** (1966) 401.
[138] Schütte, H. R.; Maier, W.; Stephan, U.: Z. Naturforsch. B **23** (1968) 1426.
[139] Scott, T. A.; Bellion, E.; Mattey, M.: Eur. J. Biochem. **10** (1969) 318.
[140] Scott, T. A.; Devonshire, A. L.: Biochem. J. **124** (1971) 949.
[141] Scott, T. A.; Glynn, J. P.: Phytochemistry **6** (1967) 505.
[142] Scott, T. A.; Hussey, H.: Biochem. J. **96** (1965) 9c.
[143] Scott, T. A.; Mattey, M.: Biochem. J. **107** (1968) 606.
[144] Skursky, L.; Burleson, D.; Waller, G. R.: J. Biol. Chem. **244** (1969) 3238.
[145] Sleeper, H. L.; Fenical, W.: J. Am. Chem. Soc. **99** (1977) 2367.
[146] Smith, R. M.: The Celastracea Alkaloids. In: Manske, R. H. F. (Ed.): The Alkaloids, Vol. 16, p. 215. Academic Press, New York 1977.
[147] Solt, M. L.; Dawson, R. F.; Christman, D. R.: Plant Physiol. **35** (1960) 887.
[148] Summons, R. E.; Ellis, J.; Gellert, E.: Phytochemistry **11** (1972) 3335.
[149] Vella, G. J.; Hill, R. E.; Spenser, I. D.: J. Biol. Chem. **256** (1981) 10469.
[150] Wagner, H.; Brüning, R.; Lotter, H.; Jones, A.: Tetrahedron Lett. **1977**, 125.
[151] Wagner, H.; Heckel, E.; Sonnenbichler, J.: Tetrahedron Lett. **1974**, 213; Tetrahedron **31** (1975) 1949.
[152] Waller, G. R.; Henderson, L. M.: J. Biol. Chem. **236** (1961) 1186.
[153] Waller, G. R.; Lee, J. L. C.: Plant Physiol. **44** (1969) 522.
[154] Waller, G. R.; Nowacki, E. K.: Alkaloid Biology and Metabolism in Plants. Plenum Press, New York/London 1978, p. 134
155] Waller, G. R.; Skursky, L.: Plant Physiol. **50** (1972) 622.
156] Waller, G. R.; Yang, K. S.; Gholson, R. K.; Hadwiger, L. A.: J. Biol. Chem. **241** (1966) 4411.
[157] Warfield, A. H.; Galloway, W. D.; Kallianos, A. G.: Phytochemistry **11** (1972) 3371.
158] Wheeler, J. W.; Olubajo, O.; Storm, C. B.; Duffield, R. M.: Science **211** (1981) 1051.
158a] Wigle, I. D.; Mestichelli, L. J. J.; Spenser, I. D.: J. Chem. Soc., Chem. Commun. **1982**, 662.

[159] Yakushijin, K.; Sugiyama, S.; Mori, Y.; Murata, H.; Furukawa, H.: Phytochemistry **19** (1980) 161.
[160] Yamada, K.; Shizuri, Y.; Hirata, Y.: Tetrahedron **34** (1978) 1915.
[161] Yang, K. S.; Gholson, R. K.; Waller, G. R.: J. Am. Chem. Soc. **87** (1965) 4184.
[162] Yang, K. S.; Waller, G. R.: Phytochemistry **4** (1965) 881.
[163] Yoshida, D.; Mitake, T.: Plant Cell Physiol. **7** (1966) 301.
[164] Yoshida, S.; Yoneyama, K.; Shiraishi, S.; Watanabe, A.; Takahashi, N.: Agric. Biol. Chem. **41** (1977) 849, 855.
[165] Zielke, H. R.; Byerrum, R. U.; O'Neal, R. M.; Burns, L. C.; Koeppe, R. E.: J. Biol. Chem. **243** (1968) 4757.

14 Alkaloids Derived from Tyrosine and Phenylalanine

H. R. Schütte and H. W. Liebisch

As noted earlier these two amino acids are derived in different ways from the precursors chorismic acid and prephenic acid (*cf.* chapter 8.2.1). Tyrosine may be converted into its more highly hydroxylated homologues, dihydroxyphenylalanine (dopa) and trihydroxyphenylalanine. Phenylalanine is preferentially converted into cinnamic acid, the parent substance of the large group of non-nitrogenous phenylpropanoids [435, 436]. Cinnamic acid can undergo hydroxylation to derivatives that are incorporated into alkaloids.

Therefore tyrosine and its higher hydroxylated derivatives act as the nitrogen-carrying C_6-C_2 or C_6-C_3 portion of a wide variety of alkaloids. The non-nitrogenous portion of these systems comes from tyrosine or phenylalanine, the latter *via* cinnamic acid, for C_6-C_1, C_6-C_2 and C_6-C_3 substances or other readily available precursors, *e.g.* formate or glyoxylate for C_1, pyruvate for C_2, mevalonate for C_5 and loganin for C_9 and C_{10} alkaloids. Thus the large group of alkaloids derived from tyrosine and phenylalanine includes simple β-phenethylamines and the reaction products of these amines with a lot of different keto-compounds such as simple tetrahydroisoquinolines, alkyltetrahydroisoquinolines and especially the different benzyltetrahydroisoquinolines [80a, 164, 194, 233, 355, 430, 431, 462].

The isoquinolines, particularly the benzylisoquinolines, can be transformed into a wide range of different structures. Some typical transformations are indicated in Fig. 14.1, which illustrates the great diversity of skeletal types that can be produced. Those of epistephanine (**1**), isoboldine (**2**) and scoulerine (**3**) bear an obvious structural relationship to their benzylisoquinoline precursor (**4**), but this relationship is less immediately obvious in the case of morphine (**5**) and chelidonine (**6**), and in protostephanine (**7**) it has been almost completely obscured. Nevertheless the relationship has now been firmly established in every case, as we shall see.

The widespread isoquinoline alkaloids occur in variable amounts particularly in the *Magnolialae, Ranalae, Aristolochiae, Papaveralae, Leguminosae, Geranialae, Rutalae, Myrtiflorae, Tubiflorae* species. Simple examples also are found in *Cactaceae* and *Chenopodiaceae* [123, 169, 366, 446, 449]. Closely related compounds are present in *Rubiaceae* and *Amaryllidaceae* species.

The most important reaction for the biosynthesis of these substances is the condensation of a phenethylamine derivative with a phenylacetaldehyde or with other

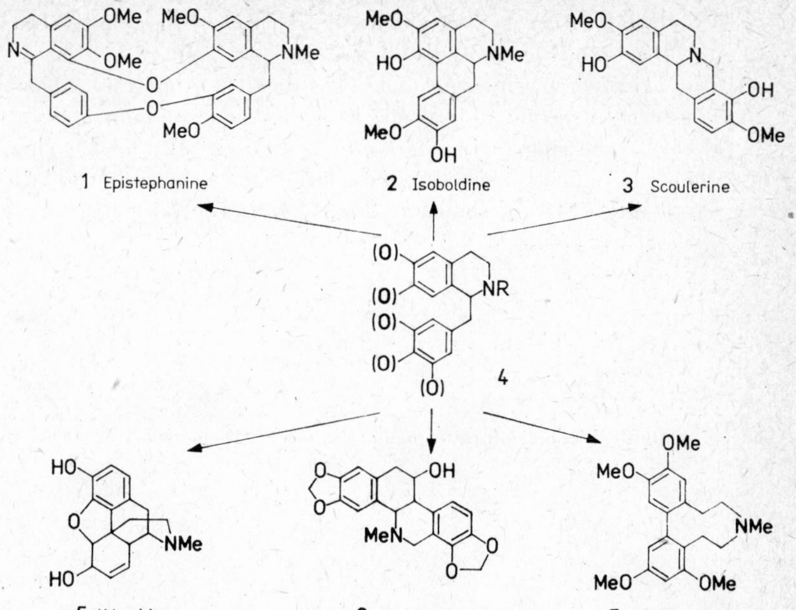

1 Epistephanine 2 Isoboldine 3 Scoulerine

4

5 Morphine 6 Chelidonine 7 Protostephanine

Fig. 14.1
Various benzylisoquinolines and derived compounds

carbonyl compounds to isoquinolines. As early as 1910, Winterstein and Trier [502] suggested that tetrahydropapaverine is synthesized in cells from β-3,4-dimethoxy-phenethylamine and 3,4-dimethoxyphenylacetaldehyde by aldehyde ammonia reaction and condensation with the proton in the *p*-position to one methoxyl group. This very simple reaction makes it clear that the simple types of the benzylisoquinolines are widespread and that more complicated derivatives such as the morphinans occur only in a few species. Such condensations have also been realized under physiological conditions of pH, temperature, and concentration. For example the compound 9 is synthesized from β-3,4-dihydroxyphenethylamine (dopamine) (8) and acetaldehyde at pH 5 in aqueous solution at 25 °C in nearly quantitative yield (Fig. 14.2).

Before going into details another key process should be introduced, that of oxidative coupling of phenolic rings. The general concept was first enunciated by early pioneers, especially Robinson [419], but the great breakthrough in our under-

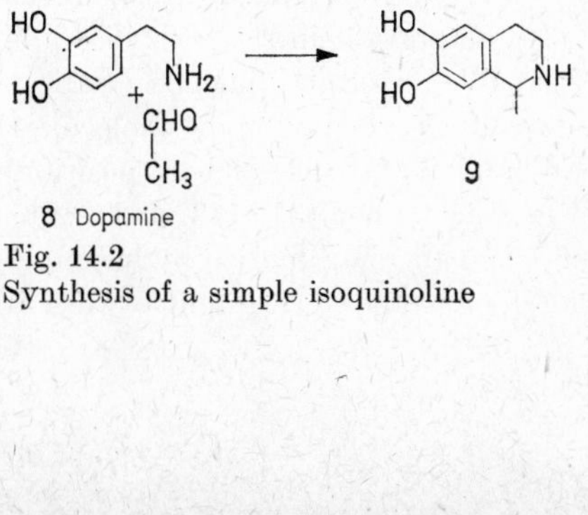

8 Dopamine 9

Fig. 14.2
Synthesis of a simple isoquinoline

standing of what is involved came when the hypothesis was placed on a sound mechanistic basis by Barton and Cohen [20]. They pointed out that if a phenolate anion **10** is oxidized to the corresponding radical **11**, the system has the potential to combine with a second radical species at any one of four sites: at the oxygen itself (**11**) or at the ring carbons *ortho* or *para* to it (**12** and **13**) (Fig. 14.3) [334]. The mesomeric radicals formed on oxidation of a benzyltetrahydroisoquinoline such as reticuline (**14**) are shown in Fig. 14.4. If the *para* hydroxy group of the

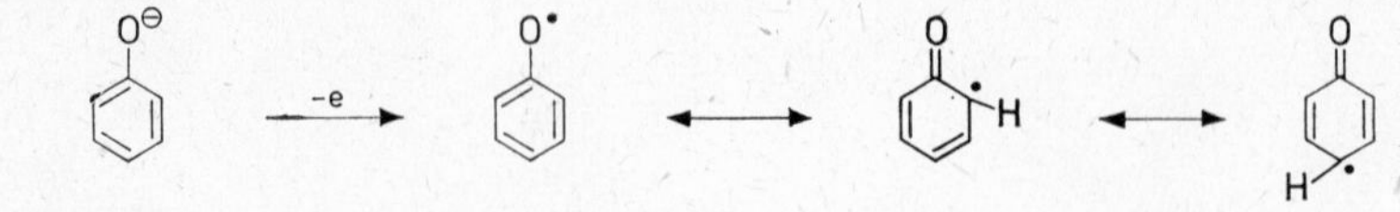

10 Phenolate Anion 11 Aroxyl Radical 12 Ketomethin radical 13

Fig. 14.3
Radical formation from phenol by loss of an electron

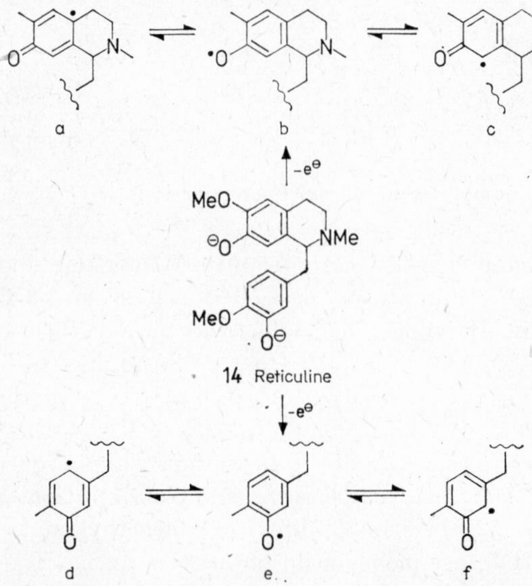

a b c

14 Reticuline

d e f

Fig. 14.4
The transformation of a tetrahydrobenzylisoquinoline to give mesomeric radicals by phenol oxidation [20]

benzylic half is free, other possibilities exist. The radical pairing reactions can be carried out intermolecularly with a second molecule (*e.g.* for bisbenzylisoquinolines) or intramolecularly. The latter is vitally important in the benzylisoquinoline family where there are two linked aryl rings capable of forming phenoxy radicals (Fig. 14.4).

Intramolecular oxidative coupling of a benzylisoquinoline molecule thus leads to the formation of a new carbon-carbon bond at the *ortho* or *para* position with respect to the free hydroxyl groups in the two aryl rings [31, 433, 462]. When there is a hydrogen at the site of coupling the ring can re-aromatize by a simple enolization. If there is a carbon substituent at the site of coupling, straightforward aromatization is not possible; the carbon skeleton may remain unchanged in subsequent steps (morphine) or it may re-aromatize either by migration of carbon (isothebaine) or by a fragmentation process (erythraline). Thus these processes greatly extend the range of skeletal types that may be formed as a consequence of oxidative coupling.

14.1 Phenylalkylamines

Phenylalkylamines can be considered as protoalkaloids of the isoquinolines. This group includes, for example, hordenine (**23**), *N*-methyltyramine (**22**), mescaline (**31**) and other simple phenethylamines (Fig. 14.6), and also ephedrine (**38**) and nor-pseudoephedrine (**39**) (Fig. 14.8). Epinephrine (**33**) and norepinephrine (**36**) (Fig. 14.7), hormones of the adrenal medulla, are characteristic compounds of mammalian systems. Biosynthesis has been investigated in detail in such systems [7, 430]. Norepinephrine has also been detected in different plants, *e.g.* in members of the

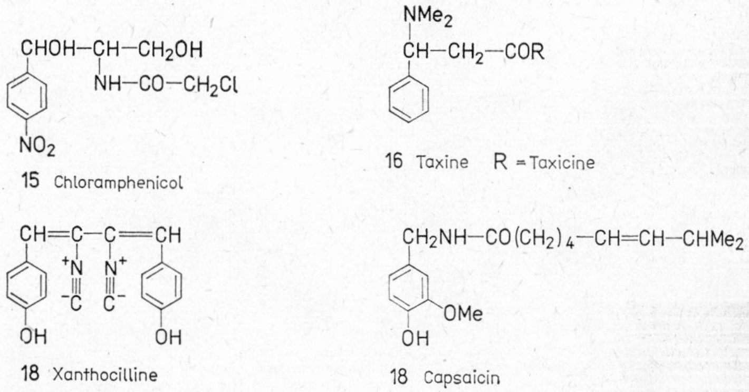

15 Chloramphenicol

16 Taxine R = Toxicine

18 Xanthocilline

18 Capsaicin

Fig. 14.5
Simple phenylalanine- or tyrosine-derived compounds

Cactaceae, Leguminosae, Musaceae, Passifloraceae, Portulacaceae, Rosaceae, Rutaceae, and *Solanaceae* [458]. Epinephrine is involved in the biogenesis of normacromerine (Fig. 14.7). Other simple phenylalanine or tyrosine derived compounds (Fig. 14.5) are the phenylpropanolamine antibiotic chloramphenicol (**15**) from *Streptomyces venezuelae* [430], the taxus ester-alkaloids, *e.g.* taxine (**16**) which contains the phenylalanine derived 3-dimethylamino-3-phenylpropionic acid [430], capsaicin (**18**), the antibiotic xanthocilline (**17**) from *Penicillium notatum* [430] and some cyanogenic glycosides and glucosinolates [158—160, 243, 434, 491].

14.1.1 Hordenine, *N*-Methyltyramine and Other Simple Phenethylamines

Hordenine (**23**) and *N*-methyltyramine (**22**) occur, for example, in roots of barley seedlings, in *Lophophora williamsii*, *Opuntia clavata* or *Citrus* species and are biosynthesized by hydroxylation of phenylalanine (**19**) to tyrosine (**20**) followed by decarboxylation and methylation with methionine (Fig. 14.6) [284, 330, 352, 358, 495]. Two distinct enzymes are responsible for the subsequent *N*-methylations [358a]. During the formation of hordenine and *N*-methyltyramine an *NIH*-shift operates [326]. Tyrosine, dopamine (8) and 3-hydroxy-4-methoxyphenethylamine were incorporated into 3,4-dimethoxyphenethylamine, the major alkaloid in *Echinocereus merkeri* [356] and also in *Lophophora williamsii* and *T. pachanoi* [341]. Perhaps the true intermediate is 3-methoxy-4-hydroxyphenethylamine (24) which has been found in *E. merkeri* [4]. The direction of methylation of dopamine determines whether the plant makes dimethoxyphenethylamine and its derivatives or whether it makes tetrahydroisoquinolines. In *Trichocereus pachanoi*, which produces 3,4-

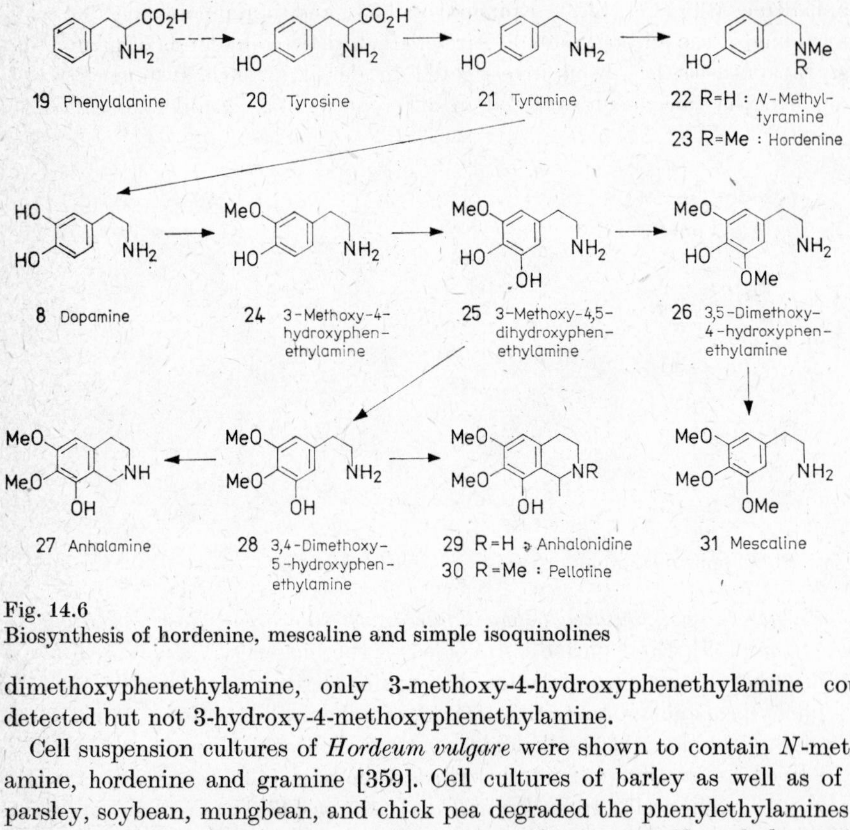

19 Phenylalanine
20 Tyrosine
21 Tyramine
22 R=H : *N*-Methyltyramine
23 R=Me : Hordenine
8 Dopamine
24 3-Methoxy-4-hydroxyphenethylamine
25 3-Methoxy-4,5-dihydroxyphenethylamine
26 3,5-Dimethoxy-4-hydroxyphenethylamine
27 Anhalamine
28 3,4-Dimethoxy-5-hydroxyphenethylamine
29 R=H : Anhalonidine
30 R=Me : Pellotine
31 Mescaline

Fig. 14.6
Biosynthesis of hordenine, mescaline and simple isoquinolines

dimethoxyphenethylamine, only 3-methoxy-4-hydroxyphenethylamine could be detected but not 3-hydroxy-4-methoxyphenethylamine.

Cell suspension cultures of *Hordeum vulgare* were shown to contain *N*-methyltyramine, hordenine and gramine [359]. Cell cultures of barley as well as of wheat, parsley, soybean, mungbean, and chick pea degraded the phenylethylamines hordenine, tyramine, and dopamine to CO_2. Intermediates were *N*-methyltyramine and *p*-hydroxyphenylacetic acid, *p*-hydroxybenzoic acid, and 3,4-dihydroxyphenylacetic acid respectively. Oxidative polymerization of phenylethylamines and intermediate phenols was also observed. Also intact plants of H. vulgare degrade hordenine to CO_2 [425a, 426, 358b]. *p*-Methoxyphenethylamine is demethylated by cell cultures of *Catharanthus roseus* and is metabolized to *p*-methoxyphenethyl-β-*D*-glucopyranoside [239]. In *Dolichothele sphaerica* (*Cactaceae*) *N*-methylphenethylamine is biosynthesised from phenylalanine *via* phenethylamine [284a].

14.1.2 Mescaline

Mescaline (**31**) is the hallucinogenic compound of the peyote cactus (*Lophophora williamsii* = *Anhalonium lewinii*). It is produced in this plant together with a number of closely related β-phenethylamines and some tetrahydroisoquinolines [123, 169]. The peyote cactus is used by Indians in Mexico and in the southern parts of the USA as a hallucinogenic drug. *Trichocereus pachanoi* is another mescaline producing cactus with only traces of other alkaloids, which is used under the name San Pedro by Indians in Peru to prepare a hallucinogenic drink called "Cimora".

Investigations of the biosynthesis of mescaline (**31**) in these two plants [323, 341–343, 358, 392, 393, 423, 424] have shown that dopamine (**8**) deriving from phenylalanine (**19**) via tyrosine (**20**) and tyramine (**21**) is *O*-methylated to give 3-methoxy-4-hydroxyphenethylamine (**24**) followed by hydroxylation to 3-methoxy-4,5-dihydroxyphenethylamine (**25**), *O*-methylation to give 3,5-dimethoxy-4-hydroxyphenethylamine (**26**) and, finally, *O*-methylation to give mescaline (**31**) (Fig. 14.6). An *O*-methyltransferase has been partially purified from *Lophophora williamsii* that transfers the methyl group of *S*-adenosyl-*L*-methionine to different hydroxy groups of dopamine and other intermediates in the biosynthesis of mescaline [29, 30]. The methylation of 3-methoxy-4,5-dihydroxyphenethylamine (**25**) apparently determines whether mescaline or the tetrahydroisoquinolines are synthesized (Fig. 14.6). Thus 3-methoxy-4,5-dihydroxyphenethylamine is a precursor for both alkaloids while the 3,4-dimethoxy-5-hydroxyderivative (**28**) is an intermediate for the tetrahydroisoquinolines [288] and the 3,5-dimethoxy-4-hydroxyderivative (**26**) is essentially only an intermediate in mescaline synthesis [287, 393]. In grafted *Trichocereus pachanoi* no translocation of mescaline has been found [408].

14.1.3 Normacromerine and Synephrine

The cactus genus *Coryphanta* has the unique ability to proliferate and accumulate β-hydroxylated phenethylamines [142, 409]. Thus normacromerine (*N*-methyl-3,4-dimethoxy-β-hydroxyphenethylamine) (**34**), a psychotropic epinephrine derivative is the major alkaloid of the cactus *Coryphantha macromeris* [285]. It is biosynthesized

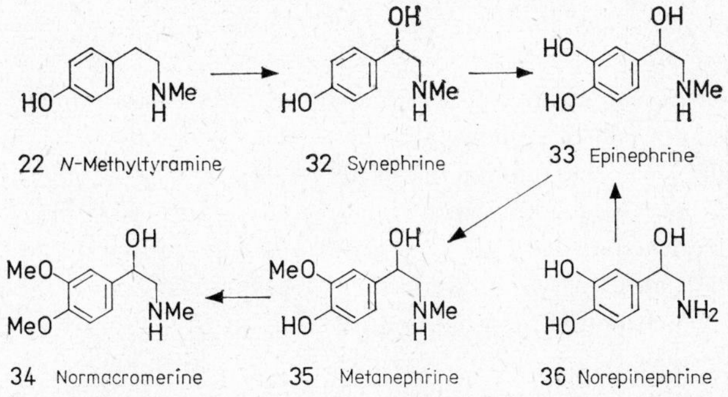

22 *N*-Methyltyramine
32 Synephrine
33 Epinephrine
34 Normacromerine
35 Metanephrine
36 Norepinephrine

Fig. 14.7
Biosynthesis of epinephrine and normacromerine in plants

by decarboxylation of tyrosine (**20**), methylation of the resultant tyramine (**21**) to *N*-methyltyramine (**22**), and hydroxylation *via* synephrine (**32**) and epinephrine (**33**) (Fig. 14.7) [282, 284, 286]. Norepinephrine (**36**) is also incorporated but seems not to be on the major biosynthetic pathway from tyrosine to normacromerine (**34**). Metanephrine (**35**) was shown to be the immediate precursor of normacromerine [283]. All of the intermediates given in Fig. 14.7 have been either detected in or isolated from extracts of *C. macromeris* including the first reported detection of epinephrine (**33**) in a member of the plant kingdom. Dopamine (**8**) is not involved in normacromerine biosynthesis. Thus the biosynthesis of these cactus alkaloids may involve a different sequence than that utilized in the mammalian biosynthesis of norepinephrine and epinephrine [7]. The β-hydroxylation of *N*-methyltyramine (**22**) to give synephrine (**32**) has been established in *Citrus* species [495].

14.1.4 Ephedrine and Norpseudoephedrine

Ephedrine (**38**) and norpseudoephedrine (**39**) are present in *Ephedra* species and have an aliphatic hydroxyl group and a C-methyl group. Ephedrine is important because of its sympaticomimetic activity. In contrast to earlier studies on the biosynthesis of ephedrine in *Ephedra distachya* (*Ephederaceae*) [430, 453], which claimed that ^{15}N labelled phenylalanine was incorporated along with its nitrogen, more detailed experiments have shown that [2-^{14}C]phenylalanine was not incorporated

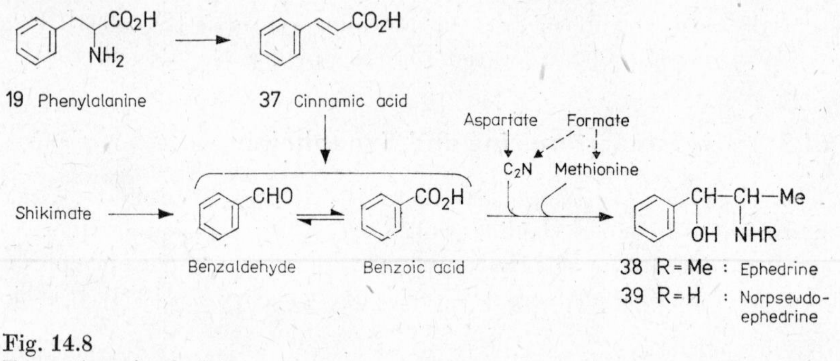

Fig. 14.8
Formation of ephedrine

into *L*-ephedrine, whereas aromatic-^{3}H and -3-^{14}C were introduced [510, 511]. These observations suggest that phenylalanine is cleaved between C-2 and C-3 and only the remaining C_6—C_1 part participates in the biosynthesis of ephedrine. The *N*-methyl is unequivocally derived from active methionine [453]. The C_2—N unit might be derived from aspartate or some closely related compounds [511]. Thus it has been suggested that C_6—C_1 compounds such as benzoic acid or benzaldehyde react with a C_2—N compound or their equivalent to give ephedrine (Fig. 14.8). Phenylalanine was also reported to be an efficient precursor for *D*-norpseudoephedrine (cathine) in *Catha edulis* [318]. The amount of ephedrine produced by *Ephedra foliata* callus culture depends on the quality of the light [358, 454].

14.1.5 Capsaicinoids

Capsaicinoids such as capsaicin (**18**) are responsible for the pungent property of red peppers. It has been reported that phenylalanine, cinnamate derivatives, and vanillylamine are incorporated *in vivo* into the vanillylamine moiety of capsaicinoids and that valine is incorporated *via* isobutyric acid and 8-methylnonanoic acid into the acyl moiety of capsaicin (**18**) and dihydrocapsaicin, the acyl constituents of which are even-no, branched chain fatty acids, while leucine was incorporated *via* isopentanoic acid and 9-methyldecanoic acid into nordihydrocapsaicin and homodihydrocapsaicin, which contain odd-no, branched chain fatty acids [77, 209a, 313, 313a, 313b, 329, 480]. Furthermore dihydrocapsaicin, capsaicin, and nordihydrocapsaicin are formed and accumulate in the fruits of the sweet pepper, *Capsicum annuum*, when the peduncle is put into an aqueous solution of vanillylamine and isocapric acid and the whole fruit aged for several days under continuous light [251]. The content of capsaicinoid reached its maximum about 25 days after flowering [481]. The fluctuation of the capsaicinoid content and composition is related to those of leucine and valine. Dihydrocapsaicin was also formed by cell-free extracts of the fruits in a reaction mixture containing vanillylamine and isocapric acid and corresponding extracts of placenta of *C. annuum* catalysed the formation of capsaicinoid from vanillylamine and different acyl-CoA compounds [210, 252]. The capsaicinoid-synthesizing enzyme activity in sweet pepper fruits was found in both mitochondrial and microsomal fractions of placenta and pericarp. The capsaicinoid synthetase was located in the vacuole [209, 209b].

14.2 Simple Isoquinoline Alkaloids

14.2.1 Pellotine, Anhalamine and Related Compounds

For the simple isoquinoline alkaloids such as pellotine (**30**), anhalonidine (**29**), and anhalamine (**27**) it has been shown that tyrosine (**20**) and dopamine (**8**) serve as specific precursors in *Lophophora williamsii* [35, 323, 342, 392]. Therefore, it is assumed that the pathway from tyrosine (**20**) to these alkaloids involves hydroxylation to 3,4-dihydroxyphenylalanine (dopa) and then decarboxylation to dopamine (**8**) (Fig. 14.6). Of the 3,4,5-trihydroxy-β-phenethylamines, 3,4-dimethoxy-5-hydroxyphenethylamine (3-demethylmescaline) (**28**) is a particularly excellent precursor of these alkaloids in peyote [36, 340, 342, 344, 393] and it cannot go on to mescaline (**31**). 3-Demethylmescaline has also been identified in the peyote cactus [3, 279, 345].

Formate and acetate have been postulated for the origin of C-1 and its substituent (C-9) in tetrahydroisoquinoline alkaloids [35, 36]. Methionine should serve as the source of C-1 in anhalamine (**27**). The identification of peyoglutam (**43**) and mescalotam (**44**) [277], compounds in peyote evidently containing an α-ketoglutarate moiety, appeared to support earlier proposals [228] regarding the carbonyl group of α-keto-acids such as pyruvic acid as a source of C-1 in tetrahydroisoquinolines (Fig. 14.9). Pyruvate has been shown to be incorporated specifically into anhalonidine (**29**) [327]. Administered *N*-acetyldemethylmescaline was deacetylated to 3-demethylmescaline (**28**) prior to its incorporation into anhalamine (**27**) and anhalonidine (**29**) [278]. Peyoxylic (**40**) and peyoruvic acid (**41**), the glyoxylic and pyruvic acid condensation products of 3-demethylmescaline, were identified in the amino acid

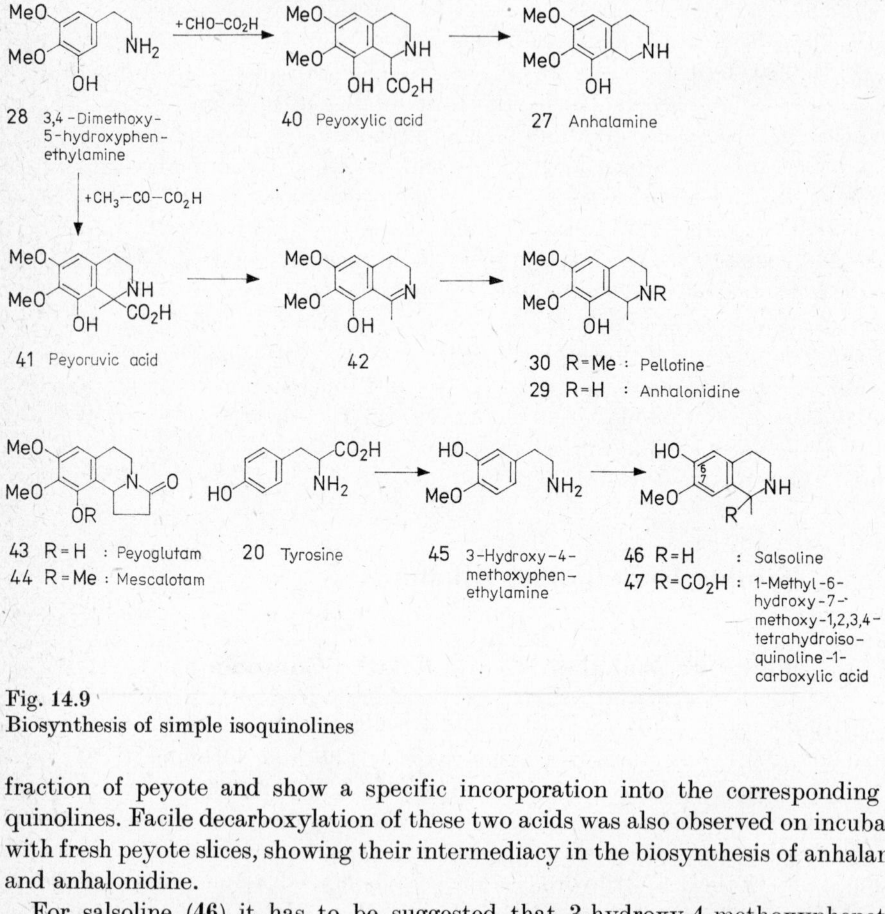

Fig. 14.9
Biosynthesis of simple isoquinolines

fraction of peyote and show a specific incorporation into the corresponding isoquinolines. Facile decarboxylation of these two acids was also observed on incubation with fresh peyote slices, showing their intermediacy in the biosynthesis of anhalamine and anhalonidine.

For salsoline (**46**) it has to be suggested that 3-hydroxy-4-methoxyphenethylamine (**45**) is a precursor (Fig. 14.9). The specific incorporation of 1-methyl-6-hydroxy-7-methoxy-1,2,3,4-tetrahydroisoquinoline-1-carboxylic acid (**47**) [356], a natural constituent of peyote, in experiments with *Echinocereus merkeri* shows that the amine should react with pyruvic acid yielding this isoquinolinic acid, which after decarboxylation and probably reduction gives rise to salsoline and possibly also to salsolidine (**46** 6-MeO) and carnegine (**46** 6-MeO and NMe) (in *Carnegiea gigantea*) [143].

Simple tetrahydroisoquinoline alkaloids are formed in human and animal metabolism subsequent to alcohol consumption by reaction of alcohol derived acetaldehyde with corresponding biogenic amines. These alkaloids affect the vegetative nervous system [129].

1-Methyl-6-hydroxy-1,2,3,4-tetrahydroisoquinoline-3-carboxylic acid occurs in *Euphorbia myrsinites* [367] and is formed from *m*-tyrosine without decarboxylation by the common tetrahydroisoquinoline synthesis [368]. Callus culture of *Stizolobium hassjoo* can convert dopa into the corresponding carboxylic acids [380, 426a].

14.2.2 Lophocerine and Pilocereine

In the case of lophocerine (**48**) and its trimeric pilocereine (**49**) from the cactus *Lophocereus schottii* it has been shown that tyrosine serves as the precursor of the C_6-C_2-N portion of the tetrahydroisoquinoline nucleus and mevalonic acid is incorporated specifically into the remaining C_5-unit [381, 382, 438]. The occurrence of lophocerine in the plant and a significantly high incorporation of this monomeric alkaloid into pilocereine substantiate its role as the natural precursor of the trimeric compound [383] by oxidative phenolic coupling (Fig. 14.10).

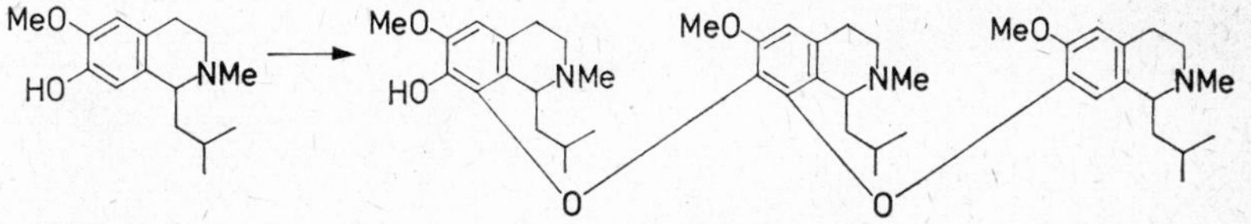

48 Lophocerine

49 Pilocereine

Fig. 14.10
Lophocerine and pilocereine

14.2.3 Ipecac Alkaloids

The ipecac alkaloids from the rhizome and roots of *Cephaelis ipecacuanha* (*Rubiaceae*) such as cephaeline (**57**) and emetine (**56**) are pharmacologically active compounds used as an emetic and as an anti-amoebic. Cephaeline has also been detected in *Alangium lamarckii* (*Alangiaceae*) along with a number of related isoquinoline alkaloids. In both plant species, neutral alkaloidal glucosides are also found, ipecoside (**52**) in *Cephaelis* and alangiside (**54**) in *Alangium*. These alkaloids bear a certain relationship to the monoterpenoid indole bases. The biosynthetic pathway involves the condensation of dopamine (**8**) and secologanin (**50**), derived from monoterpenoid metabolism, with the formation of both deacetylipecoside (**51**) and deacetylisoipecoside (**53**) [69]. In *Cephaelis ipecacuanha* as well as in *Alangium lamarckii* deacetylisoipecoside with the 1α-configuration is incorporated with retention of configuration into cephaeline (**57**) and emetine (**56**) most likely involving protoemetine (**55**) as an intermediate (Fig. 14.11) [63, 373, 374, 374a]. The 1β-epimer, deacetylipecoside (**51**), is the precursor only of the alkaloidal glucosides, ipecoside (**52**) in *C. ipecacuanha* and alangiside (**54**) in *Alangium*. Furthermore the proton at the C-7 of secologanin is retained during the formation of the metabolites with both α- and β-configurations. A similar pathway has been found for the biosynthesis of tubulosine in young *Alangium lamarckii* plants starting from tyrosine to dopamine (**8**) + secologanin (**50**) and deacetylisoipecoside (**53**) involving a tryptamine part to deoxytubulosine and tubulosine [85a].

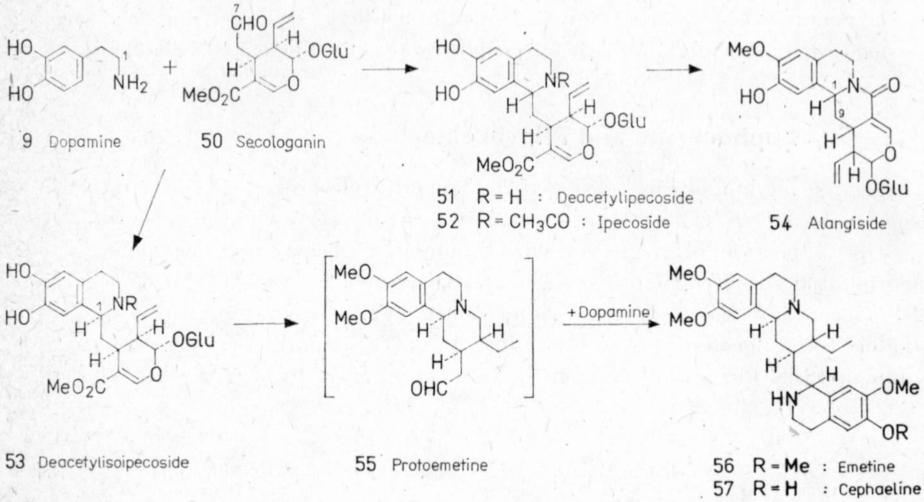

9 Dopamine
50 Secologanin
51 R = H : Deacetylipecoside
52 R = CH_3CO : Ipecoside
54 Alangiside
53 Deacetylisoipecoside
55 Protoemetine
+Dopamine
56 R = Me : Emetine
57 R = H : Cephaeline

Fig. 14.11
Biosynthesis of ipecac alkaloids

14.2.4 Cryptostyline

A unique 1-phenyl substituted isoquinoline structure is found in the cryptostyline alkaloids, *e.g.* cryptostyline-I (**61**) from *Cryptostylis erythroglossa* (*Orchidaceae*). The biosynthesis goes from tyrosine (**20**) and dopa (**62**) via dopamine (**8**) and the phenethylamines (**58** and **59**) but not (**24**). This suggests that the ring closure to the tetrahydroisoquinoline skeleton is facilitated by a *para*-hydroxy group [1, 2]. Dopamine, vanillin, and probably protocatechualdehyde (but not isovanillin) can serve as precursors for the remaining $C_6—C_1$ residue in cryptostyline-I (Fig. 14.12). Furthermore it was found that the dihydroisoquinoline (**60**) is a highly efficient precursor for cryptostyline-I, indicative of a reduction after ring closure.

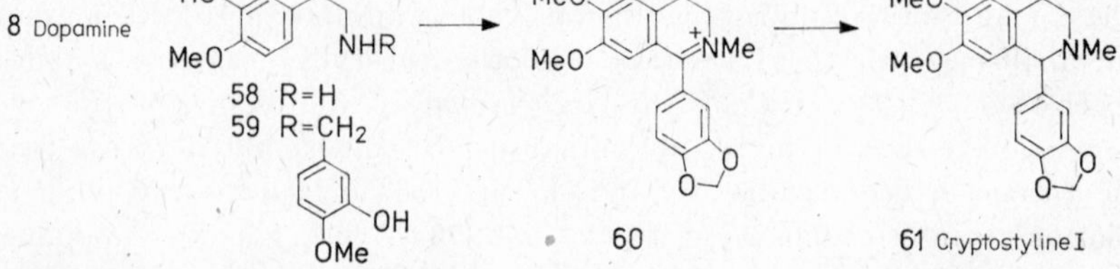

8 Dopamine
58 R = H
59 R = CH_2
60
61 Cryptostyline I

Fig. 14.12
Biosynthesis of cryptostyline I

14.3 Benzyltetrahydroisoquinolines

14.3.1 Norlaudanosoline

More important than the simple tetrahydroisoquinolines are the benzyltetrahydroisoquinolines such as norlaudanosoline (**65**) and its various methylation derivatives, *e.g.* reticuline (**66**), orientaline (**71**), laudanidine (**72**), laudanosine (**74**), and tetrahydropapaverine (**73**), which are widely distributed in the *Papaveraceae* family. Norlaudanosoline contains four OH-groups, whereas norcoclaurine (**86**) has three hydroxy groups, only one in ring C. According to earlier proposals, the norlaudanosoline system is derived biosynthetically from two units of dopa (**62**), which by decarboxylation and oxidative deamination were believed to give rise to dopamine (**8**) and 3,4-dihydroxy-phenylacetaldehyde, respectively. Formation of a Schiff's base and ring closure would give norlaudanosoline, the primary substance from which a lot of other benzylisoquinolines are derived by dehydrogenation and/or methylation (Figs. 14.13, 14.14 and 14.15).

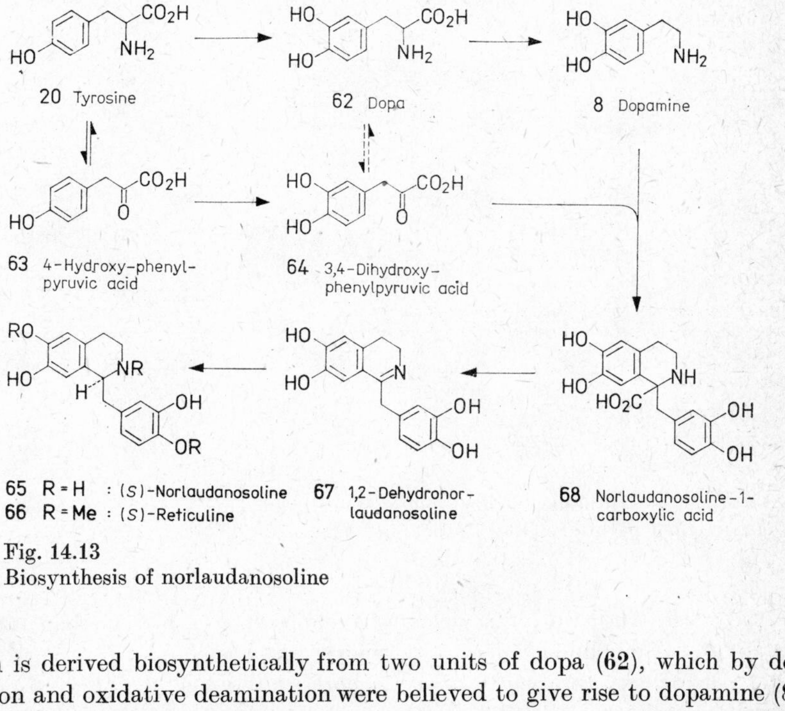

20 Tyrosine
62 Dopa
8 Dopamine
63 4-Hydroxy-phenylpyruvic acid
64 3,4-Dihydroxyphenylpyruvic acid
65 R = H : (*S*)-Norlaudanosoline
66 R = Me : (*S*)-Reticuline
67 1,2-Dehydronorlaudanosoline
68 Norlaudanosoline-1-carboxylic acid

Fig. 14.13
Biosynthesis of norlaudanosoline

In contrast to these proposals, experiments with capsules of intact *Papaver somniferum* plants and *P. orientale* latex and seedlings show that dopamine (**8**) and 3,4-dihydroxy-phenylpyruvic acid (**64**) are the reacting units. They probably both derive from dopa (**62**) by decarboxylation on the one hand, and by transamination

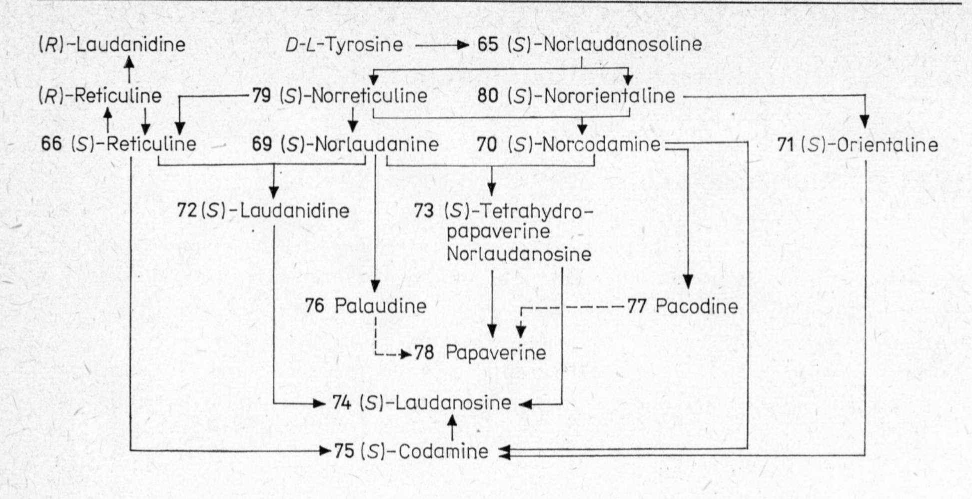

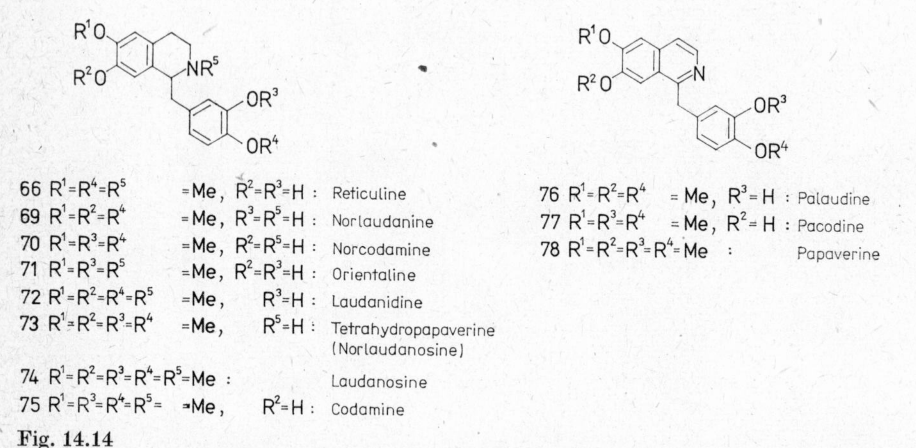

Fig. 14.14
Pathways for biosynthesis of benzylisoquinoline alkaloids in *Papaver somniferum*; broken lines designate minor pathways [130]

on the other. The two units yield norlaudanosoline carboxylic acid (**68**), the first tetrahydroisoquinoline alkaloid of the series, which is converted to norlaudanosoline (**65**) by decarboxylation *via* 1,2-dehydronorlaudanosoline (**67**) (Fig. 14.13) [58, 500]. Norlaudanosoline produced from dopa in the opium poppy is the laevorotatory enantiomer with the (*S*)-configuration.

In a cell-free system from *P. somniferum*, dopamine and 3,4-dihydroxy-phenylpyruvic acid were found to be incorporated into norlaudanosoline-1-carboxylic acid and norlaudanosoline [444].

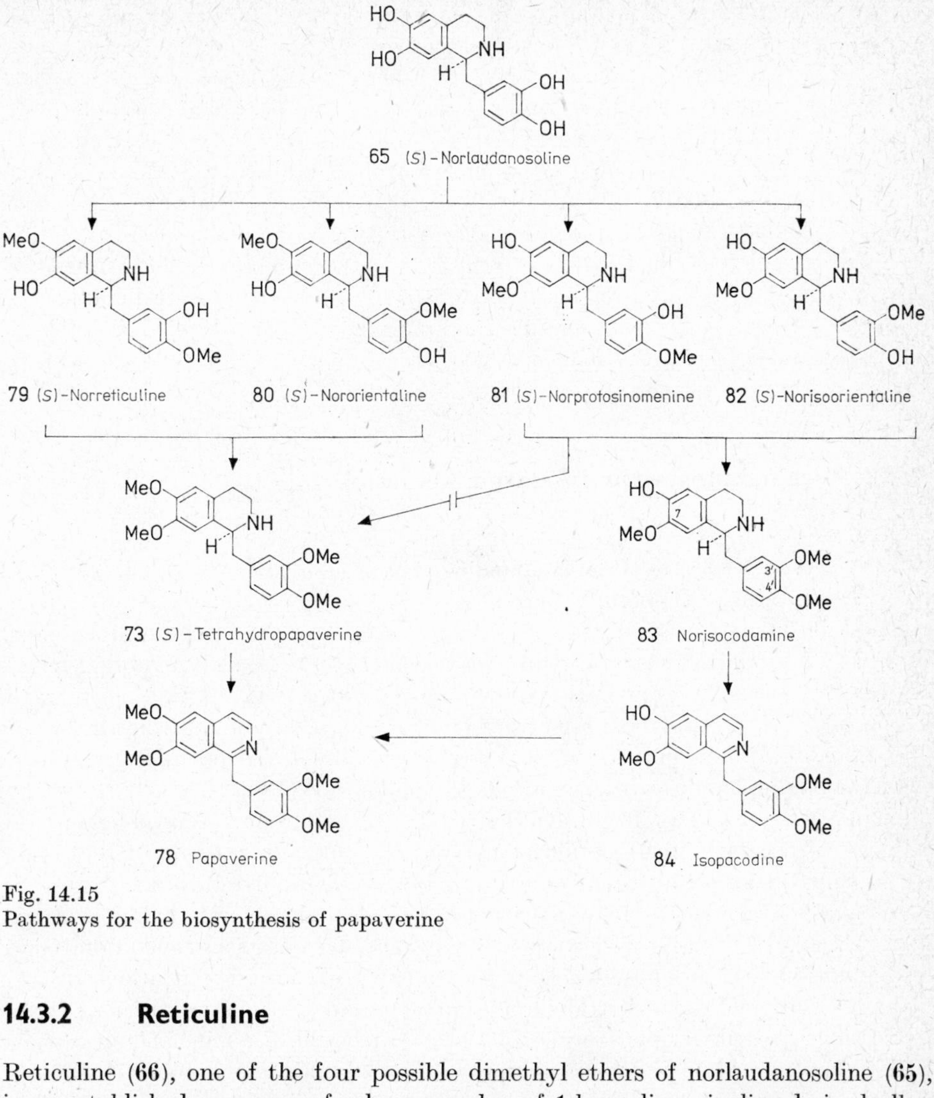

65 (*S*)-Norlaudanosoline

79 (*S*)-Norreticuline

80 (*S*)-Nororientaline

81 (*S*)-Norprotosinomenine

82 (*S*)-Norisoorientaline

73 (*S*)-Tetrahydropapaverine

83 Norisocodamine

78 Papaverine

84 Isopacodine

Fig. 14.15
Pathways for the biosynthesis of papaverine

14.3.2 Reticuline

Reticuline (**66**), one of the four possible dimethyl ethers of norlaudanosoline (**65**), is an established precursor of a large number of 1-benzylisoquinoline-derived alkaloids. It occurs in plants in (+)-, (—)-, and (±)-forms. This asymmetry seems to be generated at the norlaudanosoline stage. Racemization of reticuline, which is a necessary step for the biosynthesis of the different alkaloids, appears to be very substrate-specific. Thus (*R*)-reticuline is a specific precursor for the alkaloids of the morphine type and (*S*)-reticuline (**66**) gives rise to the tetrahydroberberine type such as (*S*)-scoulerine (**3**).

The reticuline biosynthesis was studied in detail in *Litsea glutinosa* (*Lauraceae*) [102, 489]. Tyrosine (**20**), 4-hydroxy- (**63**), and 3,4-dihydroxy-phenylpyruvic acid (**64**) participate in the formation of both portions of reticuline, whereas dopa (**62**) and dopamine (**8**) contribute only to the formation of the phenethylamine portion (Fig. 14.13). The benzylic portion is biosynthesized from 3,4-dihydroxy-phenylpyruvic acid not derived from dopa. The incorporations of these two α-keto acids into the phenethyl portion suggest that these acids could be aminated in the plant to give tyrosine (**20**) and dopa (**62**). Furthermore the intermediacy of norlaudanosoline-1-carboxylic acid (**68**), norlaudanosoline (**65**), and 1,2-dehydronorlaudanosoline (**67**) was shown, suggesting that the carboxylic acid is decarboxylated to didehydronorlaudanosoline, which is then reduced. The *O*-methylation precedes *N*-methylation and there seems to be no rigid selectivity of *O*-methylation in the biosynthesis of reticuline. *N*-Methylation is the terminal step and a specific process. A direct hydroxylation of the three hydroxy groups containing norcoclaurine (**86**) derivatives is not possible in *L. glutinosa*, showing that both aromatic building blocks must be dihydroxylated before they are joined together.

14.3.3 Laudanosine and Related Alkaloids

In *Papaver somniferum* (*S*)-laudanosine (**74**) is biosynthesized *via* (*S*)-norreticuline (**79**), (*S*)-reticuline (**66**), and (*S*)-laudanidine (**72**) or codamine (**75**) [130, 133] (Fig. 14.14). Stereospecific *N*-methylation of (*S*)-norreticuline is supported by the fact that (*R*)-norreticuline was not incorporated into laudanosine. On the other hand it can also be formed from tetrahydropapaverine (**73**). The *N*-methylation of tetrahydropapaverine shows only a low order of stereoselectivity. Both isomers gave good incorporation into laudanosine with little or no loss of tritium from the asymmetric centre. Consequently, administration of the unnatural precursor (*R*)-tetrahydropapaverine produces the unnatural (*R*)-laudanosine [130].

Laudanosine (**74**) may be biosynthesized by several routes depending on the sequence of *O*- and *N*-methylation of norlaudanosoline. Methylation of the 7-OH group shows little, if any, stereoselectivity. Both isomers of reticuline are readily methylated at the 7-position to give (*S*)- and (*R*)-laudanidine. Methylation of the 3′-OH group is very stereoselective and only the dextrorotatory enantiomers of codamine (**75**) and laudanosine (**74**) are present in the plant. The relationships between the different benzyltetrahydroisoquinolines are summarized in Fig. 14.14. Incubation of tetrahydropapaverine (**73**) with *S*-adenosylmethionine and an enzyme preparation from *Berberis aggregata* callus culture provided laudanosine (**74**) [281].

14.3.4 Papaverine and Related Alkaloids

In addition to the benzyltetrahydroisoquinolines some aromatic benzylisoquinolines exist in opium poppy. Papaverine (**78**) is one of the major benzylisoquinoline alkaloids of *Papaver somniferum*, and palaudine (**76**) and pacodine (**77**) were also formed [133]. For their biosynthesis norlaudanosoline (**65**) has to be *O*-methylated and dehydrogenated. Consistent with this hypothesis, norlaudanosoline and its precursors are effectively incorporated into papaverine by *P. somniferum* [33, 46, 52, 133].

O-Methylation of norlaudanosoline by catechol *O*-methyltransferase may conceivably produce four isomeric dimethyl ethers, norreticuline (**79**), nororientaline (**80**), norprotosinomenine (**81**), and norisoorientaline (**82**) (Fig. 14.15). Although these bases have not been isolated from opium poppy, the corresponding *N*-methyl derivatives are known to be present, namely reticuline (**66**) and orientaline (**71**). Complete *O*-methylation of the four dimethyl ethers would give tetrahydropapaverine (**73**).

All four isomeric dimethyl ethers of norlaudanosoline were specifically incorporated into papaverine; however, their pathways differ. (*S*)-Tetrahydropapaverine (**73**) is the principal immediate precursor of papaverine (**78**) in *Papaver somniferum* [130, 492] and its occurrence in opium poppy has been proved. This compound is derived preferentially from (*S*)-norreticuline (**79**) and also from (*S*)-nororientaline (**80**), but not from the other two dimethyl ethers. The dehydrogenation does not take place at the norreticuline stage, and since 1,2-dehydronorreticuline was not incorporated it is not an intermediate in a stepwise dehydrogenation to the aromatic benzylisoquinolines.

In the further methylation sequence 6-*O*-methylation must precede 7-*O*-methylation, which is consistent with the sequence of *O*-methylation of phenethylamines and tetrahydroisoquinolines in the biosynthesis of peyote alkaloids [392]. It appears that the opium poppy is not capable of methylating the hydroxy group in the 6 position once the 7-OH is methylated. The aromatization of ring *B* is a stereoselective process, and only the (*S*)-isomers of tetrahydropapaverine (**73**) and norreticuline (**79**) could serve as precursors of papaverine (**78**). Furthermore, it involves the stereospecific loss of the pro-*S*-hydrogen from C-3 of norreticuline, but the subsequent loss of the C-4 hydrogen is non-stereospecific [72].

Nevertheless the plant does not require complete methylation of all four phenolic hydroxy-groups before dehydrogenation can take place, *e.g.* after the *C*-ring is fully methylated the resulting trimethyl ether (**83**) can be dehydrogenated to the corresponding aromatic benzylisoquinoline, the 3′,4′,7-trimethylether isopacodine (**84**), which may then be methylated to give papaverine. This is the incorporation route for norprotosinomenine (**81**) and norisoorientaline (**82**). Palaudine (**76**) and pacodine (**77**) seem to be formed by dehydrogenation of the corresponding tetrahydroderivatives as shown by an efficient incorporation of norlaudanine (**69**) into palaudine (**76**). They are not important intermediates in the biosynthesis of papaverine. *P. somniferum* is able to dehydrogenate various tetrahydropapaverine analogues, *e.g.* with ethoxy groups, to the corresponding, unnatural analogues of papaverine [132].

Results similar to those for the norlaudanosoline derivatives have been found in *Annona reticulata* (*Annonaceae*) for the biosynthesis of coclaurine (**87**) [406], an established precursor of proaporphine, aporphine, and bisbenzylisoquinoline alkaloids. While tyramine, dopa, and dopamine (**8**) contribute only to the formation of the phenethylamine portion of coclaurine (**87**), tyrosine and 4-hydroxy-phenylpyruvic acid (**63**) are incorporated into both halves. Dopamine derived from dopa and 4-hydroxyphenylpyruvic acid derived from tyrosine, respectively, interact and form norcoclaurine-1-carboxylic acid (**85**) which specifically gives rise to coclaurine (**87**) *via* 1,2-didehydronorcoclaurine and norcoclaurine (**86**) (Fig. 14.16).

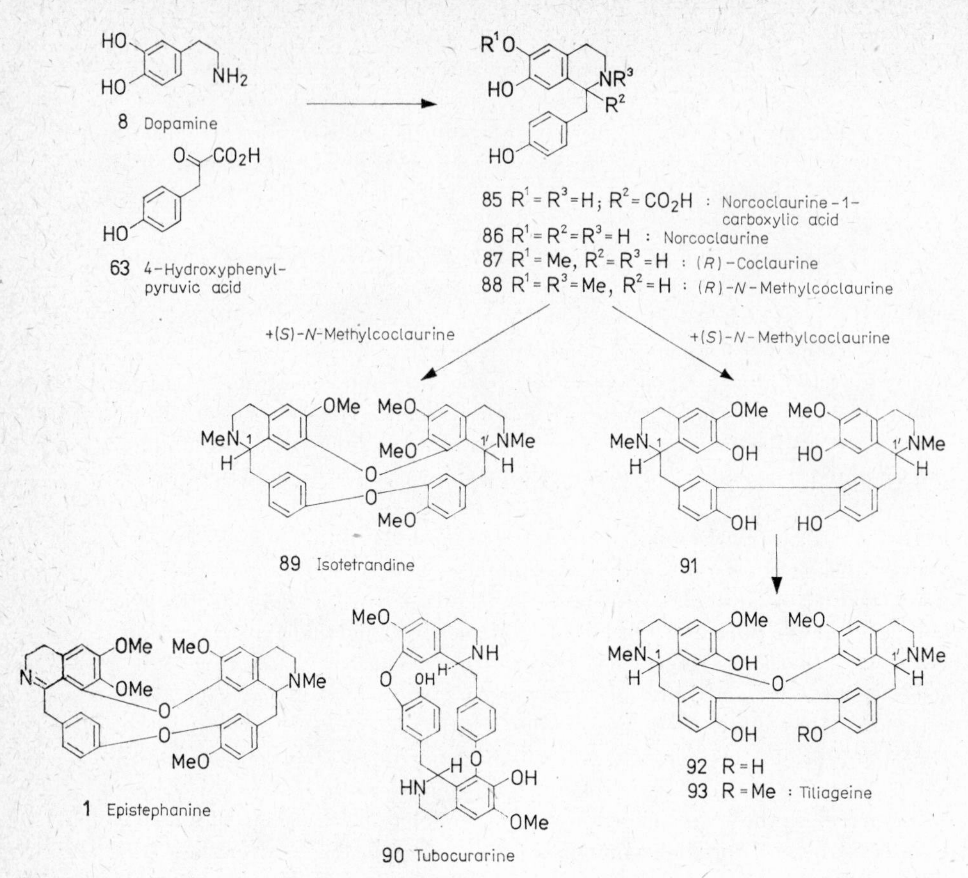

Fig. 14.16
Biosynthesis of coclaurine and different bisbenzylisoquinolines

14.4 Bisbenzylisoquinolines

Epistephanine (**1**) and tubocurarine (**90**) are representative members of different subgroups of bisbenzylisoquinoline alkaloids. About twenty such subgroups exist, and together they contain over 100 known alkaloids [224]. This large number of varieties of dimeric benzylisoquinolines comes about because of the possible variations in the joining of two benzyltetrahydroisoquinoline-derived fragments and their subsequent methylation. In all cases known, the ether bridge joining the two benzylisoquinoline-derived fragments is fixed to one benzylisoquinoline in a position dictated by the derivation of that fragment from tyrosine or its further transformation products, while the other end is fixed to a carbon ortho to that demanded by derivation from tyrosine or its further transformation products.

Earlier suggestions were that bisbenzylisoquinoline alkaloids might be formed in nature by the oxidative coupling of coclaurine derivatives [20, 31, 446]. In *Stephania japonica* it could be shown that tyrosine, coclaurine (**87**) and (*R*)-*N*-methylcoclaurine (**88**) were specifically incorporated into epistephanine (**1**) the latter without loss of the *N*-methyl group and only into the *N*-methylated half of the alkaloid (Fig. 14.16) [27]. Cycleanine and *N*-demethylcycleanine are derived from (*R*)-*N*-methylcoclaurine (88) in young *Stephania glabra* [80b].

Isotetrandine (**89**) is a bisbenzylisoquinoline alkaloid with anti-inflammatory, analgesic and hypothermic effects, having *R* and *S* configurations at C-1 and C-1′ respectively. It is biosynthesized in *Cocculus laurifolius* by intermolecular oxidative coupling of (*S*)- and (*R*)-*N*-methylcoclaurine (**88**) without interconversion of the enantiomers *via* the dehydro-*N*-methyl-coclaurinium ion (Fig. 14.16) [99]. The hydrogen atom at the asymmetric centre in *N*-methylcoclaurine is retained in the bioconversion into isotetrandine. Tetrandine, diastereoisomeric with isotetrandine (**89**), is formed from two (*S*)-*N*-normethylcoclaurine (**94**) units [88]. In a similar way oxyacanthine, the hypotensic principle of *Berberis vulgaris*, also with two phenyl ether linkages and having *S* and *R* configurations at the asymmetric centres C-1 and C-1′ respectively, is biosynthesized from (*S*)-*N*-methylcoclaurine and (*R*)-*N*-methylcoclaurine [98].

Besides the bisbenzylisoquinolines with two ether bridges, alkaloids are also known with three ether linkages, *e.g.* cocsulin (**100**) from *Cocculus laurifolius* and cocsulinin (**99**) an anti-cancer agent of *Cocculus pendulus* (*Menispermaceae*) [90]. Both alkaloids are biosynthesized in *C. laurifolius* by oxidative dimerization of two (*S*)-*N*-methylcoclaurine (**94**) units (Fig. 14.17) [87, 91], which has been isolated from *C. laurifolius*. The *O*-methyl function from one of the *N*-methylcoclaurine units is lost and the stereospecificity is maintained in the oxidative dimerization of *N*-methylcoclaurine into cocsulin. Intermolecular oxidative coupling of two (*S*)-*N*-methylcoclaurine (**94**) units could lead to a dauricine-type intermediate (**95**), which in turn can undergo intramolecular oxidative coupling to generate **97**. For the formation of the dibenzo-*p*-dioxin system this intermediate could be oxidized to the corresponding phenoxy cation **96**, which is then substituted into ring *A* to give the ion **98** which through loss of CH_3O^+ can generate cocsulin (**100**). These results strongly support the following sequence for the biosynthesis of cocsulin (**100**) and cocsulinin (**99**) in *C. laurifolius*: tyrosine → norcoclaurine → coclaurine → (*S*)-*N*-methylcoclaurine (**94**) → dimerization giving cocsulin (**100**) or (*S,S*)-*O*-methylcocsulinin and cocsulinin (**99**).

In *Tiliacora* species (*Menispermaceae*) [447, 482] tiliageine (**93**), tiliacorine and tiliacorinine are present, dimeric alkaloids of the bisbenzylisoquinoline groups in which the two benzylic "halves" are linked through a direct carbon-to-carbon bond rather than through the more common diarylether bridge. Tiliageine has only one ether bridge and the other two have additional ether bridges. It has been established that *N*-methylcoclaurine is a specific precursor of these three alkaloids in *Tiliacora racemosa* [82, 96, 100]. In tiliageine (**93**) and tiliacorine one part is derived from (*S*)- and the other from (*R*)-*N*-methylcoclaurine (**88**) in accordance with the (*S*) and (*R*) configurations at the asymmetric centres C-1 and C-1′ in these alkaloids. For tiliacorinine, the diastereomer of tiliacorine, only the (*S*)-enantiomer is incorporated in accordance with the (*S*-*S*)-configuration at both asymmetric centres. For tiliageine it is assumed that an intermolecular oxidative coupling of (*R*)- and (*S*)-*N*-methylcoclaurines can form a carbon-to-carbon bond in the benzylic halves giving the intermediate **91**, which can generate **92** by intramolecular oxidative coupling. Selective *O*-methylation of one phenolic group can finally yield tiliageine (**93**) (Fig. 14.16).

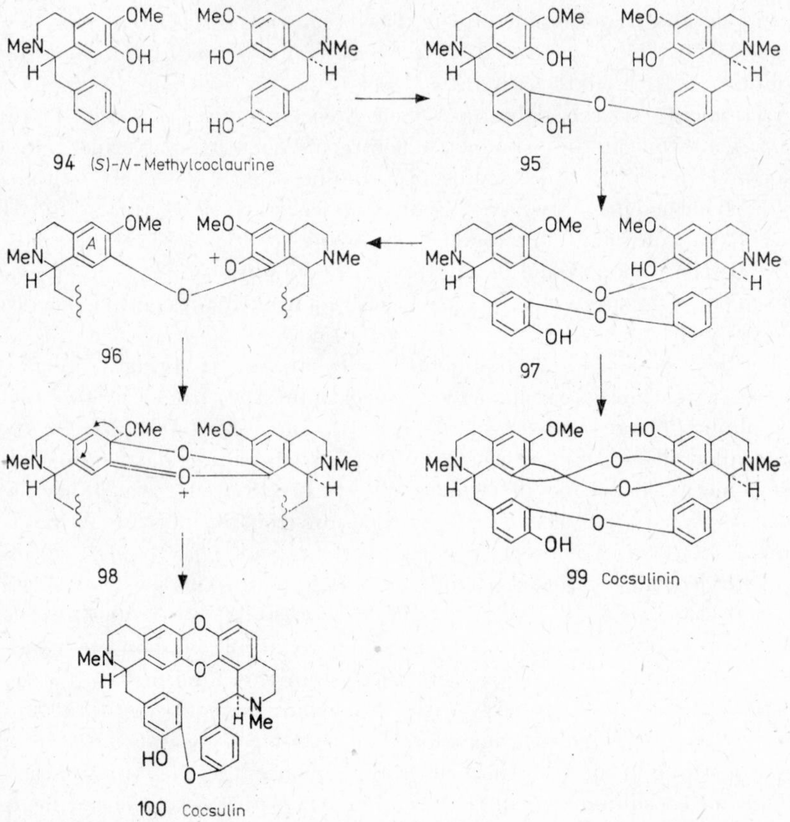

Fig. 14.17
Biosynthesis of cocsulin and cocsulinin

Besides the bisbenzylisoquinolines mentioned, which contain two benzylisoquinoline units, other dimeric alkaloids are known with only one benzylisoquinoline unit. Thus baluchistanamine (**101**) from *Berberis baluchistanica* is an isoquinolone-benzylisoquinoline, which seems to be synthesized by oxidation from an oxyacanthine-type, also found in this plant [448]. A similar oxidative phenol coupling leading to bisbenzylisoquinoline has to be suggested for the synthesis of proaporphine-benzylisoquinoline dimers such as pakistanamine from *Berberis baluchistanica*, as well as for the formation of aporphine-benzylisoquinoline dimers such as thalicarpine (**104**) and pakistanine, which occur mainly in the genus *Thalictrum* (*Ranunculaceae*) [84,

248, 350, 450, 455]. Thus in experiments with *T. minus* and *Cocculus laurifolius* (*S*)-reticuline (**66**) has been found to be specifically incorporated into thalicarpine (**104**). The biosynthetic pathway is probably *via* the bisbenzylisoquinoline (**102**) and after selective *O*-demethylation at C-4 following intramolecular oxidative coupling of one half of the molecule *via* the proaporphine-benzylisoquinoline intermediate (**103**). Dienone-phenol rearrangement can furnish the aporphine-benzylisoquinoline system such as thalicarpine (**104**) (Fig. 14.18).

The callus tissues derived from the tuber of *Stephania cepharantha* (*Menispermaceae*) can synthesize bisbenzylisoquinolines, but not the full alkaloid spectrum as in the whole plant [6]. The callus tissues lack the specific enzymes necessary for *O*-methylation and methylenedioxy group formation.

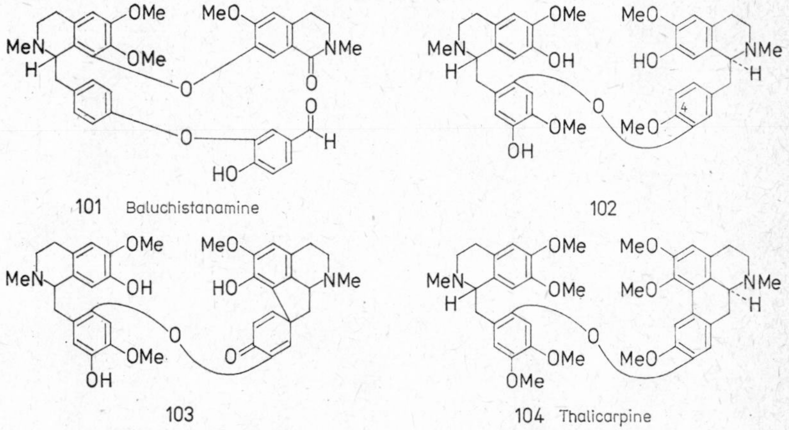

101 Baluchistanamine

102

103

104 Thalicarpine

Fig. 14.18
Mixed dimeric benzylisoquinolines and biosynthesis of thalicarpine

14.5 Aporphine Alkaloids and Derived Compounds

14.5.1 Aporphines Derived by Direct Oxidative Coupling

Aporphines are biosynthesized from phenolic tetrahydrobenzylisoquinolines either by direct oxidative coupling or by an indirect mechanism that utilizes an intermediate dienone [20]. Such a direct oxidative coupling could be carried out either by *ortho-para* or *ortho-ortho* reaction yielding compounds of isoboldine- or bulbocapnine-type (Fig. 14.19). Isoboldine (**106**) has been shown to be formed in *Papaver somniferum* by such a straightforward intramolecular oxidative coupling between the two aryl rings of reticuline (**66**) [134]. The sites of coupling both bear a hydrogen and therefore the two rings can be re-aromatized to give the biphenyl system characteristic of the aporphine alkaloids.

The same pathway operates for the synthesis of boldine (**105**), the choleretic principle of *Peumus boldus*. Reticuline and norreticuline are specific precursors of boldine in *Litsea glutinosa* (*Lauraceae*) [103] (Fig. 14.19). Parallel feedings with (*S*)-reticuline (**66**) and (*R*)-reticuline demonstrated that stereospecificity is maintained in the bioconversion of 1-benzylisoquinoline precursors into boldine (**105**) and the former is the natural precursor. Incorporation of reticuline into boldine implies that isoboldine (**106**) is an intermediate that shows specific incorporation. This indicates that a change in the methylation pattern takes place in the biosynthesis of boldine after the oxidative coupling process. Two processes could explain this transformation: a demethylation-remethylation or a migration of a methyl group

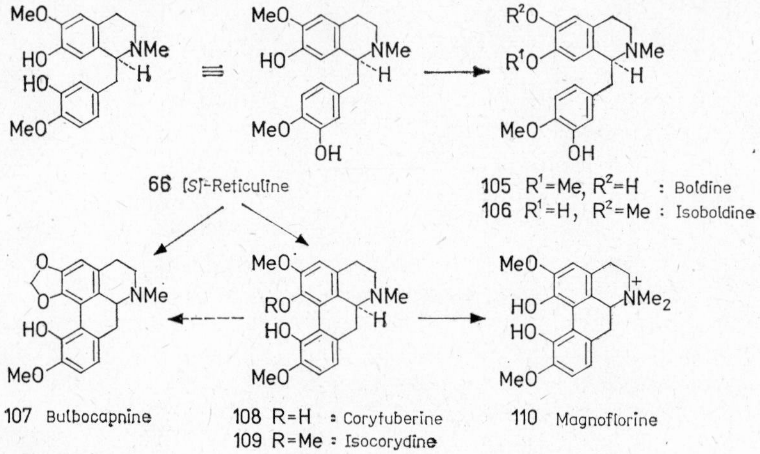

Fig. 14.19
Biosynthesis of aporphines by direct oxidative coupling

The loss of activity in the methoxy group in the biosynthesis of boldine from reticuline provided evidence against methyl migration. Similar results have been obtained in the biosynthesis of crotonosine (**135**) in *Croton linearis* [15].

Reticuline and isoboldine (**106**) have been isolated from *L. glutinosa* [229]. The young shoots of *L. glutinosa* were also found to metabolize 4′-*O*-methylnorlaudanosoline and its presence in the plants was shown by a trapping experiment. Thus the following sequence is postulated for the biosynthesis of boldine in *L. glutinosa*: 4′-*O*-methylnorlaudanosoline → norreticuline (**79**) → (*S*)-reticuline (**66**) → isoboldine (**106**) → boldine (**105**) (Fig. 14.19). In experiments with *Cocculus laurifolius* (*S*)-reticuline is specifically incorporated into laurifoline, the quaternary base of isoboldine [89].

In vitro oxidation of reticuline by reagents that favour the formation of free radicals has produced isoboldine [139, 148, 150, 195, 253, 271, 272]. However, the infrequent formation of such a product in these reactions indicates that a direct *ortho-ortho* coupling to aporphines rarely takes place. Reticuline would seem to lend itself readily to *ortho-para* coupling. It is suggested that for *ortho-ortho* coupling the substituents cause considerable steric hindrance to co-planarity, thereby making such a coupling of reticuline very difficult. Thus, in earlier experiments it was impossible to demonstrate a direct incorporation of reticuline into magnoflorine (**110**) [134] or a direct coupling for the biosynthesis of corydine (**130**), glaucine (**128**), and dicentrine (**129**) in *Dicentra eximia* [66]. On the other hand, evidence was found for an indirect mechanism involving norprotosinomenine (**81**) and dienones.

Nevertheless there is some evidence for direct *ortho-ortho* oxidative phenol coupling in nature. In larkspur (*Corydalis cava, Papaveraceae*) reticuline (**66**) was specifically incorporated into bulbocapnine (**107**) [112, 114], probably *via* corytuberine (**108**) as an intermediate. The natural occurrence of reticuline in *Corydalis cava* was demonstrated by the trapping method and by the isolation of the morphinane dienone-type alkaloid sinoacutine, which is biosynthesized unambiguously from reticuline [116]. Thus the occurrence of a dienone pathway for the bulbocapnine synthesis could be excluded. The pathway probably proceeds *via* the diradical **111** which reacts *via* the mesomeric form **112** to the stable dienone (**113**) and then yields corytuberine (**108**) by enolisation (Fig. 14.20).

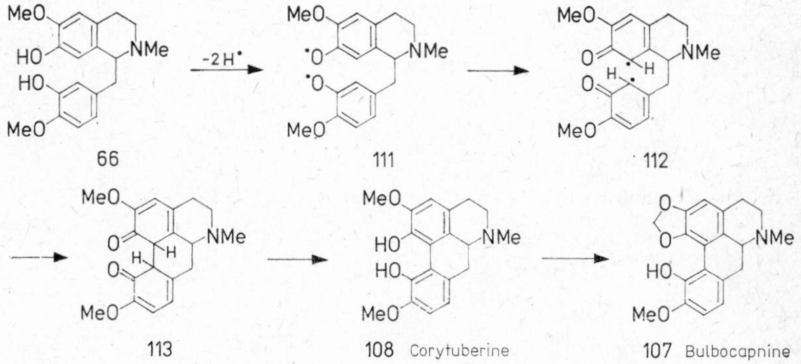

MeO HO HO MeO NMe 66 —$-2H^{\bullet}$→ MeO •O •O MeO NMe 111 → MeO O H H O MeO NMe 112

→ MeO O H H O MeO NMe 113 → MeO HO HO MeO NMe 108 Corytuberine → O O HO MeO NMe 107 Bulbocapnine

Fig. 14.20
Biosynthesis of bulbocapnine

Isocorydine (**109**) is also formed in *Annona squamosa* by direct *ortho-ortho* oxidative coupling of norreticuline and reticuline (**66**), probably *via* corytuberine (108), from which isocorydine was formed by methylation [168] (Fig. 14.19). The presence o reticuline in *A. squamosa* was confirmed by a trapping experiment.

Magnoflorine (**110**) has been shown to be derived from (*S*)-reticuline (**66**) in *Aquilegia* plants and *Cocculus laurifolius*, presumably *via* direct *ortho-ortho* oxidative phenol coupling [89, 131] (Fig. 14.19). A mechanism involving a proaporphine could be excluded.

14.5.2 Aporphines Derived *via* Proaporphines

Several aporphine alkaloids, *e.g.* anonaine (**139**), roemerine (**140**) and isothebaine (**122**), have an unusual hydroxyl substitution pattern that involves more than straightforward oxidative coupling. For these alkaloids it is supposed that a benzylisoquinoline undergoes an oxidative coupling yielding an intermediate dienone [20]. Such dienones are present in many plants and are generally referred to as proaporphines, *e.g.* orientalinone (**120**) and crotonosine (**135**). Rearrangement of such dienones with or without prior reduction may result in a large number of aporphines with different substitution patterns.

Cyclohexadienones have a strong tendency to undergo rearrangements (Fig. 14.21). In the acid-catalysed dienone-phenol rearrangement the parent dienone (**114**) is converted into the phenol (**116**) by migration of a substituent to give a carbonium ion **115** that can aromatize by loss of proton. The dienolbenzene rearrangement, a

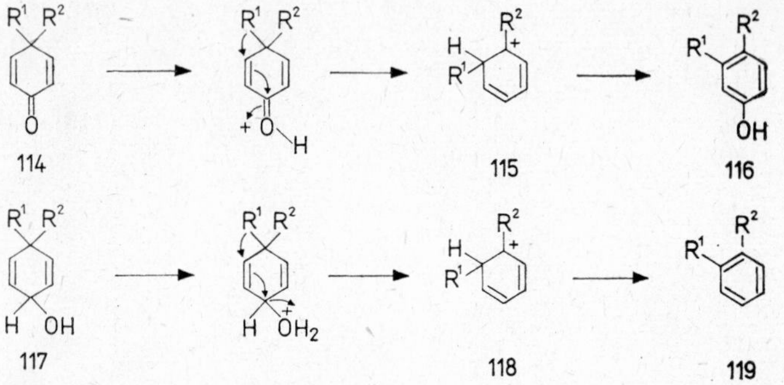

Fig. 14.21
Cyclohexadienone rearrangements

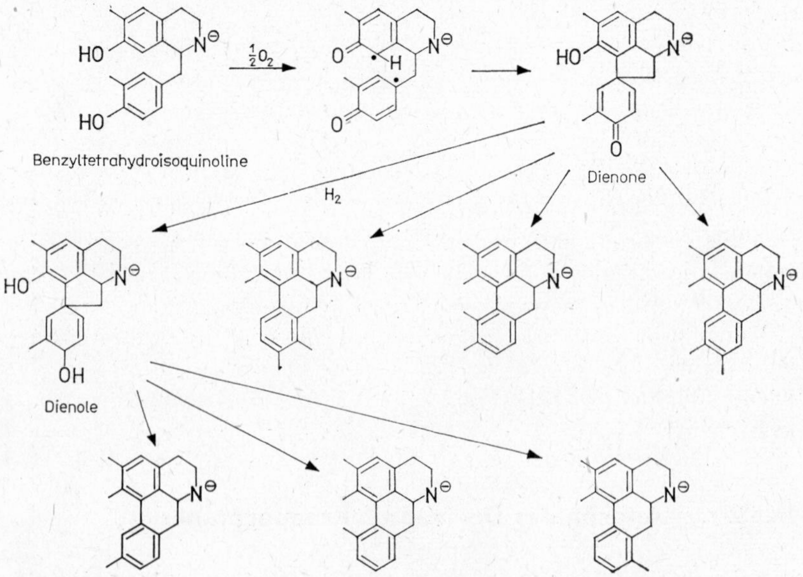

Fig. 14.22
Different possibilities for rearrangements starting from benzyltetrahydroisoquinolines

less direct process, starts with reduction of the dienol (**117**), which can now undergo an equivalent rearrangement to give carbonium ion **118**, followed by aromatization to the benzene derivative **119**. Both processes are common in alkaloid biosynthesis. The product of the first aromatization retains the oxygen of the dienone whereas in the second the oxygen is lost. In Fig. 14.22 the different possibilities of these reactions starting from benzyltetrahydroisoquinolines via a dienone intermediate are summarized.

Aporphines have also been synthesized *in vitro* by oxidative coupling to a dienone, reduction to the corresponding dienol and rearrangement to the aporphines, *e.g.* isothebaine (**122**) from orientaline (**71**) *via* orientalinone (**120**) [42, 254]. Thus a dienol-benzene rearrangement has been found in the biosynthesis of isothebaine (**122**) in *Papaver orientale* (Fig. 14.23) [40, 41]. Oxidative coupling of the benzylisoquinoline orientaline (**71**) (an isomer of reticuline) leads to a dienone, orientalinone (**120**). This is reduced to the dienol, orientalinol (**121**), which then undergoes a

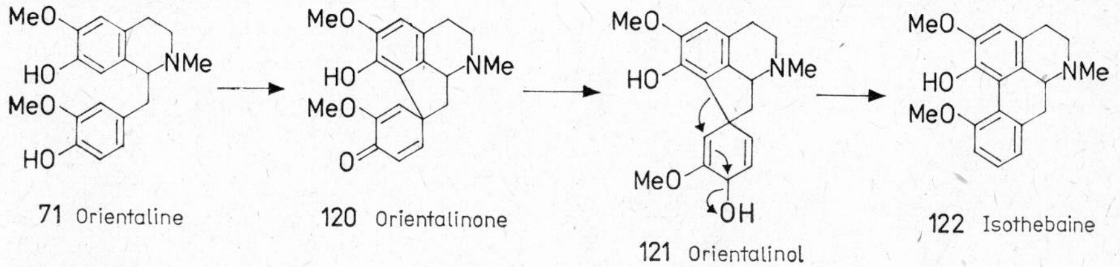

71 Orientaline — 120 Orientalinone — 121 Orientalinol — 122 Isothebaine

Fig. 14.23
Biosynthesis of isothebaine

dienol-benzene rearrangement to form the aporphine isothebaine. There is one less oxygen function in the final product than in the benzylisoquinoline; hence the pattern of hydroxyl and methoxy groups is profoundly changed in the course of the biosynthesis.

It is interesting to note the controlling effect of the methylation pattern and stereochemistry at C-1 in the tetrahydrobenzylisoquinoline system. Neither (*R*)-reticuline nor (*S*)-reticuline will serve to generate isothebaine (**122**), which comes only from (*S*)-orientaline. (*R*)-Reticuline yields only the morphinan alkaloid thebaine, which is found in the same plant (*Papaver orientale*) in which norlaudanosoline is methylated on the one hand to orientaline and on the other hand to reticuline. (*S*)-Reticuline yields only berberine and derived alkaloids.

Corydine (**130**), glaucine (**128**) and dicentrine (**129**) occur in *Dicentra eximia* [123] and their structures can readily be explained [20, 31] as involving direct coupling of the radicals formed by oxidation of reticuline (**66**) as shown (Fig. 14.19) for bulbocapnine (**107**) and isoboldine (**106**), *ortho-ortho* and *ortho-para* coupling could yield the corresponding aporphines from which the three alkaloids are derivable but reticuline and orientaline were ineffective as precursors of these alkaloids [22]. Tyrosine and dopa were incorporated specifically into the three alkaloids showing that two building blocks derivable from tyrosine are used, but only one of these can be formed from dopa as for other benzylisoquinolines. The incorporation of norlaudanosoline (**65**) and of different mono-*O*-methyl ethers preferably the 4′-*O*-methyl ether (**123**) leads to norprotosinomenine (**81**) as the di-*O*-methylated precursor of the *Dicentra* alkaloids, which shows good incorporation into the three aporphines [66] (Fig. 14.24). Oxidative coupling of norprotosinomenine or of protosinomenine (**124**) yield the dienone (**125** and **126**). Dienone phenol rearrangement of **126** gives boldine (**105**) leading to glaucine (**128**) and dicentrine (**129**) and of the dienone **125** yields aporphine **127** leading to corydine (**130**). Boldine was efficiently incorporated into glaucine and dicentrine (**129**) but not into isoboldine (**106**). The dienone **126** is

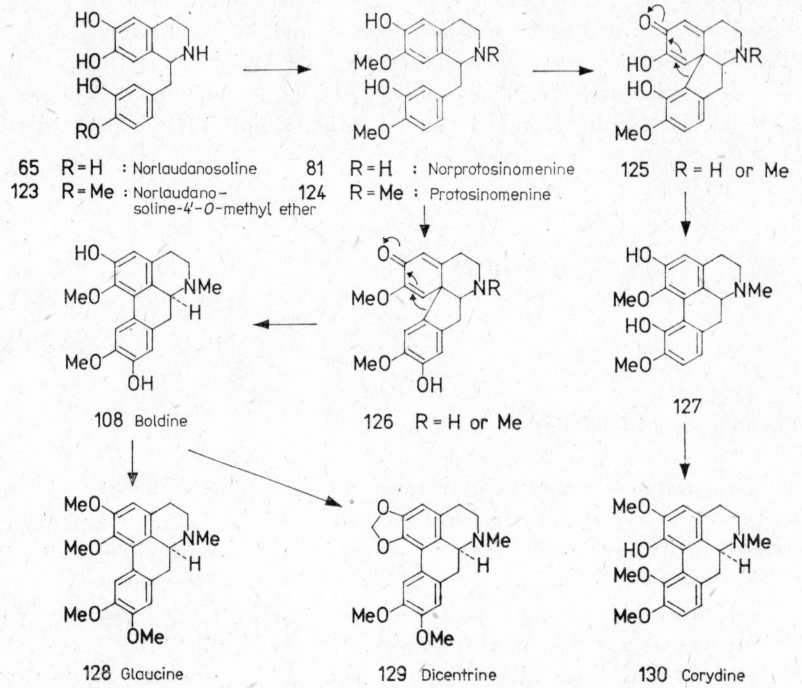

65 R = H : Norlaudanosoline
123 R = Me : Norlaudano-soline-4′-*O*-methyl ether
81 R = H : Norprotosinomenine
124 R = Me : Protosinomenine
125 R = H or Me
108 Boldine
126 R = H or Me
127
128 Glaucine
129 Dicentrine
130 Corydine

Fig. 14.24
Biosynthesis of glaucine, dicentrine, and corydine in *Dicentra eximia*

also an intermediate for the biosynthesis of the *Erythrina* alkaloids [23]. The various uses made of this skeleton in *Erythrina* (*Leguminosae*) and *Dicentra* (*Papaveraceae*) species is of chemotaxonomic interest. Boldine can be formed by two different pathways: in *Dicentra* via a dienone (Fig. 14.24) and in *Litsea* (*Lauraceae*) by direct oxidative phenol coupling (Fig. 14.19).

In *Glaucium flavum* a pathway from isoboldine (**106**) to glaucine (**128**) and oxoglaucine has been considered [146, 221b, 339]. The 1,2,9,10-substituted aporphines are synthesized first in these plants and finally the 1,2,10,11-substituted aporphines [338]; glaucine plays an important role in the regulation of aporphine synthesis [147].

14.5.3 Abnormal Aporphines

The 'abnormal aporphine' alkaloids lack oxygen function in ring *D*. According to the established biogenetic theory [20] such alkaloids, *e.g.* roemerine (**140**), anonaine (**139**), and nornuciferine-I (**138**) [94], can be biosynthesized from coclaurine derivatives with only one oxygen function in ring *C* (Fig. 14.25). *Ortho-para* oxidative coupling of a coclaurine derivate such as **94** gives the corresponding proaporphine as **132**. Reduction to the dienol **137** followed by dienol-benzene rearrangement finally yields the abnormal aporphines *e.g.* **138**. Thus tracer experiments have shown that the 'abnormal aporphine' alkaloid roemerine (**140**) in *Papaver dubium* and *Meconopsis cambrica* is biosynthesized specifically from norcoclaurine *via* coclaurine

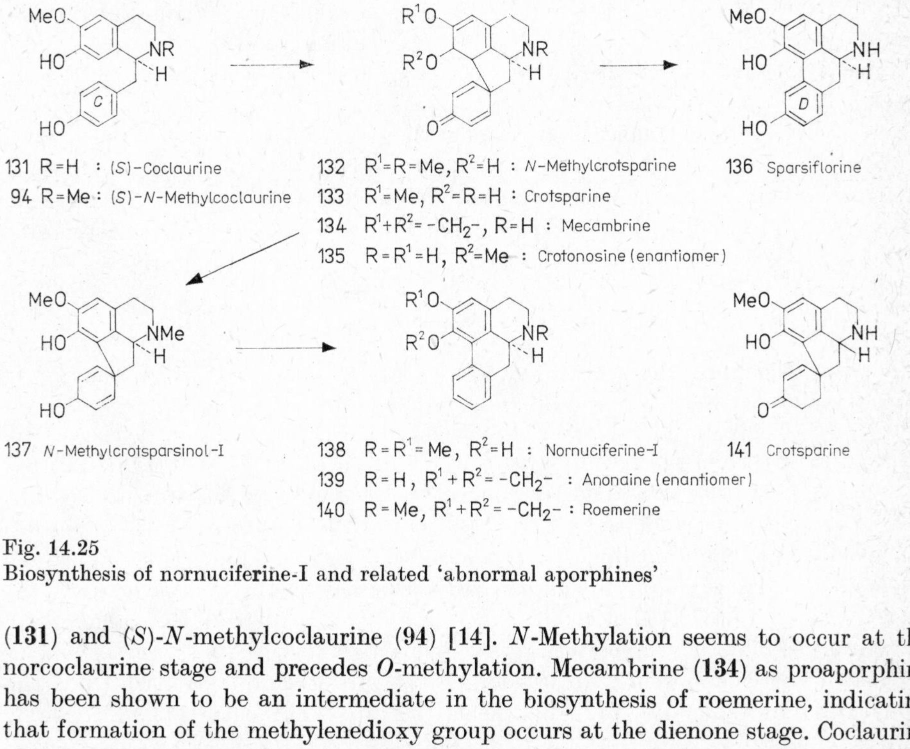

131 R=H : (*S*)-Coclaurine
94 R=Me : (*S*)-*N*-Methylcoclaurine

132 R^1=R=Me, R^2=H : *N*-Methylcrotsparine
133 R^1=Me, R^2=R=H : Crotsparine
134 R^1+R^2=$-CH_2-$, R=H : Mecambrine
135 R=R^1=H, R^2=Me : Crotonosine (enantiomer)

136 Sparsiflorine

137 *N*-Methylcrotsparsinol-I

138 R=R^1=Me, R^2=H : Nornuciferine-I
139 R=H, R^1+R^2=$-CH_2-$: Anonaine (enantiomer)
140 R=Me, R^1+R^2=$-CH_2-$: Roemerine

141 Crotsparine

Fig. 14.25
Biosynthesis of nornuciferine-I and related 'abnormal aporphines'

(**131**) and (*S*)-*N*-methylcoclaurine (**94**) [14]. *N*-Methylation seems to occur at the norcoclaurine stage and precedes *O*-methylation. Mecambrine (**134**) as proaporphine has been shown to be an intermediate in the biosynthesis of roemerine, indicating that formation of the methylenedioxy group occurs at the dienone stage. Coclaurine (**131**) and *N*-methylcoclaurine (**94**) were also incorporated into anonaine (**139**) in *Annona reticulata* and into mecambrine in *Meconopsis cambrica*, a good source for this alkaloid.

A similar pathway operates for the biosynthesis of nornuciferine-I (**138**) in *Croton sparsiflorus* (*Euphorbiaceae*). Results of feeding experiments strongly support the following sequence: tyrosine → norcoclaurine → (*S*)-coclaurine (**131**) → (*S*)-*N*-methylcoclaurine (**94**) → *N*-methylcrotsparine (**132**) → *N*-methylcrotsparinol-I (**137**) → nornuciferine-I (**138**) [85] (Fig. 14.25). The presence of *N*-methylcoclaurine (**94**) in *C. sparsiflorus* was demonstrated by trapping experiments and *N*-methylcrotsparine (**132**) has been isolated from this plant. The proaporphines, pronuciferine and stepharine are derived from (*R*)-*N*-methylcoclaurine (**88**) in young *Stephania glabra* [80b].

In addition to the proaporphine alkaloids crotsparine (**133**), *N*-methylcrotsparine (**132**), and *N*,*O*-dimethylcrotsparine, the dihydroproaporphine crotsparinine (**141**) and the aporphine sparsiflorine (**136**) occur in *Croton sparsiflorus* [94]. In contrast to the abnormal aporphine nornuciferine (**138**), sparsiflorine contains one oxygen function in ring *D*. Nevertheless coclaurine (**131**) was efficiently incorporated into sparsiflorine together with crotsparine and crotsparinine (Fig. 14.25) [83, 95]. Sparsiflorine (**136**) was produced from crotsparine by dienone-phenol rearrangement, retaining the oxygen function in ring *D*. The presence of coclaurine as key intermediate in *C. sparsiflorus* was confirmed by a trapping experiment.

Crotsparine (**133**) and crotsparinine (**141**) have opposite configurations, while sparsiflorine (**136**) and crotsparine have the same configuration at the corresponding asymmetric centres. Crotosparine and crotsparinine are biosynthesized by independent routes.

The proaporphine alkaloid crotonosine (**135**) is formed in *Croton linearis* by oxidative cyclization of coclaurine [15, 232], probably *via* the dienone (**133** opposite configuration) involving a demethylation — remethylation process. Otherwise in *Croton linearis* the dihydroproaporphine centre can be converted into crotonosine [469, 471]. It should be mentioned that this conversion varies with the sex of the plant.

14.5.4 Aristolochic Acid

Aporphines such as magnoflorine (**110**) occur in different *Aristolochia* species together with 10-nitro-1-phenanthroic acids, the aristolochic acids, such as aristolochic acid I (**143**) [123, 169]. This acid was biosynthesized by cleavage of the heterocyclic ring of an aporphine and oxidation of the nitrogen to the nitro group. Tyrosine, dopa, dopamine, noradrenaline (norepinephrine), norlaudanosoline, orientaline, and stephanine can serve as precursors for aristolochic acid I in *Aristolochia sipho* and *A. bracteata* [157, 437, 452, 459]. Its nitro group is derived from the amino group of tyrosine. From these experiments the following biosynthetic pathway can be postulated: norlaudanosoline, to orientaline (**71**), to orientalinone (**120**), to orientalinol (**121**), to stephanine (**142**), and thence to aristolochic acid I (**143**) (Fig. 14.26).

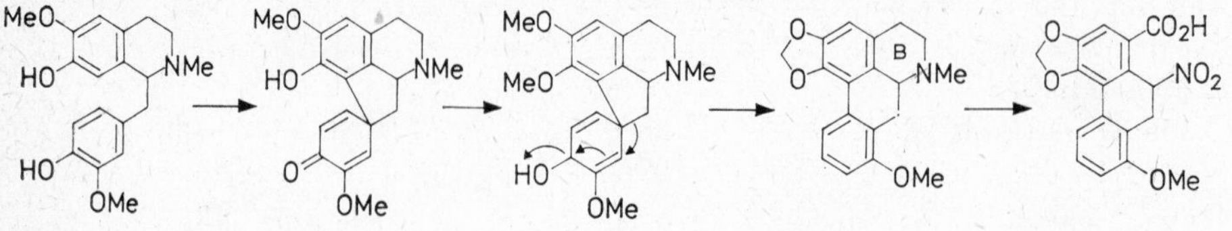

71 Orientaline → 120 Orientalinone → 121 Orientalinol → 142 Stephanine → 143 Aristolochic acid I

Fig. 14.26
Biosynthesis of aristolochic acid

It is not clear wether noradrenaline is on the normal biosynthetic pathway or wether the entry of this compound constitutes an aberrant pathway to a normal tissue component. In each case a ring *B* oxygenated compound is necessary in the transformation of the aporphine system to the aristolochic acid derivatives.

There is a possibility that an aporphine is oxidized to a 4,5-dioxoaporphine such as **144**, which can undergo net overall decarboxylation, presumably through a benzylic acid rearrangement (to **145**), to yield the corresponding aristolactam (**146**). Further oxidation would then yield the aristolochic acid skeleton (Fig. 14.27) [148]. Such 4,5-dioxoaporphines have been isolated from callus tissue of *Stephania cepharantha* [5]. Five phenanthrene derivatives have recently been found in *Aristolochia indica*, *e.g.* aristolamide, aristolinic acid methyl ester, and methyl aristolochate. These compounds must be of aporphinoidal origin [385].

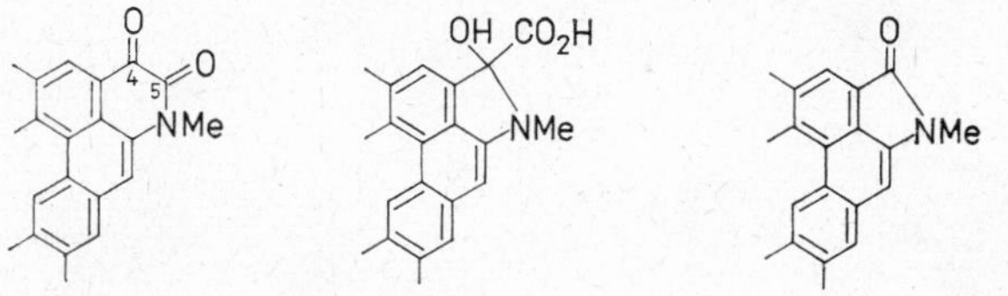

144 145 146

Fig. 14.27
Suggested conversion of the aporphine to the aristolochic acid structure

14.6 Morphinan Alkaloids

14.6.1 Thebaine, Codeine, and Morphine

A second example of straightforward phenol coupling of a benzylisoquinoline takes place in the biosynthesis of the opium alkaloids thebaine (**151**), codeine (**157**) and morphine (**154**) in *Papaver somniferum*. In different hypotheses the structural relationship of the morphinan skeleton with benzylisoquinolines has been suggested [13, 418, 419, 429]. As can be seen (Fig. 14.28) reticuline (**66**), present in *P. somniferum* [135], is again the substrate for the coupling step and (in formula **148**) the relation to the morphinan skeleton is obvious [25, 46, 105, 132c, 351]. This time the coupling takes place *ortho-para* to give the dienone salutaridine (**149**), in which only one of the two rings can re-aromatize by enolization. Because reticuline is a precursor, its progenitors tyrosine [34, 300, 319, 505] (*cf.* [431]), dopamine [47, 331], norlaudanosoline-1-carboxylic acid [58], and norlaudanosoline [46] are specifically incorporated into the opium alkaloids. A (*S*)-norlaudanosoline synthase was detected in several plants [425]. Reticuline (**66**) is produced from tyrosine, has the (*S*)-configuration and is dextrorotatory. It undergoes racemization in the plant [45], a reaction essential for the formation of the morphine alkaloids derived from (*R*)-

reticuline (**148**). This racemization proceeds by a reversible oxidation reduction mechanism *via* the 1,2-dehydroreticulinium ion (**147**) and is enzymatic and substrate specific [46, 125, 130].

In the case of the aporphine and morphinane alkaloids derived from reticuline we see the importance of the control which must be exerted by the enzymes over the mode of oxidative coupling. The same diradical is presumably made to couple in two completely different ways. Presumably the enzymes that catalyse the process

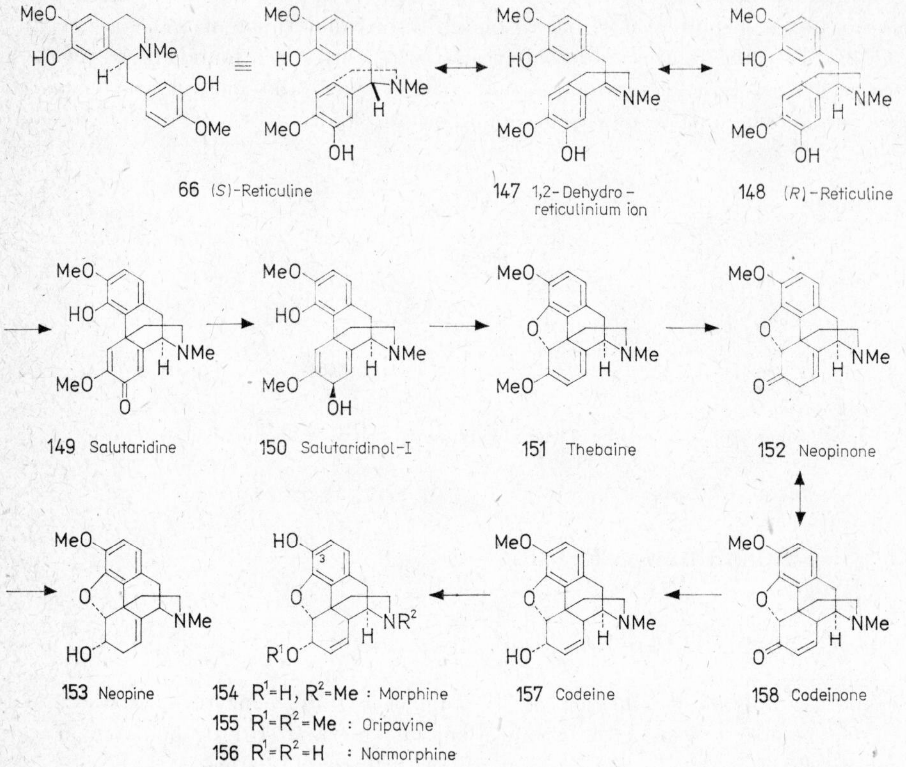

66 (*S*)-Reticuline
147 1,2-Dehydro-reticulinium ion
148 (*R*)-Reticuline
149 Salutaridine
150 Salutaridinol-I
151 Thebaine
152 Neopinone
153 Neopine
154 R^1=H, R^2=Me : Morphine
155 R^1=R^2=Me : Oripavine
156 R^1=R^2=H : Normorphine
157 Codeine
158 Codeinone

Fig. 14.28
Biosynthesis and metabolism of morphinan alkaloids

in each case direct the outcome by forcing the benzylisoquinoline to fold in the appropriate fashion. The importance of this controlling influence is brought home by the results of experiments to carry out equivalent oxidative coupling reactions *in vitro*. The usual result is a complex mixture in which the desired product is at best a minor component [16, 80, 441]. A cell-free extract from *P. somniferum* was prepared that utilize H_2O_2 to convert reticuline to salutaridine [241 a, 420].

The presence of salutaridine (**149**) in *P. somniferum* and other thebaine synthesizing *Papaver* species has been shown by trapping experiments. In the further biosynthetic pathway to morphinan alkaloids, salutaridine is reduced to the dienol salutaridinol-I (**150**) which undergoes a ring closure yielding thebaine (**151**) (Fig.

14.28). Salutaridine and salutaridinol were efficiently incorporated into thebaine, codeine, and morphine [25].

The final steps of the biosynthesis of morphine (**154**) by *P. somniferum* have been shown to involve the conversion of thebaine (**151**) into codeine (**157**), followed by 3-*O*-demethylation of codeine to morphine (Fig. 14.28). This sequence was determined using $^{14}CO_2$ exposure and precursor feeding [10, 51, 376, 394, 410, 429], which also demonstrated that both of these steps are irreversible. The first step proceeds by the cleavage of the enol ether of thebaine yielding the keto group of neopinone (**153**) and thence migration of the double bond into conjugation forming codeinone (**158**) [387]. This is then reduced to codeine (**157**).

The reduction of thebaine to codeine methyl ether and ether cleavage to codeine can be excluded. Codeine methyl ether is not incorporated into codeine or morphine [64, 115, 136]. However, codeinone is present to an extent of 5% of the amount of thebaine. It appears that the conversion of codeine methyl ether to codeine is an aberrant path, resulting from demethylation by a non-specific demethylating enzyme. The biosynthetic enol ether cleavage of thebaine proceeds by methyl cleavage with retention of the oxygen-6, probably involving an oxygenase [246]. While *O*-demethylation of the enol ether group of thebaine is an enzymatic reaction, the rearrangement of neopinone to codeinone does not require an enzyme. In aqueous solution, an equilibrium is established that favours codeinone [10]. Thus the *O*-demethylation of thebaine to morphine occurs at two steps. It is reasonable to hypothesize that these two reactions may be catalysed by different types of enzymes, since an enol ether and an aromatic ether are involved in the reactions. The alkaloids of *P. somniferum* can be formed in the latex but this is not the main site of biosynthesis [120, 183, 187, 190, 417a]. They partly occur in bound forms [189]. Cell-free extracts and immobilized cells have been prepared from *P. somniferum* which catalyse the reduction of codeinone (**158**) to codeine (**157**) [212, 212a, 241].

The unnatural thebaine analogue oripavine 3-ethyl ether, was efficiently metabolized to morphine 3-ethyl ether and morphine (**154**) [132b, 137] (*cf.* [132c]) and several unnatural codeine analogues were demethylated to the corresponding morphine analogues in the opium poppy [292].

Morphine and codeine are peculiar to *P. somniferum*, except for very small amounts in *P. setigerum*, while thebaine occurs in most other *Papaver* species. *Papaver bracteatum*, closely related to *P. orientale* and *P. pseudo-orientale*, contains mainly thebaine, which may account for 98 per cent of the alkaloid content, to the exclusion of both codeine and morphine [118, 185, 315, 378, 445]. Isolation and purification of thebaine from *P. bracteatum* are, therefore, relatively simple. In the laboratory, thebaine can be converted into codeine and other narcotic analgesics and antitussives [10, 78].

Alkaloid biosynthesis in *P. bracteatum* is not limited to any part of the plant. The highest concentration of thebaine is found in the capsule and reaches a peak 4 to 6 weeks after flowering [8, 185, 451]. In a chemical strain of *P. bracteatum* the highest amount is found in the thick storage roots [378]. As in the opium poppy, alkaloid biosynthesis in *P. bracteatum* probably takes place in the laticiferous vessels that form an anastomic network throughout the plant [186]. Feeding experiments with tyrosine and reticuline indicate that the biosynthesis of thebaine (**151**) in *P. bracteatum* occurs by the same route as in the opium poppy including racemization of reticuline (Fig. 14.28) [138, 240, 377, 379].

When codeinone (**158**) was administered to *P. bracteatum* plants or cell-free systems it was converted into codeine rapidly and efficiently, but no *O*-demethylation to morphine was detected [138, 240, 241]. The limiting step in the sequence from thebaine (**151**) to morphine (**154**) seems to be the demethylation of the enol ether group of thebaine to neopinone (**152**). The small amounts of codeine and neopine (**153**) reported [314] may best be explained by non-enzymatic hydrolysis of the enol ether group of thebaine because of acidic conditions prevailing in the plant. In *P. orientale*, thebaine appears to be converted to oripavine (**155**) [132a, 464]. These data strongly support the hypothesis that all the species investigated contain an *O*-demethylase capable of clearing the aromatic ether linkage. *P. somniferum* is unique in that it contains an enzyme system that attacts the enolic ether as well.

Morphine shows a rapid turn over in the opium poppy [184, 236]. Normorphine (**156**) has been established as an active metabolite of morphine in *P. somniferum* [361]. Feeding labelled normorphine showed that the *N*-demethylation step is irreversible. Probably the morphine degradation pathway involves an initial demethylation to normorphine, which is subsequently degraded to non-alkaloidal metabolites. The inclusion of normorphine as the final step in the alkaloid biosynthesis completes the sequence of demethylation from thebaine. It is possible that only the stem latex is involved with the morphine and normorphine activity [183]. Indeed, it was found that senescent, dying plants had almost no normorphine despite the fact that morphine concentration was still quite high.

Otherwise, morphine fed *in vivo* and *in vitro* to samples of capsules and stem latex of *P. somniferum* was converted to its *N*-oxide [188, 492a], which has been shown to occur naturally in this plant [395] and which is probably involved in *N*-demethylation.

Earlier investigations showed the presence of alkaloids in callus tissues of opium poppy. Although *P. somniferum* contains morphinan-type alkaloids as its main components, in some experiments poppy callus tissues lack such alkaloids, but are found to contain instead benzophenanthridine, protopine, and aporphine-type alkaloids [211, 249]. The ability of the poppy callus to metabolize thebaine, codeinone, codeine and morphine was investigated and the only reactions were the transformation of thebaine to neopine and the stereospecific reduction of codeinone (**158**) to codeine (**157**) [212, 485]. In other investigations the synthesis of codeine and thebaine but not of morphine by cell suspension cultures and callus tissue of *P. somniferum* is described [111, 241b, 276, 484]. Callus tissue of *P. bracteatum* synthesizes orientalidine and isothebaine [337].

14.6.2 Sinomenine and Sinoacutine

A similar pathway to that for morphine, codeine, and thebaine is seen for the biosynthesis of sinomenine (**160**), which occurs together with sinoacutine (**159**) in *Sinomenium acutum*. Tyrosine and reticuline were specifically incorporated into sinoacutine and sinomenine by *S. acutum*; sinoacutine, the enantiomer of salutaridine (**149**), is reductively converted specifically into sinomenine by this plant (Fig. 14.29) [24].

Also in *Corydalis cava* (*Papaveraceae*) the conversion of reticuline into the morphinanedienone-type alkaloid sinoacutine was demonstrated [116]. In *Croton flavens*, sinoacutine and norsinoacutine are interconvertible, sinomenine can be transformed into sinoacutine, and norsinoacutinol is used to form norsinoacutine and sinoacutine [469, 470]. *C. flavens* as well as *S. acutum*, which contain morphinandienone compounds but do not produce morphine-type alkaloids, are probably able to suppress the necessary dehydration step by oxidizing dienols to dienones.

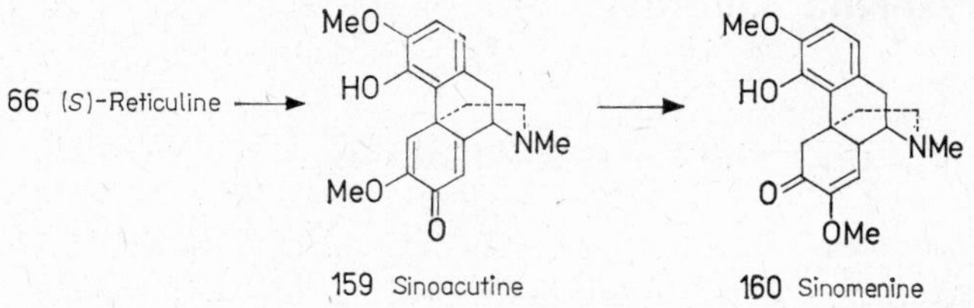

66 (*S*)-Reticuline → 159 Sinoacutine → 160 Sinomenine

Fig. 14.29
Biosynthesis of sinomenine

14.6.3 Sebiferine and Related Alkaloids

Whereas thebaine, codeine, morphine, and sinoacutine are synthesized by an *ortho-para* oxidative coupling of (*R*)- or (*S*)-reticuline, a corresponding *para-para* coupling of reticuline has been established for the biosynthesis of flavinantine (**161**) in *Croton flavens* (*Euphorbiaceae*) [470, 472].

Sebiferine (**162**) is a morphinanedienone alkaloid isolated initially from the stem bark of *Litsea sebifera* (*Lauraceae*) [456]. It is identical with *O*-methylflavinantine. Also the biosynthesis of this alkaloid starts from norreticuline and reticuline giving isosalutaridine by *para-para* oxidative coupling.

In experiments with *Cocculus laurifolius* (*Menispermaceae*) both enantiomers of reticuline were incorporated with nearly the same efficiency. This establishes the

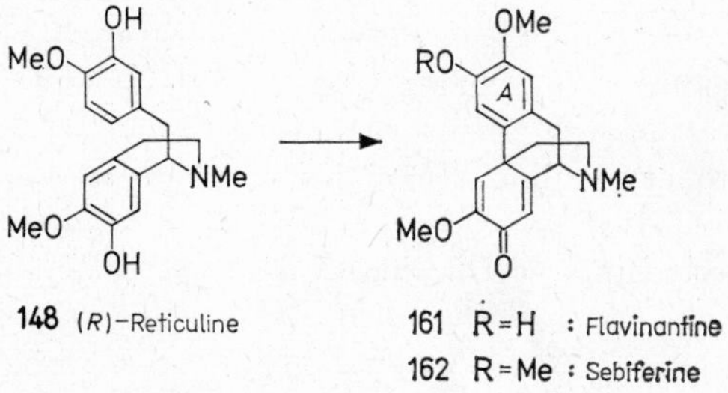

148 (*R*)-Reticuline → 161 R=H : Flavinantine; 162 R=Me : Sebiferine

Fig. 14.30
Biosynthesis of sebiferine

presence of a highly active oxidation—reduction system involving the 1,2-didehydro-derivative, as has been observed in poppy (Fig. 14.28), and supports the following sequence for the biosynthesis of sebiferine in *C. laurifolius*: norreticuline → (*S*)-reticuline → dehydroreticuline → (*R*)-reticuline → sebiferine (**162**) (Fig. 14.30) [93]. The enzyme systems present in *C. laurifolius* are also capable of converting flavinantine into sebiferine. A similar biosynthetic pathway is suggested for amurine (having a methylenedioxy group in ring *A* of sebiferine) and its dienol, nudaurine, from *Papaver amurense* [170].

14.7 Erythrina Alkaloids

14.7.1 Erythraline and Related Alkaloids

Dienones which are established precursors for some aporphines, are also intermediates for the *Erythrina* alkaloids, such as erythraline (**172**), which occur mainly in *Erythrina* species. It can be seen that bond migration is not the only secondary reaction that can lead to modification of the carbon skeleton of a dienone produced by oxidative coupling.

In experiments with *Erythrina crista galli* and *E. berteroana* the immediate and exclusive benzylisoquinoline precursor for the *Erythrina* alkaloids was found to be (*S*)-*N*-norprotosinomenine (**81**) [18, 23, 28], although there is an *in vitro* synthesis of the erythrina skeleton starting from norreticuline [196]. (*S*)-*N*-Norprotosinomenine is converted to the dienone (**126**) by *para-para* oxidative phenol coupling, which is also found as an intermediate in the biosynthetic pathway for the aporphine alkaloids glaucine and dicentrine (Fig. 14.24). This dienone undergoes cleavage to a nine-membered ring system **163** rather than migration of a C—C-bond.

The imine so generated is reduced to give the dibenzazonine intermediate **164**, which is efficiently incorporated into erythraline (**172**) and erythratine (**169**). The dibenzazonine is subjected to further modification *via* a recyclization and oxidation leading to erysodienone (**165**); the (*S*)-antipode, which has been related to the natural alkaloids is the *in vivo* progenitor of the *Erythrina* alkaloids (Fig. 14.31) [19]. *In vitro,* this oxidation proceeds *via* a planar diphenoquinone, the diquinone of **164**, but *in vivo* such an intermediacy has not been demonstrated [17].

Erythratinone (**166**), erysodine (**170**), and erysopine are all well incorporated into erythraline (**172**), suggesting the pathway shown (Fig. 14.31) for the conversion of erysodienone into erythraline. In addition the synthetic compounds erysotinone (**167**) and erysotine (**168**), which have not yet been isolated from natural sources, are efficiently converted into erythraline by *E. crista galli*. In a survey of different *Erythrina* species for corresponding alkaloids some possible intermediate bases have been found [21].

Furthermore the aromatic *Erythrina* alkaloids, such as erysodienone (**165**), erysodine (**170**), and erysovine, are shown to be specifically incorporated into the lactone alkaloids α- and β-erythroidines (**175**) with retention of the 17-hydrogen atom. This probably involves the intermediates **173** and **174** (Fig. 14.31) [19, 324]. At present it is not possible to decide whether the lactone alkaloids are formed by an intradiol (C-15—C-16) or extradiol (C-16—C-17) cleavage. Both processes would lead to retention of the 17-hydrogen atom and loss of C-16. The cleavage of aromatic rings by higher plants is rare but has been proven unambiguously for the betalains.

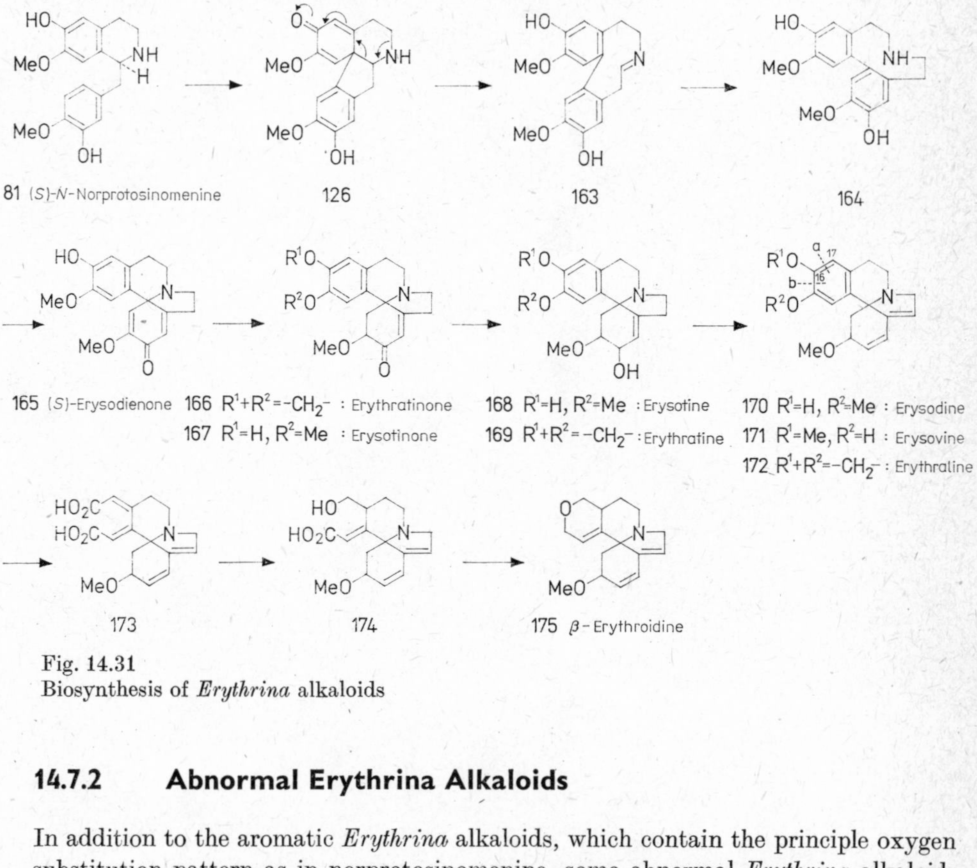

81 (S)-N-Norprotosinomenine
126
163
164
165 (S)-Erysodienone
166 $R^1+R^2=-CH_2-$: Erythratinone
167 $R^1=H, R^2=Me$: Erysotinone
168 $R^1=H, R^2=Me$: Erysotine
169 $R^1+R^2=-CH_2-$: Erythratine
170 $R^1=H, R^2=Me$: Erysodine
171 $R^1=Me, R^2=H$: Erysovine
172 $R^1+R^2=-CH_2-$: Erythraline
173
174
175 β-Erythroidine

Fig. 14.31
Biosynthesis of *Erythrina* alkaloids

14.7.2 Abnormal Erythrina Alkaloids

In addition to the aromatic *Erythrina* alkaloids, which contain the principle oxygen substitution pattern as in norprotosinomenine, some abnormal *Erythrina* alkaloids are known with fewer oxygen functions, such as cocculidine (**185**) and cocculine (**186**), the hypotensive principles, as well as isococculidine (**181**), the neuromuscular blocking principle of *Cocculus laurifolius*, and coccuvine (**184**) [104, 413]. These structures can also be formed in nature from norprotosinomenine (**81**) or its precursors [81, 97, 101]. The (*S*)-enantiomer is especially used, yielding at first the same dienone (**126**) as for the normal *Erythrina* alkaloids. In the bioconversion into the abnormal *Erythrina* alkaloids a reduction of the dienone is involved. This yields the dienol (**176**) that can lose one of the oxygen functions of the precursor by a dienol-benzene rearrangement during the formation of the corresponding imine

(**177**) (Fig. 14.32). The next steps are similar to those for the normal *Erythrina* alkaloids. Feeding protosinomenine revealed that the plants cannot convert protosinomenine into *N*-norprotosinomenine. The 7-methoxy group and the C-1 hydrogen atom (at the asymmetric centre) in norprotosinomenine (**81**) are retained during the bioconversion into cocculidine (**185**).

The key dienone **180**, an erysodienone derivative, can be formed from a dibenz-[*d,f*]azonine intermediate **178** as shown in **179** by recyclization. The occurrence of dibenz [*d,f*]azonine bases such as laurifinine in *C. laurifolius* [386] indirectly supports the intermediacy of these bases in the biosynthesis of the abnormal *Erythrina*

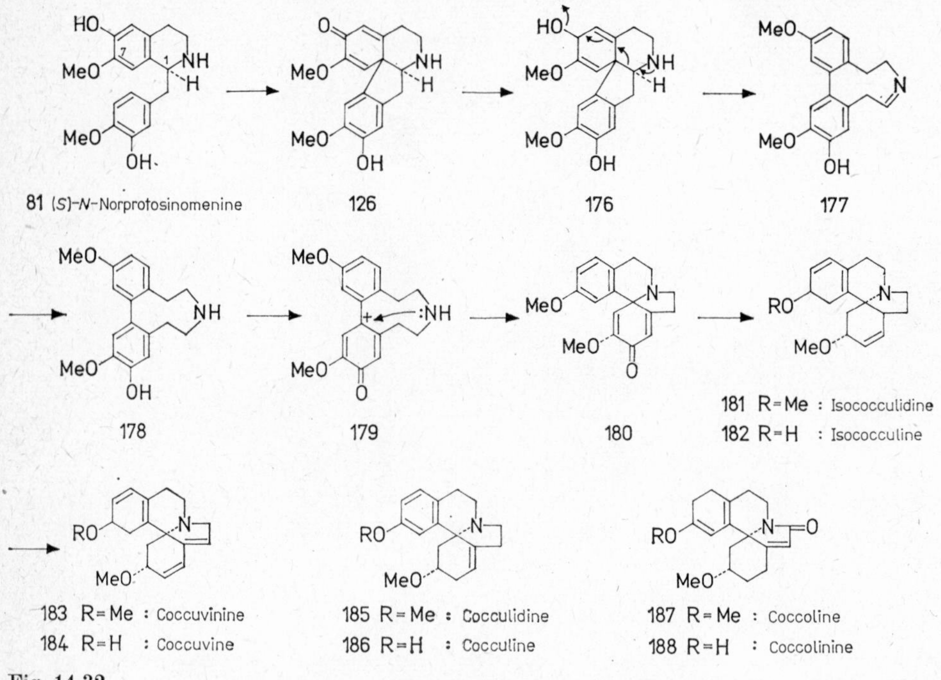

Fig. 14.32
Formation of "abnormal" *Erythrina* alkaloids

alkaloids isococculidine, cocculidine, and cocculine. Laurifinine (the $N-CH_3$-derivative of **178**) is biosynthesised in *C. laurifolius* according the pathway outlined in Fig. 14.32 [82a].

The dienone (**180**) is a precursor of the different abnormal *Erythrina* alkaloids in *C. laurifolius* [81] (Fig. 14.32). On the one hand it can be converted into cocculidine (**185**) and — by demethylation — into cocculine (**186**) which have a double bond at C-1—C-2, and on the other hand it yields isococculidine (**181**). This can be transformed into isococculine (**182**), coccoline (**187**), coccolinine (**188**), coccuvinine (**183**), and coccuvine (**184**). The monodemethylated isococculine is only transformed into coccuvine and coccolinine with a conjugated diene system (C-1—C-7 and C-2—C-3).

The diene alkaloids coccuvinine and coccuvine can be converted to coccoline and coccolinine. The 4-*O*-methylated alkaloids can be 4-demethylated but no further methylation of the demethyl compounds was observed.

14.8 Protostephanine and Hasubanonine

Stephania japonica produces a wide variety of alkaloids [490], among them hasubanonine (**200**) and the rare base protostephanine (**198**). These were the first natural examples of the hasubanane and dibenz [*d,f*]azonine skeletons to be characterized. For the biosynthesis of these unusual systems many pathways have been suggested [31, 123]. Most probably an oxidative coupling process is involved, which seems to follow an unusual course. Tracer experiments with *S. japonica,* have given positive incorporations of labelled tyrosine, dopa, tyramine, and dopamine into hasubanonine and protostephanine. The results show that both alkaloids are built from two different C_6-C_2 units derivable from tyrosine [59]. Dopa, tyramine, and dopamine as a phenethylamine unit generate only ring-*C* of these alkaloids with its attached ethanamine residue.

The suggestion that to generate a trioxygenated aromatic ring oxidation must occur early before the C_6-C_2 units are joined was confirmed by an efficient incorporation of the amines **189** and **190** into hasubanonine and protostephanine yielding again specifically ring *C* and its ethanamine side chain. No other methylation pattern of **189** can be used. Therefore the first isoquinoline derivative to be formed should have in its phenethylamine unit the same pattern of hydroxyl and methoxy groups as these intermediates. From feeding experiments with a large number of different benzylisoquinolines it has to be suggested that the phenethylamine **190** is converted into the benzylisoquinoline **191** by reaction with 4-hydroxyphenylpyruvic acid (**63**) derived from tyrosine and following decarboxylation in the usual manner (Fig. 14.33). Both the secondary and tertiary amines of **191** and of the benzylisoquinolines **192** and **193** were efficiently incorporated [60, 61, 67]. The time of *N*-methylation does not seem to be critical, indicating that possibly two interconnected pathways operate in parallel at these stages of the biosynthesis. In contrast the order of the hydroxylation and methylation by which the functionality of the aryl ring is modified seems to be precisely defined because no other benzylisoquinoline derivatives were incorporated.

The next step in the biosynthesis of protostephanine may be an intramolecular oxidative coupling (*para-para*) of the intermediate **193** in the configuration of **193a**. One has to consider the possibility that the benzylisoquinoline enters the phenol coupling with two free hydroxyl groups in one of its aryl rings. Therefore it is suggested that a nucleophilic addition to an *ortho*-quinone **194** is involved rather than the usual coupling of a biradical. It should be noted that in the other examples considered so far there has been only one free hydroxyl group in each aryl ring of the substrate. The remaining oxygen functions have always been protected as methoxy group to avoid side reactions. The *ortho*-quinone may be converted to the

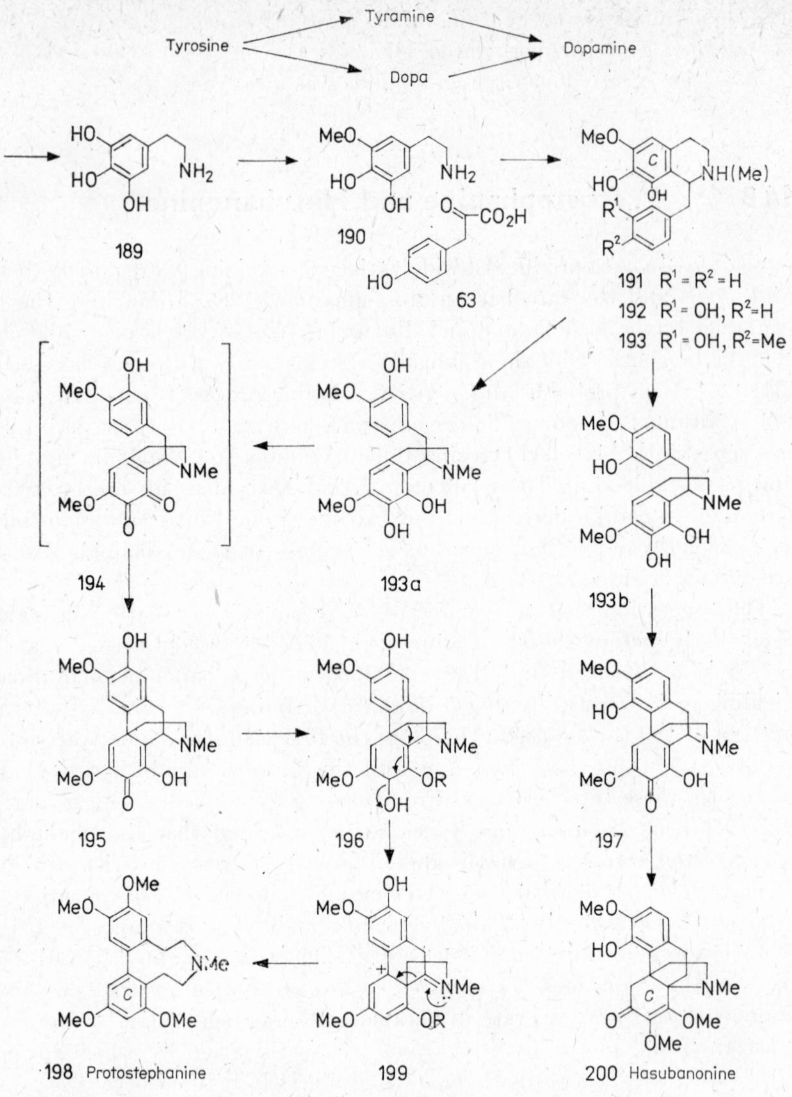

Fig. 14.33
Biosynthesis of hasubanonine and protostephanine

intermediate dienone **195**, which yields protostephanine (**198**) by reduction to the dienol **196** following a migration to **199**, a subsequent fragmentation recalling the erythraline biosynthesis and methylation [462].

The biosynthesis of hasubanonine (**200**) in *Stephania japonica* shows the same unusual features as for protostephanine (Fig. 14.33). Also in this case the benzylisoquinoline (**193**) with two free hydroxy groups in one aryl ring is a necessary intermediate that seems to undergo oxidative phenol coupling. In this case the configuration **193b** (*ortho-para*) yields the dienone **197** followed by migration of a C—N-bond. As far as is known these cases are unique in requiring two phenolic hydroxyl groups in one of the rings undergoing oxidative coupling.

14.9 Cularine, Dibenzopyrrocoline, and Pavine-Type

Oxidative coupling of an unusual hydroxylated benzylisoquinoline (**201**) yields alkaloids of the cularine type, for example cularine (**203**) [115a, 122a]. A biogenetic-type synthesis has led to the suggestion that the biosynthesis takes place *via* the dienone (**202**) as an intermediate (Fig. 14.34) [270].

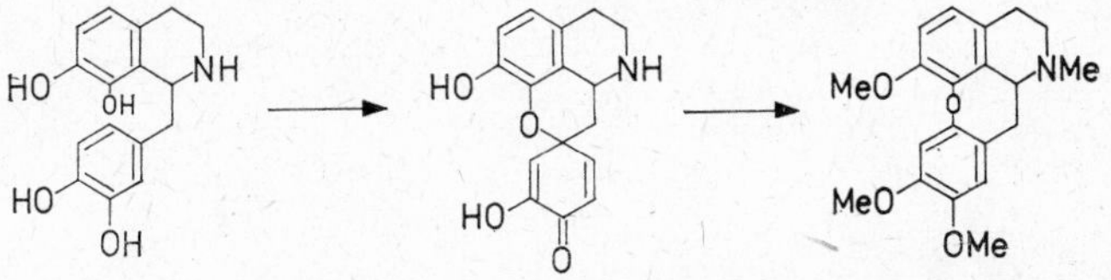

HO NH OH HO OH → HO NH O HO O → MeO NMe O MeO OMe

201 202 203 Cularine

Fig. 14.34
Biosynthesis of cularine

An oxidative coupling of the laudanosoline system **204** with the nitrogen leads to the dibenzopyrrocoline alkaloids (Fig. 14.35) of which cryptaustoline (**206**) and cryptowolline (**207**) occur naturally. *O*-Methylcryptaustoline could be synthesized by oxidation of laudanosoline (**204**) with horseradish peroxidase and hydrogen peroxide to furnish the tetraphenol (**205**) in 81% yield. Subsequent *O*-methylation afforded *O*-methylcryptaustoline [139].

HO HO NMe OH OH → R^1O HO + N Me OR^2 OR^3

204 Laudanosoline

205 $R^1 = R^2 = R^3 = H$
206 $R^1 = R^2 = R^3 = Me$: Cryptaustoline
207 $R^1 = Me, R^2 + R^3 = -CH_2-$: Cryptowolline

Fig. 14.35
Formation of the dibenzopyrrocoline-system

A ring closure with the C-3 or C-4 position of a corresponding benzylisoquinoline (*e.g.* reticuline **66**), most probably by oxidative coupling, yields the pavine alkaloids such as munitagine (**212**) and argemonine (**211**). These both occur in *Argemone* species such as *A. munita* and the isopavine type such as amurensine (**209**) are formed in different *Papaver* species [428]. A dihydroisoquinoline (**208**) has been considered as an intermediate of the isopavines and pavines [140, 172]. *Para*-coupling would yield bisnorargemonine (**210**) and argemonine, whereas *ortho* coupling would give munitagine (Fig. 14.36) [22, 465].

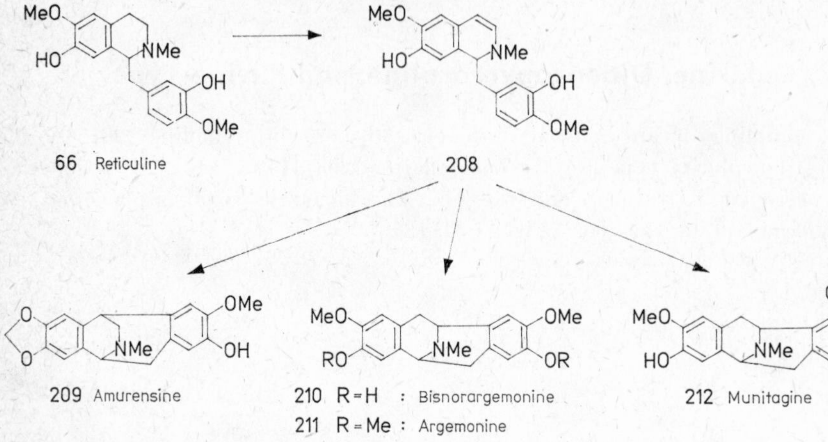

66 Reticuline

208

209 Amurensine

210 R = H : Bisnorargemonine

211 R = Me : Argemonine

212 Munitagine

Fig. 14.36
Formation of the pavine- and isopavine-system

14.10 Protoberberine and Related Compounds

14.10.1 Protoberberines

Berberine (**224**) exemplifies a group of alkaloids based on the benzylisoquinoline skeleton, the distinguishing feature of which is an extra carbon atom appearing as a bridge at C-8. This is the 'berberine bridge'. It has been suggested [419] that several important classes of alkaloids are formed in nature by modification of the protoberberine (berbine) skeleton. Examples are narcotine, a representative of the phthalideisoquinoline class, the benzophenanthridines such as chelidonine, and corydaline, and rhoeadines.

Berberine (**224**), together with hydrastine and canadine (**220**), was shown to be constituted in *Hydrastis canadensis* from two C_6-C_2 units derived from tyrosine. One of them is incorporated via dopamine as for other benzylisoquinolines [245, 365]. Both radioactive methionine and formate label the methylenedioxy- and methoxy groups of berberine and, more significantly, C-8 as well [225]. In experiments with *Fagara coco* one fifth of the formate incorporated is located in the C-8 of palmatine (**226**) [457].

It has been discovered that the berberine bridge arises by oxidative cyclization involving the *N*-methyl group of a 1-benzylisoquinoline in a manner analogous to the formation of a methylenedioxy group. (*S*)-Reticuline (**66**) is incorporated intact into berberine (224) with *Hydrastis canadensis* [22] and *Berberis japonica* [50], into stylopine (217) with *Chelidonium majus* [48], into tetrahydropalmatine (218) and palmatine (226) with *Cocculus laurifolius* [86], and into stepholidine, corydalmine, capaurine and corynoxidine with *Stephania glabra* [80b]. On the other hand (*R*)-reticuline is a precursor of (*R*)-sinactine in *Cocculus laurifolius* [86a]. Tetrahydropalmatine is a precursor of palmatine. C-8 of these alkaloids is formed by oxidative cyclization involving the *N*-methyl group of (*S*)-reticuline (66) (Fig. 14.37). The methylendioxy groups of stylopine arise from the ortho-hydroxy-methoxy-system of reticuline. In *Berberis stolonifera* callus and in *B. aggregata* the methylene dioxy group of berberine (224) is opened in the formation of jatrorrhizine (224 $R^1 = Me$, $R^2 = H$) [76a, 224a].

The product of oxidative ring closure at the *N*-methyl group of (*S*)-reticuline should be (*S*)-scoulerine (214), which is an effective precursor of (*S*)-stylopine (217), tritium is not lost from the chiral centre (C-14) in the biochemical conversion when the precursor reticuline was tritiated at C-1. By trapping experiments it was found that a small pool of scoulerine is present in *C.* majus plants, a result that is supported by its rapid turnover into the alkaloids further along the biosynthetic route (*e.g.* stylopine (217) and chelidonine (230)). Thus the pathway (*S*)-reticuline (66) → (*S*)-scoulerine → (*S*)-stylopine is defined [48], most probably via cheilanthifoline (215). Reticuline is incorporated *via* scoulerine into sinactin [86a].

Even rat liver homogenates convert reticuline into coreximine (232) and scoulerine (214) [271, 274]. Laudanosine was similarly biotransformed into xylopinine (233) and tetrahydropalmatine (218) [273]. The conversion of reticuline to scoulerine took place in an enzyme preparation of *Macleaya cordata* suspension cultures [121, 417]. Incubation of reticuline with $[^{14}CH_3]$*S*-adenosylmethionine and a crude enzyme preparation from *Berberis aggregata* callus culture resulted in the formation of labelled palmatine (226) and columbamine (225) [281]. Cell suspension cultures of 36 different species of *Berberis* contain protoberberine alkaloids [238].

The pathway to berberine (224) proceeds *via* scoulerine (214), isocorypalmine (219), and canadine (220) [76]. Probably a stereospecific enzymic process operates at C-8 for the aromatization of canadine to berberine. Experiments with stereospecific 13-T labelled canadine in *Corydalis ophiocarpa* established that hydroxylation of the C-13 benzylic methylene, which affords ophiocarpine (221), removes the *pro-13R* hydrogen and corresponds to an overall retention of configuration in the hydroxylation process [264].

In a similar way (*S*)-reticuline (66) and its precursors are incorporated into protopine (222) in *Dicentra spectabilis, Argemone* species and *Chelidonium majus* [22, 76] (Fig. 14.37). In this case, (*S*)-scoulerine (214), cheilanthifoline (215), (*S*)-stylopine (217), and stylopine methochloride (216) are also precursors of protopine in *Chelidonium majus* (*Papaveraceae*) and *Corydalis incisa* (*Fumariaceae*) [49, 74, 76, 483, 486]. The indications are that the cis B/C fused stylopine *N*-metho salt rather than the trans analogue is a precursor for protopine. Dihydroprotopine (where CHOH replaces CO) does not seem to be an intermediate.

Isocorypalmine (219) has been shown to be a precursor of the closely related allocryptopine (where methoxyls replace the $O-CH_2-O$ in ring *D* of protopine).

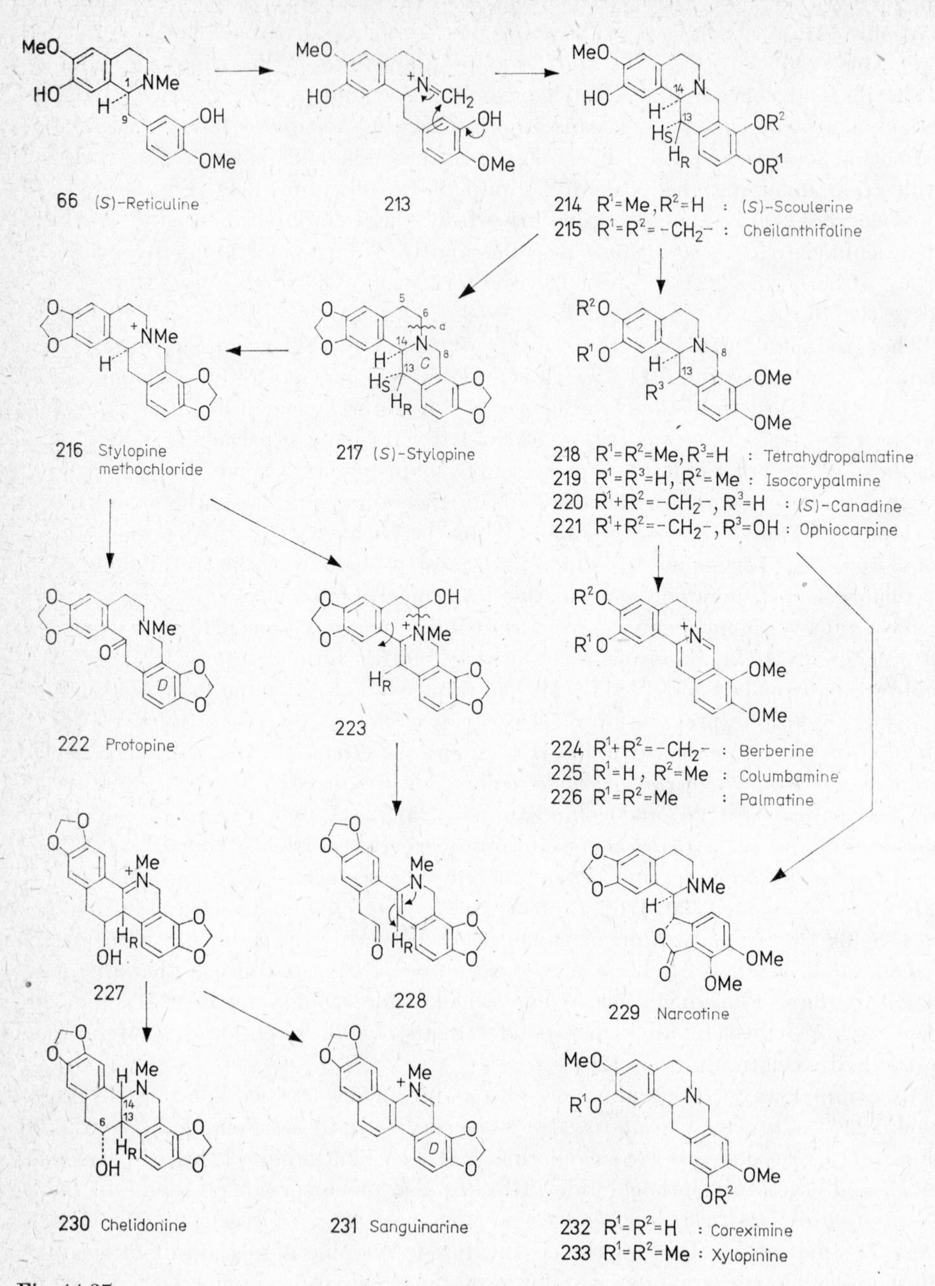

Fig. 14.37
Biosynthesis of protoberberines, and related alkaloids

A reasonable biosynthetic pathway to allocryptopine thus only requires a switch by *O*-methylation at the scoulerine stage of the sequence given for protopine to generate isocorypalmine and so forwards *via* canadine (**220**) to allocryptopine [76]. In *Corydalis bulbosa* tetrahydropalmatine seems to be the first and corydaline and bulbocapnine later products of alkaloid biosynthesis [493].

Corydaline (**240**) and the spirobenzylisoquinoline ochotensimine (**238**), with an additional C_1 group, can be seen as yet further variants on the protoberberine theme (Figs. 14.38 and 14.39). Spirobenzylisoquinoline alkaloids have been found only within the plant family *Fumariaceae*, preferably within the genera *Fumaria* and *Corydalis* [407]. They occur both with and without exocyclic methylenes, *e.g.* ocho-

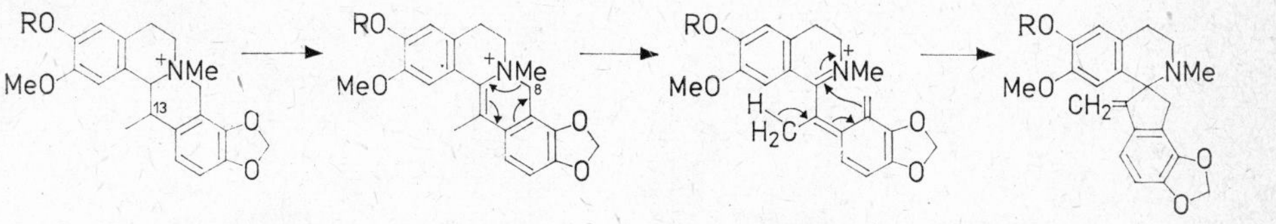

234 235 236 237 R=H : Ochotensine
238 R=Me : Ochotensimine

Fig. 14.38
Possible pathway for the biosynthesis of ochotensimine and ochotensine

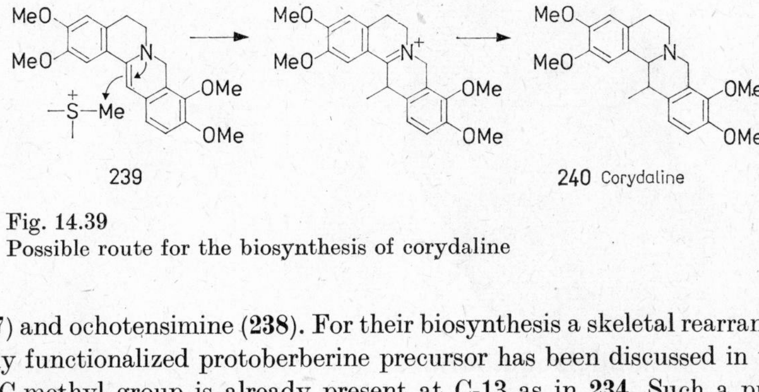

239 240 Corydaline

Fig. 14.39
Possible route for the biosynthesis of corydaline

tensine (**237**) and ochotensimine (**238**). For their biosynthesis a skeletal rearrangement of a suitably functionalized protoberberine precursor has been discussed in which a preformed C-methyl group is already present at C-13 as in **234**. Such a precursor would be closely related to the alkaloid corydaline (**240**), which occurs simultaneously with ochotensine (**237**) in *Corydalis solida*. Oxidation of the tetrahydroprotoberberine (to **235**) provides the driving force for fragmentation of the *N*—C-8 bond leading to the intermediate **236**, which cyclizes to give ochotensimine (**238**) or ochotensine (**237**) (Fig. 14.38).

In studies with *Corydalis solida* and *C. ochotensis* (*Fumariaceae*) tyrosine and methionine are incorporated into these alkaloids along the protoberberine pathway. Methionine supplies the OCH_3-groups, the bridge-C, and also the additional C_1 group of the protoberberine corydaline (**240**) and the spirobenzylisoquinoline ochotensimine (**238**) [244, 245]. The incorporation of reticuline (**66**) into corydaline shows a more obvious correlation to the protoberberines [113]. In accordance with a postulated $\Delta^{13,14}$-didehydroprotoberberine precursor such as **239** the tetradehydroderivative palmatine (**226**) (but not tetrahydropalmatine) is incorporated into corydaline (**240**) in *C. solida* (Fig. 14.39) [245]. The stage in the pathway of norlaudanosoline to corydaline at which the enamine functionality is introduced is unknown. The protopine analogue corycavine (**246**) was shown to be biosynthesized in *Corydalis incisa* plants from tetrahydrocorysamine (**243**) *via* its *N*-methosalt (Fig. 14.40) [483].

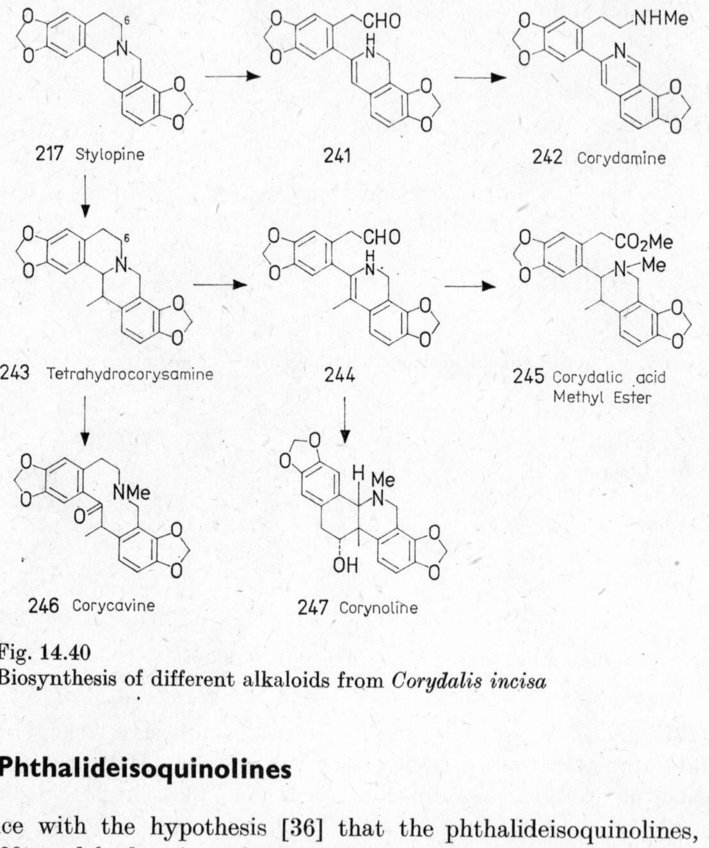

217 Stylopine
241
242 Corydamine
243 Tetrahydrocorysamine
244
245 Corydalic acid Methyl Ester
246 Corycavine
247 Corynoline

Fig. 14.40
Biosynthesis of different alkaloids from *Corydalis incisa*

14.10.2 Phthalideisoquinolines

In accordance with the hypothesis [36] that the phthalideisoquinolines, such as narcotine (**229**) and hydrastine, also arise by oxidative modification of the tetrahydroprotoberberine skeleton, its precursors tyrosine, tyramine, dopamine, methionine, and formate were incorporated specifically into hydrastine and narcotine in *Hydrastis canadensis* and *Papaver somniferum* [56, 57, 214, 225, 365]. Of particular note was the labelling of the lactone carbonyl groups by C_1 sources, especially the methyl group of methionine. The specific incorporation of norlaudanosoline and (*S*)-reticuline (**66**) established the derivation of narcotine (**229**) from an 1-benzylisoquinoline system with the *N*-methyl group becoming the lactone carbonyl (Fig. 14.37). The incorporation of (*S*)-reticuline occurs with retention of half of the hydrogen at C-9. Finally, the protoberberines, (*S*)-scoulerine (**214**), isocorypalmine (**219**), and canadine (**220**), were shown to be precursors of narcotine in opium poppies (Fig. 14.37) [49, 57, 76]. Tritium at position 8 in canadine is lost as expected. The conversion of scoulerine into narcotine in *P. somniferum* involves removal of the *pro-13S* hydrogen atom [75]. Introduction of the oxygen atom at C-13 thus proceeds with overall retention of configuration. The phthalideisoquinoline hemiacetal egenine may be an intermediate in phthalideisoquinoline biogenesis [221 a].

14.10.3 Benzophenanthridines

Chelidonine (**230**) is the most abundant member of the benzophenanthridine class of alkaloids. It occurs in *Chelidonium majus* together with more highly oxidized members, *e.g.* sanguinarine (**231**). Stylopine (**217**) is also present in this plants and it has been suggested [419] that the benzophenanthridine bases arise by modification of the tetrahydroprotoberberine (berbine) skeleton (Fig. 14.37). The necessary changes are C—N fission at a (in **217**) and formation of a new C—C bond between C-6 and C-13. The alkaloids in *Chelidonium majus* latex seem to be stored in the vacuolar sap of the vesicles, though only 70% were found in the vesicle fraction after centrifugation, possibly due to the greater fragility of the latter [353].

Earlier work [320, 331] has shown the incorporation of [2-^{14}C]tyrosine and [1-^{14}C]-dopamine into chelidonine (**230**) and sanguinarine (**231**); the former precursor labelled C-6 and C-14 of chelidonine and the latter C-6 only. The carbon skeleton is thus built from two C_6—C_2 units derivable from tyrosine. Furthermore it has been shown with *C. majus* plants that (*S*)-reticuline (**66**) is an effective and specific precursor of chelidonine [76] giving a labelling pattern qualitatively and quantitatively in agreement with biosynthetic conversion of the tetrahydroprotoberberine skeleton (*e.g.* stylopine (**217**)) into the benzophenanthridines **230** and **231** (Fig. 14.37).

Direct evidence for the protoberberine-benzophenanthridine relationship came from feeding the plants with labelled (*S*)-scoulerine (**214**), (*S*)-stylopine (**217**) and stylopine methochloride (**216**) which are well incorporated into chelidonine, sanguinarine (**231**), and chelerythrine (in ring *D* of sanguinarine two OCH_3 groups in place of O—CH_2—O) [49, 57, 76, 331]. The C-8 of stylopine is unaffected during the further transformation into chelidonine. Tritium at C-9 of (*S*)-reticuline is retained through to stylopine at C-13 whereas there is significant loss in its subsequent conversion into chelidonine [76, 483]. During the biosynthetic pathway from scoulerine *via* stylopine to chelidonine there is a stereospecific removal of the *pro-13S* hydrogen suggesting that the enamine system (*e.g.* **223**) is generated by a *cis*-dehydrogenation or after hydroxylation with retention of configuration at C-13 of stylopine by elimination of water in an overall *cis*-manner [75].

Complete loss of tritium from the chiral centre of scoulerine (**214**) as it is transformed into chelidonine (**230**) has been observed. This tritium loss is in keeping with the suggested intermediate **223** (Fig. 14.37), which allows the new carbon-carbon bond to be formed by enamine chemistry. The conversion of stylopine into chelidonine involves fission of the C-6-*N*-bond with loss of at least one hydrogen atom from C-6. The corresponding oxidation of the methylene group at C-6 to the aldehyde level **223** occurs with stereospecific removal of the hydrogen atom in *si*-space (*pro-S*) [74] as it happens with the hydrogen removal from C-13 and C-14 of stylopine (**217**) during the biosynthesis of chelidonine (**230**) [75]. C-5 is not involved in the conversion of stylopine into chelidonine. Stylopine methochloride (**216**) is incorporated without degradation indicating that it lies on the pathway to chelidonine. Thus the biosynthetic pathway to chelidonine is as shown in Fig. 14.37. In cell suspension cultures of *Macleaya cordata* the following pathway is defined: 7,8,13,13a-tetrahydrocoptisine (**217**) → cis-*N*-methyl-7,8,13,13a-tetrahydrocoptisinium salt (**216**) → protopine (**222**) → sanguinarine (**231**) → chelirubine (**231**-type) → macarpine (**231**-type) [483a]. Suspension cultures of *Eschscholtzia californica* accumulate the dihydro forms of sanguinarine (**231**), chelirubine, macarpine, and chelerythrine, all of which are known to be constituents of the *Eschscholtzia* plant [79a].

Corydalis incisa, a plant also known to produce benzophenanthridines, contains alkaloids of the corydaline-type together with secoprotoberberine alkaloids. These are postulated as intermediates in the benzophenanthridine biosynthesis. It has been demonstrated with this plant that stylopine (**217**) can be converted into corydamine (**242**), probably *via* the aldehyde **241** formed by cleavage of the C-6-*N* bond, and also into corydalic acid methyl ester (**245**) and the chelidonine analogue corynoline (**247**) *via* tetrahydrocorysamine (**243**) and the aldehyde **244** by a corresponding oxidative fission of its C-6 to *N* bond [487, 509] (Fig. 14.40).

14.10.4 Rhoeadine-Type Alkaloids

Rhoeadine (**250**) is present in *Papaver rhoeas* and arises by extensive modification of scoulerine (**214**). During this biotransformation a stereospecific loss of the *pro-13S* hydrogen of the tetrahydroberberine occurs [73]. It seems likely that scoulerine

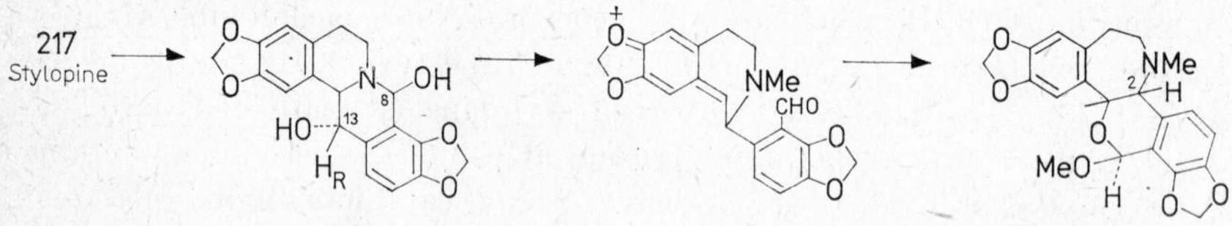

Fig. 14.41
Formation of rhoeadine

is converted first into stylopine (**217**), which is oxidized to the 8,13-dihydroxy compound (**248**). *N*-Methylation and rearrangement to **249** then provide the correct skeleton from which rhoeadine (**250**) can arise (Fig. 14.41).

In other experiments with *P. rhoeas* $[N\text{-}^{13}CH_3]$stylopine methochloride and $[13\text{-}T_2]$protopine were incorporated into rhoeadine [488]. All the tritium of the latter was found at C-2 of rhoeadine.

The tetrahydrobenzazepine alkaloid alpinigenine (**254**) belongs to the trans series of rhoeadine-type bases [223, 354]. It originates in *Papaver bracteatum* from two molecules of tyrosine [122]. In addition six C_1-units were incorporated into the four methoxyls, the aminomethyl, and the hemiacetal C-14 after feeding of [S-$^{14}CH_3$]-methionine. Tetrahydropalmatine (**218**) and its methiodide (**251**) were established as specific precursors of alpinigenine [421, 422] showing that this tetrahydrobenzazepine skeleton is derived from the tetrahydroprotoberberine ring system (Fig. 14.42). The rearrangement involves reactions at carbons-8, -13 and -14 and at the nitrogen. C-8 of tetrahydropalmatine yields C-14 of alpinigenine and C-1 and C-2 of the latter are derived from C-14 and C-13 of the tetrahydroprotoberberine system. As mentioned above (Fig. 14.37) tetrahydropalmatine (**218**) is synthesized from norlaudanosoline, (S)-reticuline (**66**), and (S)-scoulerine (**214**). In accordance with this pathway only (S)-reticuline is a precursor of alpinigenine whereas the (R)-enantiomer is used in the biosynthesis of thebaine in *P. bracteatum*.

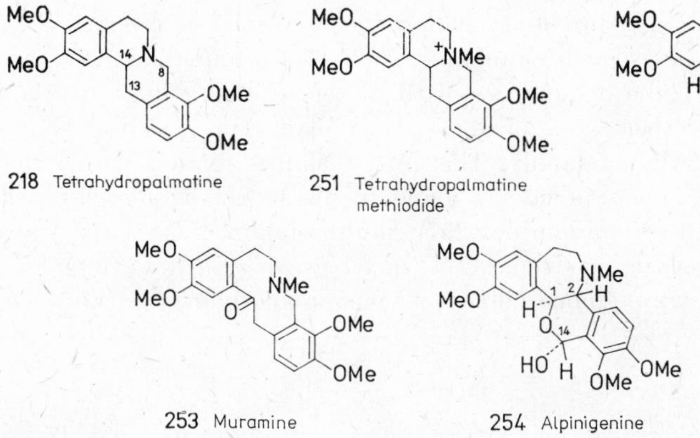

218 Tetrahydropalmatine

251 Tetrahydropalmatine methiodide

252

253 Muramine

254 Alpinigenine

Fig. 14.42
Biosynthetic pathway for alpinigenine

With [8,13,14-^{3}H]tetrahydropalmatine (**218**) and its methiodide (**251**) it has been shown that during transformation into alpinigenine (**254**) one of the two C-8 and one of the two C-13 hydrogens are retained at C-14 and C-2 of alpinigenine respectively. This shows that a phthalideisoquinoline intermediate is not involved in the reactions — as has been discussed in an earlier biogenetic proposal-because this would implicate the total loss of tritium from C-8. One hydrogen of C-8, one of C-13 and the hydrogen of C-14 are lost during the biosynthetic pathway. Therefore compounds lacking hydrogen at C-14 must be potential intermediates. According to this proposal the protopine-type alkaloid muramine (**253**), a natural base of *Papaver nudicaule* [124], could be shown to be an efficient and specific precursor of alpinigenine [422]. Therefore it is postulated that an N-methylation step (to **251**) is followed by hydroxylation at C-14 (**252**), the carbon atom adjacent to the quaternary nitrogen. Possibly a non-enzymatic step could complete the fission of the C-14-N bond yielding the protopine-type alkaloid **253**. The occurrence of tetrahydropalmatine, its methiodide, and muramine has been established in *P. bracteatum* by trapping experiments.

Some plants of *Papaver bracteatum* (thebain-type) contain alpinigenine only during their early stages of development but this alkaloid is not detectable in older plants [117]. Old plants are, however, able to form alpinigenine after feeding of tetrahydropalmatine [119]. Apparently one of the early steps in the biosynthetic pathway is blocked in old plants and a new formation of alpinigenine is triggered by feeding of tetrahydropalmatine.

14.11 1-Phenethylisoquinolines

14.11.1 Simple 1-Phenethylisoquinolines

During the course of investigations on the biosynthesis of colchicine (**276**) it was shown that 1-phenethylisoquinolines such as autumnaline (**267**) are important intermediates. These homologues of the 1-benzylisoquinolines are the central core of an important new class of alkaloids of many types, commonly isolated from plants of the *Liliaceae* family. This relationship is reminiscent of the vast group of 1-benzylisoquinoline alkaloids found in plants of the *Papaveraceae*. Thus the C-homoaporphines floramultine (**257**), multifloramine (**258**), and kreysigine (**259**) were found alongside colchicine (**276**) in *Kreysigia multiflora* [39] (Fig. 14.43). The minor alkaloid kreysiginone (**261**) is a homoproaporphine and kreysiginine (**262**) is

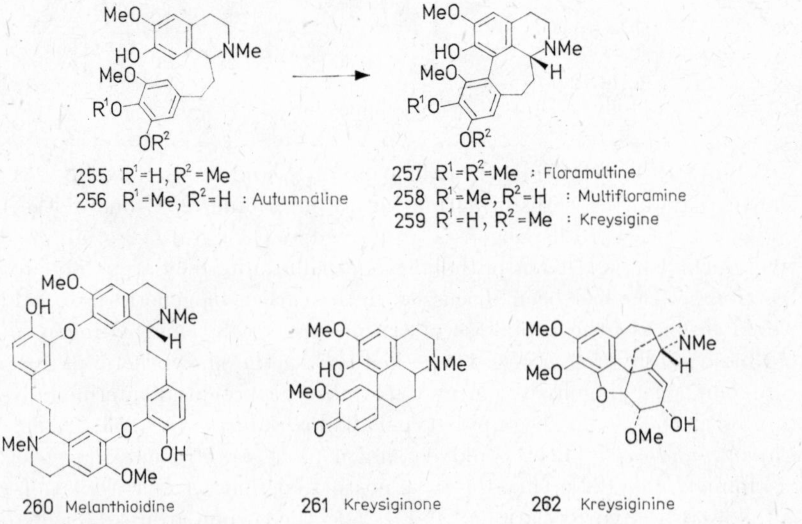

255 R^1=H, R^2=Me
256 R^1=Me, R^2=H : Autumnaline

257 R^1=R^2=Me : Floramultine
258 R^1=Me, R^2=H : Multifloramine
259 R^1=H, R^2=Me : Kreysigine

260 Melanthioidine

261 Kreysiginone

262 Kreysiginine

Fig. 14.43
Different phenethylisoquinoline derivatives and biosynthesis of homoaporphines

related to the morphine group [68, 230]. Melanthioidine (**260**) from *Androcymbium melanthioides* resembles the bisbenzylisoquinolines and may be synthesized by oxidative phenol coupling of two phenethylisoquinoline units [54]. The synthesis of androcymbine (**269**) as a precursor of colchicine from a 1-phenethylisoquinoline is to be compared with the biogenesis of salutaridine as a progenitor of morphine from reticuline. A series of homoerythrina alkaloids has been isolated from the Australian

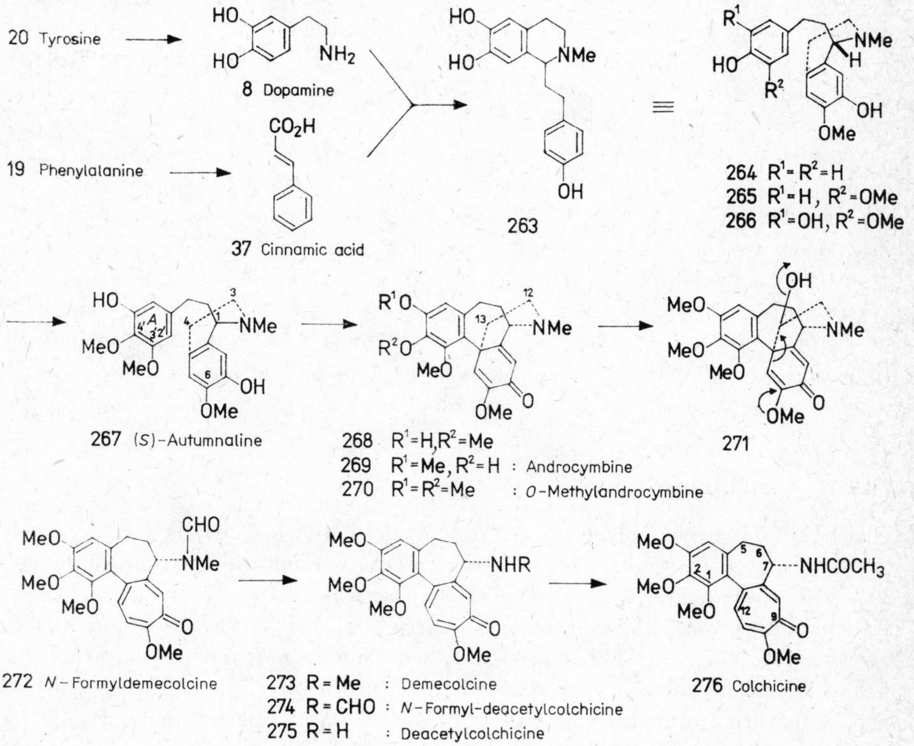

Fig. 14.44
Biosynthesis of colchicine

plant *Schelhammera pedunculata* (*Liliaceae*) [268, 317] and also the alkaloids of *Cephalotaxus harringtonia* are modified homoerythrina bases [403]. The parent 1-phenethylisoquinoline system has been found in *Colchicum cornigerum* as autumnaline (**267**) [70]. Some structural types familiar in the 1-benzylisoquinoline series (*e.g.* protoberberines) are not present in the above list.

Research on colchicine biosynthesis has identified autumnaline (**267**) as a key intermediate and *trans*-cinnamic acid and dopamine as primary precursors (Fig. 14.44) [53]. Therefore these substances are also progenitors of autumnaline. The sequence to this phenethylisoquinoline was examined by feeding a set of 1-phenethylisoquinolines with various degrees of oxygenation and of *O*- and *N*-methylation.

Since neither the oxygenated cinnamic acids nor hydrocinnamic acid are precursors of colchicine, it was suggested that the carboxyl group of cinnamic acid must be modified, probably to give cinnamaldehyde, prior to reduction of the double bond or hydroxylation of the aromatic ring, and that sequential oxidation of the C_6-C_3 unit might take place at the isoquinoline level. This was confirmed by efficient incorporation of [9-^{14}C]1-phenethylisoquinoline (**264**, **265** and **266**). Studies with multiply labelled **264** and **265** revealed that these precursors retain their *O*-methyl groups during intact bioconversion into colchicine *via* autumnaline. *N*-Methylation occurs early in the sequence. As the isoquinoline lacking an oxygen function in the phenethyl part was not incorporated, the timing of the initial oxygenation of the C_6-C_3 unit has still to be established.

In an investigation with different 1-phenethylisoquinolines applied to *Kreysigia multiflora* plants it has been shown that autumnaline (**267**) is an efficient and specific precursor for the homoaporphines floramultine (**257**), multifloramine (**258**), and kreysigine (**259**) (Fig. 14.43) [38]. The incorporation of the base **265** results presumably from its conversion into autumnaline as happens in *Colchicum autumnale* plants [53]. It is very probable that a direct *ortho-para* phenol coupling operates. In contrast to this, it is clear for the homoaporphines that *para-para* coupling is involved for the tropolones co-occurring as colchicine (**276**) [11]. The results also indicate that the presence of small amounts of kreysiginone in *K. multiflora* is not connected with the main pathway to the homoaporphines.

14.11.2 Colchicine

In colchicine (**276**) the 1-phenethylisoquinoline skeleton (as **263**) is so extensively modified that it is difficult to discern the essential structural relation. Moreover, when research was begun on colchicine biosynthesis no naturally occurring 1-phenethylisoquinolines were known. Earlier research with *Colchicum autumnale* and *C. byzantinum* [37, 237, 321, 332] had established that ring A and carbon atoms -5, -6 and -7 of colchicine (**276**) are derived in nature from cinnamic acid, which is generated in turn from phenylalanine. Tyrosine cannot replace phenylalanine for this part of the molecule. [3-^{14}C]Tyrosine was used to form colchicine carrying its radioactivity in the tropolone ring at C-12. This ring system thus appears to be formed by ring expansion of a C_6-C_1 unit provided by tyrosine [43]. Following [4'-^{14}C]tyrosine incorporation, the isolated colchicine was found to be labelled at C-9 [322].

The incorporation of phenylalanine but not of tyrosine into the highly oxygenated ring A is surprising but relates to the situation for the biosynthesis of *Amaryllidaceae* alkaloids, where phenylalanine is converted into trans-cinnamic acid before oxygenation of the aromatic ring. Acetate, a precursor of fungal tropolones [79] is incorporated only into the *N*-acetyl group of colchicine [37, 333]. Tyramine and dopamine were both incorporated with satisfactory efficiency into colchicine and they presumably contribute the same C_6-C_1 unit as that provided by tyrosine [53]. Support is thus given for decarboxylation of the related amino acids before combination of the nitrogenous unit with that derived from cinnamic acid.

It has been suggested that the colchicine skeleton is biosynthesized from a dienone skeleton [43], which is formed by oxidative phenol coupling of a suitably substituted 1-phenethylisoquinoline. Such a dienone was found as androcymbine (**269**) [55], which occurs alongside colchicine in *Androcymbium melanthioides* [247, 400]. The conviction that androcymbine is of biosynthetic importance for colchicine was strengthened by the close botanical relationship between *C. autumnale* and *A. melanthioides*; both species fall into the sub-family *Wurmbaeoideae* of the *Liliaceae* family. On the basis of such a biosynthetic connection between androcymbine (**269**) and colchicine (**276**) a pathway can be considered making use of a 1-phenethylisoquinoline, *e.g.* autumnaline (**267**). By oxidative coupling and subsequent *O*-methylation this could generate *O*-methylandrocymbine (**270**). Hydroxylation of **270** to form structure **271**, as occurs at a late stage in the biosynthesis of haemanthamine, could provide a starting point for a homoallylic ring expansion possibly *via* the *O*-phosphate. Such a process would generate the colchicine skeleton specifically labelled at C-12 from [3-^{14}C]tyrosine (Fig. 14.44).

In accordance with these suggestions labelled *O*-methylandrocymbine (**270**) and autumnaline (**267**) were administered to *C. autumnale* or *C. byzantinum* plants. Efficient and specific incorporations into both demecolcine (**273**) and colchicine (**276**) were observed [53], supporting the suspected biosynthetic relationship between colchicine and a 1-phenethylisoquinoline system (Fig. 14.44). Demecolcine was found to be a precursor of colchicine [11].

The oxidative coupling step (**267** $\rightarrow$ **268**) produces a new carbon-carbon bond between the positions *para* to the phenolic hydroxy groups. The trioxygenated ring could in principle be rotated to allow *ortho-para* coupling. By feeding different multiply labelled autumnalines (**267**), it was shown [11] that the aryl-hydrogen in position 2′ of ring A is lost as expected and that the *N*-atom and the 6-*O*-methyl group of autumnaline are completely retained. The 3′- and 4′-*O*-methyl group correspond to the *O*-methyl groups at C-1 and C-2 in colchicine proving that autumnaline cyclizes exclusively by *para-para* coupling. The *N*-methyl group of autumnaline is retained in demecolcine (**273**) but is absent in colchicine (**276**). (*1S*)-Autumnaline, whose configuration corresponds to that at C-7 in colchicine was incorporated 180 times better than the (*1R*)-enantiomer. The specific incorporation of autumnaline into colchicine was also demonstrated using [1-^{13}C]-autumnaline [71].

Incorporation and trapping experiments support the sequence demecolcine (**273**) $\rightarrow$ *N*-formyldeacetyl colchicine (**274**) $\rightarrow$ deacetylcolchicine (**275**) $\rightarrow$ colchicine (**276**). The first tropolone intermediate found during the transformation of *O*-methylandrocymbine is *N*-formyldemecolcine (**272**). It contains all the carbon atoms of *O*-methyl-androcymbine and is a precursor of demecolcine. The formyl group of *N*-formyldemecolcine (**272**) corresponds to the C-12 of *O*-methylandrocymbine (**270**) and to the C-3 of autumnaline (**267**). By feeding stereospecifically labelled autumnalines (**267**) to *Colchicum autumnale* it was found that the *pro-3S*- and *pro-4S*-hydrogens are removed and the *pro-3R*- and *pro-4R*-hydrogens retained during the conversion to *N*-formyldemecolcine (**272**) [355]. Most probably these stereospecific reactions occur during oxidative attack at C-12 and C-13 of the androcymbine skeleton prior to the fragmentation of the C-12—C-13 bond. These stereochemical results are consistent with either hydroxylation with retention of configuration at both C-12 and C-13 leading to a diol or cis dehydrogenation to an enamine. These intermediates would then be converted into *N*-formyldemecolcine (**272**). Demecolcine (**273**) can be alkylated to the naturally occurring alkaloids *N*-methyldemecolcine and speciosine (*N*-*o*-hydroxybenzyldemecolcine), but cannot be converted into *N*-formyldemecolcine.

14.11.3 Schelhammeridine and Cephalotaxus Alkaloids

Conifers of the genus *Cephalotaxus* (*Cephalotaxaceae*) contain a group of alkaloids of unique structure. The most abundant member of this group is cephalotaxine (**283**) [404], which is accompanied in nature by small quantitaties of related alkaloids [391, 403] such as cephalotaxinone (**285**), demethylcephalotaxinone (**287**) and demethylcephalotaxine (**286**), as well as by several cephalotaxine esters including deoxyharringtonine (**284**), isoharringtonine, harringtonine, and homoharringtonine. Some of these possess significant antitumour activity [405]. The alkaloid esters are also known to be potent inhibitors of eukaryotic protein synthesis [198]. In *C. harringtonia,* the alkaloids derived from the cephalotaxine skeleton occur simultaneously with a number of bases possessing the homoerythrina skeleton such as 3 epi-schelhammericine [401]. Schelhammeridine (**280**) and schelhammericine are the major alkaloids of *Schelhammera pedunculata* (*Liliaceae*) [193, 269]. The occurrence of both structural types (**280** and **283**) together in *Cephalotaxus* species [402] led to the suggestion that they are divergent products from an initially common biosynthetic pathway.

Schelhammeridine (**280**) is structurally related to erythraline. It is thus reasonable to expect schelhammeridine to be biosynthesized from a 1-phenethylisoquinoline (as **277**), a system commonly produced by plants of the *Liliaceae* family. In accordance with this hypothesis, tyrosine is incorporated by *Schelhammera pedunculata* specifically (and exclusively) into the C_6-C_2 residue (ring *A*) of schelhammeridine (**280**). Dopamine follows tyrosine on the biosynthetic pathway and the C_6-C_3 unit of this alkaloid (ring *B*) is derived naturally from phenylalanine *via* cinnamic acid [65, 390]. It is suggested that the phenethylisoquinoline (**277**) is converted in similar reactions as those mentioned for *Erythrina* alkaloids (Fig. 14.31) into the amine **278** the homologue of the corresponding *Erythrina* amine **164**. Probably the dienone **279**, the homologue of erysodienone, is an intermediate formed by ring closure as in **278** (Fig. 14.45).

In the case of cephalotaxine (**283**) it is predicted that this alkaloid is generated *via* a 1-phenethyltetrahydroisoquinoline derivative (**277**) [388]. In experiments with *C. harringtonia* it has been shown that tyrosine leads to exclusive labelling of ring *A* and the atoms C-10 and -11 of cephalotaxine (**283**) and that phenylalanine is specifically incorporated *via* cinnamic acid into the ring *D* and the atoms C-6, -7 and -8 of this alkaloid [389, 390, 440]. The carboxylic group of these acids gives C-8. These results provided substantial evidence that cephalotaxine is a member of the family of phenethylisoquinoline alkaloids that includes the *Schelhammera* and

Colchicum alkaloids. The suggested intermediate **278** for the *Schelhammera* alkaloids could undergo a ring closure according b yielding the dienone **281**, which could be rearranged to the dienone **282**. This could undergo ring contraction to give cephalotaxine (Fig. 14.45). Ring *D* of this alkaloid may be derived from the aromatic ring of phenylalanine by ring contraction and loss of one carbon atom. Incorporation experiments with [*p*-^{14}C]- and [*m*-^{14}C]-phenylalanine have shown that one of the

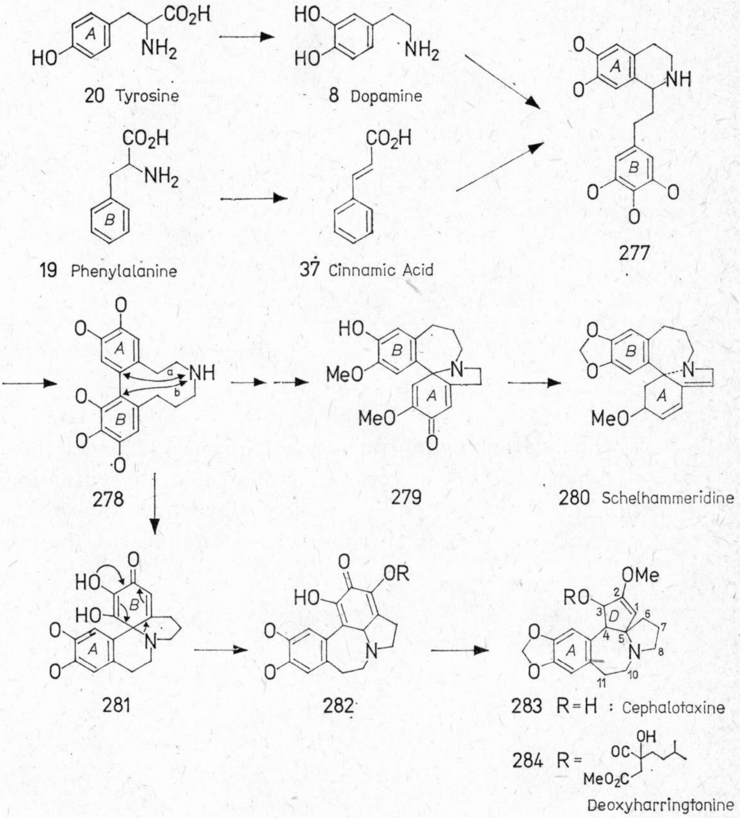

Fig. 14.45
Biosynthesis of schelhammeridine and cephalotaxine

meta carbon atoms of phenylalanine is lost during the conversion of this amino acid into cephalotaxine. The remaining *meta* carbon atom is located at C-2 of the alkaloid. These results are consistent with the occurrence of a benzilic acid rearrangement of the hypothetical dienone **282**.

Investigations on metabolic interrelationships between the *Cephalotaxus* alkaloids cephalotaxine (**283**), cephalotaxinone (**285**), demethylcephalotaxine (**286**) and demethylcephalotaxinone (**287**) have shown that cephalotaxine and cephalotaxinone are interconvertible in *Cephalotaxus* and that demethylation of cephalotaxinone takes place readily (Fig. 14.46).

The key to the elucidation of the biosynthesis of the uncommon acyl portions of the antileukemic *Cephalotaxus* ester alkaloids was provided by the recognition that the parent diacid linked to cephalotaxine in deoxyharringtonine, the deoxyharring-

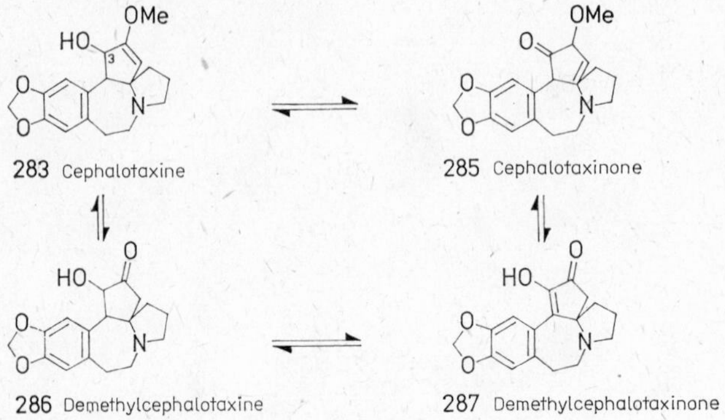

283 Cephalotaxine

285 Cephalotaxinone

286 Demethylcephalotaxine

287 Demethylcephalotaxinone

Fig. 14.46
Metabolic interrelationships between *Cephalotaxus* alkaloids

tonic acid (**293**), bears a close resemblance to a diacid intermediate involved in the biosynthesis of leucine from valine in microorganisms [222]. On the basis of this resemblance, a hypothesis for deoxyharringtonic acid biosynthesis is shown in Fig. 14.47 [215]. This predicts that 3-carboxy-3-hydroxy-5-methyl-hexanoic acid (**289**) should be an intermediate in the biosynthesis of deoxyharringtonic acid (**293**)

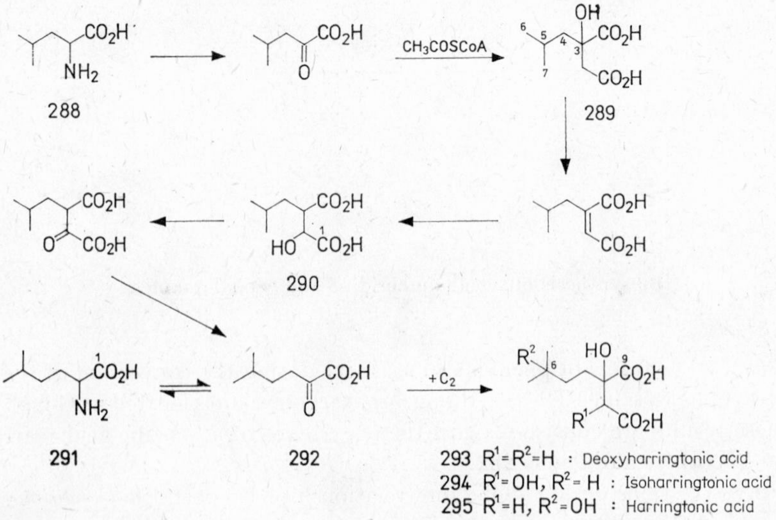

$CH_3COSCoA$

288

289

290

$+C_2$

291

292

293 $R^1=R^2=H$: Deoxyharringtonic acid
294 $R^1=OH$, $R^2=H$: Isoharringtonic acid
295 $R^1=H$, $R^2=OH$: Harringtonic acid

Fig. 14.47
Biosynthesis of the acids in the cephalotaxine esters

and that atoms C-3 to C-8 of **289** should be derived from *L*-leucine (**288**). The presence of diacid **289** in *Cephalotaxus* plants was proved by isotopic trapping. Furthermore the hypothesis predicts that 2-hydroxy-3-carboxy-5-methylhexanoic acid (**290**) lies on the biosynthetic pathway to **293**.

[1-^{14}C]-(**290**) was specifically incorporated into **295**. On the basis of this hypothesis (Fig. 14.47) 2-oxo-5-methylhexanoic acid (**292**) would be expected to be the immediate precursor of **293**. If this is so, then homoleucine (**291**) should be specifically incorporated into **293** as the result of the facile interconversion between α-amino acids and the corresponding α-keto acids. It was demonstrated that homoleucine is a highly efficient and specific precursor of **293** and it was established that isoharringtonic acid (**294**) and harringtonic acid (**295**) are derived from deoxyharringtonic acid (**293**) by direct hydroxylation. Homoharringtonic acid could be formed by homologation of deoxyharringtonic acid (**293**) as suggested in Fig. 14.47 with subsequent hydroxylation. Moreover an attempt might be made to determine whether deoxyharringtonic acid (**293**) is hydroxylated at C-6 before or after it is linked to cephalotaxine. Most probably deoxyharringtonine (**284**) is directly converted to harringtonine without prior deacylation.

Tissue cultures of *C. harringtonia* synthesize the same alkaloids as intact plants although at much lower concentrations. This fact is important for the production of the antitumour alkaloids [167]. A complex mixture of new cephalotaxus alkaloids is produced in another callus line [166].

The alkaloid composition of *C. harringtonia* is influenced by environmental factors on plant growth [165]. Moreover the alkaloids of this plant show a catabolic turnover. In young plants grown in a controlled environment, the concentration of cephalotaxine esters increases. Physiological stress (pruning) causes hydrolysis of part of the stored alkaloid esters to free cephalotaxine (**283**) and also the oxidation of cephalotaxine to 11-hydroxycephalotaxine and drupacine (11,2-epoxycephalotaxine) and of homoerythrina alkaloids to epoxy derivatives. It is suggested that to prevent autotoxicity high concentrations of alkaloid esters are hydrolysed to free cephalotaxine since this is only 1/1000 as toxic as its esters. When the cephalotoxine levels tend to rise too high then further metabolism occurs.

14.12 Amaryllidaceae Alkaloids and Mesembrine

14.12.1 Amaryllidaceae Alkaloids

The alkaloids of this group occur almost exclusively in the family of *Amaryllidaceae*. The different structures could be derived from the same precursor, norbelladine (**298**), which contains a phenethylamine and a C_6-C_1 part. By simple or multiple oxidative coupling norbelladine yielded the galanthamine type (**302**) (Fig. 14.48) (*para-ortho*-coupling), the lycorine type (**313**) (*ortho-para*-coupling), and the haemanthamine type (**323**) (*para-para*-coupling), which could be precursors for other alkaloids of this group.

In earlier experiments with *Narcissus pseudonarcissus*, *Galanthus elwesii*, and *Nerine bowdenii* it was shown that labelled norbelladine (**298**) and the corresponding amines **296** and **297** are specifically incorporated into galanthamine (**302**). Norbelladine as well as *O*-methylnorbelladine (**307**) are specific precursors of galanthine (**311**), haemanthamine (**323**), lycorine (**313**), norpluviine (**312**), crinamine, and belladine (**299**) [26, 32, 44, 298] (*cf*. [432]). The occurrence of *O*-methylnorbelladine and *N*,*O*-dimethylnorbelladine (**296**) in daffodils was established by trapping experiments. Tyrosine (**20**) is incorporated *via* tyramine (**21**) only into the phenethylamine part of norbelladine (**298**) and the corresponding alkaloids (Fig. 14.49) [26, 32, 44, 476]. Thus the C-2 of tyrosine corresponds to carbon α to the nitrogen in the C_6-C_2-unit of galanthamine (**302**), to C-5 of lycorine (**313**), and to C-12 of haemanthamine (**323**). The C-3 of this amino acid corresponds to C-11 of haemanthamine (**323**) and haemanthidine (**324**) and the corresponding hydroxylated carbon of tazettine (**328**)

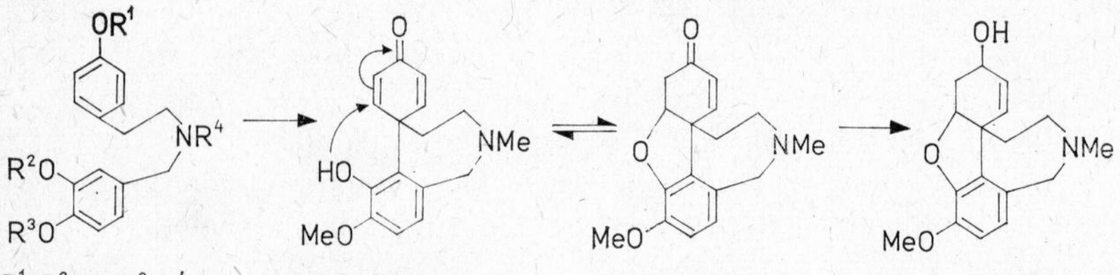

296 $R^1=R^2=H$, $R^3=R^4=Me$
297 $R^1=R^2=R^3=H$, $R^4=Me$
298 $R^1=R^2=R^3=R^4=H$: Norbelladine
299 $R^1=R^2=R^3=R^4=Me$: Belladine
300
301 Narwedine
302 Galanthamine

Fig. 14.48
Biosynthesis of galanthamine

(in experiments with *Sprekelia formosissima* and *Haemanthus natalensis* [255, 498]) and to the C-4 of lycorine (**313**) [32, 476]. These results show that the C_6-C_1 unit yielding the ring *A* in these alkaloids is not derived from tyrosine.

[3-^{14}C]Phenylalanine (**19**) can be incorporated into ring *A* and the benzylic carbon of lycorine (**313**), haemanthamine (**323**), and belladine (**299**) but not into the rings *C* and *D* [32, 476, 477]. The incorporation of phenylalanine proceeds by way of cinnamic acid (**37**) and the hydroxylated derivatives *p*-coumaric acid (**303**) and caffeic acid (**304**) with subsequent cleavage of two carbon atoms from the latter to generate a C_6-C_1 precursor such as protocatechualdehyde (**305**) or isovanillin (**306**). The cinnamic acids and benzaldehyde derivatives were incorporated specifically into lycorine (**313**) (Fig. 14.49), belladine (**299**), and haemanthamine (**323**) [26, 32, 473, 476–479, 512]. The presence of phenylalanine ammonia-lyase in *Amaryllidaceae* plants has been demonstrated [478]. The biochemical formation of the C_6-C_1 unit involves loss of the hydrogen atom from C-3 of cinnamic acid as has been found in biosynthetic experiments for haemanthamine and oduline ($-O-CH_2-O-$ instead of two $O-Me$ in lycorenine (**315**)) [496].

From incorporation experiments in King Alfred daffodils (*Narcissus pseudonarcissus*) with different norbelladine derivatives the following pathway for the biosynthesis of galanthamine has been established: Norbelladine (**298**) → *N*-methylnorbelladine (**297**) → *N,O*-dimethylnorbelladine (**296**) → galanthamine (**302**) (Fig. 14.48) [26]. The dienone (**300**) and narwedine (**301**) are probable intermediates.

The biosynthetic pathway for lycorine (**313**) starts with *O*-methylnorbelladine (**307**), which yields norpluviine (**312**) by oxidative coupling as the first pyrrolophenanthridine alkaloid, probably *via* **308** and **309** as intermediates. Norpluviine was found as precursor for galanthine (**311**), narcissidine (**314**), and lycorine (**313**), but lycorine cannot be transformed into norpluviine (Fig. 14.49) [32].

In Twink daffodils conversion of [3′,5′-$^3H_2$; 1-^{14}C]*O*-methylnorbelladine (**307**) into lycorine (**313**) involves retention at C-2 of **313** of half the tritium originally present in the precursor. This is the net result of two stereospecific processes: firstly

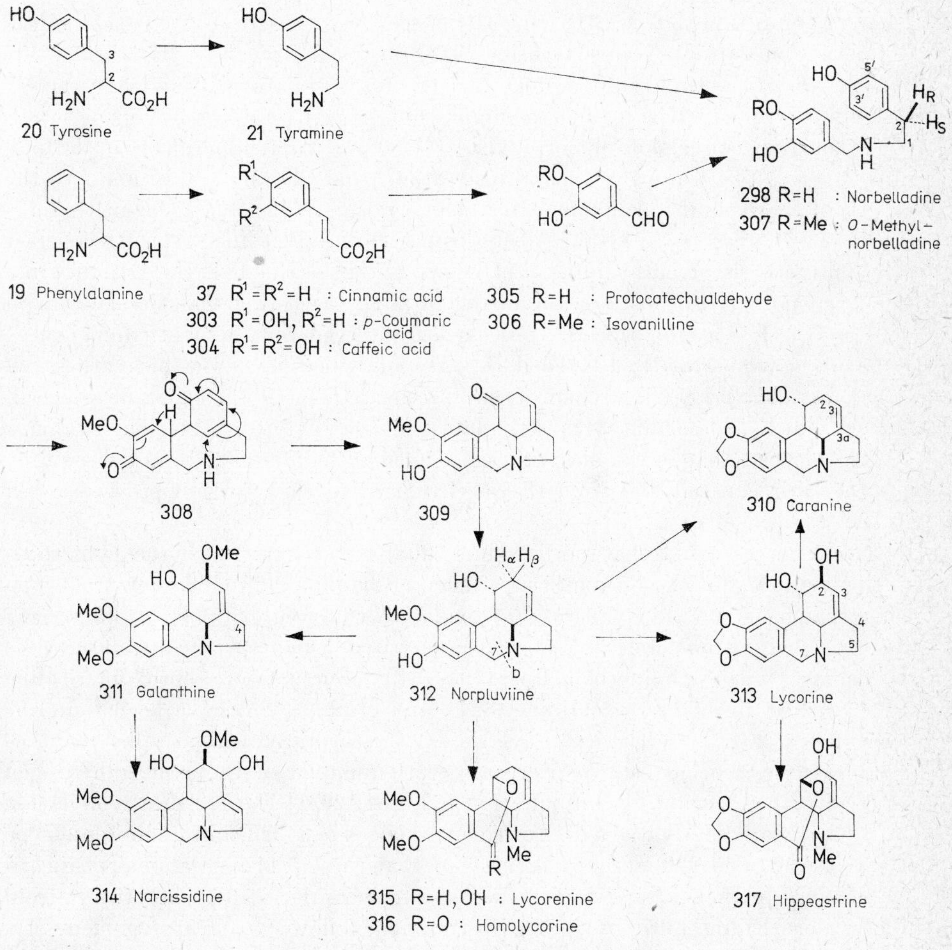

20 Tyrosine
21 Tyramine
298 R=H : Norbelladine
307 R=Me : O-Methyl-norbelladine
19 Phenylalanine
37 $R^1=R^2=H$: Cinnamic acid
303 $R^1=OH, R^2=H$: p-Coumaric acid
304 $R^1=R^2=OH$: Caffeic acid
305 R=H : Protocatechualdehyde
306 R=Me : Isovanilline
308
309
310 Caranine
311 Galanthine
312 Norpluviine
313 Lycorine
314 Narcissidine
315 R=H,OH : Lycorenine
316 R=O : Homolycorine
317 Hippeastrine

Fig. 14.49
Biosynthesis of lycorine, narcissidine, lycorenine, and related alkaloids

protonation to form the C-2 methylene of norpluviine (**312**) bearing the ^3H label in the β-configuration, and secondly hydroxylation of the next intermediate, caranine (**310**), with complete inversion. This possibly proceeds through a β-3,3a-epoxide as intermediate [141]. The same must occur at least partly in the conversion of [2β-^{3}H]-caranine (**310**) into [2-^{3}H]lycorine (**313**) in *Zephyranthes candida* [499]. In *Clivia miniata* the conversion of *O*-methylnorbelladine (**307**) into lycorine (**313**) occurs also with the intermediacy of norpluviine (**312**) and caranine (**310**) without the participation of a 2-oxo-derivative but with almost complete loss of the tritium originally present *ortho* to the phenolic hydroxy-group of the C_6—C_2-fragment of **307**. This shows a different stereochemical course of the hydroxylation process [206]. In experiments with stereospecifically labelled caranine (**310**) and norpluviine (**312**) it was shown that hydroxylation at C-2 takes place with removal of the hydrogen having the β-configuration without inversion. The protonation of the intermediate leading from **307** to **312** and **310** is seen to occur from the α-side of the molecule, as in daffodils, since lycorine (**313**) obtained from 3′,5′-^{3}H labelled **307** can be devoid of tritium at C-2 only if in the intermediates **312** and **310** the ^3H originally present in **307** is in the β-configuration.

The biological conversion of norpluviine (**312**) into narcissidine (**314**) in daffodil plants involves the removal of a hydrogen atom from the allylic position β to the nitrogen of the lycorane skeleton. After incorporation of [2-^{3}H, 1-^{14}C]*O*-methylnorbelladine (**307**), there is no loss of tritium from C-4 of galanthine (**311**). By contrast in the biosynthesis of narcissidine (**314**) there is the loss of half the tritium from the corresponding positions β to the nitrogen [201], suggesting that the conversion of **307** into **314** is a stereospecific process, as observed for haemanthamine (**323**). By feeding asymmetrically 2-labelled *O*-methylnorbelladine in the biosynthesis of **314** the loss of a *pro-4S* hydrogen of the intermediate with the lycorane skeleton has been shown. Galanthine (**311**) is the latest intermediate for (**314**). The evidence therefore suggests that in the biosynthesis of haemanthamine (**323**) and narcissidine (**314**) the hydrogen removal from the methylene β to the nitrogen proceeds with opposite stereochemistry.

It has been discussed that norbelladine (**298**) is methylated to the *O*-methylnorbelladine (**307**) for the biosynthesis of haemanthamine (**323**) and then yields the haemanthamine-structure by *para-para* oxidative coupling [26, 44]. From *Nerine bowdenii* a methyltransferase has been characterized which specifically methylates norbelladine to *O*-methylnorbelladine [191, 349]. Furthermore noroxomaritidine (**319**), normaritidine (**322**), oxocrinine (**321**), and crinine (**325**) have been found as intermediates [204, 208]. The hydroxylation at the saturated carbon in haemanthamine biosynthesis occurs stereospecifically with removal of the *pro-2R* hydrogen atom and with retention of configuration [62, 201, 293]. The chemical conversion of 11-hydroxylated derivatives of the crinane nucleus (*e.g.* **323**) into montanine-type derivatives (**329** and **330**) supported the view that these 5,11-methanomorphanthridine alkaloids produced by several *Haemanthus* species derive biosynthetically from *O*-methylnorbelladine (**307**) by *para-para*-coupling, followed by rearrangement of a haemanthamine-like intermediate. Thus in *Haemanthus coccineus* the intermediacy of **307** was established in the biosynthesis of manthine (**329**) and montanine (**330**)

[200]. During the conversion a *pro-2S* of **307** is lost. This suggests that if a haemanthamine-like intermediate is involved, there is a different stereochemical course of hydroxylation at C-11 of the crinane skeleton to that already observed in daffodil plants.

Furthermore some hemiacetal- and lactonalkaloids were derived by cleavage of a corresponding bond (b) from the norpluviine- (**312**) of haemanthamine-type. Thus norbelladine (**298**) is incorporated into homolycorine (**316**) [12], *O*-Methylnor-

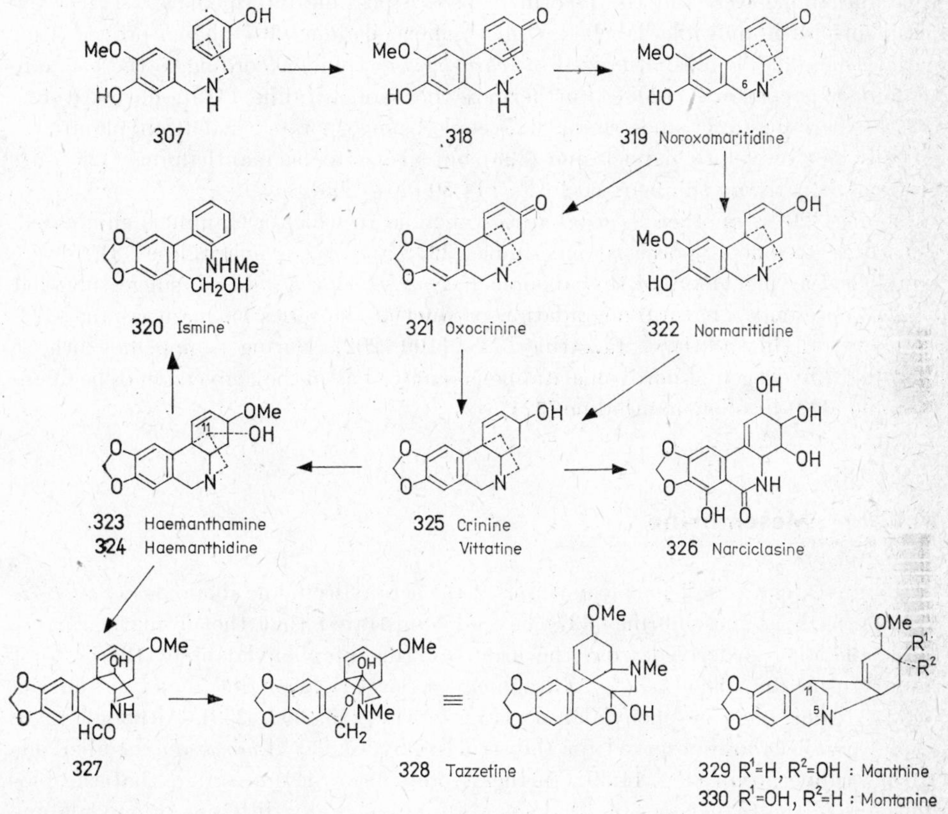

Fig. 14.50
Biosynthesis of haemanthamine and related alkaloids

belladine (**307**) is incorporated into clivonine, a derivative of hippeastrine (**317**), (the ester component of clivimine in *Clivia miniata* [206]) and into oduline (derivative of lycorenine (**315**); King Alfred daffodils) [62]. Finally lycorine (**313**) is incorporated into hippeastrine (**317**) (in *Hippeastrum vittatum*) [474]. The biological conversion of norpluviine (**312**) into lycorenine (**315**) involves the loss of the *pro-7R* hydrogen of the former [207]. The hydrogen that is lost is the one introduced in protonation during incorporation of 3,4-dihydroxybenzaldehyde (**305**) into the aromatic C_6-C_1 unit of norpluviine (**312**) [203]. This result confirms the stereospecificity of both protonation and hydroxylation processes. Similar results are obtained for the conversion of haemanthamine (**323**) into haemanthidine (**324**). In a similar way tazettine (**328**) is derived from haemanthamine (**323**) *via* haemanthidine (**324**) by cleavage of the bond b (in **324**) yielding the tautomeric aldehyde **327** [497].

Many *Amaryllidaceae* plants contain the lactam narciclasine (**326**) in addition to haemanthamine (**323**), norpluviine (**312**), and lycorine (**313**). Narciclasine was derived from the latter by elimination of two carbon atoms. Theoretically the haemanthamine-type and the lycorine-type are possible intermediates. Tracer experiments with multiple labelled *O*-methylnorbelladine (**307**) have proved that narciclasine (**326**) is biosynthesized by *para-para*-coupling according to the haemanthamine-type pathway [208]. Furthermore noroxomaritidine (**319**), normaritidine (**322**), oxocrinine (**321**), and crinine (**325**) could be incorporated in daffodil plants specifically not only into narciclasine (**326**) but also into haemanthamine (**323**) suggesting the pathway summarized in Fig. 14.50 [204, 205].

Ismine (**320**) is another *Amaryllidaceae* alkaloid in which a C_2 unit is eliminated. According to one hypothesis oxocrinine (**321**) and noroxomaritidine (**319**) were established as precursors of this alkaloid in *Sprekelia formosissima*, suggesting that ismine represents a natural degradation product of the series haemanthamine (**323**) → haemanthidine (**324**) → tazettine (**328**) [199, 202]. During ismine biosynthesis the *pro-6R* hydrogen of noroxomaritidine is removed as in the conversion of haemanthamine (**323**) into haemanthidine (**324**).

14.12.2 Mesembrine

Earlier investigations of the biosynthesis of the octahydroindole alkaloids in *Sceletium strictum* such as mesembrine (**337**) have demonstrated that the aromatic ring of these alkaloids is derived from the aromatic ring of phenylalanine (**19**) but not from tyrosine (**20**). The $C_6—C_2—N$ fragment of the perhydroindole moiety is derived from tyrosine but not phenylalanine (Fig. 14.51) [256, 257, 259]. Although these results parallel those observed for the structurally similar *Amaryllidaceae* alkaloids of the crinine group (Fig. 14.50), further studies have demonstrated that the biosynthesis of the two classes of alkaloids are fundamentally different. Phenylalanine is incorporated into mesembrine, with the loss of the C_3 side-chain but without loss of the protons *ortho* to the side chain. This rules out the possibility that the biosynthesis proceeds *via* an extension of the crinine pathway since this would require the loss of one of the two *ortho* protons [258]. Carbon-1′ of phenylalanine corresponds to the C-1′ position in mesembrine. These results have been interpreted as indicating the probable intermediacy of a dienone (**335**) which undergoes a Michael addition to generate the octahydroindole ring system.

A systematic characterization of the immediate post-amino acid metabolites beyond tyrosine and phenylalanine in the pathway to the mesembrine alkaloids has established that the incorporation of phenylalanine into the 3a-aryl octahydroindole ring system occurs *via* the intermediacy of cinnamic acid (**37**), 4′-hydroxy-cinnamic acid (**303**) and 4′-hydroxydihydrocinnamic acid (4′-phloretic acid) (**331**). 3′,4′-Dioxygenated cinnamic acids are not involved as intermediates showing that the 3′-aryl oxygen substituent is introduced at a late state [263]. Tyrosine is incorporated *via* tyramine (**21**) and *N*-methyltyramine (**22**) [261]. During the incorporation of tyrosine and its derivatives a stereospecific protonation at C-7 of the mesembrine skeleton occurs from the β-face.

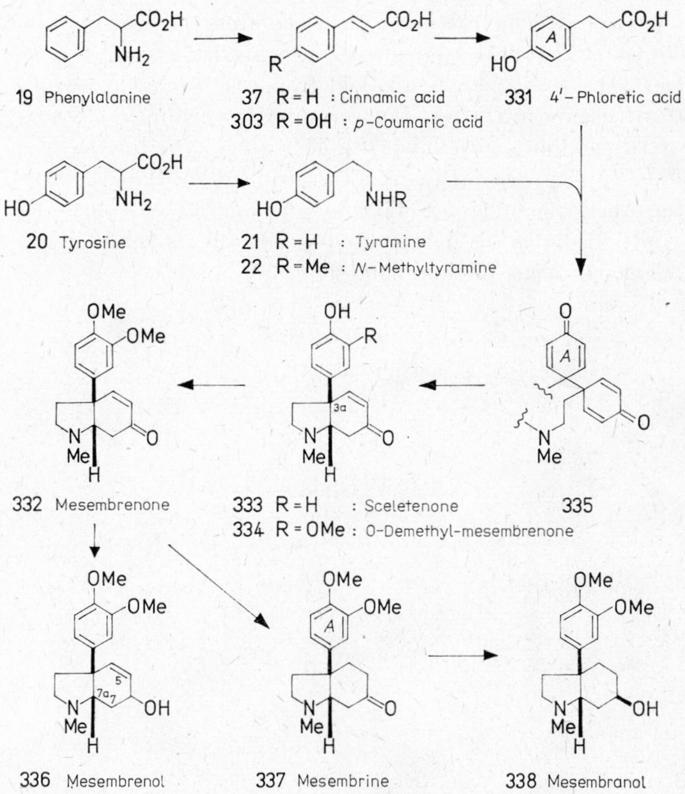

19 Phenylalanine

37 R = H : Cinnamic acid
303 R = OH : *p*-Coumaric acid

331 4′-Phloretic acid

20 Tyrosine

21 R = H : Tyramine
22 R = Me : *N*-Methyltyramine

332 Mesembrenone

333 R = H : Sceletenone
334 R = OMe : O-Demethyl-mesembrenone

335

336 Mesembrenol

337 Mesembrine

338 Mesembranol

Fig. 14.51
Biosynthesis of mesembrine and related alkaloids

The suggestion that the 3′-aryl oxygen substituent is introduced at a late stage is supported by the bioconversion in *S. strictum* of sceletenone (**333**) to mesembrenol and other related alkaloids. This established the following sequence (Fig. 14.51): sceletenone → 4′-*O*-demethylmesembrenone (**334**) → mesembrenone (**332**). Mesembrenone is converted into mesembrine (**337**), mesembrenol (**336**) and mesembranol (**338**) [260, 262, 263]. The late stages of biosynthesis involve the sequential reduction of the cyclohexenone chromophore. These results also provide evidence for the formation of non-phenolic alkaloids from phenolic alkaloids.

14.13 Other Phenylalanine and Tyrosine-Derived Alkaloids

14.13.1 Annuloline

There are some phenylpropane-derived alkaloids that do not belong to any group yet mentioned. They can be synthesized only by phenylpropane units such as annuloline (**341**) or in combination with other units such as gliotoxin (**345**), tylophorine (**366**), shihunine (**374**), tenellin (**376**), or securinine (*cf.* chapter 12.1.4.1.).

Annuloline (**341**) is an oxazole alkaloid from the annual ryegrass *Lolium multiflorum* [280]. It occurs in the roots of germinating seedlings and is synthesized between the sixth and fourteenth day [**439**]. This alkaloid is formed from two phenylalanine units (**19**) [384]. One of these is transformed *via* tyrosine (**20**) into tyramine (**21**) and the other *via* cinnamic (**37**) and *p*-coumaric acid (**303**) into caffeic acid (**304**). These intermediates are incorporated specifically into annuloline most probably as shown in Fig. 14.52 *via* **339** and **340**. Methylation seems to occur at a late stage.

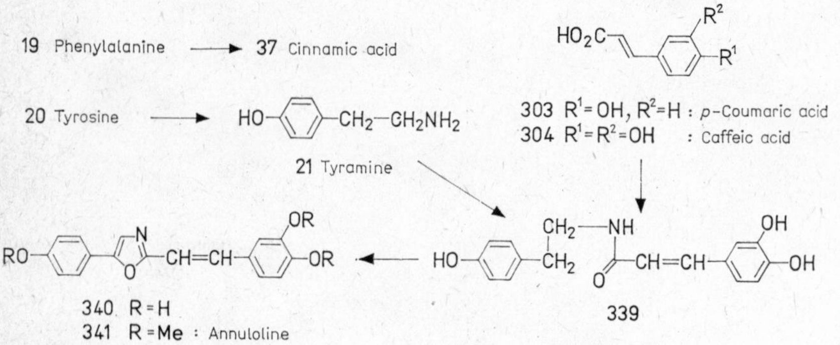

Fig. 14.52
Biosynthesis of annuloline

14.13.2 Gliotoxin

The epipolythiodioxopiperazines constitute a group of toxic fungal metabolites exemplified by the first known member, gliotoxin (**345**) [265, 296, 427]. Phenylalanine (**19**) rather than tryptophan provides the reduced indole nucleus of gliotoxin in cultures of *Trichoderma viride* (Fig. 14.53) [475]. The carbon skeleton of the amino acid has been incorporated intact including the side chain. C-1 of phenylalanine corresponds to position 1 and its nitrogen giving the *N*-5 of gliotoxin [126, 127]. The other portion of the dioxopiperazine ring is derived from serine [501]. *m*-Tyrosine cannot be an obligatory intermediate between phenylalanine and gliotoxin [128, 501]. All five aromatic hydrogens are retained in the incorporation of phenylalanine into gliotoxin [145]. Also *m*-tritiated phenylalanine was incorporated with full retention of tritium, half of which appears at C-6 in gliotoxin (**345**) [266, 267]. The formation of the cyclohexadienol ring of gliotoxin is thought to involve an arene oxide intermediate probably of the **344** type.

Cyclo-*L*-phenylalanyl-*L*-seryl (**342**) fed to *Penicillium terlikowskii* and *T. viride* shows an intact incorporation into gliotoxin [144, 348]. Of the four stereo-isomers only the *LL*-isomer was incorporated efficiently [294].

The enzymes catalysing the transformation of **342** into gliotoxin (**345**) seem not to have high substrate specificities. The analogue of **342**, cyclo-*L*-alanyl-*L*-phenylalanyl can be transformed by *T. viride* to 3a-deoxygliotoxin (**348**) as efficiently as the natural process [295].

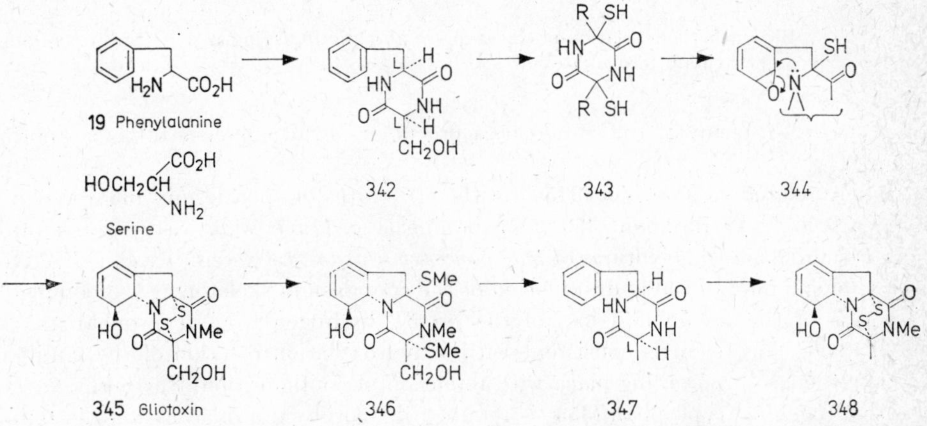

Fig. 14.53
Biosynthesis and metabolism of gliotoxin and related compounds

The biosynthetic introduction of sulphur into gliotoxin is not yet clear. It is suggested that cysteine is a better source of the sulphur bridge than methionine. A dehydrodioxopiperazine is suggested for the sulphur acceptor. Consecutive introduction of two thiol groups would give **343** suitable for oxidative cyclization to an epidisulphide or methylation to a dimethylthio compound. Bisdethiobismethylthiogliotoxin (**346**) has been identified as a minor metabolite of *Gliocladium deliquescens* and is formed apparently irreversible from gliotoxin (**345**) (Fig. 14.53) [297].

An interesting structural resemblance to gliotoxin (**345**) is shown in the aranotin (**349**) group of metabolites. Phenylalanine is a much more efficient precursor than *m*-tyrosine and dopa for bisdethiodimethylthioacetylaranotin (BDA) (**351**) in *Aspergillus aureus* [128] and has contributed to both halves of the symmetrical structure. All of the aromatic and both of the methyleneprotons of phenylalanine were retained upon incorporation into acetylaranotin (**350**) by *Aspergillus terreus*. The presence of oxepin rings in the aranotin structures is best explained by the intervention of an arene oxide capable of rapid valence tautomerism (Fig. 14.54). The close biosynthetic relationship between the aranotin and gliotoxin ring systems is conveniently displayed in the structure of apoaranotin (**352**), a metabolite of *A. aureus*. Cyclo-*L*-

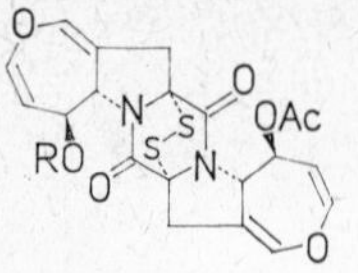

349 R = H : Aranotin
350 R = Ac : Acetylaranotin 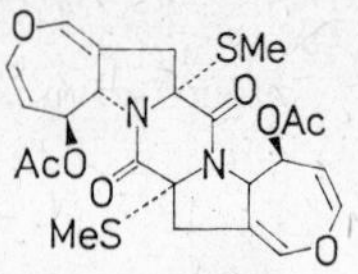
351 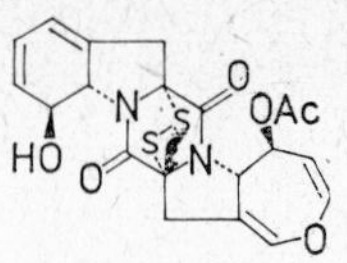
352

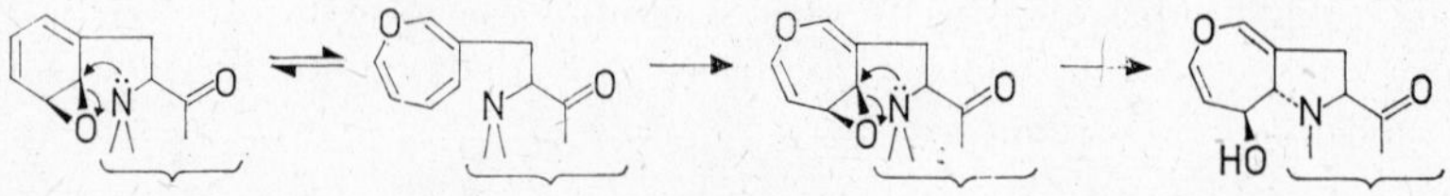

Fig. 14.54
Biosynthetic derivation of the aranotin and gliotoxin ring system from a common arene oxide intermediate

phenylalanyl-*L*-phenylalanyl should be a precursor for all members of the aranotin group.

Closely related to gliotoxin (**345**) are the structures of sporidesmin (**355**) and its co-metabolites. Tryptophan (**353**) and alanine (Fig. 14.55) were incorporated into sporidesmin (**355**) using cultures of *Pithomyces chartarum*. *L*-Cysteine was an effective source of sulphur. In the hydroxylation of the sporidesmin skeleton at C-3 complete retention of the *pro-3S* and loss of the *pro-3R* hydrogen from tryptophan takes place [299]. This result is consistent with a hydroxylation reaction of the familiar monooxygenase type taking place with retention of configuration. Furthermore the C-2′-hydrogen of tryptophan (**353**) is retained in sporidesmin (**355**) as for gliotoxin. Cyclo-*L*-alanyl-*L*-tryptophanyl (**354**) has shown good incorporation into sporidesmin.

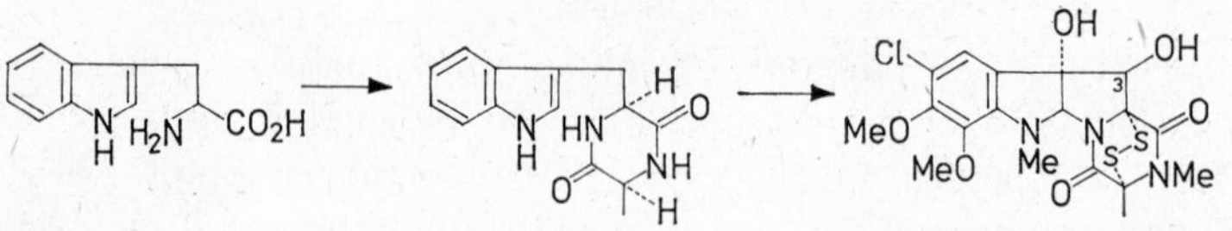

353 Tryptophan
354
355 Sporidesmin

Fig. 14.55
Biogenesis of sporidesmin

14.13.3 Tylophorine and Related Bases

Tylophora asthmatica contains some phenanthroindolizidine alkaloids such as tylophorine (**366**), tylophorinine (**368**), and tylophorinidine (**369**). These alkaloids are derived from phenylalanine (**19**) *via* cinnamic acid as the precursors of ring *A*, C-10, and C-6′. Tyrosine (**20**) *via* dopa yields ring *B*, C-9, and C-7′ and ornithine the pyrrolidine part [92, 369—371] (Fig. 14.56). Different 2-phenacylpyrrolidines (**358**—**360**) and benzoylacetic acids (**357**) have been established as intermediates

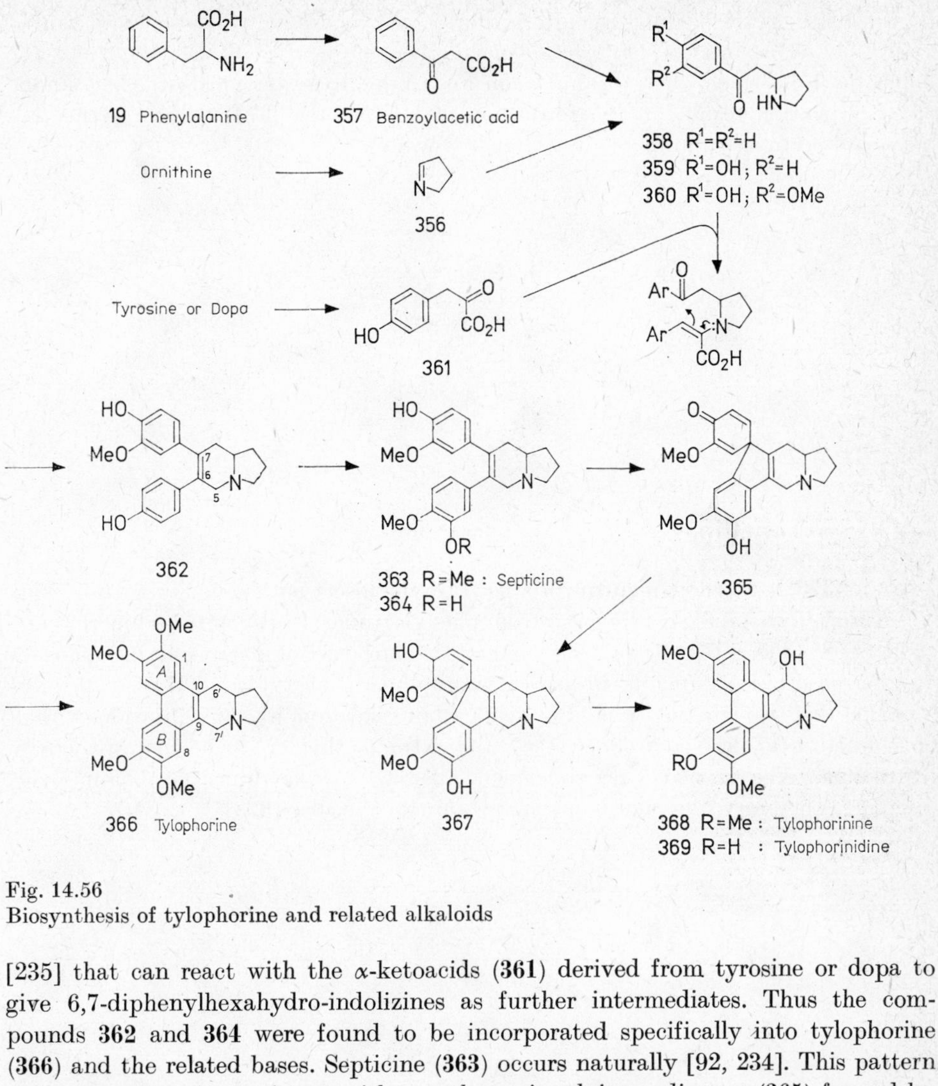

Fig. 14.56
Biosynthesis of tylophorine and related alkaloids

[235] that can react with the α-ketoacids (**361**) derived from tyrosine or dopa to give 6,7-diphenylhexahydro-indolizines as further intermediates. Thus the compounds **362** and **364** were found to be incorporated specifically into tylophorine (**366**) and the related bases. Septicine (**363**) occurs naturally [92, 234]. This pattern of incorporation is consistent with a pathway involving a dienone (**365**) formed by phenoloxidation, which gives either tylophorine (**366**) by dienone phenol re-arrangement, or tylophorinine (**368**) and tylophorinidine (**369**) *via* reduction to the dienol **367** followed by dienol benzene re-arrangement.

14.13.4 Shihunine

Shihunine (**374**) is a phthalidopyrrolidine from *Dendrobium* species. In the orchid *Dendrobium pierardii* it was shown that *o*-succinylbenzoic acid (**370**) is a precursor [325]. This is formed from shikimic acid and α-ketoglutaric acid and is a known intermediate in the biosynthesis of several naphthoquinones and anthraquinones. In Fig. 14.57 it is suggested that *o*-succinylbenzoic acid (**370**) is reduced to the aldehyde **371**, which on transamination and *N*-methylation yields the ketoamine **372**. Cyclization results in the formation of the betaine (**373**), which is the predominant state of shihunine (**374**) in polar solvents.

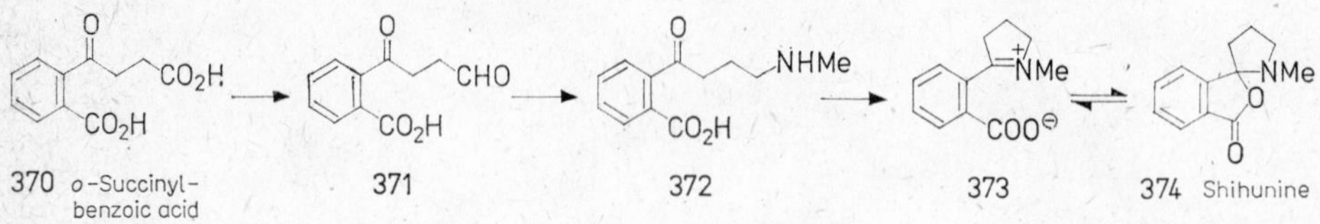

370 *o*-Succinylbenzoic acid
371
372
373
374 Shihunine

Fig. 14.57
Biogenesis of shihunine

14.13.5 Tenellin

Tenellin (**376**) is a yellow pigment produced by the insect pathogenic fungi *Beauveria tenella* and *B. bassiana*. It is derived from phenylalanine (**19**), acetate and methionine (Fig. 14.58) [357, 507]. It was shown that the carboxyl of phenylalanine migrates and becomes C-4 of tenellin. Probably α-formylphenylacetyl-CoA (**375**) is an intermediate that would yield the required carbon skeleton by condensation with a poly-β-keto acid derived from acetate, the extra methyl group being provided by methionine. The involved rearrangement of the phenylalanine side chain, reminiscent of the biosynthesis of tropic acid, is intramolecular [328].

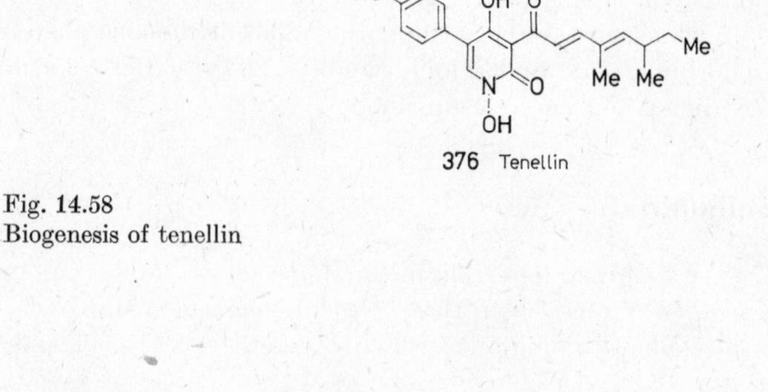

19 Phenylalanine
375
Methionine
376 Tenellin

Fig. 14.58
Biogenesis of tenellin

14.14 Betalains

14.14.1 Structure and Occurrence

The betalains comprise two groups of water-soluble vacuolar pigments, isolated from numerous species of the *Centrospermae*. These are the red-violet betacyanins and the yellow-orange betaxanthins (*cf.* [307, 346, 347, 396, 415]).

Some 50 different betacyanins are derived from betanidin (**377**). They have the 2(*S*):15(*S*) configuration and occur together with the 15(*R*) epimers (isobetacyanins). Betanidin (**377**) is found most frequently as betanin (**378**), its 5-*O*-*β*-*D*-glucopyrano-

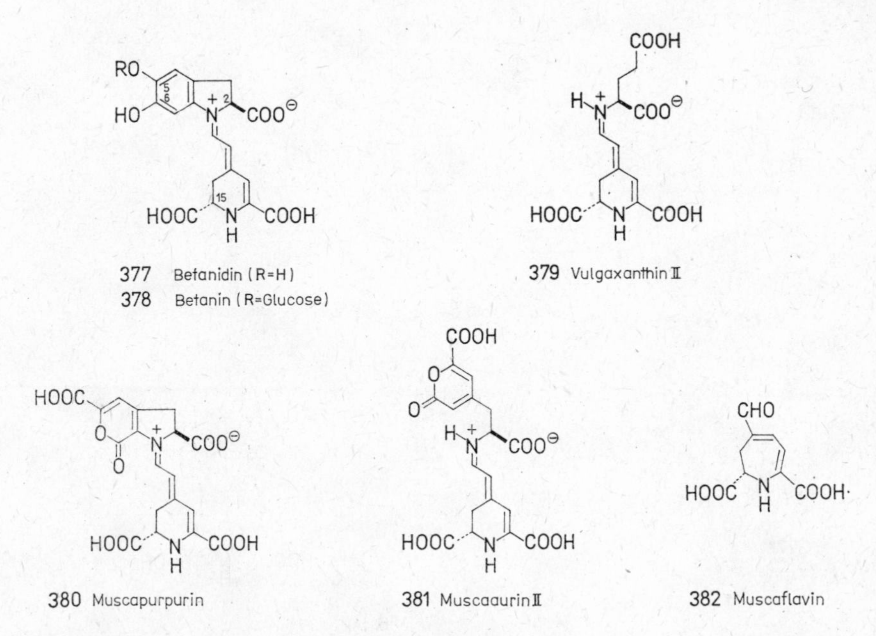

Fig. 14.59
Structures of typical betalains and *Amanita* pigments

side. Other betacyanins are also glycosides such as amaranthin (**386**). Some are acylated with one or more of the following acids: sulphuric, malonic, 3-hydroxy-3-methylglutaric, citric, *p*-coumaric, ferulic, caffeic, and sinapic acids. Unlike the anthocyanins the aglycon can also be acylated.

Betaxanthins exhibit the same dihydropyridine moiety as betacyanins but the cyclodopa (**383**) unit is displaced by an amine or amino acid. Vulgaxanthin II (**379**) contains a residue derived from glutamic acid while indicaxanthin (**385**) is similarly related to proline (Figs. 14.59 and 14.60). Other moieties of betaxanthins are glutamine, aspartic acid, methionine sulphoxide, tyramine, dopamine, and dopa.

With the exception of the *Caryophyllaceae* and *Molluginaceae* both groups of betalains have been detected only in species of centrospermous families (*cf.* [347, 396, 415]) where they displace the anthocyanins. Betalains and anthocyanins are mutually exclusive and are never detected together [289]. Recently, quite another

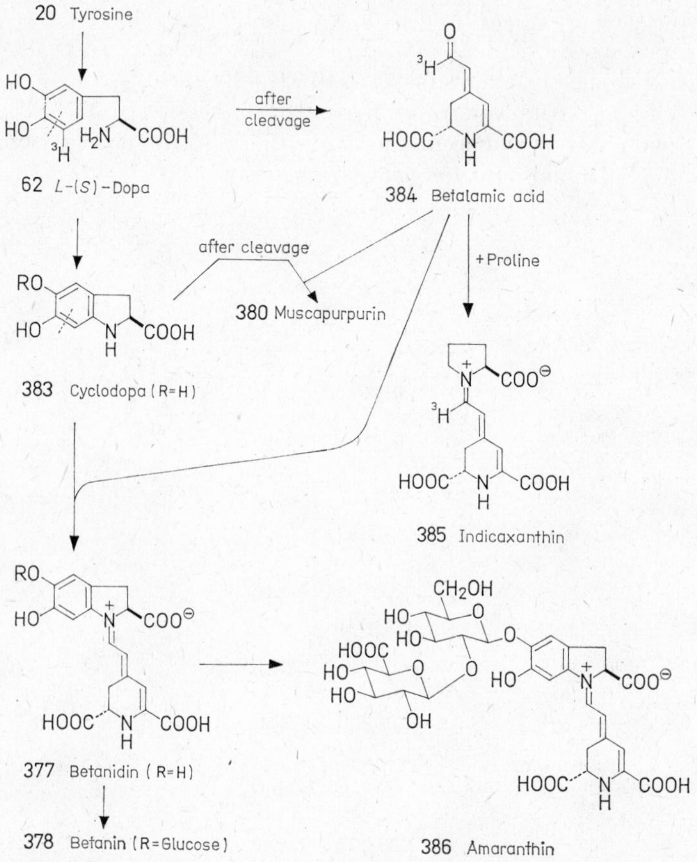

Fig. 14.60
Biosynthesis of betalains

source of betaxanthins has become known. From the red cap of the fly agaric (*Amanita muscaria*) a number of pigments has been isolated [372]: the purple muscapurpurin (**380**), the orange muscaaurins I to VII and the yellow muscaflavin (**382**). Muscapurpurin (**380** or an isomeric structure) and muscaaurin II (**381**) both contain the unusual amino acid stizolobic acid, the former in a cyclized form. Some muscaaurins are identical to known betaxanthins (*i.e.* vulgaxanthin).

14.14.2 Biosynthesis of the Betalains

At the time that the structure was elucidated for betanin (**378**) a proposal for its biosynthesis was elaborated [508]. According to this betanidin (**377**) originates from two molecules of dopa (**62**), one of which has to undergo ring cleavage and recyclization to form the dihydropyridine moiety (Fig. 14.60). Incorporation of labelled tyrosine and dopa (**62**) into the betacyans betanin (**378**) [242, 335, 336, 360, 362] and amaranthin (**386**) [213, 467, 466] and also into the betaxanthins indicaxanthin (**385**) [362] and vulgaxanthin II (**379**) [335] substantiated this hypothesis. Betalamic acid (**384**) is the most likely intermediate for both types of pigments. Its presence in *Centrospermae* has been demonstrated [416] as has its derivation from labelled tyrosine [110]. To gain a better understanding of the ring cleavage mechanism leading from dopa to betalamic acid (**384**) [3,5-^{3}H, 1-^{14}C]tyrosine (**20**) was administered to fruits of *Opuntia decumbens* [192] and *O. ficus-indica* [250] and the remaining tritium label was found to be attached at the aldehyde function of the betalamic acid unit of the betalains (Fig. 14.60). The radiocarbon of carboxyl labelled tyrosine (**20**) and dopa (**62**) was introduced in the carboxyl group at the asymmetric centre of the betalamic acid (**384**) moiety [151, 171]. The heterocyclic nitrogen is derived from the amino group of tyrosine (**20**) [336]. The incorporation of labelled betanidin (**377**) into betanin (**378**) and amaranthin (**386**) in *O. dillenii* fruits [442] and *O. plumosa* seedlings [443] pointed to the glucosylation at a late stage, although even [^{14}C] cyclodopa (**383**) and its glucoside (**383**, R = glucose) gave rise to labelled pigments. As the latter results were obtained with a yellow variety of *O. plumosa*, the question of induced metabolites arises, also observed following administration of excessive dopa (**62**) to *Amaranthus* [219]. Incorporation of labelled tyrosine (**20**) into both betacyanins and betaxanthins has been realized also with illuminated callus cultures of *Portulaca grandiflora* [335].

Preliminary feeding experiments with *Amanita muscaria* showed the label of tyrosine (**20**) to be associated with the yellow pigments [372].

14.14.3 Regulation of Betalain Synthesis

Numerous compounds and environmental conditions have been found to interfere with the control of betalain synthesis (Table 14.1). Some species of *Centrospermae* are known to have an absolute requirement of light for pigment synthesis (*i.e. Amaranthus tricolor*) whereas others form betalains even in the dark (*i.e. Beta vulgaris*) [399, 504]. This resulted in light being the most intensively investigated environmental factor. In seedlings of *A. tricolor* an increased amaranthin (**386**) formation was observed after irradiation with red (660 nm) or blue (440 nm) light [155, 305, 306]. The light-induced pigment synthesis could be reversed by far red (730 nm) treatment in *A.* species [155, 305, 306, 397], *Chenopodium rubrum* [494], and *Celosia* species [217, 221]. Such behaviour is consistent with photobiological processes controlled by phytochrome. According to the hypothesis on the mode of action of phytochrome [363, 364], light-enhanced betalain synthesis by P_{fr}, the

Table 14.1
Factors Effecting Pigment Formation in *Centrospermae*

Factors	Response	References
Light (440, 660 nm)	enhanced betalain synthesis	9, 155, 216–221, 213, 301–303, 305, 306, 312, 335, 397, 398, 412, 494, 503, 504, 506
Temperature	influence on betalain synthesis	177, 178, 303
Ions	influence on betalain synthesis	174, 180, 182, 310, 335, 466
Precursors (phenylalanine, tyrosine, dopa)	increased pigment synthesis (except leaf disks and callus cultures), induced betaxanthin synthesis	197, 213, 219, 226, 302, 303, 316, 375
Cytokinins	enhanced pigment synthesis in light and dark	9, 161, 173, 175, 176, 178a, 179, 180, 216 to 218, 221, 275, 301 to 303, 311, 399, 414
Purines, nucleotides, pteridines	enhanced pigment synthesis	179–182, 218, 411
Gibberellins	inhibition of pigment formation (except cytokinin induced)	107–109, 153, 156, 227, 290, 291, 316, 460, 461, 468
Abscisic acid	inhibition of light- and cytokinin-induced betalain formation	108, 227, 468
Fusicoccin	enhanced pigment synthesis	106, 173
Inhibitors of nucleic acid and protein synthesis	counteract the light- and cytokinin-induced betalain formation	110, 216, 217, 304, 308, 309, 398, 399, 411, 468, 503
Growth regulators (CCC, Phosfon D)	counteract gibberellin action	291, 467
Inhibitors of cyclic photophosphorylation	counteract light-induced betalain synthesis	216, 217, 220
Phenolic compounds	inhibition of cytokinin-induced betalain formation	149, 216, 231

active phytochrome, is mediated through the activation or derepression of potentially active genes (Fig. 14.61). This has been substantiated by the lowering of the light-stimulating effect on the pigment synthesis by inhibitors of the nucleic acid and protein synthesis (Table 14.1). This photocontrol of betalain formation seems to be focussed on the synthesis of the betalamic acid (**384**) moiety [217, 219], although there are other sites in the biogenetic route that may be influenced by light [197]. Furthermore, cyclic photophosphorylation seems to be involved [220], which may supply the system with energy-rich substances.

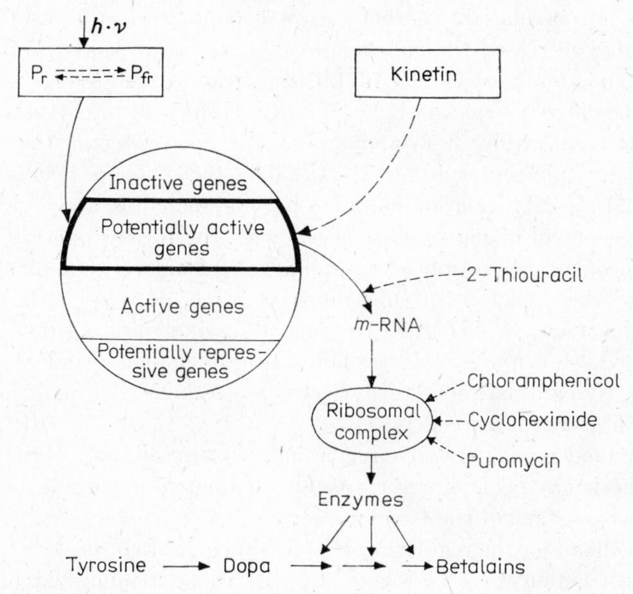

Fig. 14.61
Regulation of betalain synthesis

Light-induced synthesis of amaranthin (**386**) was reported to be stimulated by kinetin [302, 303, 397, 398]. As cytokinins, especially kinetin, are able to induce betalain formation even in the dark [399], these compounds were considered to mimic the effect of light. Cytokinin-enhanced pigment formation is annihilated by the action of inhibitors of nucleic acid and protein synthesis and thus the mode of action can be described as gene activation. Although light increases the cytokinin content in *A. caudatus* [312], such phytohormones seem unlikely to be involved in the sequence by which the response to light is expressed. These factors seem to operate independently of each other [301].

As a result of the enhanced amaranthin (**386**) synthesis in *Amaranthus* under the influence of kinetin, an optical bioassay for cytokinins has been worked out [311] and further developed [107, 109, 161, 414]. Under reproducible physiological conditions [175, 176] this assay is superior to other biological assays for cytokinins and is especially rapid.

A wide range of purine bases, nucleosides, and cyclic nucleotides have been shown to induce amaranthin (**386**) [179]. All of them, including cyclic AMP [179, 218, 411], may act as analogues of cytokinins. Certain results [218] contradict the assumption that cyclic AMP is a "second messenger" of light action.

Gibberellins are another type of plant hormones that inhibit light-induced synthesis of amaranthin (**386**) in *A. caudatus* [156, 460, 468]. This inhibitory phenomenon led to the development of a bioassay for gibberellins [290] that, with respect to sensitivity and speed, is superior over other gibberellin bioassays. From this inhibitory response a regulation of the betalain biosynthesis through endogenous gibberellins may be postulated. Synthetic growth regulators such as CCC or AMO 1618 in *A. caudatus* decreased the endogenous gibberellin levels and simultaneously enhanced the pigment synthesis [291, 467]. Stimulation by retardants was reversed by application of gibberellic acid (GA_3). GA_3 had no effect on the cytokinin-induced dark synthesis of amaranthin (**386**) and kinetin dit not overcome the gibberellin-induced inhibition of pigment synthesis in the light [467]. These results point to a more complex nature of regulation because both gibberellins and cytokinins are implicated in the control of pigment synthesis with the former operating at a site different from the gene activation level. Inhibition of amaranthin (**386**) synthesis was attributed to the depletion of substrates (tyrosine, dopa) and their diversion to other metabolic pathways [461, 467]. Recently, brassinolide as a growth regulating steroidal lactone was shown to inhibit betalain synthesis similar to and synergistically with GA_3 [348a]. By contrast, betalains affect the metabolism of plant hormones in as far as they inhibit auxin oxidase [463].

The effect of added precursors (phenylalanine, tyrosine, dopa) on the regulation of betalain synthesis gave contradictory results that may, at least in part, depend on the developmental stage of the tissues used.

A further mechanism for the regulation of betalain content is based on the enzymic hydrolysis of both betacyanins and betaxanthins to betalamic acid (**384**) [179a, 453a]. Numerous other factors have been found to influence the formation of betalains (Table 14.1), pointing once more to complex regulation mechanisms. Among these factors the control of betalain levels with temperature [106, 177, 178, 303] may be discussed in connection with the autumnal reddening of *Kochia scoparia* plants.

Some of the factors mentioned in Table 14.1 (*i.e.* ions, precursors, cytokinins, light) are very important in the formation of pigments in cell cultures of *Beta* and *Portulaca* [154, 162, 163, 180—182, 335, 375, 429a].

14.15 References

[1] Agurell, S.; Granelli, I.; Leander, K.; Lüning, B.; Rosenblom, J.: Acta Chem. Scand. **B28** (1974) 239.

[2] Agurell, S.; Granelli, I.; Leander, K.; Rosenblom, J.: Acta Chem. Scand. **B28** (1974) 1175.

[3] Agurell, S.; Lundström, J.: Chem. Commun. **1968**, 1638.

[4] Agurell, S.; Lundström, J., Masoud, A.: J. Pharm. Sci. **58** (1969) 1413.
[5] Akasu, M.; Itokawa, H.; Fujita, M.: Phytochemistry **14** (1975) 1673.
[6] Akasu, M.; Itokowa, H.; Fujita, M.: Phytochemistry **15** (1976) 471.
[7] Axelrod, J.: Angew. Chem. **83** (1971) 827.
[8] Aynehchi, Y.; Jaffarian, S.: Lloydia **36** (1973) 427.
[9] Bamberger, E.; Mayer, A. M.: Science **131** (1960) 1094.
[10] Barber, R. B.; Rapoport, H.: J. Med. Chem. **19** (1976) 1175.
[11] Barker, A. C.; Battersby, A. R.; McDonald, E.; Ramage, R.; Clements, J. H.: Chem. Commun. **1967**, 390.
[12] Barton, D. H. R.: Proc. Chem. Soc. **1963**, 293.
[13] Barton, D. H. R.: In: Festschrift K. Mothes. Gustav Fischer Verlag, Jena 1965, p. 67.
[14] Barton, D. H. R.; Bhakuni, D. S.; Chapman, G. M.; Kirby, G. W.: J. Chem. Soc. C **1967**, 2134.
[15] Barton, D. H. R.; Bhakuni, D. S.; Chapman, G. M.; Kirby, G. W.; Haynes, L. J.; Stuart, K. L.: J. Chem. Soc. C **1967**, 1295.
[16] Barton, D. H. R.; Bhakuni, D. S.; James, R.; Kirby, G. W.: J. Chem. Soc. C **1967**, 128.
[17] Barton, D. H. R.; Boar, R. B.; Widdowson, D. A.: J. Chem. Soc. C **1970**, 1208.
[18] Barton, D. H. R.; Boar, R. B.; Widdowson, D. A.: J. Chem. Soc. C **1970**, 1213.
[19] Barton, D. H. R.; Bracho, R. D.; Potter, C. J.; Widdowson, D. A.: J. Chem. Soc. Perkin I **1974**, 2278.
[20] Barton, D. H. R.; Cohen, T.: In: Festschrift A. Stoll, Birkhäuser-Verlag, Basel 1957, p. 117.
[21] Barton, D. H. R.; Gunatilaka, A. A. L.; Letcher, R. M.; Lobo, A. M. F. T.; Widdowson, D. A.: J. Chem. Soc. Perkin I **1973**, 874.
[22] Barton, D. H. R ; Hesse, R. H.; Kirby, G. W.: J. Chem. Soc. **1965**, 6379.
[23] Barton, D. H. R.; James, R.; Kirby, G. W.; Turner, D. W.; Widdowson, D. A.: J. Chem. Soc. C **1968**, 1529.
[24] Barton, D. H. R.; Kirby, A. J.; Kirby, G. W.: J. Chem. Soc. C **1968**, 929.
[25] Barton, D. H. R.; Kirby, G. W.; Steglich, W.; Thomas, G. M.; Battersby, A. R.; Dobson, T. A.; Ramuz, H.: J. Chem. Soc. **1965**, 2423.
[26] Barton, D. H. R.; Kirby, G. W.; Taylor, J. B.; Thomas, G. M.: J. Chem. Soc. **1963**, 4545.
[27] Barton, D. H. R.; Kirby, G. W.; Wiechers, A.: J. Chem. Soc. C **1966**, 2313.
[28] Barton, D. H. R.; Potter, C. J.; Widdowson, D. A.: J. Chem. Soc. Perkin I **1974**, 346.
[29] Basmadjian, G. P.; Hussain, S. F.; Paul, A. G.: Lloydia **41** (1978) 375.
[30] Basmadjian, G. P.; Paul, A. G.: Lloydia **34** (1971) 91.
[31] Battersby, A. R.: In: Battersby, A. R.; Taylor, W. I. (Eds.): Oxidative Coupling of Phenols. Marcel Dekker, New York 1967.
[32] Battersby, A. R.; Binks, R.; Breuer, S. W.; Fales, H. M.; Wildman, W. C.; Highet, R. J.: J. Chem. Soc. **1964**, 1595.
[33] Battersby, A. R.; Binks, R.; Francis, R. J.; McCaldin, D. J.; Ramuz, H.: J. Chem. Soc. **1964**, 3600.
[34] Battersby, A. R.; Binks, R.; Harper, B. J. T.: J. Chem. Soc. **1962**, 3534.
[35] Battersby, A. R.; Binks, R.; Huxtable, R.: Tetrahedron Lett. **1967**, 563.
[36] Battersby, A. R.; Binks, R.; Huxtable, R.: Tetrahedron Lett. **1968**, 6111.
[37] Battersby, A. R.; Binks, R.; Reynolds, J. J.; Yeowell, D. A.: J. Chem. Soc. **1964**, 4257.
[38] Battersby, A. R.; Böhler, P.; Munro, M. H. G.; Ramage, R.: J. Chem. Soc. Perkin I **1974**, 1399.
[39] Battersby, A. R.; Bradbury, R. B.; Herbert, R. B.; Munro, M. H. G.; Ramage, R.: J. Chem. Soc. Perkin I **1974**, 1394.
[40] Battersby, A. R.; Brocksom, T. J.; Ramage, R.: Chem. Commun. **1969**, 464.
[41] Battersby, A. R.; Brown, R. T.; Clements, J. H.; Iverach, G. G.: Chem. Commun. **1965**, 230.

[42] Battersby, A. R.; Brown, T. H.; Clements, J. H.: J. Chem. Soc. **1965**, 4550.
[43] Battersby, A. R.; Dobson, T. A.; Foulkes, D. M.; Herbert, R. B.: J. Chem. Soc. Perkin I **1972**, 1730.
[44] Battersby, A. R.; Fales, H. M.; Wildman, W. C.: J. Am. Chem. Soc. **83** (1961) 4098.
[45] Battersby, A. R.; Foulkes, D. M.; Binks, R.: J. Chem. Soc. **1965**, 3323.
[46] Battersby, A. R.; Foulkes, D. M.; Hirst, M.; Parry, G. V.; Staunton, J.: J. Chem. Soc. C **1968**, 210.
[47] Battersby, A. R.; Francis, R. J.: J. Chem. Soc. **1964**, 4078.
[48] Battersby, A. R.; Francis, R. J.; Hirst, M.; Ruveda, E. A.; Staunton, J.: J. Chem. Soc. Perkin I **1975**, 1140.
[49] Battersby, A. R.; Francis, R. J.; Hirst, M.; Southgate, R.; Staunton, J.: Chem. Commun. **1967**, 602.
[50] Battersby, A. R.; Francis, R. J.; Hirst, M.; Staunton, J.: Proc. Chem. Soc. **1963**, 268.
[51] Battersby, A. R.; Harper, B. J. T.: Tetrahedron Lett. **1960**, No. 27, 21.
[52] Battersby, A. R.; Harper, B. J. T.: J. Chem. Soc. **1962**, 3526.
[53] Battersby, A. R.; Herbert, R. B.; McDonald, E.; Ramage, R.; Clements, J. H.: J. Chem. Soc. Perkin I **1972**, 1741.
[54] Battersby, A. R.; Herbert, R. B.; Mo, L.; Santavy, F.: J. Chem. Soc. C **1967**, 1739.
[55] Battersby, A. R.; Herbert, R. B.; Pijewska, L.; Santavy, F.; Sedmera, P.: J. Chem. Soc. Perkin I **1972**, 1736.
[56] Battersby, A. R.; Hirst, M.: Tetrahedron Lett. **1965**, 669.
[57] Battersby, A. R.; Hirst, M.; McCaldin, D. J.; Southgate, R.; Staunton, J.: J. Chem. Soc. C **1968**, 2163.
[58] Battersby, A. R.; Jones, R. C. F.; Kazlauskas, R.: Tetrahedron Lett. **1975**, 1873.
[59] Battersby, A. R.; Jones, R. C. F.; Kazlauskas, R.; Poupat, C.; Thornber, C. W.; Ruchirawat, S.; Staunton, J.: Chem. Commun. **1974**, 773.
[60] Battersby, A. R.; Jones, R. C. F.; Kazlaukas, R.; Thornber, C. W.; Ruchirawat, S.; Staunton, J.: J. Chem. Soc. Perkin I **1981**, 2016.
[61] Battersby, A. R.; Jones, R. C. F.; Minta, A.; Ottridge, A. P.; Staunton, J.: J. Chem. Soc. Perkin I **1981**, 2030.
[62] Battersby, A. R.; Kelsey, J. E.; Staunton, J.; Suckling, K. E.: J. Chem. Soc. Perkin I **1973**, 1609.
[63] Battersby, A. R.; Lewis, N. G.; Tippett, J. M.: Tetrahedron Lett. **1978**, 4849.
[64] Battersby, A. R.; Martin, J. A.; Brochmann-Hanssen, E.: J. Chem. Soc. C **1967**, 1785.
[65] Battersby, A. R.; McDonald, E.; Milner, J. A.: Tetrahedron Lett. **1975**, 3419.
[66] Battersby, A. R.; McHugh, J. L.; Staunton, J.; Todd, M.: Chem. Commun. **1971**, 985.
[67] Battersby, A. R.; Minta, A.; Ottridge, A. P.; Staunton, J.: Tetrahedron Lett. **1977**, 1321.
[68] Battersby, A. R.; Munro, M. H. G.; Bradbury, R.; Santavy, F.: Chem. Commun. **1968**, 695.
[69] Battersby, A. R.; Parry, R. J.: Chem. Commun. **1971**, 901.
[70] Battersby, A. R.; Ramage, R.; Cameron, A. F.; Hannaway, C.; Santavy, F.: J. Chem. Soc. C **1971**, 3514.
[71] Battersby, A. R.; Sheldrake, P. W.; Milner, J. A.: Tetrahedron Lett. **1974**, 3315.
[72] Battersby, A. R.; Sheldrake, P. W.; Staunton, J.; Summers, M. C.: Bioorg. Chem. **6** (1977) 43.
[73] Battersby, A. R.; Staunton, J.: Tetrahedron **30** (1974) 1707.
[74] Battersby, A. R.; Staunton, J.; Summers, M. C.; Southgate, R.: J. Chem. Soc. Perkin I **1979**, 45.
[75] Battersby, A. R.; Staunton, J.; Wiltshire, H. R.; Bircher, B. J.; Fuganti, C.: J. Chem. Soc. Perkin I **1975**, 1162.
[76] Battersby, A. R.; Staunton, J.; Wiltshire, H. R.; Francis, R. J.; Southgate, R.: J. Chem. Soc. Perkin I **1975**, 1147.
[76a] Beecher, C. W. W.; Kelleher, W. J.: Tetrahedron Lett. **24** (1983) 469.
[77] Bennet, D. J.; Kirby, G. W.: J. Chem. Soc. C **1968**, 442.

[78] Bentley, K. W.; Hardy, D. G.: J. Am. Chem. Soc. **89** (1967) 3267, 3273, 3281.
[79] Bentley, R.: Biochim. Biophys. Acta **29** (1958) 666.
[79a] Berlin, J.; Forche, E.; Wray, V.; Hammer, J.; Hoesel, W.: Z. Naturforsch. **38c** (1983) 346.
[80] Beyerman, H. C.; Lie, T. S.; Maat, L.; Bosman, H. H.; Buurman, E.; Bijsterveld, E. J. M.; Sinnige, H. J. M.: Rec. Trav. Chim. Pays-Bas **95** (1976) 24.
[80a] Bhakuni, D. S.: Curr. Sci. **52** (1983) 897.
[80b] Bhakuni, D. S.; Gupta, S., Jain, S.: Tetrahedron **39** (1983) 4003.
[81] Bhakuni, D. S.; Jain, S.: Tetrahedron **36** (1980) 2153.
[82] Bhakuni, D. S.; Jain, S.: J. Chem. Soc. Perkin I **1981**, 1598.
[82a] Bhakuni, D. S.; Jain, S.: Tetrahedron **37** (1981) 3171.
[83] Bhakuni, D. S.; Jain, S.: Tetrahedron **37** (1981) 3175.
[84] Bhakuni, D. S.; Jain, S.: Tetrahedron **38** (1982) 729.
[85] Bhakuni, D. S.; Jain, S.; Chaturvedi, R.: Tetrahedron **35** (1979) 2323.
[85a] Bhakuni, D. S.; Jain, S.; Chaturvedi, R.: J. Chem. Soc. Perkin I **1983**, 1949.
[86] Bhakuni, D. S.; Jain, S.; Gupta, S.: Tetrahedron **36** (1980) 2491.
[86a] Bhakuni, D. S.; Jain, S.; Gupta, S.: Tetrahedron **39** (1983) 455.
[87] Bhakuni, D. S.; Jain, S.; Singh, A. N.: J. Chem. Soc. Perkin I **1978**, 380.
[88] Bhakuni, D. S.; Jain, S.; Singh, A. N.: Phytochemistry **19** (1980) 234.
[89] Bhakuni, D. S.; Jain, S.; Singh, R. S.: Tetrahedron **36** (1980) 2525.
[90] Bhakuni, D. S.; Joshi, P. P.: Tetrahedron **31** (1975) 2575.
[91] Bhakuni, D. S.; Labroo, V. M.; Singh, A. N.; Kapil, R. S.: J. Chem. Soc. Perkin I **1978**, 121.
[92] Bhakuni, D. S.; Mangla, V. K.: Tetrahedron **37** (1981) 401.
[93] Bhakuni, D. S.; Mangla, V. K.; Singh, A. N.; Kapil, R. S.: J. Chem. Soc. Perkin I **1978**, 267.
[94] Bhakuni, D. S.; Satish, S.; Dhar, M. M.: Phytochemistry **9** (1970) 2573.
[95] Bhakuni, D. S.; Satish, S.; Uprety, H.; Kapil, R. S.: Phytochemistry **13** (1974) 2767.
[96] Bhakuni, D. S.; Singh, A. N.: Tetrahedron **34** (1978) 1409.
[97] Bhakuni, D. S.; Singh, A. N.: J. Chem. Soc. Perkin I **1978**, 618.
[98] Bhakuni, D. S.; Singh, A. N.; Jain, S.: J. Chem. Soc. Perkin I **1978**, 1318.
[99] Bhakuni, D. S.; Singh, A. N.; Jain, S.: Tetrahedron **36** (1980) 2149.
[100] Bhakuni, D. S.; Singh, A. N.; Jain, S.; Kapil, R. S.: Chem. Commun. **1978**, 226.
[101] Bhakuni, D. S.; Singh, A. N.; Kapil, R. S.: Chem. Commun. **1977**, 211.
[102] Bhakuni, D. S.; Singh, A. N.; Tewari, S.; Kapil, R. S.: J. Chem. Soc. Perkin I **1977**, 1662.
[103] Bhakuni, D. S.; Tewari, S.; Kapil, R. S.: J. Chem. Soc. Perkin I **1977**, 706.
[104] Bhakuni, D. S.; Uprety, H.; Widdowson, D. A.: Phytochemistry **15** (1976) 739.
[105] Bhatia, S. K.; Prakash, O.; Kapil, R. S.: Indian J. Chem. **20B** (1981) 701.
[106] Bianco-Colomas, J.; Bulard, C.: Physiol. Veg. **19** (1981) 1.
[107] Biddington, N. L.; Thomas, T. H.: Planta **111** (1973) 183.
[108] Biddington, N. L.; Thomas, T. H.: Physiol. Plant. **40** (1977) 312.
[109] Bigot, M. C.: C. R. Acad. Sci, Ser. D **266** (1968) 349.
[110] Birnbaum, D.; Köhler, K. H.: Biochem. Physiol. Pflanzen **161** (1970) 521.
[111] Blanarik, P.; Pospisilova, Z.; Blanarikova, V.: Farm. Obz. **50** (1981) 575; C. A. **96** (1982) 31692.
[112] Blaschke, G.: Arch. Pharm. **301** (1968) 432.
[113] Blaschke, G.: Arch. Pharm. **301** (1968) 439.
[114] Blaschke, G.: Arch. Pharm. **303** (1970) 358.
[115] Blaschke, G.; Parker, H. I.; Rapoport, H.: J. Am. Chem. Soc. **89** (1967) 1540.
[115a] Blaschke, G.; Scriba, G.: Tetrahedron Lett. **24** (1983) 5855.
[116] Blaschke, G.; Waldheim, G.; Schantz, M. v.; Peura, P.: Arch. Pharm. **307** (1974) 122.
[117] Böhm, H.: Planta Med. **15** (1967) 215.
[118] Böhm, H.: Planta Med. **19** (1970) 93.
[119] Böhm, H.: Biochem. Physiol. Pflanzen **162** (1971) 474.

[120] Böhm, H.; Olesch, B.; Schulze, C.: Biochem. Physiol. Pflanzen **163** (1972) 126.
[121] Böhm, H.; Rink, E.: Biochem. Physiol. Pflanzen **168** (1975) 69.
[122] Böhm, H.; Rönsch, H.: Z. Naturforsch. **23b** (1968) 1552.
[122a] Boente, J. M.; Castedo, L.; Cuadros, R.; Rodriquez de la Lera, A.; Saa, J. M.; Suan, R.; Vidal, M. C.: Tetrahedron Lett. **24** (1983) 2303.
[123] Boit, H.-G.: Ergebnisse der Alkaloid-Chemie bis 1960. Akademie-Verlag, Berlin 1961.
[124] Boit, H.-G.; Flentje, H.: Naturwissenschaften **47** (1960) 180.
[125] Borkowski, P. R.; Horn, J. S.; Rapoport, H.: J. Am. Chem. Soc. **100** (1978) 276.
[126] Bose, A. K.; Das, K. G.; Funke, P. T.; Kugajevsky, I.; Shukla, O. P.; Khanchandani, K. S.; Suhadolnik, R. J.: J. Am. Chem. Soc. **90** (1968) 1038.
[127] Bose, A. K.; Khanchandani, K. S.; Tavares, R.; Funke, P. T.: J. Am. Chem. Soc. **90** (1968) 3593.
[128] Brannon, D. R.; Mabe, J. A.; Molloy, B. B.; Day, W. A.: Biochem. Biophys. Res. Commun. **43** (1971) 588.
[129] Bringmann, G.: Naturwissenschaften **66** (1979) 22.
[130] Brochmann-Hanssen, E.; Chen, C.-H.; Chen, C. R.; Chiang, H.-C.; Leung, A. Y.; McMurtrey, K.: J. Chem. Soc. Perkin I **1975**, 1531.
[131] Brochmann-Hanssen, E.; Chen, C.-H.; Chiang, H.-C.; McMurtrey, K.: Chem. Commun. **1972**, 1269.
[132] Brochmann-Hanssen, E.; Chen, C.-Y.; Linn, E. E.: Lloydia **43** (1980) 736.
[132a] Brochmann-Hanssen, E.; Cheng, C. Y.: J. Nat. Prod. **45** (1982) 434.
[132b] Brochmann-Hanssen, E.; Cheng, C. Y.: J. Nat. Prod. **45** (1982) 437.
[132c] Brochmann-Hanssen, E.; Cheng, C. Y.: J. Nat. Prod. **47** (1984) 175.
[132d] Brochmann-Hanssen, E.; Cheng, C. Y.; Chiang, H. C.: J. Nat. Prod. **45** (1982) 629.
[133] Brochmann-Hanssen, E.; Fu, C.-C.; Leung, A. Y.; Zanati, G.: J. Pharm. Sci. **60** (1971) 1672.
[134] Brochmann-Hanssen, E.; Fu, C.-C.; Misconi, L. Y.: J. Pharm. Sci. **60** (1971) 1880.
[135] Brochmann-Hanssen, E.; Nielsen, B.: Tetrahedron Lett. **1965**, 1271.
[136] Brochmann-Hanssen, E.; Nielsen, B.; Aadahl, G.: J. Pharm. Sci. **56** (1967) 1207.
[137] Brochmann-Hanssen, E.; Okamoto, Y.: J. Nat. Prod. **43** (1980) 731.
[138] Brochmann-Hanssen, E.; Wunderly, S. W.: J. Pharm. Sci. **67** (1978) 103.
[139] Brossi, A.; Ramel, A.; O'Brien, J.; Teitel, S.: Chem. Pharm. Bull. (Tokyo) **21** (1973) 1839.
[140] Brown, D. W.; Dyke, S. F.; Hardy, G.; Sainsbury, M.: Tetrahedron Lett. **1969**, 1515.
[141] Bruce, I. T.; Kirby, G. W.: Chem. Commun. **1968**, 207.
[142] Bruhn, J. G.; Agurell, S.; Lindgren, J. E.: Acta Pharm. Suecica **12** (1975) 199.
[143] Bruhn, J. G.; Svenson, U.; Agurell, S.: Acta Chem. Scand. **24** (1970) 3775.
[144] Bu'Lock, J. D.; Leigh, C.: Chem. Commun. **1975**, 628.
[145] Bu'Lock, J. D.; Ryles, A. P.: Chem. Commun. **1970**, 1404.
[146] Buzuk, G. N.; Lovkova, M. Y.; Grinkevich, N. I.: Istvest. Akad. Nauk SSSR, Ser. Biol. **1980**, 546.
[147] Buzuk, G. N.; Lovkova, M. Y.; Grinkevich, N. I.: Istvest. Akad. Nauk SSSR, Ser. Biol. **1981**, 458.
[148] Castedo, L.; Suau, R.; Mourino, A.: Tetrahedron Lett. **1976**, 501.
[149] Challice, J. S.: Biol. Plant. **19** (1977) 212.
[150] Chan, W. W. C.; Maitland, P.: J. Chem. Soc. C **1966**, 753.
[151] Chang, C.; Kimler, L.; Mabry, T. J.: Phytochemistry **13** (1974) 2777.
[152] Chang, C. P.; Mabry, T. J.: Biochem. Syst. **1** (1973) 185.
[153] Chhabra, N.; Singh, O. S.: Indian J. Plant Physiol. **18** (1975) 189.
[154] Colomas, J.; Barthe, P.; Bulard, C.: Z. Pflanzenphysiol. **87** (1978) 341.
[155] Colomas, J.; Bulard, C.: Planta **124** (1975) 245.
[156] Colomas, J.; Layton, J. L.; Bulard, C.: C. R. Acad. Sci. Ser. D **277** (1973) 173.
[157] Comer, F.; Tiwari, H. P.; Spenser, I. D.: Canad. J. Chem. **47** (1969) 481.
[158] Conn, E. E.: Naturwissenschaften **66** (1979) 28.
[159] Conn, E. E.: Ann. Rev. Plant Physiol. **31** (1980) 433.

[160] Conn, E. E.: Encycl. Plant Physiol. **8** (1980) 461.
[161] Conrad, K.: Biochem. Physiol. Pflanzen **165** (1974) 531.
[162] Constabel, F.: Naturwissenschaften **54** (1967) 175.
[163] Constabel, F.; Nassif-Makki, H.: Ber. Dtsch. Bot. Ges. **84** (1971) 629.
[164] Dalton, D. R.: The Alkaloids. Marcel Dekker, New York/Basel 1979.
[165] Delfel, N. E.: Phytochemistry **19** (1980) 403.
[166] Delfel, N. E.: Planta Med. **39** (1980) 168.
[167] Delfel, N. E.; Rothfus, J. A.: Phytochemistry **16** (1977) 1595.
[168] Dewan, O. P.; Bhakuni, S.; Kapil, R. S.: J. Chem. Soc. Perkin I **1978**, 622.
[169] Döpke, W.: Ergebnisse der Alkaloidchemie 1960–1968. Akademie-Verlag, Berlin 1976.
[170] Döpke, W.; Flentje, H.; Jeffs, P. W.: Tetrahedron **24** (1968) 4459.
[171] Dunkelblum, E.; Miller, H. E.; Dreiding, A. S.: Helv. Chim. Acta **55** (1972) 642.
[172] Dyke, S. F.; Ellis, A. C.: Tetrahedron **27** (1971) 3803.
[173] Elliott, D. C.: Plant Sci. Letters **15** (1979) 251.
[174] Elliott, D. C.: Plant Physiol. **63** (1979) 264.
[175] Elliott, D. C.: Plant Physiol. **63** (1979) 269.
[176] Elliott, D. C.: Plant Physiol. **63** (1979) 274.
[177] Elliott, D. C.: Plant Physiol. **63** (1979) 277.
[178] Elliott, D. C.: Plant Physiol. **64** (1979) 521.
[178a] Elliott, D. C.: Physiol. Plant. **59** (1983) 428.
[179] Elliott, D. C.; Murray, A. W.: Biochem. J. **146** (1975) 333.
[179a] Elliott, D. C.; Schultz, C. G.; Cassar, R. A.: Phytochemistry **22** (1983) 383.
[180] Endress, R.: Biochem. Physiol. Pflanzen **169** (1976) 87.
[181] Endress, R.: Phytochemistry **16** (1977) 1549.
[182] Endress, R.: Biochem. Physiol. Pflanzen **175** (1980) 17.
[183] Fairbairn, J. W.; Djoté, M.: Phytochemistry **9** (1970) 739.
[184] Fairbairn, J. W.; El-Masry, S.: Phytochemistry **6** (1967) 499.
[185] Fairbairn, J. W.; Hakim, F.: J. Pharm. Pharmacol. **25** (1973) 353.
[186] Fairbairn, J. W.; Hakim, F.; Dickenson, P. B.: J. Pharm. Pharmacol., Suppl. **25** (1973) 113P.
[187] Fairbairn, J. W.; Hakim, F.; El-Kheir, Y.: Phytochemistry **13** (1974) 1133.
[188] Fairbairn, J. W.; Handa, S. S.; Gurkan, E.; Phillipson, J. D.: Phytochemistry **17** (1978) 261.
[189] Fairbairn, J. W.; Steele, M. J.: Phytochemistry **19** (1980) 2317.
[190] Fairbairn, J. W.; Steele, M. J.: Phytochemistry **20** (1981) 1031.
[191] Fales, H. M.; Mann, T.; Mudd, S. H.: J. Am. Chem. Soc. **85** (1963) 2025.
[192] Fischer, N.; Dreiding, A. S.: Helv. Chim. Acta **55** (1972) 649.
[193] Fitzgerald, J. S.; Johns, S. R.; Lamberton, J. A.; Sioumis, A. A.: Austral. J. Chem. **22** (1969) 2187.
[194] Fodor, G. B.: In: Bell, E. A.; Charlwood, B. V. (Eds.): Encyclopedia of Plant Physiology, Vol 8, p. 92. Springer-Verlag, Heidelberg/New York 1980.
[195] Franck, B.; Blaschke, G.; Schlingloff, G.: Angew. Chem. **75** (1963) 957.
[196] Frank, B.; Teetz, V.: Angew. Chem. **83** (1971) 409.
[197] French, C. J.; Pecket, R. C.; Smith, H.: Phytochemistry **13** (1974) 1505.
[198] Fresno, M.; Jimenez, A.; Vazquez, D.: Eur. J. Biochem. **72** (1977) 323.
[199] Fuganti, C.: Tetrahedron Lett. **1973**, 1785.
[200] Fuganti, C.; Ghiringhelli, D.; Grasselli, P.: Chem. Commun. **1973**, 430.
[201] Fuganti, C.; Ghiringhelli, D.; Grasselli, P.: Chem. Commun. **1974**, 350.
[202] Fuganti, C.; Mazza, M.: Chem. Commun. **1970**, 1466.
[203] Fuganti, C., Mazza, M.: Chem. Commun. **1971**, 1196.
[204] Fuganti, C.; Mazza, M.: Chem. Commun. **1971**, 1388.
[205] Fuganti, C.; Mazza, M.: Chem. Commun. **1972**, 239.
[206] Fuganti, C.; Mazza, M.: Chem. Commun. **1972**, 936.
[207] Fuganti, C.; Mazza, M.: J. Chem. Soc. Perkin I **1973**, 954.
[208] Fuganti, C.; Staunton, J.; Battersby, A. R.: Chem. Commun. **1971**, 1154.

[209] Fujiwake, H.; Suzuki, T.; Iwai, K.: Plant Cell Physiol. **21** (1980) 1023.
[209a] Fujiwake, H.; Suzuki, T.; Iwai, K.: Agric. Biol. Chem. **46** (1982) 2591.
[209b] Fujiwake, H.; Suzuki, T.; Iwai, K.: Agric. Biol. Chem. **46** (1982) 2685.
[210] Fujiwake, H.; Suzuki, T.; Oka, S.; Iwai, K.: Agric. Biol. Chem. **44** (1980) 2907.
[211] Furuya, T.; Ikuta, A.: Syono, K.: Phytochemistry **11** (1972) 3041.
[212] Furuya, T.; Nakano, M.; Yoshikawa, T.: Phytochemistry **17** (1978) 891.
[212a] Furuya, T.; Yoshikawa, T.; Taira, M.: Phytochemistry **23** (1984) 999.
[213] Garay, A. S.; Towers, G. H. N.: Canad. J. Bot. **44** (1966) 231.
[214] Gear, J. R.; Spenser, I. D.: Nature **191** (1961) 1393.
[215] Gitterman, A.; Parry, R. J.; Dufresne, R. F.; Sternbach, D. D.; Cabelli, M. D.: J. Am. Chem. Soc. **102** (1980) 2074.
[216] Giudici de Nicola, M.; Amico, V.: Phytochemistry **11** (1972) 1011.
[217] Giudici de Nicola, M.; Amico, V.: Phytochemistry **12** (1973) 353.
[218] Giudici de Nicola, M.; Amico, V.; Piatelli, M.: Phytochemistry **14** (1975) 989.
[219] Giudici de Nicola, M.; Amico, V.; Sciuto, S.; Piatelli, M.: Phytochemistry **14** (1975) 479.
[220] Giudici de Nicola, M.; Piatelli, M.: Phytochemistry **12** (1973) 2163.
[221] Giudici de Nicola, M.; Piatelli, M.; Castrogiovanni, V.; Molina, C.: Phytochemistry **11** (1972) 1005.
[221a] Goezler, B.; Goezler, T.; Shamma, M.: Tetrahedron **39** (1983) 577.
[221b] Grinkevich, N. L.; Lovkova, M. Ya.; Buzuk, G. N.: Rastit. Resur **18** (1982) 367.
[222] Gross, S. R.; Burns, R. O.; Umbarger, E.: Biochemistry **2** (1963) 1046.
[223] Guggisberg, A.; Hesse, M.; Schmid, H.; Böhm, H.; Rönsch, H.; Mothes, K.: Helv. Chim. Acta **50** (1967) 621.
[224] Guha, K. P.; Mukherjee, B.; Mukherjee, R.: J. Nat. Prod. **42** (1979) 1.
[225] Gupta, R. N.; Spenser, I. D.: Canad. J. Chem. **43** (1965) 133.
[226] Guruprasad, K. N.; Laloraya, M. M.: Planta **130** (1976) 185.
[227] Guruprasad, K. N.; Laloraya, M. M.: Biochem. Physiol. Pflanzen **175** (1980) 582.
[228] Hahn, G.; Rumpf, F.: Ber. Dtsch. Chem. Ges. **71** (1938) 2141.
[229] Hart, N. K.; Johns, S. R.; Lamberton, J. A.; Loder, J. W.; Moorhouse, A.; Sioumis, A. A.; Smith, T. K.: Austral. J. Chem. **22** (1969) 2259.
[230] Hart, N. K.; Johns, S. R.; Lamberton, J. A.; Saunders, J. K.: Tetrahedron Lett. **1968**, 2891.
[231] Hayashi, H.; Koshimizu, K.: Agric. Biol. Chem. **43** (1979) 113.
[232] Haynes, L. J.; Stuart, K. L.; Barton, D. H. R.; Bhakuni, D. S.; Kirby, G. W.: Chem. Commun. **1965**, 141.
[233] Herbert, R. B.: In: Barton, D. H. R.; Ollis, W. D. (Eds.): Comprehensive Organic Chemistry. Vol. 5, 1045. Pergamon Press, Oxford 1979.
[234] Herbert, R. B.; Jackson, F. B.: Chem. Commun. **1977**, 955.
[235] Herbert, R. B.; Jackson, F. B.; Nicolson, I. T.: Chem. Commun. **1976**, 865; J. Chem. Soc. Perkin I **1984**, 825.
[236] Heydenreich, K.; Pfeifer, S.: Sci. Pharm. **30** (1962) 164.
[237] Hill, R. D.; Unrau, A. M.: Canad. J. Chem. **43** (1965) 709.
[238] Hinz, H.; Zenk, M. H.: Naturwissenschaften **68** (1981) 620.
[239] Hiraoka, N.; Carew, D. P.: J. Nat. Prod. **44** (1981) 285.
[240] Hodges, C. C.; Horn, J. S.; Rapoport, H.: Phytochemistry **16** (1977) 1939.
[241] Hodges, C. C.; Rapoport, H.: Phytochemistry **19** (1980) 1681.
[241a] Hodges, C. C.; Rapoport, H.: Biochemistry **21** (1982) 3729.
[241b] Hodges, C. C.; Rapoport, H.: J. Nat. Prod. **45** (1982) 481.
[242] Hörhammer, L.; Wagner, H.; Fritzsche, W.: Biochem. Z. **339** (1964) 398.
[243] Hösel, W.; Nahrstedt, A.: Arch. Biochem. Biophys. **203** (1980) 753.
[244] Holland, H. L.; Castillo, M.; MacLean, D. B.; Spenser, I. D.: Canad. J. Chem. **52** (1974) 2818.
[245] Holland, H. L.; Jeffs, P. W.; Capps, T. M.; MacLean, D. B.: Canad. J. Chem. **57** (1979) 1588.

[246] Horn, J. S.; Paul, A. G.; Rapoport, H.: J. Am. Chem. Soc. **100** (1978) 1895.
[247] Hrbek, J.; Santavy, F.: Coll. Czech. Chem. Commun. **27** (1962) 255.
[248] Hussain, S. F.; Siddiqui, M. T.; Shamma, M.: Tetrahedron Lett. **21** (1980) 4573.
[249] Ikuta, A.; Syono, K.; Furuya, T.: Phytochemistry **13** (1974) 2175.
[250] Impellizzeri, G.; Piatelli, M.: Phytochemistry **11** (1972) 2499.
[251] Iwai, K.; Lee, K.-R.; Kobashi, M.; Suzuki, T.; Oka, S.: Agric. Biol. Chem. **42** (1978) 201.
[252] Iwai, K.; Suzuki, T.; Lee, K.-R.; Kobashi, M.; Oka, S.: Agric. Biol. Chem. **42** (1977) 1877.
[253] Jackson, A. H.; Martin, J. A.: J. Chem. Soc. C **1966**, 2061.
[254] Jackson, A. H.; Martin, J. A.: J. Chem. Soc. C **1966**, 2222.
[255] Jeffs, P. W.: Proc. Chem. Soc. **1962**, 80.
[256] Jeffs, P. W.; Archie, W. C.; Farrier, D. S.: J. Am. Chem. Soc. **89** (1967) 2509.
[257] Jeffs, P. W.; Archie, W. C.; Hawks, R. L.; Farrier, D. S.: J. Am. Chem. Soc. **93** (1971) 3752.
[258] Jeffs, P. W.; Campbell, H. F.; Farrier, D. S.; Ganguli, G.; Martin, N. H.; Molina, G.: Phytochemistry **13** (1974) 933.
[259] Jeffs, P. W.; Campbell, H. F.; Farrier, D. S.; Molina, G.: Chem. Commun. **1971**, 228.
[260] Jeffs, P. W.; Capps, T.; Johnson, D. B.; Karle, J. M.; Martin, N. H.; Rauckman, B.: J. Org. Chem. **39** (1974) 2703.
[261] Jeffs, P. W.; Johnson, D. B.; Martin, N. H.; Rauckman, B. S.: Chem. Commun. **1976**, 82.
[262] Jeffs, P. W.; Karle, J. M.: Chem. Commun. **1977**, 60.
[263] Jeffs, P. W.; Karle, J. M.; Martin, N. H.: Phytochemistry **17** (1978) 719.
[264] Jeffs, P. W.; Scharver, J. D.: J. Am. Chem. Soc. **98** (1976) 4301.
[265] Johne, S.; Gröger, D.: Pharmazie **32** (1977) 1.
[266] Johns, N.; Kirby, G. W.: Chem. Commun. **1971**, 163.
[267] Johns, N.; Kirby, G. W.; Bu-Lock, J. D.; Ryles, A. P.: J. Chem. Soc. Perkin I **1975**, 383.
[268] Johns, S. R.; Kowala, C.; Lamberton, J. A.; Sioumis, A. A.; Wunderlich, J. A.: Chem. Commun. **1968**, 1102.
[269] Johns, S. R.; Lamberton, J. A.; Sioumis, A. A.: Austral. J. Chem. **22** (1969) 2219.
[270] Kametani, T.; Kikuchi, T.; Fukumoto, K.: Chem. Commun. **1967**, 546.
[271] Kametani, T.; Ohta, Y.; Takemura, M.; Ihara, M.; Fukumoto, K.: Heterocycles **6** (1977) 415; Bioorg. Chem. **6** (1977) 249.
[272] Kametani, T.; Satoh, Y.; Takemura, M.; Ohta, Y.; Ihara, M.; Fukumoto, K.: Heterocycles **5** (1976) 175.
[273] Kametani, T.; Takemura, M.; Ihara, M.; Takahashi, K.; Fukumoto, K.: J. Am. Chem. Soc. **98** (1976) 1956.
[274] Kametani, T.; Takemura, M.; Takahashi, K.; Ihara, M.; Fukumoto, K.: Heterocycles **3** (1975) 139.
[275] Kaminek, M.; Paces, V.; Corse, J.; Challice, J. S.: Planta **145** (1979) 239.
[276] Kamo, K. K.; Kimoto, W.; Hsu, A. F.; Mahlberg, P. G.; Bills, D.: Phytochemistry **21** (1982) 219.
[277] Kapadia, G. J.; Fales, H. M.: Chem. Commun. **1968**, 1688.
[278] Kapadia, G. J.; Subba Rao, G.; Leete, E.; Fayez, M. B. E.; Vaishnav, Y. N.; Fales, H. M.: J. Am. Chem. Soc. **92** (1970) 6943.
[279] Kapadia, G. J.; Vaishnav, Y. N.; Fayez, M. B. E.: J. Pharm. Sci. **58** (1969) 1157.
[280] Karimoto, R. S.; Axelrod, B.; Wolinsky, J.; Schall, E. D.: Phytochemistry **3** (1964) 349.
[281] Kelleher, W. J.; Rother, A.; Wellmann, E.; Grisebach, H.: Planta Med. **40** (1980) 127.
[282] Keller, W. J.: Lloydia **41** (1978) 37.
[283] Keller, W. J.: J. Pharm. Sci. **68** (1979) 85.
[284] Keller, W. J.: Phytochemistry **19** (1980) 413.
[284a] Keller, W. J.: Phytochemistry **21** (1982) 2851.

[285] Keller, W. J.; McLaughlin, J. L.; Brady, L. R.: J. Pharm. Sci. **62** (1973) 408.
[286] Keller, W. J.; Spitznagle, L. A.; Brady, L. R.; McLaughlin, J. L.: Lloydia **36** (1973) 397.
[287] Khanna, K. L.; Rosenberg, H.; Paul, A. G.: Chem. Commun. **1969**, 315.
[288] Khanna, K. L.; Takido, M.; Rosenberg, H.; Paul, A. G.: Phytochemistry **9** (1970) 1811.
[289] Kimler, L.; Mears, J.; Mabry, T. J.; Rösler, H.: Taxon **19** (1970) 875.
[290] Kinsman, L. T.; Pinfield, N. J.; Stobart, A. K.: Planta **127** (1975) 149.
[291] Kinsman, L. T.; Pinfield, N. J.; Stobart, A. K.: Planta **127** (1975) 207.
[292] Kirby, G. W.; Massey, S. R.; Steinreich, P.: J. Chem. Soc. Perkin I **1972**, 1642.
[293] Kirby, G. W.; Michael, J.: J. Chem. Soc. Perkin I **1973**, 115.
[294] Kirby, G. W.; Patrick, G. L.; Robins, D. J.: J. Chem. Soc. Perkin I **1978**, 1336.
[295] Kirby, G. W.; Robins, D. J.: Chem. Commun. **1976**, 354.
[296] Kirby, G. W.; Robins, D. J.: In: P. Steiyn (Ed.): The Biosynthesis of Mycotoxins. A Study in Secondary Metabolism. Academic Press, New York 1980, p. 301.
[297] Kirby, G. W.; Robins, D. J.; Sefton, M. A.; Talekar, R. R.: J. Chem. Soc. Perkin I **1980**, 119.
[298] Kirby, G. W.; Tiwari, H. P.: J. Chem. Soc. C **1966**, 676.
[299] Kirby, G. W.; Varley, M. J.: Chem. Commun. **1974**, 833.
[300] Kleinschmidt, G.; Mothes, K.: Z. Naturforsch. **14b** (1959) 52; Arch. Pharmaz. **293** (1960) 948.
[301] Kochhar, V. K.; Kochhar, S.; Mohr, H.: Ber. Dtsch. Bot. Ges. **94** (1981) 27.
[302] Köhler, K. H.: Naturwissenschaften **52** (1965) 561.
[303] Köhler, K. H.: Ber. Dtsch. Bot. Ges. **80** (1967) 403.
[304] Köhler, K. H.: Phytochemistry **11** (1972) 127.
[305] Köhler, K. H.: Phytochemistry **11** (1972) 133.
[306] Köhler, K. H.: Biol. Zbl. **92** (1973) 307.
[307] Köhler, K. H.: Pharmazie **28** (1973) 18.
[308] Köhler, K. H.: Biochem. Physiol. Pflanzen **168** (1975) 113.
[309] Köhler, K. H.; Birnbaum, D.: Biochem. Physiol. Pflanzen **161** (1970) 511.
[310] Köhler, K. H.; Birnbaum, D.: Biol. Zbl. **89** (1970) 201.
[311] Köhler, K. H.; Conrad, K.: Biol. Rdschau **5** (1966) 36.
[312] Köhler, K. H.; Dörfler, M.; Göring, H.: Biol. Plant. **22** (1980) 128.
[313] Kopp, B.; Jurenitsch, J.: Planta Med. **43** (1981) 272.
[313a] Kopp, B.; Jurenitsch, J.: Sci. Pharm. **50** (1982) 150.
[313b] Kopp, B.; Jurenitsch, J.; Woeginger, R.; Haunold, E.: Sci. Pharm. **51** (1983) 274.
[314] Kuppers, F. J. E. M.; Salemink, C. A.; Bastart, M.; Paris, M.: Phytochemistry **15** (1976) 444.
[315] Lalezari, I.; Nasseri, P.; Asgharian, R.: J. Pharm. Sci. **63** (1974) 1331.
[316] Laloraya, M. M.; Srivastav, H. N.; Guruprasad, K. N.: Planta **128** (1976) 275.
[317] Langlois, N.; Das, B.; Portier, P.; Lacombe, L.: Bull. Soc. chim. France **1970**, 3535.
[318] Leete, E.: Chem. & Ind. **1958**, 1088.
[319] Leete, E.: J. Am. Chem. Soc. **81** (1959) 3948.
[320] Leete, E.: J. Am. Chem. Soc. **85** (1963) 473.
[321] Leete, E.: J. Am. Chem. Soc. **85** (1963) 3666.
[322] Leete, E.: Tetrahedron Lett. **1965**, 333.
[323] Leete, E.: J. Am. Chem. Soc. **88** (1966) 4218.
[324] Leete, A.; Ahmad, A.: J. Am. Chem. Soc. **88** (1966) 4722.
[325] Leete, E.; Bodem, G. B.: J. Am. Chem. Soc. **98** (1976) 6321.
[326] Leete, E.; Bowman, R. M.; Maneul, M. F.: Phytochemistry **10** (1971) 3029.
[327] Leete, E.; Braunstein, J. D.: Tetrahedron Lett. **1969**, 451.
[328] Leete, E.; Kowanko, N.; Newmark, R. A.; Vining, L. C.; McInnes, A. G.; Wright, J. L. C.: Tetrahedron Lett. **1975**, 4103.
[329] Leete, E.; Louden, M. C. L.: J. Am. Chem. Soc. **90** (1968) 6837.
[330] Leete, E.; Marion, L.: Canad. J. Chem. **31** (1953) 126.
[331] Leete, E.; Murrill, S. J. B.: Tetrahedron Lett. **1964**, 147.

[332] Leete, E.; Nemeth, P. E.: J. Am. Chem. Soc. **82** (1960) 6055.
[333] Leete, E.; Nemeth, P. E.: J. Am. Chem. Soc. **83** (1961) 2192.
[334] Liebisch, H. W.: In: Mothes, K.; Schütte, H. R. (Eds.): Biosynthese der Alkaloide. Deutscher Verlag der Wissenschaften, Berlin 1969, p. 101.
[335] Liebisch, H. W.; Böhm, H.: Pharmazie **36** (1981) 218.
[336] Liebisch, H. W.; Matschiner, B.; Schütte, H. R.: Z. Pflanzenphysiol. **61** (1969) 269.
[337] Lockwood, G. B.: Phytochemistry **20** (1981) 1463.
[338] Lovkova, M. Y.; Buzuk, G. N.; Grinkevich, N. I.: Isvest. Akad. Nauk SSSR, Ser. Biol. **1** (1980) 111.
[339] Lovkova, M. Y.; Buzuk, G. N.; Grinkevich, N. I.: Prikl. Biochim. Mikrobiol. **16** (1980) 760.
[340] Lundström, J.: Acta Pharm. Suecica **8** (1971) 485.
[341] Lundström, J.: Acta Chem. Scand. **25** (1971) 3489.
[342] Lundström, J.; Agurell, S.: Tetrahedron Lett. **1968**, 4437.
[343] Lundström, J.; Agurell, S.: Abh. Dtsch. Akad. Wiss. Berlin, 1971, Vol. b Symposiumsbericht „Biochemie und Physiologie der Alkaloide", Halle, 25.—28. 6. 69. Akademie-Verlag, Berlin 1972, p. 89
[344] Lundström, J.; Agurell, S.: Tetrahedron Lett. **1969**, 3371.
[345] Lundström, J.; Agurell, S.: Acta Pharm. Suecica **8** (1971) 261.
[346] Mabry, T. J.: In: Bell, E. A.; Charlwood, B. V. (Eds.): Encyclopedia of Plant Physiology. N. S., Vol. 8, p. 513; Springer-Verlag, Berlin/Heidelberg/New York 1980.
[347] Mabry, T. J.; Kimler, L.; Chang, C.: The Betalains — Structure, Function, and Biogenesis, and the Plant Order Centrospermae. In: Runeckles, V. C.; Tso, T. C. (Eds.): Recent Advances in Phytochemistry, Vol. 5, p. 105; New York **1972**.
[348] MacDonald, J. C.; Slater, G. P.: Canad. J. Biochem. **53** (1975) 475.
[348a] Mandava, N. B.; Sasse, J. M.; Yopp, J. H.: Physiol. Plant. **53** (1981) 453.
[349] Mann, J. D.; Fales, H. M.; Mudd, S. H.: J. Biol. Chem. **238** (1963) 3820.
[350] Marekov, N.; Sidjimov, A.: Tetrahedron Lett. **22** (1981) 2311.
[351] Martin, R. O.; Warren, M. E.; Rapoport, H.: Biochemistry **6** (1967) 2355.
[352] Massicot, J.; Marion, L.: Canad. J. Chem. **35** (1957) 1.
[353] Matile, P.; Jans, B.; Rickenbacher, R.: Biochem. Physiol. Pflanzen **161** (1970) 447.
[354] Maturová, M.; Potesilova, H.; Santavy, F.; Cross, A. D.; Hanus, V.; Dolejs, L.: Coll. Czech. Chem. Commun. **32** (1967) 419.
[355] McDonald, E.: In: Grethe, G. (Ed.): Isoquinolines. John Wiley & Sons, London 1981, p. 275.
[356] McFarlane, I. J.; Slaytor, M.: Phytochemistry **11** (1972) 235.
[357] McInnes, A. G.; Smith, D. G.; Walter, J. A.; Vining, L. C.; Wright, J. L. C.: Chem. Commun. **1974**, 282.
[358] McLaughlin, J. L.; Paul, A. G.: Lloydia **30** (1967) 91.
[358a] Meyer, E.: Plant Cell Reports **1** (1982) 236.
[358b] Meyer, E.: Phytochemistry **22** (1983) 1381.
[359] Meyer, E.; Barz, W.: Planta Med. **33** (1978) 336.
[360] Miller, H. E.; Rösler, H.; Wohlpart, A.; Wyler, H.; Wilcox, M. E.; Frohofer, H.; Mabry, T. J.; Dreiding, A. S.: Helv. Chim. Acta **51** (1968) 1470.
[361] Miller, R. J.; Jolles, C.; Rapoport, H.: Phytochemistry **12** (1973) 597.
[362] Minale, L.; Piatelli, M.; Nicolaus, R. A.: Phytochemistry **4** (1965) 593.
[363] Mohr, H.: Z. Pflanzenphysiol. **54** (1966) 63.
[364] Mohr, H.: Naturwissenschaften **68** (1981) 193.
[365] Monkovic, I.; Spenser, I. D.: Canad. J. Chem. **43** (1965) 2017.
[366] Mothes, K.; Romeike, A.: In: Ruhland, W. (Ed.): Handbuch der Pflanzenphysiologie. Vol. 8. Springer-Verlag, Berlin/Göttingen/Heidelberg 1958, p. 989.
[367] Müller, P.; Schütte, H. R.: Z. Naturforsch. **23b** (1968) 491.
[368] Müller, P.; Schütte, H. R.: Biochem. Physiol. Pflanzen **162** (1971) 234.
[369] Mulchandani, N. B.; Iyer, S. S.; Badheka, L. P.: Phytochemistry **8** (1969) 1931.
[370] Mulchandani, N. B.; Iyer, S. S.; Badheka, L. P.: Phytochemistry **10** (1971) 1047.

[371] Mulchandani, N. B.; Iyer, S. S.; Badheky, L. P.: Phytochemistry **15** (1976) 1697.
[372] Musso, H.: Tetrahedron **35** (1979) 2843.
[373] Nagakura, N.; Höfle, G.; Coggiola, D.; Zenk, M. H.: Planta Med. **34** (1978) 381.
[374] Nagakura, N.; Höfle, G.; Zenk, M. H.: Chem. Commun. **1978**, 896.
[374a] Nagakura, N.; Rüffer, M.; Zenk, M. H.: J. Chem. Soc., Perkin I, **1979**, 2308.
[375] Nassif-Makki, H.; Constabel, F.: Z. Pflanzenphysiol. **67** (1972) 201.
[376] Neubauer, D.: Planta Med. **12** (1964) 43.
[377] Neubauer, D.: Arch. Pharm. **298** (1965) 737.
[378] Neubauer, D.; Mothes, K.: Planta Med. **11** (1963) 387.
[379] Nordal, A.; Paulsen, B. S.; Wold, J. K.: Acta Pharm. Suecica **14** (1977) 37.
[380] Obata-Sasamoto, H.; Komamine, A.; Saito, K.: Z. Naturforsch. **36C** (1981) 921.
[381] O'Donovan, D. G.; Barry, E.: J. Chem. Soc. Perkin I **1974**, 2528.
[382] O'Donovan, D. G.; Horan, H.: J. Chem. Soc. C **1968**, 2791.
[383] O'Donovan, D. G.; Horan, H.: J. Chem. Soc. C **1969**, 1737.
[384] O'Donovan, D. G.; Horan, H.: J. Chem. Soc. C **1971**, 331.
[385] Pakrashi, S. C.; Ghosh-Dastidar, P.; Basu, S.; Achari, B.: Phytochemistry **16** (1977) 1103.
[386] Pande, H.; Bhakuni, D. S.: J. Chem. Soc. Perkin I **1976**, 2197.
[387] Parker, H. I.; Blaschke, G.; Rapoport, H.: J. Am. Chem. Soc. **94** (1972) 1276.
[388] Parry, R. J.: Recent Adv. Phytochem. **13** (1979) 55.
[389] Parry, R. J.; Chang, M. N. T.; Schwab, J. M.; Foxman, B. M.: J. Am. Chem. Soc. **102** (1980) 1099.
[390] Parry, R. J.; Schwab, J. M.: J. Am. Chem. Soc. **97** (1975) 2555.
[391] Paudler, W. W.; McKay, J.: J. Org. Chem. **38** (1973) 2110.
[392] Paul, A. G.: Lloydia **36** (1973) 36.
[393] Paul, A. G.; Khanna, K. L.; Rosenberg, H.; Takido, M.: Chem. Commun. **1969**, 838.
[394] Pfeifer, S.; Heydenreich, K.: Naturwissenschaften **48** (1961) 222.
[395] Phillipson, J. D.; Handa, S. S.; El-Dabbas, S. W.: Phytochemistry **15** (1976) 1297.
[396] Piatelli, M.: Betalains. In: Goodwin, T. W. (Ed.): Chemistry and Biochemistry of Plant Pigments. 2nd Ed. Vol. 1, p. 560; 1976.
[397] Piatelli, M.; Giudici de Nicola, M.; Castrogiovanni, V.: Phytochemistry **8** (1969) 731.
[398] Piatelli, M.; Giudici de Nicola, M.; Castrogiovanni, V.: Phytochemistry **9** (1970) 785.
[399] Piatelli, M.; Giudici de Nicola, M.; Castrogiovanni, V.: Phytochemistry **10** (1971) 289.
[400] Pijewska, L.; Kaul, J. L.; Joshi, R. K.; Santavy, F.: Coll. Czech. Chem. Comm. **32** (1967) 158.
[401] Powell, R. G.: Phytochemistry **11** (1972) 1467.
[402] Powell, R. G.; Mikolajczak, K. L.; Weisleder, D.; Smith, C. R.: Phytochemistry **11** (1972) 3317.
[403] Powell, R. G.; Mikolajczak, K. L.: Phytochemistry **12** (1973) 2987.
[404] Powell, R. G.; Weisleder, D.; Smith, C. R.; Wolff, I. A.: Tetrahedron Lett. **1969**, 4081.
[405] Powell, R. G.; Weisleder, D.; Smith, C. R.: J. Pharm. Sci. **61** (1972) 1227.
[406] Prakash, O.; Bhakuni, D. S.; Kapil, R. S.: J. Chem. Soc. Perkin I **1979**, 1515.
[407] Preisner, R. M.; Shamma, M.: J. Nat. Prod. **43** (1980) 305.
[408] Pummangura, S.; McLaughlin, J. L.; Schifferdecker, R. C.: J. Nat. Prod. **45** (1982) 214.
[409] Ranieri, R. L.; McLaughlin, J. L.; Arp, G. K.: Lloydia **39** (1976) 172.
[410] Rapoport, H.; Stermitz, F. R.; Baker, D. R.: J. Am. Chem. Soc. **82** (1960) 2765.
[411] Rast, D.; Skrivanova, R.; Bachofen, R.: Phytochemistry **12** (1973) 2669.
[412] Rast, D.; Skrivanova, R.; Wohlpart, A.: Ber. Schweiz. Bot. Ges. **82** (1972) 213.
[413] Razakov, R.; Yunusov, S. Y.; Nasyrov, S. M.; Chekhlov, A. N.; Adrianov, V. G.; Struchkov, Y. T.: Chem. Commun. **1974**, 150.
[414] Reda, F.; Rasmussen, O.: Biol. Plant. **17** (1975) 368.
[415] Reznik, H.: Ber. Dtsch. Bot. Ges. **88** (1975) 179.
[416] Reznik, H.: Z. Pflanzenphysiol. **87** (1978) 95.
[417] Rink, E.; Böhm, H.: FEBS-Letters **49** (1975) 396.
[417a] Roberto, M. F.; McCarthy, D.; Kutchan, T. M.; Coscia, C. J.: Arch. Biochem. Biophys. **222** (1983) 599.

[418] Robinson, R.: J. Chem. Soc. **1936**, 1079.
[419] Robinson, R.: The Structural Relations of Natural Products. Clarendon Press, Oxford 1955.
[420] Robinson, T.; Nagel, W.: Phytochemistry **21** (1982) 535.
[421] Rönsch, H.: Eur. J. Biochem. **28** (1972) 123.
[422] Rönsch, H.: Phytochemistry **16** (1977) 691.
[423] Rosenberg, H.; Khanna, K. L.; Takido, M.; Paul, A. G.: Lloydia **32** (1969) 334.
[424] Rosenberg, H.; McLaughlin, J. L.; Paul, A. G.: Lloydia **30** (1967) 100.
[424a] Rueffer, M.; Ekundayo, O.; Nagakura, N.; Zenk, M. H.: Tetrahedron Lett. **24** (1983) 2643.
[425] Rueffer, M.; El-Shagi, H.; Nagakura, N.; Zenk, M. H.: FEBS-Letters **129** (1981) 5.
[425a] Russo, C. A.; Burton, C.; Gros, E. C.: Phytochemistry **22** (1983) 71.
[426] Russo, C. A.; Gros, E. G.: Phytochemistry **20** (1981) 1763; **21** (1982) 609; **22** (1983) 1839.
[426a] Saito, K.; Obata-Sasamoto, H.; Hatanaka, S.-I.; Noguchi, H.; Sankawa, U.; Komamine, A.: Phytochemistry **21** (1982) 474.
[427] Sammes, P. G.: Progr. Chem. org. nat. Prod. **32** (1975) 51.
[428] Santavy, F.; Maturova, M.; Hruban, L.: Chem. Commun. **36** (1966) 144.
[429] Schöpf, C.: Naturwissenschaften **39** (1952) 241.
[429a] Schröder, W.; Böhm, H.: Plant Cell Report **3** (1984) 14.
[430] Schütte, H. R.: In: Mothes, K.; Schütte, H. R. (Eds.): Biosynthese der Alkaloide. Deutscher Verlag der Wissenschaften, Berlin 1969, p. 344.
[431] Schütte, H. R.: In: Mothes, K.; Schütte, H. R. (Eds.): Biosynthese der Alkaloide. Deutscher Verlag der Wissenschaften, Berlin 1969, p. 367.
[432] Schütte, H. R.: In: Mothes, K.; Schütte, H. R. (Eds.): Biosynthese der Alkaloide. Deutscher Verlag der Wissenschaften, Berlin 1969, p. 420.
[433] Schütte, H. R.: Fortschr. Bot. **34** (1972) 165.
[434] Schütte, H. R.: Fortschr. Bot. **35** (1973) 103.
[435] Schütte, H. R.: Progr. Bot. **40** (1978) 126; **41** (1979) 93.
[436] Schütte, H. R.: Biosynthese von niedermolekularen Naturstoffen. Gustav Fischer Verlag, Jena 1982.
[437] Schütte, H. R.; Orban, U.; Mothes, K.: Eur. J. Biochem. **1** (1967) 70.
[438] Schütte, H. R.; Seelig, G.: Liebigs Ann. Chem. **730** (1969) 186.
[439] Schunack, W.; Rochelmeyer, H.: Planta Med. **13** (1965) 1.
[440] Schwab, J. M.; Chang, M. N. T.; Parry, R. J.: J. Am. Chem. Soc. **99** (1977) 2368.
[441] Schwartz, M. A.; Mami, I. S.: J. Am. Chem. Soc. **97** (1975) 1239.
[442] Sciuto, S.; Oriente, G.; Piatelli, M.: Phytochemistry **11** (1972) 2259.
[443] Sciuto, S.; Oriente, G.; Piatelli, M.; Impellizzeri, G.; Amico, V.: Phytochemistry **13** (1974) 947.
[444] Scott, A. I.; Lee, S. L.; Hirata, T.: Heterocycles **11** (1978) 159.
[445] Shafiee, A.; Lalezari, I.; Nasseri-Nouri, P.; Asgharian, R.: J. Pharm. Sci. **64** (1975) 1570.
[446] Shamma, M.: The Isoquinoline Alkaloids. Academic Press, New York, and Verlag Chemie, Weinheim 1972.
[447] Shamma, M.; Foy, J. E.; Govindachari, T. R.; Viswanathan, N.: J. Org. Chem. **1976**, 1293.
[448] Shamma, M.; Foy, J. E.; Miana, G. A.: J. Am. Chem. Soc. **96** (1974) 7809.
[449] Shamma, M.; Moniot, J. L.: Isoquinoline Alkaloids Research. 1972—1977. Plenum Press, New York/London 1978.
[450] Shamma, M.; Moniot, J. L.; Yao, S. Y.; Miana, G. A.; Ikram, M.: J. Am. Chem. Soc. **95** (1973) 5742.
[451] Sharghi, N.; Lalezari, I.: Nature **213** (1967) 1244.
[452] Sharma, V.; Jain, S.; Bhakuni, D. S.; Kapil, R. S.: J. Chem. Soc. Perkin I **1982**, 1153.
[453] Shibata, S.; Imaseki, I.; Yamazaki, M.: Chem. Pharm. Bull. (Tokyo) **7** (1959) 449.
[453a] Shih, C. C.; Wiley, R. C.: J. Food. Sci. **47** (1982) 164.
[454] Shukla, R. M.: Indian Drugs **17** (1980) 392.

[455] Sidjimov, A. K.; Marekov, N. L.: Phytochemistry **21** (1982) 871.
[456] Sivakumaran, M.; Gopinath, K. W.: Indian J. Chem. **14B** (1976) 150.
[457] Skerl, A. R.; Gros, E. G.: Phytochemistry **10** (1971) 2719.
[458] Smith, T. A.: Phytochemistry **16** (1977) 9.
[459] Spenser, I. D.; Tiwari, H. P.: Chem. Commun. **1966**, 55.
[460] Srivastav, H. N.; Laloraya, M. M.: Nature, New Biology **243** (1973) 224.
[461] Srivastav, H. N.; Laloraya, M. M.: Plant Biochem. J. **3** (1976) 134.
[462] Staunton, J.: Planta Med. **36** (1979) 1.
[463] Stenlid, G.: Phytochemistry **15** (1976) 661.
[464] Stermitz, F. R.; Rapoport, H.: J. Am. Chem. Soc. **83** (1961) 4045.
[465] Stermitz, F. R.; Seiber, J. N.: J. Org. Chem. **31** (1966) 2925.
[466] Stobart, A. K.; Hendry, G. A. F.; Ei-Hussein, S.; Kinsman, L. T.: Z. Pflanzenphysiol. **96** (1980) 217.
[467] Stobart, A. K.; Kinsman, L. T.: Phytochemistry **16** (1977) 1137.
[468] Stobart, A. K.; Pinfield, N. J.; Kinsman, L. T.: Planta **94** (1970) 152.
[469] Stuart, K. L.; Graham, L.: Chem. Commun. **1971**, 392.
[470] Stuart, K. L.; Graham, L.: Phytochemistry **12** (1973) 1967.
[471] Stuart, K. L.; Graham, L.: Phytochemistry **12** (1973) 1973.
[472] Stuart, K. L.; Teetz, V.; Frank, B.: Chem. Commun. **1969**, 333.
[473] Suhadolnik, R. J.: Lloydia **27** (1964) 315.
[474] Suhadolnik, R. J.: Abh. Dtsch. Akad. Wiss. Berlin, Klasse Chem., Geol., Biol. **1966**, No. 3, 369.
[475] Suhadolnik, R. J.; Chenoweth, R. G.: J. Am. Chem. Soc. **80** (1958) 4391.
[476] Suhadolnik, R. J.; Fischer, A. G.; Zulalian, J.: J. Am. Chem. Soc. **84** (1962) 4348.
[477] Suhadolnik, R. J.; Fischer, A. G.; Zulalian, J.: Proc. Chem. Soc. **1963**, 132.
[478] Suhadolnik, R. J.; Fischer, A. G.; Zulalian, J.: Biochem. Biophys. Res. Commun. **11** (1963) 208.
[479] Suhadolnik, R. J.; Zulalian, J.: Proc. Chem. Soc. **1963**, 216.
[480] Suzuki, T.; Kawada, T.; Iwai, K.: Plant Cell Physiol. **22** (1981) 23.
[481] Suzuki, T.; Kawada, T.; Iwai, K.: Agric. Biol. Chem. **45** (1981) 535.
[482] Tackie, A. N.; Dwuma-Badu, D.; Ayim, J. S. K.; Dabra, T. T.; Knapp, J. E.; Slatkin, D. J.; Schiff, P. L., Jr.: Lloydia **38** (1975) 210.
[483] Takao, N.; Iwasa, K.; Kamigauchi, M.; Sugiura, M.: Chem. Pharm. Bull. (Tokyo) **24** (1976) 2859.
[483a] Takao, N.; Kamigauchi, M.; Okada, M.: Helv. Chim. Acta **66** (1983) 473.
[484] Tam, W. H. J.; Constabel, F.; Kurz, W. G. W.: Phytochemistry **19** (1980) 486.
[485] Tam, W. H. J.; Kurz, W. G. W.; Constabel, F.; Chatson, K. B.: Phytochemistry **21** (1982) 253.
[486] Tani, C.; Tagahara, K.: Chem. Pharm. Bull. (Japan) **22** (1974) 2457.
[487] Tani, C.; Tagahara, K.: J. Pharm. Soc. (Japan) **97** (1977) 87.
[488] Tani, C.; Tagahara, K.: J. Pharm. Soc. (Japan) **97** (1977) 93.
[489] Tewari, S.; Bhakuni, D. S.; Kapil, R. S.: Chem. Commun. **1975**, 554.
[490] Thornber, C. W.: Phytochemistry **9** (1970) 157.
[491] Underhill, E. W.: In: Bell, E. A.; Charlwood, B. V. (Eds.): Encyclopedia of Plant Physiology. Vol. 8. Springer-Verlag, Berlin/Heidelberg/New York 1980, p. 493.
[492] Uprety, H.; Bhakuni, D. S.; Kapil, R. S.: Phytochemistry **14** (1975) 1535.
[492a] Vágújfalvi, D.; Petz-Stifter, M.: Phytochemistry **21** (1982) 1533.
[493] Verzar-Petri, G.; Minh-Hoang, P. J.; Szarvas, T.: Acta Pharm. Hung. **49** (1979) 168.
[494] Wagner, E.; Cumming, B. G.: Canad. J. Bot. **48** (1970) 1.
[495] Wheaton, T. A.; Stewart, I.: Phytochemistry **8** (1969) 85.
[496] Wightman, R. H.; Staunton, J.; Battersby, A. R.; Hanson, K. R.: J. Chem. Soc. Perkin I **1972**, 2355.
[497] Wildman, W. C.; Fales, H. M.: J. Am. Chem. Soc. **86** (1964) 294.
[498] Wildman, W. C.; Fales, H. M.; Battersby, A. R.: J. Am. Chem. Soc. **84** (1962) 681.
[499] Wildman, W. C.; Heimer, N. E.: J. Am. Chem. Soc. **89** (1967) 5265.

[500] Wilson, M. L.; Coscia, C. J.: J. Am. Chem. Soc. **97** (1975) 431.
[501] Winstead, J. A.; Suhadolnik, R. J.: J. Am. Chem. Soc. **82** (1960) 1644.
[502] Winterstein, E.; Trier, G.: Die Alkaloide. Bornträger-Verlag, Berlin 1910, S. 307.
[503] Wohlpart, A.; Black, S. M.: Phytochemistry **12** (1973) 1325.
[504] Wohlpart, A.; Mabry, T. J.: Plant Physiol. **43** (1968) 457.
[505] Wold, J. K.; Paulsen, B. S.; Nordal, A.: Acta Pharm. Suecica **14** (1977) 403.
[506] Woodhead, S.; Swain, T.: Phytochemistry **13** (1974) 953.
[507] Wright, J. L. C.; Vining, L. C.; McInnes, A. G.; Smith, D. G.; Walter, J. A.: Canad. J. Biochem. **55** (1977) 678.
[508] Wyler, H.; Mabry, T. J.; Dreiding, A. S.: Helv. Chim. Acta **46** (1963) 1745.
[509] Yagi, A.; Nonaka, G.; Nakayama, S.; Nishioka, I.: Phytochemistry **16** (1977) 1197.
[510] Yamasaki, K.; Sankawa, U.; Shibata, S.: Tetrahedron Lett. **1969**, 4099.
[511] Yamasaki, K.; Tamaki, T.; Uzawa, S.; Sankawa, U.; Shibata, S.: Phytochemistry **12** (1973) 2877.
[512] Zulalian, J.; Suhadolnik, R. J.: Proc. Chem. Soc. **1964**, 422.

15 Alkaloids Derived from Tryptophan

D. Gröger

Tryptophan-derived alkaloids, which are found in higher plants and microorganisms, are the largest group — about one quarter — of known alkaloids. Most of them were isolated and structurally determined during the last two decades. This was done with the aid of classic chemical degradation and modern analytical techniques (mass spectrometry, nuclear magnetic resonance, X-ray crystallography). Many indole alkaloids are biologically active and some are extremely useful therapeutic agents in human medicine, *e.g. Catharanthus, Curare,* ergot, *Rauwolfia, Vinca* alkaloids.

According to the chromophore seven groups of indole alkaloids have been found [94]. Most indole alkaloids possess an indole or indoline chromophore (Fig. 15.1)

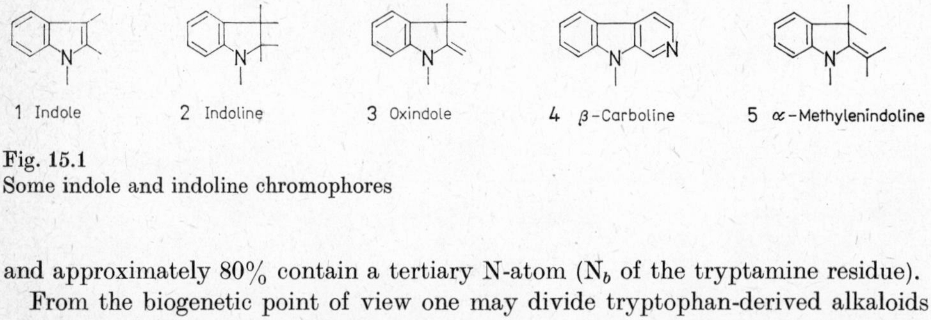

1 Indole 2 Indoline 3 Oxindole 4 β-Carboline 5 α-Methylenindoline

Fig. 15.1
Some indole and indoline chromophores

and approximately 80% contain a tertiary N-atom (N_b of the tryptamine residue).

From the biogenetic point of view one may divide tryptophan-derived alkaloids into different groups: indolealkylamines, simple β-carboline bases, alkaloids derived from tryptophan plus a C_5-unit (ergolines) and alkaloids built up from tryptamine and a monoterpene moiety. The vast majority of indole alkaloids belong to the latter group, which are found solely in vascular plants. Some tryptophan-derived alkaloids do not contain an indole nucleus (camptothecin, *Cinchona* alkaloids). Many aspects of indole alkaloid chemistry and biochemistry are covered in recent volumes of "The Alkaloids" [142] and other excellent compilations [94, 165, 214].

15.1 Simple Indole Bases

The compounds mentioned in this chapter are so-called proto-alkaloids, which contain no heterocyclic system except the indole chromophore. A selection of them is listed in Fig. 15.2. The biosynthesis and occurrence of simple tryptophan derivatives and indolyl amines have been reviewed [82, 195–197].

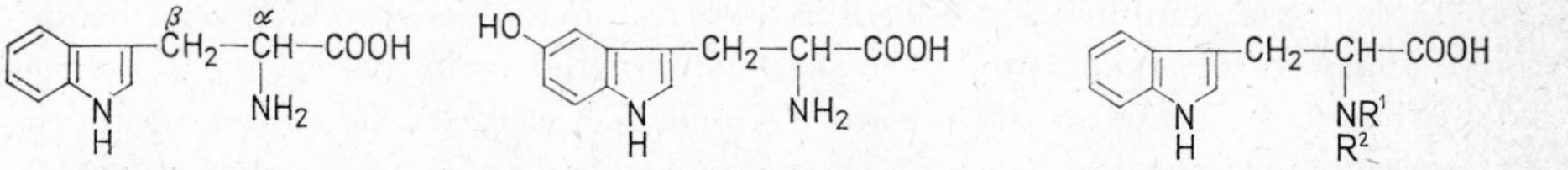

6 Tryptophan
7 5-Hydroxytryptophan
8 R^1=H, R^2=Me : *N*-Methyltryptophan ; Abrine
9 R^1, R^2=Me : *N,N*-Dimethyltryptophan

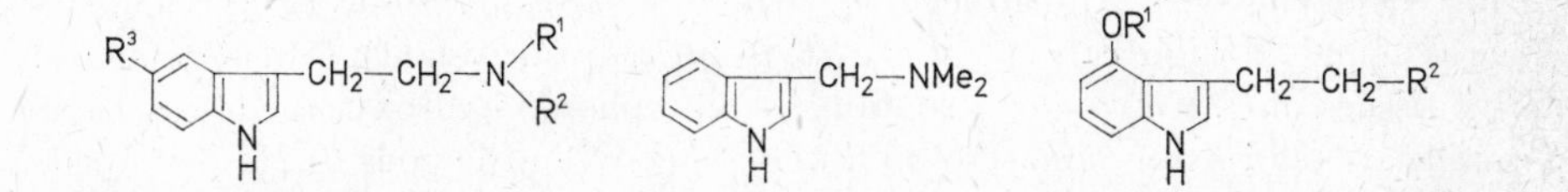

10 R^1=H, R^2=H, R^3=H : Tryptamine
11 R^1=Me, R^2=H, R^3=H : *N*-Methyltryptamine
12 R^1=Me, R^2=Me, R^3=H : *N,N*-Dimethyltryptamine
13 R^1=H, R^2=H, R^3=OH : 5-Hydroxytryptamine ; Serotonin
14 R^1=Me, R^2=H, R^3=OH : 5-Hydroxy-*N*-methyltryptamine
15 R^1, R^2=Me, R^3=OH : 5-Hydroxy-*N,N*-dimethyltryptamine ; Bufotenine
16 Gramine
17 R^1=H, R^2=NMe_2 : Psilocin
18 R^1=PO_3H_2, R^2=NMe_2 : Psilocybin

Fig. 15.2
Simple tryptophan derivatives and indolylalkylamines

5-Hydroxytryptophan (**7**) is a constituent of *Panaeolus* species, while N_b-methyl-*L*-tryptophan (abrine) (**8**) is a specific compound of the seeds of *Abrus precatorius*.

The indolyl amines occur in some fungi and a wide variety of higher plants that belong to 25 plant families. The *Gramineae* and *Leguminosae* families are especially rich in indole amines. 5-Hydroxytryptamine (serotonin) (**13**) is a well-known neurohormone common in the animal and plant kingdoms. This biologically highly active substance is also found in some edible fruits (*e.g.* bananas, plums, pineapples, tomatoes and nuts).

Typical constituents of hallucinatory snuffs and decoctions used by some ethnic groups in South America are dimethyltryptamine (**12**), bufotenine (**15**) and 5-methoxy-*N,N*-dimethyltryptamine. Species of *Banisteriopsis*, *Piptadenia* and *Virola* as well as *Mimosa hostilis* are used in the preparation. The active principles of the hallucinogenic mushrooms used by Mexican Indians are psilocin (**17**) and psilocybin (**18**). These unique compounds were first isolated by Hofmann et al. [96]. 3-Dimethylaminomethyl-indole (gramine) (**16**) is a well-known constituent of sprouting barley and other *Gramineae*. It has also been found outside this family in some species of *Acer* and some lupins (*Leguminosae*). *Phalaris arundinacia* and *Phalaris tuberosa*, which are major pasture grasses in some parts of North America and Australia can be highly toxic to sheep causing "Phalaris staggers". Apparently this is due to the content of **12**, **15**, **16** and 5-methoxy-*N,N*-dimethyltryptamine.

Using seedlings of *P. tuberosa* a pyridoxalphosphate-dependent tryptophan decarboxylase has been purified 20-fold. The decarboxylase activity of the seedlings reached a maximum after 4 days. The decarboxylation of **6** was inhibited at a concentration of 1 mM by a number of indole derivatives including **12** which is one of the major alkaloids synthesized by the plant [34]. This might be important in the control of alkaloid biosynthesis. The metabolic pathways leading to *N*,*N*-dimethyltryptamine (**12**) and 5-methoxy-*N*,*N*-dimethyltryptamine in *P. tuberosa* have been studied by a combination of feeding and trapping experiments. It was demonstrated that five of six possible pathways constructed from alternative sequences of decarboxylation, hydroxylation and *O*- and *N*-methylation could lead to the end-product. Apparently, studies at the enzymatic level are necessary to establish the pathway operative in nature [35].

In seeds of the West African legume *Griffonia simplicifolia* 5-hydroxytryptophan (**7**) was found in high concentrations (6 to 10% fresh weight). All metabolically active tissues of *G. simplicifolia* contain *L*-tryptophan-5-hydroxylase. Furthermore, a variety of indoles inter alia serotonin were detected in the pods (0.2% dry weight) of this particular plant [59]. The main indole compound found in the seeds of *Piptadenia peregrina* is bufotenine (**15**). The sequence leading to **15** starting from tryptophan was elucidated by incubating different potential ^{14}C-labelled precursors with tissues of etiolated stems. The following pathway seems plausible: tryptophan (**6**) → tryptamine (**10**) → serotonin (**13**) → 5-hydroxy-*N*-methyltryptamine (**14**) → bufotenine (**15**). Serotonin formation was studied in seeds of *Juglans regia* [84a]. The first step is the decarboxylation of (**6**) giving (**10**) catalyzed by the adaptive tryptophan decarboxylase. This enzyme is obviously the control element of serotonin biosynthesis in *Juglans regia* and *Peganum harmala* cell cultures [178a]. Finally a constitutive tryptamine-5-hydroxylase converts (**10**) into serotonin. Most surprising is the formation of serotonin from tryptamine. In mammals and other plants (**13**) is formed by the decarboxylation of 5-hydroxytryptophan (**7**) [60].

Psilocybe-, *Stropharia*-, *Conocybe*- and *Panaeolus*-species contain psychotropic indolyl amines (**17, 18**). Some *Psilocybe* species produce psilocybin (**18**) in emerged and submerged culture. According to feeding experiments with various labelled precursors and comparison of the specific incorporation rates the following pathway is assumed: tryptophan (**6**) → tryptamine (**10**) → *N*-methyltryptamine (**11**) → *N*,*N*-dimethyltryptamine (**12**) → psilocin (**17**) → psilocybin (**18**) [5, 42].

The side-chain nitrogen of the proto-alkaloid gramine (**16**) is separated from the indole ring by only one carbon atom compared with tryptamine (**10**). Gramine was the first alkaloid whose biosynthesis was studied with the aid of radioactive tracer techniques. In a classic paper [41] it was demonstrated that label from [β-^{14}C]-tryptophan appeared specifically at the side-chain methylene of gramine. Also [β-^{14}C, β-^{3}H]tryptophan showed specific incorporation [157]. The isotopes were confined with the same ratio as in the precursor in the methylene group of the side-chain of **16**. 3-Aminomethyl-indole (**20**) and 3-Methylaminomethyl-indole (**21**) are both present as "minor alkaloids" in barley and are efficiently incorporated into gramine (**16**) after administration in labelled form [74]. An intriguing question is the origin of the non-indole nitrogen in **16**. There is reliable evidence [84] that the dimethylamino group of **16** is not derived from the general metabolic nitrogen pool. According to experiments with [β-^{14}C,α-^{15}N]tryptophan, it can be concluded that the α-amino group of tryptophan (**6**) provides the side-chain nitrogen of gramine [74]. Furthermore, because the α-C-atom of **6** is not transferred to **16**, intact incorporation of the α-C-atom and the amino-*N* of tryptophan as *N*-CH_3 unit in the course of gramine formation can be excluded (Fig. 15.3) [83].

An attractive hypothesis was put forward for the conversion of the tryptophan side-chain into gramine involving 2,3 cleavage of the amino acid side chain of an enzyme-bound tryptophan-pyridoxal phosphate intermediate [221]. However, experimental evidence supporting this idea is still lacking.

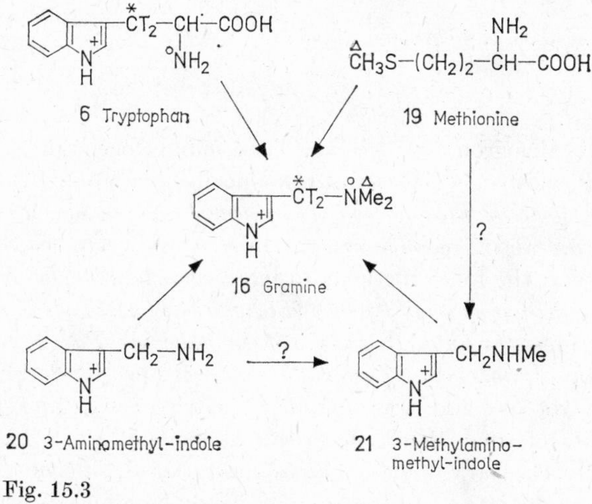

Fig. 15.3
Biosynthesis of gramine

Recently it was found [121] that tryptophan serves as specific precursor for gramine in *Lupinus hartwegii* seedlings, i.e. not in the *Gramineae*. [β-^{14}C]Tryptophan was specifically incorporated into **16** and indole-3-aldehyde after feeding to 8-week-old plants of *L. hartwegii*. Gramine is possibly being metabolized to indole-3-aldehyde in the maturing plant.

In barley the main site of gramine biosynthesis is in the basal region of the seedling leaf. The proto-alkaloid is then transported to the tip of the leaf where it accumulates [181]. In leaves of some barley cultivars which were grown under high-temperature stress a massive gramine accumulation was observed [87a].

15.2 Simple β-Carboline Alkaloids

"Simple" β-carboline alkaloids are derivatives of the tricyclic pyrido (3,4-*b*)indole, possessing alkyl substituents at C-1. The β-carbolines differ in the state of oxidation of the pyridyl ring. Thus there are aromatic (**22**), dihydro- (**23**) and tetrahydro-β-carbolines (**24**) (Fig. 15.4).

β-carbolines were first isolated from *Peganum harmala* [73] — the so-called Syrian rue — which is mainly found in arid zones. This led to the use of several trivial names for a variety of β-carboline alkaloids based on the root-"harm". For the 1,2,3,4-tetrahydro-β-carbolines the generic name "tryptoline" has been proposed.

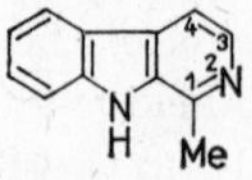

22 Harman
23 Harmalan 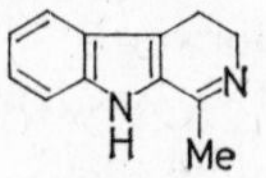
24 Tetrahydroharman = Eleagnine 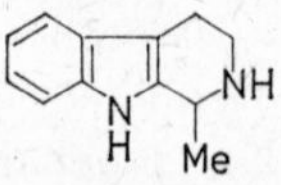

Fig. 15.4
Basic ring systems of β-carboline alkaloids

About 60 alkaloids of this group are known. They mainly occur in higher plants from 25 plant families, most of which are taxonomically unrelated. Especially rich in β-carbolines are *Apocynaceae, Eleagnaceae, Gramineae, Leguminosae, Malpighiaceae, Passifloraceae, Rubiaceae* and *Zygophyllaceae*. While in many species only one or two alkaloids are present, the following contain more than four β-carbolines: *Carex brevicollis* (*Cyperaceae*), *Eleagnus angustifolius* (*Eleagnaceae*), *Desmodium gangeticum* (*Leguminose*), *Banisteriopsis argenta, B. caapi* (*Malpighiaceae*), *Passiflora incarnata* (*Passifloraceae*) *Peganum harmala* (*Zygophyllaceae*). Ruine is 8-glucosyloxy-7-methoxy-1-methyl-β-carboline and was found in *Peganum harmala* [153]. The psychoactive properties of some South American hallucinogens are based on the content of β-carbolines. The bark of *Banisteriopsis caapi* or *B. inebrians* (*Malpighiaceae*) is used in the preparation of the famous drink "ayahuasca", "caapi" or "yajé".

Several *Virola* species (*Myristicaceae*) are the source of the native hallucinogenic snuff "epena" or "yakee" and contain, in addition to simple tryptamine derivatives, tetrahydro-β-carbolines which act as monoamine oxidase inhibitors [183]. 2-Methyl- and 1,2-dimethyl-6-methoxy-tetrahydro-β-carboline were first detected in *Virola* species [4].

The formation of β-carbolines by pyrolysis of tryptophan, *e.g.* tobacco smoke, is well documented [172]. Mammalian tissues also contain 1,2,3,4-tetrahydro-β-carbolines [56, 140]. A comprehensive review on β-carbolines has been published recently [8].

The incorporation of tryptophan into the indolyl part of simple β-carbolines has been demonstrated using different plants. Root cultures [75] of *Peganum harmala* gave radioactive labelled harmine (**26**) and harmol when [β-^{14}C]tryptophan was fed. Similar results with *Peganum harmala* were obtained by Liljegren [125]. Administration of [α-^{14}C]tryptophan to *Eleagnus angustifolia* specifically labelled C-3 of eleagnine (**24**) (tetrahydroharman) [156a].

Studies with *Carex brevicollis* [110] and *Passiflora edulis* [193] have provided evidence for the biosynthetic route from tryptophan to β-carboline alkaloids. It was shown that the alanine side-chain of tryptophan (**6**), except the carboxyl group, is incorporated into harmine (**26**) after administration of [β-^{14}C, $^{15}N_\beta$]tryptophan. Essentially the same results were obtained with [β-^{14}C, $^{15}N_\beta$]tryptamine [200]. [Me-^{14}C]methionine gives the methyl group of the OMe moiety in harmine [200].

The remaining C_2-unit to complete the β-carboline skeleton may be derived from acetate or pyruvate (**27**) as demonstrated for eleagnine (**24**) [155, 156] and harmine (**26**) [200]. Pyruvate is more plausible as immediate precursor because it was incorporated into harmine with much less scatter than that observed with acetate [200] (Fig. 15.5). The final proof comes from experiments with 1-methyl-1,2,3,4-

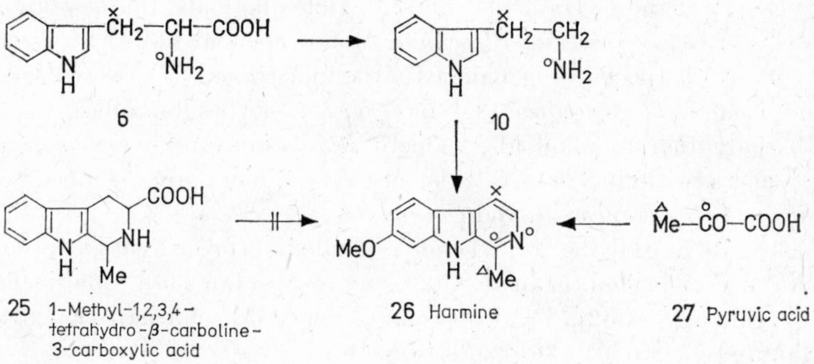

6

10

25 1-Methyl-1,2,3,4-tetrahydro-β-carboline-3-carboxylic acid

26 Harmine

27 Pyruvic acid

Fig. 15.5
Biosynthesis of harmine in *Peganum harmala*

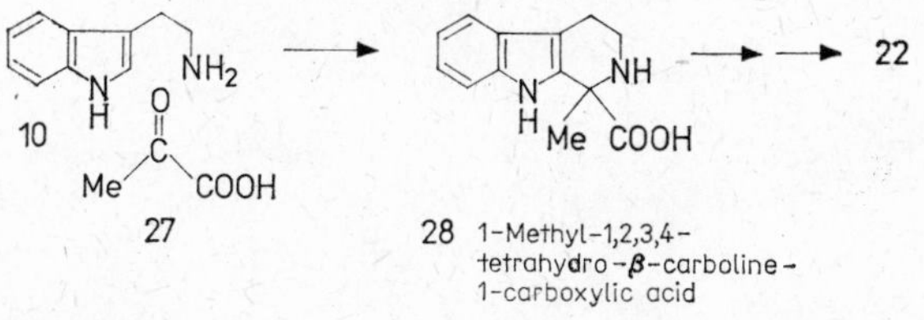

10

27

28 1-Methyl-1,2,3,4-tetrahydro-β-carboline-1-carboxylic acid

22

Fig. 15.6
Biosynthesis of harman in *Passiflora edulis*

tetrahydro-β-carboline-1-carboxylic acid (**28**), which was very efficiently transformed into harman (**22**) (Fig. 15.6) in *Passiflora edulis* and into eleagnine in *Eleagnus angustifolia*; it is also a natural constituent of *Passiflora edulis* [93, 93a].

An alternative pathway leading to harman in *P. edulis* was proposed based on feeding and trapping experiments [144, 193]. Thus N-acetylation of tryptamine is followed by cyclo-dehydration to 3,4-dihydro-β-carbolines (*e.g.* harmalan) (**23**) which in turn is oxidized to harman (**22**). However, this pathway is not operative in the formation of tetrahydroharman in *Eleagnus*. So far most results favour the assumption that Schiff base intermediates are necessary in β-carboline alkaloid biosynthesis.

15.3 Monoterpenoid Indole Alkaloids

The indole bases of this group contain two structural elements: tryptamine (**10**) with the indole nucleus and a C_9- or C_{10} unit derived from secologanin (**44**). Approximately 1100 indole alkaloids of this type are known. The overwhelming majority of monoterpenoid indole alkaloids are restricted to three plant families: *Loganiaceae*, *Apocynaceae*, and *Rubiaceae*, which belong to the order *Gentianales* (*Contortae*). Reviews on the distribution of indole alkaloids in the plant kingdom are available [165, 224]. Especially useful with regard to chemotaxonomical problems are the sophisticated contributions by Kisakürek and Hesse [107, 108].

Four genera of the *Loganiaceae* contain indole alkaloids, the best known of which are species of *Gelsemium* and *Strychnos*. Monoterpenoid indole bases occur only in one subfamily of *Apocynaceae*, namely in the four tribes of *Plumerioideae*: *Carrisseae*, *Tabernaemontaneae*, *Alstonieae* and *Rauwolfieae*. Two subfamilies of the *Rubiaceae* are especially rich in alkaloids, namely *Rubioideae* and *Cinchonoideae* to which belong genera with indole alkaloids. Most prominent are members of *Cephaelis*, *Mitragyna*, *Adina*, *Cinchona*, and *Corynanthe*.

Recently [107, 108] the iridoid indole alkaloids were classified in eight different skeleton types: corynanthean (C)- (**29**), vincosan (D)- (**30**), vallesiachotaman (V)- (**31**), strychnan (S)- (**32**), aspidospermatan (A)- (**33**), eburnan (E)- (**34**), plumeran (P)- (**35**), and ibogan (I)- (**36**) types (Fig. 15.7).

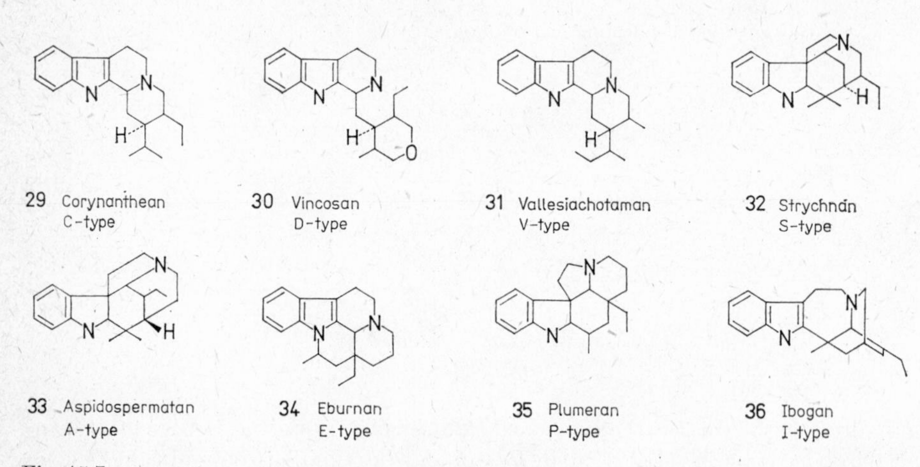

Fig. 15.7
Various skeletal types of monoterpenoid indole alkaloids

The C-, D-, V-, S-, and A-types possess a non-rearranged secologanin moiety and the E-, P- and I-types a rearranged secologanin unit [107]. The latter three more "complex" types occur exclusively in *Plumerioideae*, a subfamily of *Apocynaceae*. The C-, D-, and V-type indole alkaloids have been found in three plant families. The strychnan (S)- and aspidospermatan (A)-type alkaloids occur only in *Apo-*

cynaceae and *Loganiaceae*. Chemotaxonomically *Rubiaceae* are comparable with *Apocynaceae* with regard to the "evolutionary stage" and both should be placed higher than the *Loganiaceae* [108].

15.3.1 Early Work on Monterpene Indole Alkaloid Biosynthesis

There is no other group of alkaloids whose biosynthesis has been so intensively studied during the last two decades than the monoterpenoid indole bases. Until 1975 this was achieved by the use of radio-tracers administered to whole differentiated plants at various stages of development. Most of the experimental work was carried out with *Catharanthus roseus* plants and seedlings. With the aid of these "classical methods" the basic building blocks leading to the bewildering variety of structures in complex indole alkaloids were clarified. The first insight into the stereoselectivity of indole alkaloid interconversion was gained: These *in vivo* feeding experiments paved the way for more crucial investigations with cell-free preparations. This type of experimental approach based on enzymes and their enzymatically produced intermediates is well suited for the elucidation of the ultimate pathways and reaction mechanisms in complex indole alkaloid biosynthesis [202, 227] (*vide infra*).

For the historical background and the earlier work on indole alkaloid biosynthesis the reader is referred to comprehensive reviews [49, 162, 184].

The incorporation of tryptophan (**6**) into all three major classes of indole alkaloids was shown by Leete's group [119, 123, 225]. [α-^{14}C]Tryptophan specifically labelled reserpine and serpentine in *Rauwolfia serpentina* and vindoline (**47**) in *Catharanthus roseus*. The label from [β-^{14}C]tryptophan was located at the predicted site in ibogaine.

Recently [122] *D,L*-[2-^{14}C, α-.β-$^{13}C_2$]tryptophan was fed to *Catharanthus roseus*. The specific incorporation deduced from the intensity of the satellites relative to the singlet peaks of the contiguous ^{13}C atoms located at C-5 and C-6 of vindoline was the same as that of ^{14}C located in the indole nucleus. Thus the intact incorporation of tryptophan into **47** is established. [Ar-^{3}H]tryptamine was also a specific precursor for some indole bases [21, 115].

The biosynthetic origin of the remaining nine or ten skeletal carbon atoms remained an intriguing question for a long time. As suggested in the early 1960s [215, 221] this "C_9-C_{10}" unit is derived by fission of the five-membered ring of a cyclopentanoid monoterpene. Mevalonic acid (**37**) and geraniol (**38**) are intermediates on the route to the cyclopentane system (**39**) [15]. The fission of this ring system with rotation around a single bond gives the *Corynanthe* type (**40**) and further rearrangements should generate the *Aspidosperma* (**41**) and *Iboga* skeletons (**42**) (Fig. 15.8). The ingenious experimental work to test this hypothesis was essentialy done by the groups of Arigoni, Battersby and Scott [49]. The evidence rests mainly upon the administration of "mevalonic acids" labelled at various positions or geraniol preparations. Unambiguous degradation procedures revealed that the sites of labelling in alkaloids of the three main types are as outlined in Fig. 15.8.

It is remarkable that during formation of *Corynanthe-*, *Iboga-* and *Aspidosperma* type alkaloids an equilibration of C-2 and C-6 of one mevalonate unit of the monoterpene moiety was observed. The first proof of such an incorporation pattern of mevalonic acid into a terpenoid compound was found by Schmid's group with regard to the iridoid plumieride [226]. The search for the appropriate cyclopentane system (**39**) involved in indole alkaloid biosynthesis was stimulated by the isolation

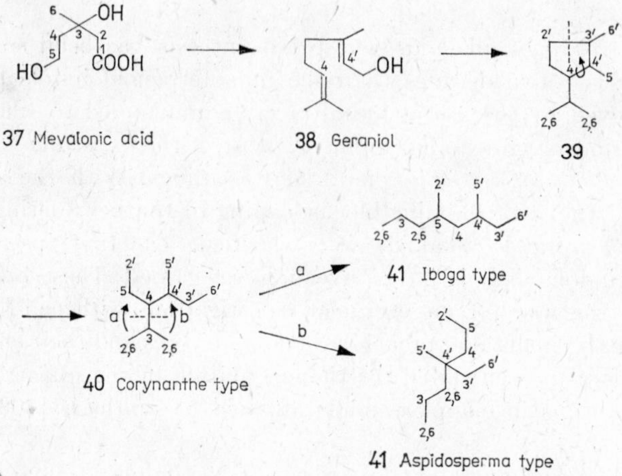

37 Mevalonic acid

38 Geraniol

39

40 Corynanthe type

41 Iboga type

41 Aspidosperma type

Fig. 15.8
Origin of the C_9–C_{10} unit in indole alkaloid biosynthesis

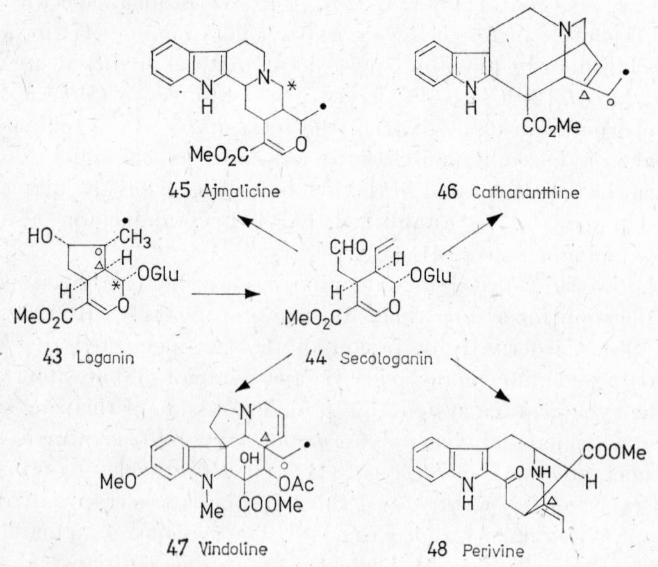

Fig. 15.9
Incorporation of secologanin into different types of monoterpenoid indole alkaloids

of ipecoside (**49**) [24]. Actually four iridoid glucosides were tested as potential candidates: verbenalin, genepin, monotropeine and loganin [17]. Loganin (**43**) proved to be the essential compound in indole alkaloid biosynthesis. This iridoid is transformed to secologanin (**44**), which in turn is also specifically incorporated into various types of indole bases (Fig. 15.9) [20, 128].

Some nitrogenous glycosides that contain a tryptamine and secologanin unit were found in the late 1960s, *e.g.* cordifoline and adifoline.

Most important were the findings of strictosidine [194] (**50**) and the C-3 epimers of **50** vincoside (**51**) and isovincoside (**52**) [21]. At first there was much controversy about the correct stereochemistry of C-3 of both glycosides. Now it is well established that vincoside (**51**) bears a hydrogen in β-configuration at C-3. In earlier experiments [19, 21] specific incorporation of only vincoside into the three major classes of indole alkaloids was found. Therefore an inversion should occur at the carbon during the latter stages of biosynthesis with the C-3 hydrogen atom in α-configuration, but *vide infra* (Fig. 15.10)!

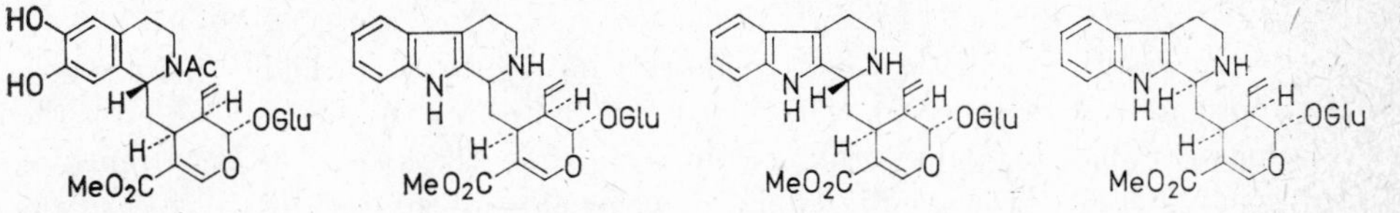

49 Ipecoside   50 Strictosidine   51 Vincoside   52 Isovincoside

Fig. 15.10
Various nitrogenous glucosides

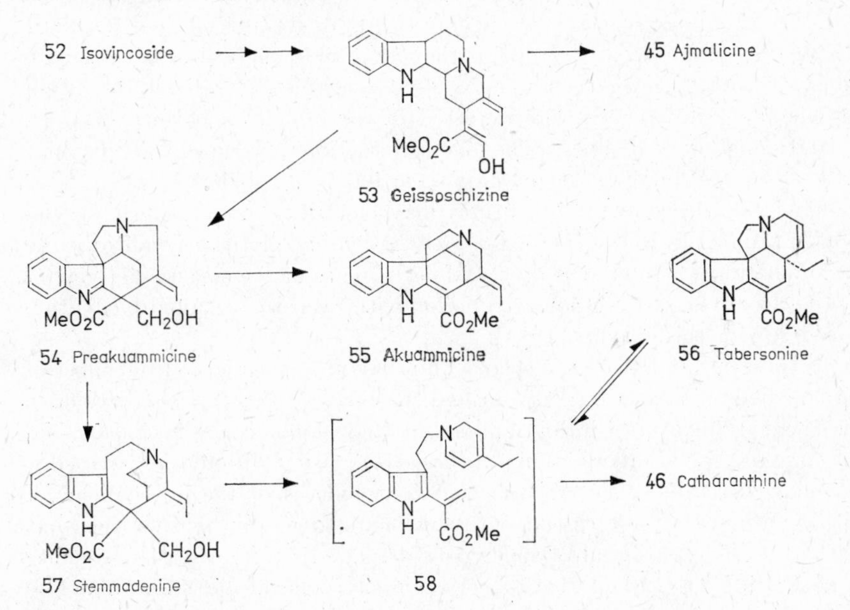

Fig. 15.11
Possible interconversions of complex indole alkaloids

Many efforts have been made to determine the stages beyond "vincoside" leading to the rearranged skeletons, mainly using convential techniques. The *in vivo* transformations of differentially labelled alkaloids were followed and the sequential formation of alkaloids in germinating *C. roseus* seeds were analysed [184, 185].

Chemical *in vitro* transformations of alkaloids served to explain some mechanistic aspects of the observed rearrangements.

Summarising all the results the following pathway (Fig. 15.11) seems to be established: *Corynanthe* → *Corynanthe/Strychnos* → *Aspidosperma* → *Iboga* type alkaloids [191]. But one should keep in mind that the ultimate evidence for most of the transformations and rearrangements of monoterpenoid indole alkaloids in nature is still not available because not all enzymes responsible for those reactions have been detected.

15.3.2 Biosynthetic Routes to the Precursors of Monoterpenoid Indole Alkaloids

The biosynthesis of aromatic amino acids and its regulation in higher plants have been thoroughly reviewed [72]. Surprisingly there are few studies available on the tryptophan branch in plant forming complex indole alkaloids. The occurrence of tryptophan decarboxylase activity in homogenates of *Catharanthus roseus* was reported by Scott and Lee [187]. This particular enzyme was purified to apparent homogeneity and represents a soluble protein with a molecular weight of 115000 Da [153a]. Cell lines resistant to 5-methyltryptophan were selected from wild-type cells of different *C. roseus* suspension cultures [179]. Resistant cell lines accumulated up to 30 times more free tryptophan than the wild-type cells. Anthranilate synthetase in normal *C. roseus* cell lines is very sensitive to feedback inhibition by *L*-tryptophan. No differences were detected in the inhibition pattern between wild type and resistant cell lines. The over-production of tryptophan did not stimulate the accumulation of either tryptamine or indole alkaloids.

Indole alkaloid synthesis in tissue cultures may be enhanced by using a special "production" medium. In a cell line of *C. roseus* grown under these conditions the alkaloid accumulation was preceded by a 12-fold increase of the specific activity of tryptophan decarboxylase. A regulatory function with regard to alkaloid biosynthesis was considered for this enzyme [108a].

Some steps from mevalonic acid to secologanin have been elucidated but most of the mechanistic details have still to be clarified. Reviews on this topic are available [129, 202]. The steric course of iridoid and secoiridoid monoterpene formation [50] as well as complex indole alkaloid biosynthesis [23] using differentially labelled mevalonic acids were investigated. One may conclude that the mechanism of isoprenoid biosynthesis in these cases from mevalonate to geranyl pyrophosphate is similar to that of cholesterol and other terpenoids. The first step of this sequence is the hydroxylation of C-10 of the isomeric monoterpene alcohols geraniol (**38**) and nerol (**59**) to **60** and **61**, which are precursors of loganin and indole alkaloids [18, 55].

An O_2, NADPH and cytochrome P_{450} dependent mono-oxygenase that oxidizes **38** and **59** has been described [132, 143, 146, 198a]. This enzyme was found only in *Catharanthus roseus* plants and tissue cultures and in other indole alkaloids as well as monoterpenoid glucosides producing plants. Apparently, this monoterpene hydroxylase is specific for the pathway leading to secologanin. Catharanthine inhibits this membrane-bound cytochrome P_{450} dependent mono-oxygenase in *C. roseus* [143], indicating a feedback control of indole alkaloid biosynthesis. 10-Hydroxycitronellol, citronellal and 10-hydroxylinalool are not incorporated into **43** proving that 2,3-reduction of geraniol does not occur before loganin synthesis. The route from **60** and **61** to the cyclopenta [c]pyran-system in loganin is still being investigated. A trialdehyde (**65**) was proposed [55] as a potential intermediate that could be cyclized by a Michael-type condensation leading to **66** and **67**, which are in equilibrium [27, 28]. Some further enzyme activities of a more non-specific nature have been found in *C. roseus* cell-free extracts [129]. They catalysed the oxidation of **38** and **59** to geranial and neral, respectively, in the presence of oxidized pyridine nucleotides. Feeding experiments using mixtures of **62a/b**, **63a/b** and **64a/b** gave good incorporations into loganin (**43**), catharanthine (**46**) and vindoline (**47**) [27, 28]. This might indicate that the pathway outlined in Fig. 15.12 is operative in *C. roseus*.

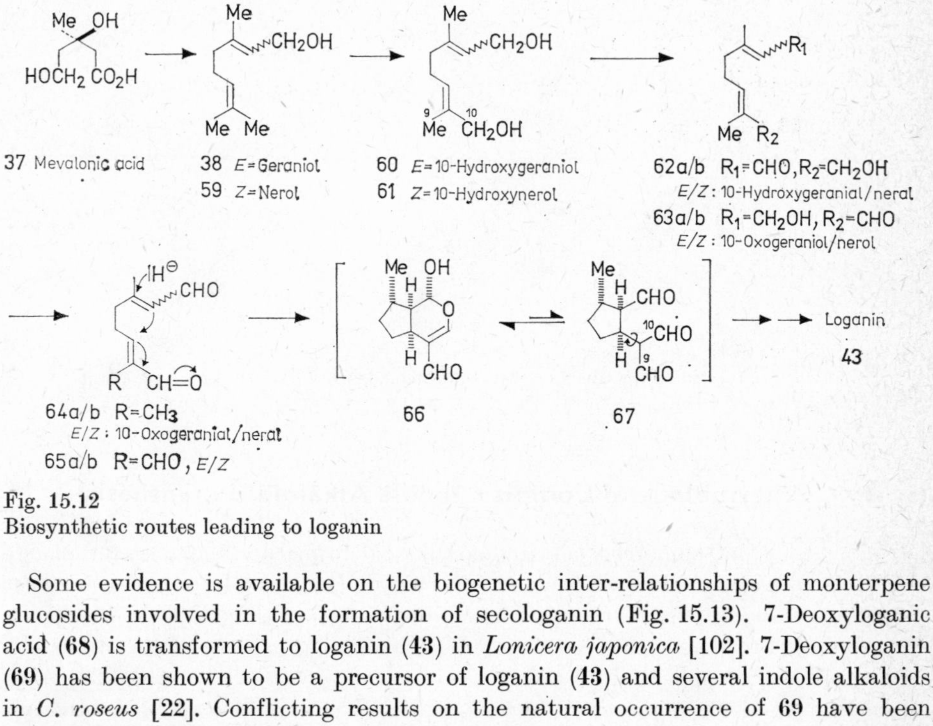

Me OH
HOCH2 CO2H
37 Mevalonic acid
Me
CH2OH
Me Me
38 E=Geraniol
59 Z=Nerol
Me
CH2OH
9 10
Me CH2OH
60 E=10-Hydroxygeraniol
61 Z=10-Hydroxynerol
R1
Me R2
62a/b R1=CHO,R2=CH2OH
E/Z: 10-Hydroxygeranial/neral
63a/b R1=CH2OH, R2=CHO
E/Z: 10-Oxogeraniol/nerol
H⊖
CHO
R CH=O
64a/b R=CH3
E/Z: 10-Oxogeranial/neral
65a/b R=CHO, E/Z
Me OH
H
O
H
CHO
66
Me
H
CHO
10 CHO
H 9
CHO
67
Loganin
43

Fig. 15.12
Biosynthetic routes leading to loganin

Some evidence is available on the biogenetic inter-relationships of monterpene glucosides involved in the formation of secologanin (Fig. 15.13). 7-Deoxyloganic acid (**68**) is transformed to loganin (**43**) in *Lonicera japonica* [102]. 7-Deoxyloganin (**69**) has been shown to be a precursor of loganin (**43**) and several indole alkaloids in *C. roseus* [22]. Conflicting results on the natural occurrence of **69** have been reported [22, 129] and further studies are needed to clarify this point. Considerable work in this area has been done by Coscia's group [85, 86, 130, 131].

In *Catharanthus roseus* they found loganic acid (**70**) and secologanic acid (**71**), which were methylated *in vivo* giving loganin (**43**) and secologanin (**44**) respectively. The reverse processes were also demonstrated, *e.g.* the conversion of **43** to **70**. An enzyme was found in seedlings of *C. roseus* catalysing the O-methylation of **70** and **71**. In the presence of dithiothreitol the S-adenosyl-*L*-methionine: loganic acid methyltransferase showed enhanced activity. The methyltransferase activity is drastically reduced by sulphhydryl group inhibitors. Geniposidic acid and 7-deoxy-loganic acid (**68**) were not converted to the corresponding methyl esters. Apparently hydroxylation at the C-7 position of iridoids precedes carboxyl group methylation. The ring cleavage mechanism leading to the seco-iridoid compounds **44** and **71** is not completely understood. 10-Hydroxyloganin (**72**) and the corresponding 7-epi derivative are not precursors in secologanin biosynthesis [27, 28].

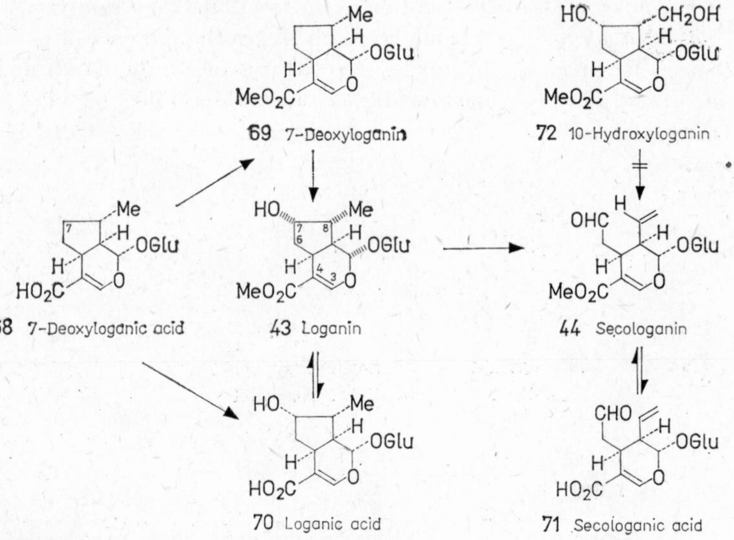

69 7-Deoxyloganin

72 10-Hydroxyloganin

68 7-Deoxyloganic acid

43 Loganin

44 Secologanin

70 Loganic acid

71 Secologanic acid

Fig. 15.13
Inter-relationships of monoterpene glucosides involved in the formation of secologanin

15.3.3 Enzymology of Complex Indole Alkaloid Biosynthesis

The enzymology of monoterpenoid indole alkaloid formation and relevant *in vivo* studies on this topic since 1975 will be discussed. Remarkable progress has been made due to the use of cell-free systems mainly from *C. roseus* suspension cultures. The screening procedure for the selection of high-yielding cell lines was considerably facilitated by the application of radioimmunoassay methods (RIA) [10, 228]. Most information available at present is on enzymes involved in indole alkaloid formation of the *Corynanthe*-type. Several useful reviews relating to enzymatic approaches in indole alkaloid biosynthesis have been published [188, 202, 227].

The first cell-free system from seedlings and callus cultures of *C. roseus* was isolated in Scott's laboratory [187]. This crude enzyme preparation catalysed the formation of ajmalicine (**45**) and geissoschizine (**53**) from tryptamine (**10**) and secologanin (**44**) in the presence of NADPH, thiols and Tris buffer. Geissoschizine

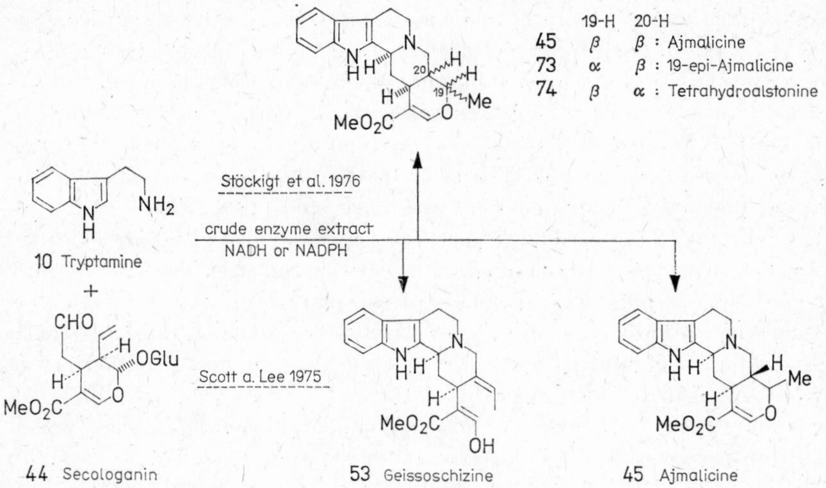

	19-H	20-H	
45	β	β	Ajmalicine
73	α	β	19-epi-Ajmalicine
74	β	α	Tetrahydroalstonine

Stöckigt et al. 1976
crude enzyme extract
NADH or NADPH
Scott a. Lee 1975
10 Tryptamine
44 Secologanin
53 Geissoschizine
45 Ajmalicine

Fig. 15.14
First enzyme-catalysed reactions starting from tryptamine and secologanin leading to *Corynanthe*-type alkaloids

(**53**) was converted into ajmalicine and several unknown metabolites. Subsequently Zenk's group [208, 216] using cell-free extracts from cell suspension cultures of *C. roseus* demonstrated the formation of ajmalicine (**45**), 19-epiajmalicine (**73**) and tetrahydroalstonine (**74**). They found that the ajmalicine synthesizing complex is strongly dependent on the presence of a reduced pyridine nucleotide (NADH or NADPH) in the incubation mixture. The overall enzymology of this reaction (Fig. 15.14) was determined with regard to pH optimum, time course, substrate specificities, Km values and inhibitors. Among the inhibitors tested, δ-*D*-gluconolactone, a known inhibitor of β-glucosidases was found to be very effective.

Considerable efforts were made to elucidate in more detail the pathway to ajmalicine (**45**) starting from **10** and **44**. New intermediates were found as well as specific enzymes catalysing their formation.

The first pathway specific intermediate, which occupies a central position in indole alkaloid biosynthesis, is 3α(*S*)-strictosidine (isovincoside) (**52**) [209, 210]. Using cell-free extracts from *Catharanthus*, *Rhazia stricta* cells and other cell suspensions of *Apocynaceae* plants exclusively **52** is formed from tryptamine and secologanin in the presence of δ-*D*-gluconolactone. These results clearly rule out the participation of the 3β(*R*)-isomer of **52**, called vincoside (**51**), in alkaloid formation. The strictosidine synthase catalyses a Pictet-Spengler-type reaction between the aldehyde function of secologanin and the amino group of **10**. This enzyme was purified about 50-fold to near homogeneity [218]. The apparent molecular weight of the enzyme was 34000. The pH optimum was 6.8, apparent Km values for tryptamine and secologanin were 2.3 mM and 3.4 mM respectively. The same enzyme was purified 740-fold to homogeneity and characterized [148]. Mizumaki et al. claimed that end-products in the biosynthetic pathway of indole alkaloids such as ajmalicine, vindoline and catharanthine do not inhibit the synthase activity. The enzyme was detected in suspension cultures of 15 different species belonging to 9 different genera of *Plumerioideae*, a subfamily of *Apocynaceae* [217, 218]. Strictosidine synthase from *C. roseus* cell cultures was partially purified and immobilized. [164f]. Apparently the next step in *Corynanthe*-type alkaloid biosynthesis is the removal of the glucose moiety of **52**. The participation of two non-specific glucosidases in an ajmalicine synthetase complex was suggested [189]. Findings in 1980 [92] led to the discovery of two highly specific glucoalkaloid β-glucosidases called strictosidine-β-*D*-glucosidase I and II, disproving earlier proposals. The strictosidine-specific glucosidases are present in members of the plant family *Apocynaceae* but not in plants devoid of indole alkaloids. They catalyse the removal of glucose from strictosidine leading to highly reactive intermediates that might be formed in the sequence: **52** $\rightarrow$ **75** $\rightarrow$ **76** $\rightarrow$ **77** (Fig. 15.15).

The paramount role of strictosidine (**52**) in alkaloid biosynthesis was further substantiated by *in vivo* studies [45, 151]. **52** gives rise to *Corynanthe* (3α- and 3β)

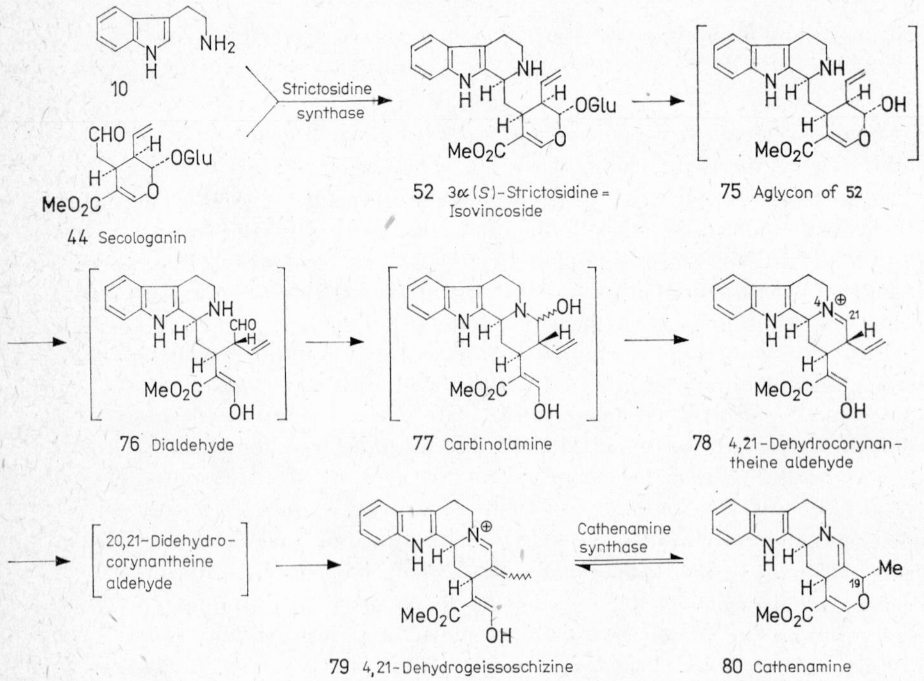

Fig. 15.15
Possible biogenetic pathway leading to cathenamine

series, *Iboga*, and *Aspidosperma* type alkaloids as well as strychnine, gelsemine and probably quinine (**108**). The conversion of strictosidine into 3β-alkaloids proceeds with loss of hydrogen at C-3, while it is retained in the formation of the 3α-series.

Indirect evidence was provided that 4,21-dehydrocorynantheine aldehyde (**78**) is involved in ajmalicine biosynthesis. In the presence of $NaBH_4$ in the incubation mixture enzyme preparations convert **52** to sitsirikine and 16-episitsirikine, which are directly derived from **78**.

The isolation of a rare indole alkaloid, 4,21-dehydrogeissoschizine (**79**), from *Guettarda exima* (*Rubiaceae*) and its biomimetic conversions were reported in 1979 [104]. **79** may be considered as an important branch-point in the biosynthesis of *Iboga*-, *Aspidosperma*-, and *Corynanthe*-type alkaloids. This assumption is based *inter alia* on the *in vitro* conversion of 4,21-dehydrogeissoschizine (**79**) in the presence of NADPH into ajmalicine (**45**) and its isomers **73** and **74**. Omission of NADPH led to the formation of cathenamine (**80**) in 74% yield. Some ingenious experiments proved that **79** is indeed an obligatory intermediate on the route from **10** and **44** to heteroyohimbine alkaloids [177]. The enzyme responsible for the conversion of **79** → **80** was named cathenamine synthase. Previously [204, 208] cathenamine (**80**) was a major product when crude cell-free preparations of *C. roseus* were incubated using **10** and **44** as substrates in the absence of either NADH or NADPH. Cathenamine was transformed in the presence of reduced pyridine nucleotides using cell-free extracts of *C. roseus* into ajmalicine (**45**) and its isomers (**73**, **74**). These experiments proved that **80** is a natural intermediate on the route to ajmalicine and that stereochemical control may occur at the cathenamine stage. The latter assumption is further supported by the fact that ajmalicine is not transformed to **73** or **74**.

An interconversion of the enamine and iminium forms of cathenamine was achieved by chemical methods. Both compounds (**80**, **82**) were also enzymatically formed using cell-free extracts of *Rauwolfia verticillata* [91]. The hitherto unknown 19-epi-cathenamine, the 19α-H(*R*) isomer of **80**, was isolated in addition to cathenamine after incubation of strictosidine with β-glucosidase [44]. Nothing is known about the enzymes catalysing the penultimate steps from **82** and **83** leading to ajmalicine/tetrahydroalstonine and 19-epi-ajmalicine respectively (Fig. 15.16).

The role of geissoschizine (**53**) as direct precursor of ajmalicine (**45**) is still controversial. In a time-dependent study [118, 188] a mixture of [aryl-^{3}H]geissoschizine, [2-^{14}C]tryptamine, and secologanin was incubated. The results show an increasing incorporation of tryptamine into geissoschizine and the latter into ajmalicine. Tetrahydroalstonine was also found to be doubly labelled, but only traces of 19-epi-ajmalicine were detectable. Furthermore [^{14}C]tryptamine was incorporated into **45**, **73** and **74** to the extent of 11.9%, 0.35% and 4.5% respectively, whilst [^{3}H,^{14}C]-cathenamine gave 3.8%, 7.1% and 1.3%. Tryptamine and geissoschizine form ajmalicine and its isomers in a "natural" ratio, while cathenamine behaves "abnormally" in forming 19-epi-ajmalicine, which has never been reported as a natural alkaloid of *C. roseus* [188]. Scott et al. seem to favour a stepwise mechanism of ajmalicine formation from **53**. It involves proton attack at C-20 on the **si** face (→ ajmalicine) or at the **re** face (tetrahydroalstonine) (Fig. 15.17).

The same problem was investigated using enzyme preparations synthesizing as

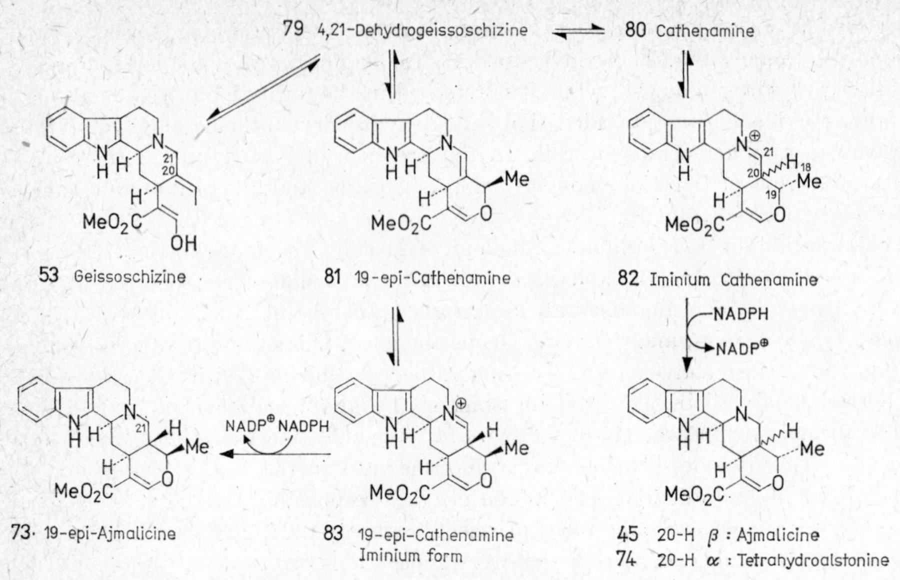

Fig. 15.16
Late steps in the formation of 3-α heteroyohimbine alkaloids [202, 227]

main product, besides **45** and **74**, 19-epi-ajmalicine [203]. The following experiments were the most conclusive:

a) By incubation of **53** in the presence of NADPD 19-epi-ajmalacine (**73**) was formed possessing one D-Atom at the α-position of C-21;
b) Geissoschizine (**53**) labelled with deuterium at C-21-α gave rise after incubation to 19-epi-ajmalicine without the label.

These results are only compatible with a pathway leading *via* 4,21-dehydrogeissoschizine to **73** as outlined in Fig. 15.16 and proposed earlier [201]. The conversion of **53** → **79** is catalysed by geissoschizine dehydrogenase in a $NADP^+$-dependent reaction. This enzyme shows an extremely low specific activity and plays no crucial role in the biosynthesis of heteroyohimbine-type alkaloids [164a]. 17-OH-19-epi-Cathenamine, which is converted under biomimetic conditions to **73**, is not involved in the biosynthetic formation of **73** but undergoes enzymatic reduction to 17-OH-19-epi-ajmalicine [206].

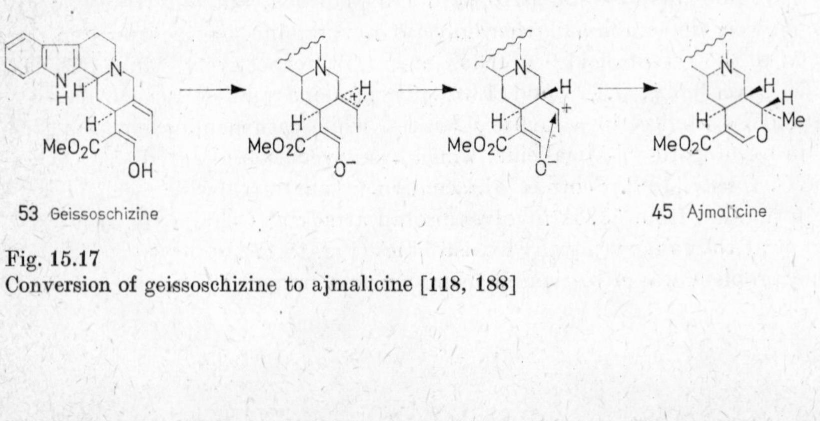

Fig. 15.17
Conversion of geissoschizine to ajmalicine [118, 188]

A further enzymatic conversion of geissoschizine (**53**) was reported for cell-free extracts of *C. roseus*, which reduced its C-16 double bond forming (16*R*)-isositsirikine [95]. The latter alkaloid might be involved in the biosynthesis of polyneuridine and akuammidine.

Most interesting indole alkaloids with a non-rearranged secologanin skeleton are members of the sarpagine/ajmaline group. Especially rich in these indole alkaloids are *Rauwolfia* species. Apparently in the first reactions of this pathway [164 b—e, 202a, 205a] (**79**) is converted into 4,5-dehydrogeissoschizine which in turn is transformed to polyneuridine aldehyde bearing the characteristic C-5—C-16 bond. A highly specific polyneuridine aldehyde esterase removes a C_1-unit from the latter compound ultimately giving 16-epivellosimine. This alkaloid may be regarded as a branch-point leading to the sarpagan-route and alternatively to the ajmalan-route. 16-Epivellosimine is readily isomerized to vellosimine which serves as substrate for the NADPH-specific vellosimine reductase producing 10-deoxysarpagine. Apparently in the final reaction leading to sarpagine a hydroxyl group is introduced at C-10.

In the presence of *Rauwolfia* enzymes and acetyl-CoA 16-epivellosimine is converted to vinorine. Thus the ajmalan skeleton is formed by a stereochemically controlled addition of C-17 at the nucleophilic C-7 carbon giving vinorine with a β-orientated acetoxy group. The next step should be a not yet clarified hydroxylation giving vomilenine. The latter one is converted in the presence of an enzyme preparation from *Rauwolfia* cells and NADPH/S-adenosylmethionine as cofactors into ajmaline.

15.3.4 Formation of Dimeric Indole Alkaloids

A smaller group of indole alkaloids are composed of several monomeric units. The so-called bis-indole alkaloids are formed from monomeric units that may be identical (*e.g.* toxiferine I) or not (*e.g.* geissospermine). In tubulosine a tryptamine unit is combined with a skeleton typical of *Ipecacuanha* bases. Unique in alkaloid chemistry are the quadrigemines A and B [163]. Both alkaloids are constructed of four N_b-methyltryptamine units. For the older literature the reader is referred to [71].

Many efforts were made to clarify the biosynthetic pathway of the antitumour agents of *C. roseus*, namely vinblastine (VLB) (**84**) and vincristine (VLC) (**85**) (Fig. 15.18). Catharanthine (**46**) and vindoline (**47**), the two major alkaloids of *C. roseus*, have long been assumed to be potential precursors of dimeric alkaloids. Several attempts to demonstrate significant incorporation of labelled **46** and **47** into VLB ended with negative results or with extremely low incorporation [98, 182]. However, [OMe-^{3}H]catharanthine and [O-Ac-^{14}C]vindoline fed in a mixture to apical cuttings of *C. roseus* regiospecifically label VLB despite low incorporation [89]. Based on the outstanding synthetic work in this field [173] anhydrovinblastine (**86**) is now regarded as a key intermediate in the biosynthesis of VLB and other dimeric *C. roseus* alkaloids. **86** gave excellent incorporation into leurosine (**87**) in 6-week-old *C. roseus* cuttings [186] and in cell-free extracts of *C. roseus* [211, 212]. The same reaction was also catalysed by horseradish peroxidase. 3′,4′-Anhydro-VLB was transformed by crude enzyme preparations of *C. roseus* to vinblastine in a

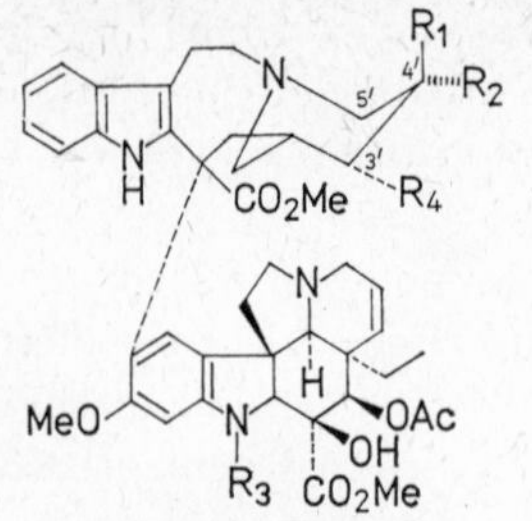

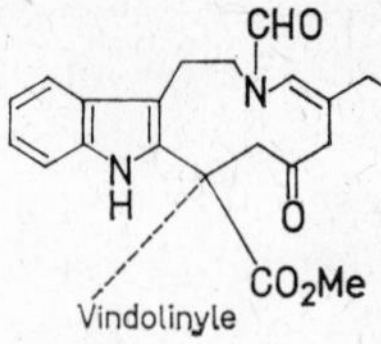

84 R_1=OH, R_2=Et, R_3=Me, R_4=H : VLB
85 R_1=OH, R_2=Et, R_3=CHO, R_4=H : VLC
86 R_1=Et, R_3=Me, R_4=H, $\Delta^{3',4'}$: Anhydro VLB
87 R_1=Et, R_2→R_4=epoxide, R_3=Me : Leurosine
88 R_1=Et, R_2=OH, R_3=Me, R_4=H : Leurosidine
89 R_1=H, R_2=Et, R_3=Me, R_4=H : Deoxy VLB
90 R_1=Et, R_2=H, R_3=Me, R_4=H : Deoxyleurosidine

91 Catherine

Fig. 15.18
Dimeric indole alkaloids of *Catharanthus roseus*

yield of 1.3 to 1.9% [31] and to leurosine and catharine (**91**). The latter alkaloid was also obtained by using leurosine as substrate [211]. Recently the biotransformation of **86** into leurosine and catharine (**91**) was reported [113a, 114] in suspension cultures of an alkaloid-lacking cell line of *C. roseus*. The basic buildingstones of VLB and related alkaloids, catharanthine (**46**) and vindoline (**47**), are efficiently incorporated into anhydrovinblastine. The total radioactivity incorporations are as high as 2.6% under *in vivo* conditions [186]. A cell-free system from *Catharanthus roseus* plants was capable to form vindoline (**47**) using tryptamine and secologanin (**44**) as substrates [114a]. Later on these results could not be confirmed by another laboratory [202b]. Cell-free extracts converted both labelled (**46**) and (**47**) into anhydrovinblastine in a range of 0.36% to 0.54% [211]. The latter compound is, in turn, incorporated and converted to the natural alkaloids leurosine (**87**), catharine (**91**) and vinblastine (**84**) [114a]. Leurosine (**87**) is no precursor of vinblastine (**84**) [188], indicating that there is no reductive ring opening of the 3′,4′-epoxide leading to **84**. [Aryl-^{3}H]deoxyleurosidine is incorporated into vinblastine by intact plants [87]. It is feasible that a Polonovski-type elimination of the *N*-oxide of **90** could lead to an enamine which is then oxidized to **84**.

The pathway from **46** and **47** to vinblastine (**84**) is still only partially understood. One intriguing question is which steps need specific enzymes and which occur by spontaneous chemical reactions or non-specific enzymes. Certainly further enzymatic work is needed in this fascinating area.

15.3.5 Alkaloids Related to Monoterpene Indole Alkaloids

In apparicine (**94**) and uleine (**95**) the two-carbon side-chain of tryptamine is absent. Uleine completely lacks the "tryptamine" bridge and apparicine is characterized by having only one carbon atom between the indole nucleus and N_b. Obviously both alkaloids contain a rearranged skeleton of the akuammicine type. The biosynthetic work so far reported was done by Kutney's group [112]. Tryptophan serves as precursor of the indole part including its β-C-atom, but with the loss of the α-carbon of the side-chain. Stemmadenine (**57**) was found to act as an advanced intermediate for **94**. This implies that the loss of the original α-atom of tryptophan (**6**) occurs at a late stage. The secodine (**92**) incorporation may proceed *via* dehydrosecodine, which is transformed into stemmadenine. The crucial step in the conversion of **57** into apparicine could be a Polonovski-type fragmentation. There was some evidence that in uleine biosynthesis tryptophan is also involved and the β-C-atom of this amino acid is retained as the N-CH_3-group of this alkaloid (Fig. 15.19).

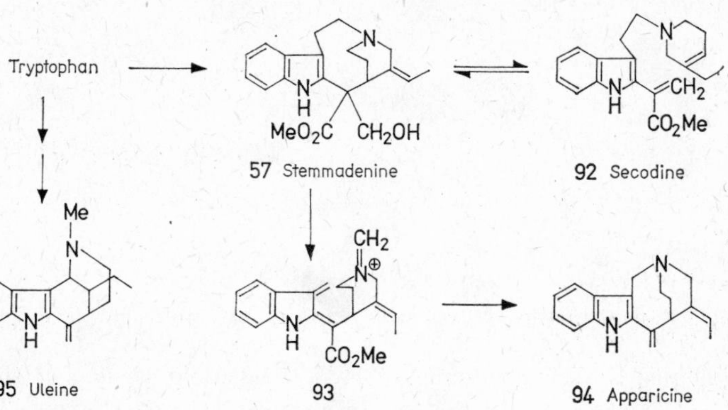

Tryptophan

57 Stemmadenine

92 Secodine

95 Uleine

93

94 Apparicine

Fig. 15.19
Possible biosynthetic route to apparicine and uleine

Camptothecin (**100**) is the first example of a new heterocyclic ring system and was isolated in 1966 from *Camptotheca acuminata* (*Nyssaceae*). It was suggested that this pyrrolo [3,4-*b*]quinoline alkaloid could be derived biogenetically from tryptophan and a monoterpene unit [222]. This assumption was supported by the incorporation of tryptophan, tryptamine, some monoterpene precursors (loganin, strictosidine (**52**)), and most important strictosamide (**96**) into camptothecin [100, 192]. A more detailed picture emerged after feeding [5-^{13}C]strictosamide. The proton noise decoupled spectrum of **100** showed clearly that only the resonance of C-5 had been significantly enhanced [101]. Surprising results were obtained using [14-^{3}H, 5-^{14}C]strictosamide (**96**) as precursor. It was expected that **96** would lose ca. one-half of its ^{3}H at carbon atom 14 during oxidative formation of the pyridone ring of **100**. The experiments showed, however, only a 5 to 9% decrease in the $^{3}H:^{14}C$ ratio. NMR studies revealed that the label was solely confined to the C-14, excluding any intramolecular ^{3}H migration. The most plausible explanation for this result is loss of hydrogen from C-14 during oxidation of the *D* ring of **96** to the pyridone ring of **100**, which might be both non-stereospecific and subject to a significant kinetic isotope effect discriminating against tritium removal [99, 101].

The role of strictosamide as penultimate precursor for **100** is now well established but biosynthetic events between **96** and camptothecin remain unclear. The mechanism of formation of the pyrrolo [3,4-*b*]quinoline ring system *in vivo* could proceed by reduction of a ketolactam (**97**) derived from strictosamide to **98** followed by ring closure to quinoline (**99**) *via* stepwise ionic or concerted electrocyclic processes [99] (Fig. 15.20).

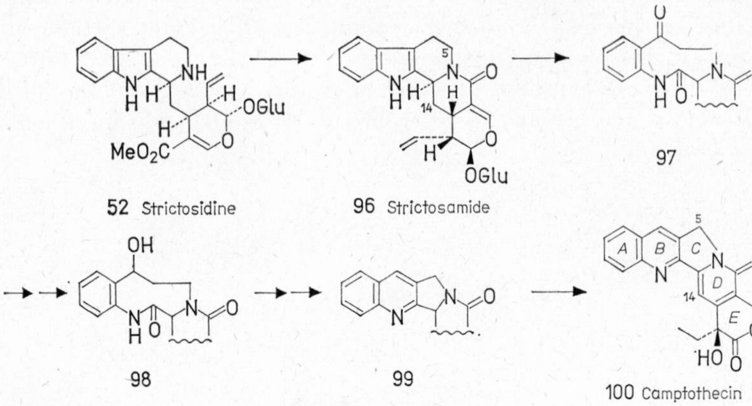

52 Strictosidine

96 Strictosamide

97

98

99

100 Camptothecin

Fig. 15.20
Proposed pathway from strictosidine to camptothecin

In 1829 just at the beginning of modern alkaloid chemistry the anti-malaria drug quinine (**108**) was isolated in a crystalline state. Quinine and related bases are quinoline derivatives found mainly in *Cinchona* and *Remijia* species of *Rubiaceae*. The minor alkaloids of the cinchonamine-type are localized in leaves and possess indole and quinuclidine moieties. It was suggested that the latter type might act as precursor for the quinoline bases.

In the 1960s the participation of tryptophan in quinine biosynthesis was demonstrated [120, 124]. [α-^{14}C]Tryptophan was incorporated into **108** labelling carbon atom 2 specifically. Feeding experiments with [1-^{15}N, 2-^{14}C]tryptophan gave identical specific incorporation for both ^{15}N and ^{14}C. Only the nitrogen in the quinoline nucleus was enriched with ^{15}N and the ^{14}C was located at C-9.

Specific incorporation of [2-^{14}C]- and [3-^{14}C]geraniol into **108** proved that the C_9 unit of quinine is of terpenoid origin [16, 124]. This was subsequently confirmed by administration of [7-^{3}H]loganin to *C. ledgeriana* plants. The isolated quinine contained all the tritium at C-8′ [25]. The glucoside sweroside also gave labelled quinine, apparently by biological conversion into secologanin [103]. A mixture of [aryl-^{3}H]-vincoside and isovincoside (**51**, **52**) led to the formation of radioactive cinchonine, chinchonidine (**107**) and quinine [26]. Heavily labelled *Cinchona* alkaloids were obtained using [6-^{14}C]strictosidine = isovincoside as precursor, but no incorporation of vincoside (**51**) was observed [151]. Corynantheal (**101**) and cinchonaminal (**102**) are key intermediates in the post-strictosidine pathway leading to *Cinchona* alkaloids [26]. Unexpectedly, corynantheine-aldehyde gave no incorporation suggesting that loss of the methoxycarbonyl group of **52** occurs at an early stage in the formation of the *Corynanthe* skeleton. A mixture of [1-^{3}H]tryptamine and [1-^{14}C]tryptamine gave results showing that 50% of the tritium at C-1 of tryptamine was lost during the alkaloid formation. Apparently the cleavage between C-5 and N-4 in corynantheal leading to cinchonaminal is a stereospecific process. The ring expansion to quinoline may proceed *via* the indolenine (**103**) and the recyclisation of **104** giving cinchonidinone (**105**). The latter is a natural product in *Cinchona*, which is converted to cinchonidine (**107**) and quinine (**108**) (Fig. 15.21). Incorporation of **105** into cinchonine was also obtained, indicating an isomerization at the 9-ketoquinoline stage. A more exact understanding of single steps from strictosidine to *Cinchona* bases, which involves intriguing skeletal rearrangements, will be greatly facilitated by use of cell-free systems.

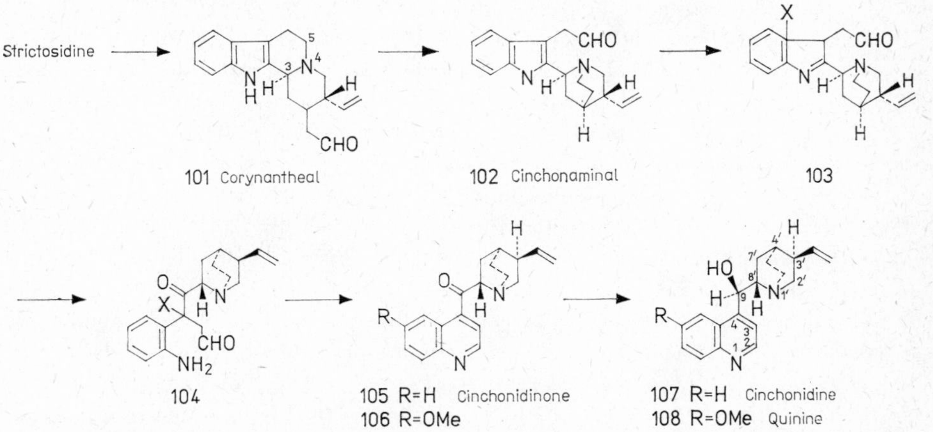

Strictosidine → 101 Corynantheal → 102 Cinchonaminal → 103 → 104 → 105 R=H Cinchonidinone, 106 R=OMe → 107 R=H Cinchonidine, 108 R=OMe Quinine

Fig. 15.21
Possible biogenetic sequence leading from strictosidine to *Chinchona* alkaloids

15.3.6 Production of Indole Alkaloids by Plant Cell Cultures

Plant cell cultures offer the fascinating possibility to produce natural compounds according to methods used in industrial microbiology. The biological and biotechnological aspects of this new field have been explored, especially during the last decade, and comprehensive reviews on this topic have been published [7, 14].

Tissue cultures of *Peganum harmala* are able to form simple β-carbolines in trace amounts [145, 175]. Reserpine was found in very low quantities in tissue cultures of *Alstonia constricta* [46]. Ajmaline and serpentine have been reported to occur in callus tissue of *Rauwolfia* [220]. The work on cell cultures of *Catharanthus roseus* until 1975 has been reviewed [47].

A major breakthrough in this area was achieved by the outstanding work of Zenk's group [228]. They developed methods to obtain substantial amounts of *Corynanthe* type alkaloids using *C. roseus* suspension cultures. Based on this procedure strains were selected with a remarkably high yield of serpentine and ajmalicine equal to 1.3% of the total cell dry weight. Different strains were obtained: S-strain, yielding solely serpentine (162 mg/l) and the A-strain, yielding 77 mg/l serpentine plus 264 mg/l ajmalicine. The alkaloid content was stimulated three-fold by the addition of *L*-tryptophan (500 mg/l) to the serpentine-producing strain. However, a variant of this particular strain behaved in a completely different way. Here alkaloid biosynthesis was severely inhibited at all concentrations of *L*-tryptophan.

Alginate-entrapped cells produce ajmalicine and its isomers from tryptamine and secologanin in a recirculated packed bed reactor [43].

Subsequently the potential of *Catharanthus* cell cultures to produce alkaloids was investigated by Zenk et al. [228] as well as in the laboratories of Reinhard [7, 14], Scott [116, 190] and by the joint efforts of two Canadian groups [113]. Several carotenoid inducers are effective in promoting alkaloid formation including catharanthine in *C. roseus* cell culture [116].

Some results of the Canadian groups should be mentioned: Out of 458 cell lines 312 accumulated alkaloids. The alkaloid production varied with the cell line and age of the subculture and ranged from 0.1 to 1.5% of cell dry weight. *Corynanthe-*, *Strychnos-*, and *Aspidosperma*-type alkaloids were found. The alkaloids occurred in a variety of combinations. Some alkaloids were detected for the first time in tissue cultures of *C. roseus, e.g.* vallesiachotomine, hörhammericine, lochnericine and some tabersonine derivatives. Of particular interest is the cell line 200 GW, which accumulates catharanthine 0.005% of cell dry weight and strictosidine lactam as major compound (60% of crude alkaloid mixture). Alkaloid profiles of callus derived from the original explants (hypocotyls) as well as callus derived from regenerated shoots are almost identical [48a]. A procedure for cryopreservation of *C. roseus* cells cultured *in vitro* has been developed [104a].

Eighteen different indole alkaloids belonging to the major types *Corynanthe*, *Strychnos*, *Aspidosperma* and *Iboga* were isolated in varying amounts from cell suspension cultures of *C. roseus* and *C. ovalis*. Especially notable is the detection of akuammigine, iso-sitsirikine and apparicine [207]. Antirhin, possessing a vallesiachotomine-like structure was found besides other alkaloids in the methanol extract of *C. roseus* tissue cultures [109a]. Cell suspension cultures of *Stemmadenia tomentosa* synthesize alkaloids belonging to the *Aspidosperma-*, *Strychnos-* and *Iboga*-type, whereas *Voacanga africana* cell cultures accumulate only Aspidosperma-type constituents [204a].

Recently cell lines of *Rauwolfia serpentina* were established producing twelve alkaloids, representatives of the ajmaline, sarpagine, yohimbine, and heteroyohimbine types [205]. The most abundant base was vomilenine (57 mg/l), which was found for the first time in *R. serpentina* cells. Vinorine, a typical constituent of *Vinca minor*, was detected for the first time in a *Rauwolfia* species. The medically important alkaloids ajmaline and reserpine occurred only in trace amounts (1.7 mg/l).

The biosynthetic capability of tissue cultures derived from plants containing monoterpene indole alkaloids is now well documented.

As in other plant species, the alkaloid pattern of cultured cells often differs remarkably from those found in the intact plant both quantitatively and qualitatively. So far about twenty alkaloids have been detected in cell suspension cultures of

Catharanthus roseus compared with approximately 90 present in the plant. Substantial amounts of catharanthine and vindoline as well as of cytotoxic dimeric indole alkaloids have not yet been found in cultured cells of *Catharanthus*. Nevertheless, recent advances in this area are promising and will certainly lead to new biotechnological applications.

15.4 Ergot Alkaloids

The fungus *Claviceps purpurea* is the oldest known producer of mycotoxins. Ergot poisoning in man, known as "holy fire" or "St. Anthony's fire", caused epidemics in many regions of central Europe during the Middle Ages. Ergotism is now practically eliminated in contrast to other mycotoxicoses. Nevertheless there is still a continuous and growing interest in these alkaloids due to their fascinating chemical properties and their importance as medicinally useful agents [62, 76, 79]. Different species of the ascomycete *Claviceps* are the main source of ergot alkaloids. The hosts for the parasitic ergot fungi are rye plants and various wild grasses. Various clavine alkaloids occur in some filamentous fungi outside the genus *Claviceps, e.g.* species of *Aspergillus, Penicillium* and *Rhizopus*. Even in one family of higher plants, some genera of the bindweed (*Convolvulaceae*), ergot alkaloids have been detected. Among them are peptide ergot alkaloids, ergosine (**118**) and its epimer ergosinine and cycloclavine representing a novel structural type of clavines, so far not found in fungi.

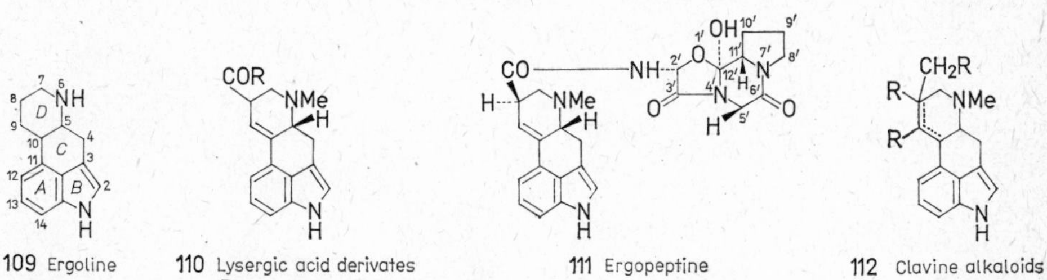

109 Ergoline

110 Lysergic acid derivates
R = Tripeptide or smaller units

111 Ergopeptine

112 Clavine alkaloids
R = H or OH

Fig. 15.22
General formulas of ergoline derivatives

The base of ergot alkaloids is the tetracyclic ergoline ring system (**109**), which is a partly hydrogenated indolo [4,3-*fg*]quinoline or in a few cases a slight modification of it. The naturally occurring alkaloids are mostly derivatives of 6,8-dimethyl-$\Delta^{8,9}$ or $\Delta^{9,10}$ ergolene, which are conventionally divided into two major classes: (a) lysergic acid derivatives (**110**) and (b) clavine alkaloids or clavines (Fig. 15.22). In the lysergic acid amides the amide portion can be a smaller peptide (Fig. 15.24) or a simpler alkylamide (Fig. 15.23). The basic skeleton of the ergot peptide alkaloids is called ergopeptine (**111**), containing the lysergic acid moiety and a peptide part

No.	Compound	R
113	Ergonovine	R = –NH–C(Me)(H)–CH$_2$OH
114	Lysergic acid amide	R = –NH$_2$
115	Lysergic acid α-hydroxyethylamide	R = –NH–CH(Me)–OH
116	$\Delta^{8,9}$-Lysergic acid or Paspalic acid ($\Delta^{9,10}$-shifted to $\Delta^{8,9}$-position)	R = –OH

Fig. 15.23
Simple lysergic acid derivatives

No.	Compound	R_1	R_2	R_3
117	Ergotamine	H	H	$-CH_2-C_6H_5$
118	Ergosine	H	H	$-CH_2CH(CH_3)_2$
119	Ergocristine	Me	Me	$-CH_2-C_6H_5$
120	α-Ergokryptine	Me	Me	$-CH_2CH(CH_3)_2$
121	β-Ergokryptine	Me	Me	$-CH(CH_3)CH_2CH_3$
122	Ergocornine	Me	Me	$-CH(CH_3)_2$
123	Ergostine	H	Me	$-CH_2-C_6H_5$

Fig. 15.24
Peptide-type ergot alkaloids

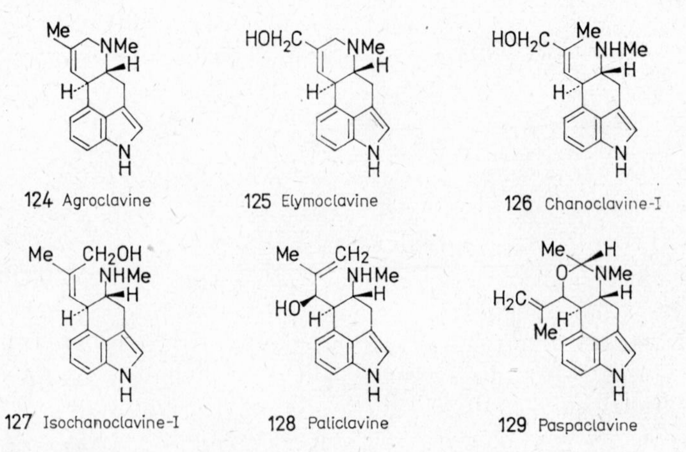

Fig. 15.25
Some clavine and secoclavine alkaloids

reduced to the tricyclic ring system with all hetero atoms and the four asymmetric centres in their natural configurations [199]. The clavines do not contain the carboxyl group but a group of lower oxidation state at position 17 (**112**). The double bond may be in the 8,9 or 9,10 position or may be missing altogether. In the 6,7-secoergolines or chanoclavines (**126**) the ring *D* is not closed. Several *N*-demethyl- or nor-analogues of normal clavines are found in nature. The minor alkaloids of *C. paspali* paliclavine (**128**) and its mixed *O,N*-acetal called paspaclavine (**129**) have unusual structures (Fig. 15.25).

15.4.1 Formation of the Ergoline Ring Systems

Most of the work on the biosynthesis of ergot alkaloids during the past 20 years has been done using the tracer technique and ergot strains producing saprophytic alkaloids. The major building blocks and some intermediates on the pathway to lysergic acid are now well established. Furthermore the steric course of various reactions has been determined and some enzymes involved in ergoline formation have been detected. Nevertheless the picture that emerged is still far from complete, *e.g.* formation of ergopeptines is scarcely understood. Many surprising results unique in alkaloid biochemistry were obtained and the efforts in this area were therefore called the "Story of the Unexpected" [63].

The present results will be summarized and some special problems discussed in greater detail. Comprehensive reviews (including the older literature) are available [62, 64].

It was suggested that the ergoline ring system is biosynthetically constructed from tryptophan, an isoprene unit derived from mevalonic acid and a methyl group from methionine [40, 149]. Subsequently in the late 1950s and early 1960s various laboratories proved that this assumption was correct (Fig. 15.26).

L-Tryptophan (**6**) is incorporated intact into ergolines (**110**, **112**) except for the loss of the carboxyl carbon atom. The hydrogens 5, 6, 7 of **6** are retained during ergoline biosynthesis but hydrogen at position 4 is lost [37, 166]. Almost complete retention of the α-hydrogen and the amino nitrogen of *L*-tryptophan was observed during ergoline formation [67]. The configuration at position 5 in ergolines (**125**) is

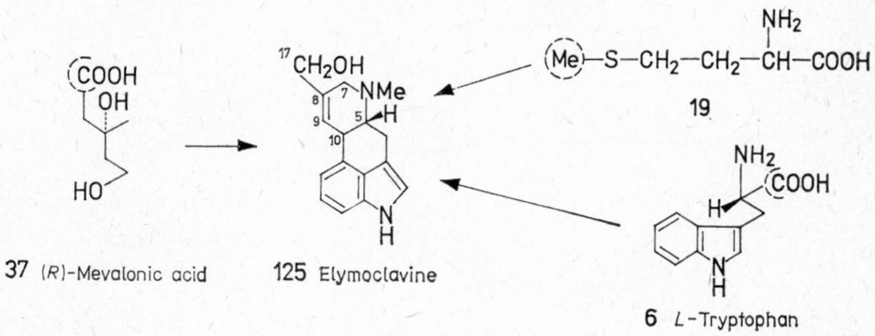

Fig. 15.26
Main precursors of ergoline biosynthesis

the opposite of the configuration of the α-C-Atom of **6**. An inversion of configuration at this particular carbon atom with retention of the original hydrogen takes place during ring *C* formation of ergoline derivatives.

The five carbon isoprene unit derived from mevalonate is involved in ergoline synthesis. Label from [2-^{14}C]mevalonate (**37**) is almost specifically incorporated into C-17 of clavines but a little scrambling (C-7) was always observed [32, 39].

(3*R*)-Mevalonate is the precursor of ergolines [66]. The same isomer of **37** is incorporated into squalene in rat liver.

Experiments with *L*-[methyl-^{14}C, T]methionine showed that the methyl group is transferred intact to the N-6 of ergolines [33].

The biosynthetic sequence chanoclavine → agroclavine → elymoclavine → lysergic acid derivatives, first proposed by Rochelmeyer [176], was experimentally confirmed by several laboratories [3, 6, 150]. This oxidative sequence proved to be the main biogenetic route. Other observed conversions may be regarded as secondary transformations (Fig. 15.27). The first proof that chanoclavine-I (**126**) and not isochanoclavine-I (**127**) is an obligate intermediate in ergoline biosynthesis was obtained in 1966 [58, 80]. It is surprising that **127**, which from its stereochemistry seems to be a potential precursor, is not involved in the formation of tetracyclic ergolines.

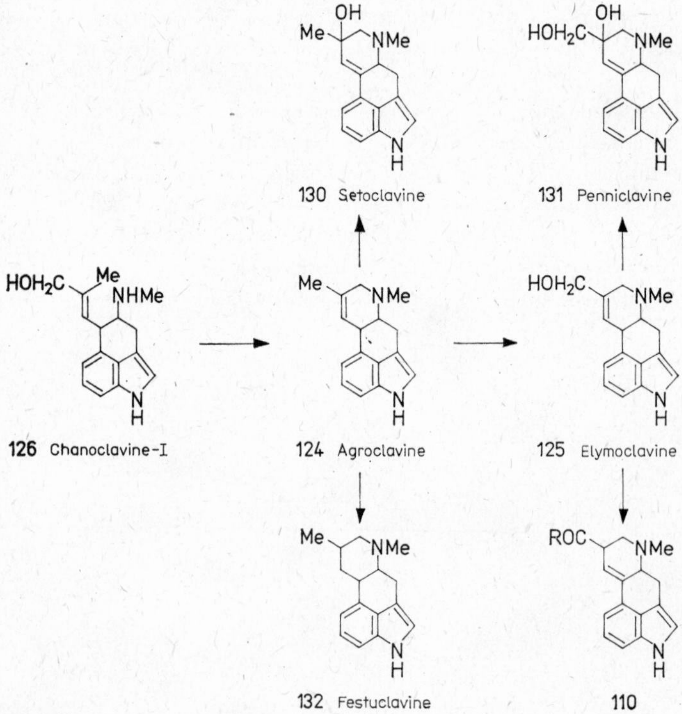

130 Setoclavine

131 Penniclavine

126 Chanoclavine-I

124 Agroclavine

125 Elymoclavine

132 Festuclavine

110

Fig. 15.27
Biogenetic inter-relationships of clavine alkaloids

The first pathway-specific step in the assembly of the ergoline ring is the isoprenylation in 4-position of tryptophan, although this position is not particularly susceptible to electrophilic attack. The proposed intermediate γ,γ-dimethylallyltryptophan (DMAT) (**134**) was synthesized by Plieningers' group. DMAT was much more efficiently incorporated into clavines compared with the "Weygand"-precursor **133** [167, 168, 223]. Experiments with double-labelled **134** bearing tritium in the alanine side-chain and ^{14}C in the dimethylallyl-group supported earlier results [169] (Fig. 15.28). Later an enzyme was isolated from *Claviceps* [77, 117], which catalyses the condensation of *L*-tryptophan with dimethylallyl pyrophosphate giving DMAT.

The apparently "normal" labelling pattern of tetracyclic clavines at C-17 during the transformation of [2-^{14}C]mevalonic acid into clavines is the fortuitous result of two cis-trans isomerizations at the allylic double bond (Fig. 15.29). This most unexpected outcome rests upon the following experiments. One cis-trans isomerization occurs between chanoclavine-I (**126**) and agroclavine (**124**). Feeding [17-^{14}C]- or [7-^{14}C]chanoclavine-I [58, 66] and degrading **124** and **125** showed that the hydro-

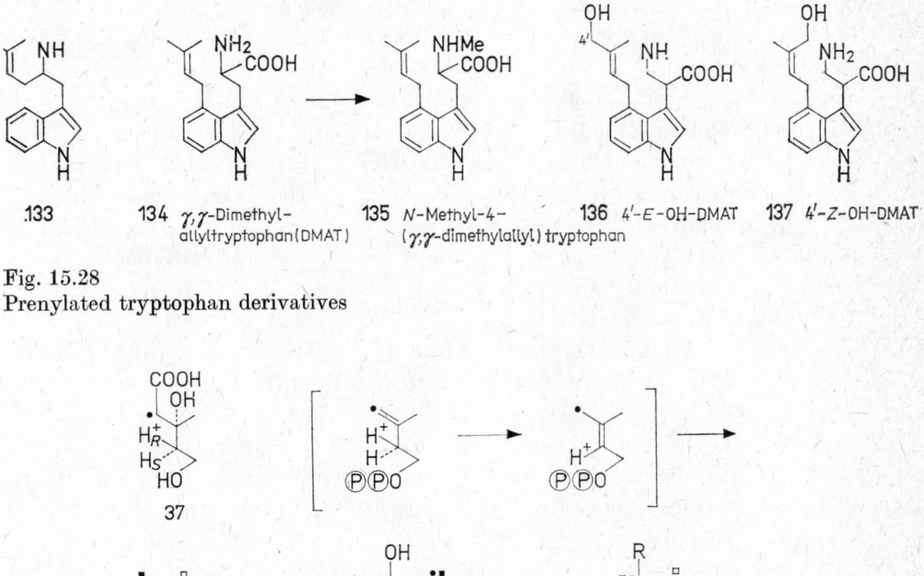

133
134 γ,γ-Dimethylallyltryptophan (DMAT)
135 *N*-Methyl-4-(γ,γ-dimethylallyl) tryptophan
136 4'-*E*-OH-DMAT
137 4'-*Z*-OH-DMAT

Fig. 15.28
Prenylated tryptophan derivatives

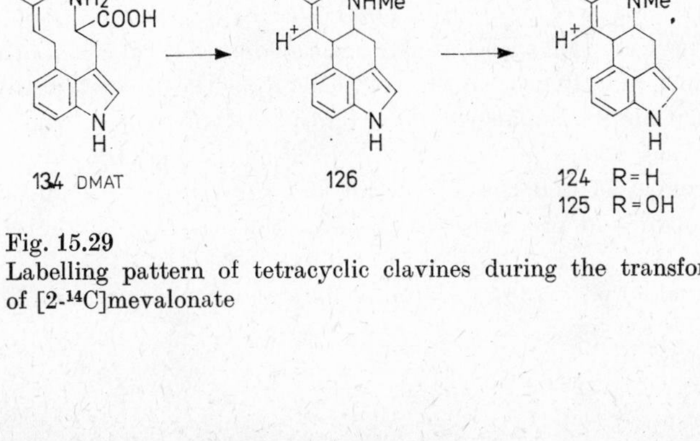

37
134 DMAT
126
124 R=H
125 R=OH

Fig. 15.29
Labelling pattern of tetracyclic clavines during the transformation of [2-^{14}C]mevalonate

xymethyl group of chanoclavine-I becomes C-7 of agroclavine and elymoclavine and that the methyl group becomes C-17. The location of the label of C-2 of mevalonate in the methyl group of **126** suggested another cis-trans isomerization [57]. This was also supported by the finding that the isopentenyl pyrophosphate isomerase reaction takes the normal steric course in *Claviceps*. The *pro-4S* hydrogen of mevalonate was lost in the formation of elymoclavine [61]. Feeding experiments with [*Z*-^{14}Me]DMAT showed unequivocally that isomerization takes place between DMAT and chanoclavine-I [160]. Further confirmation was achieved by tracer experiments using [4′-(*E*)-^{13}C]4-(3-methyl-2-butenyl)]tryptophan [170]. The ^{13}C-enriched C-atom of the E-methyl group of DMAT labelled during ergoline formation the C-methyl group of chanoclavine and C-17 of agroclavine, elymoclavine and lysergic acid.

For a long time it was believed that *N*-methylation of ergoline takes place after formation of ring *C*. Attractive hypotheses were put forward proposing the involvement of pyridoxal phosphate in the decarboxylation and *C*-ring closure steps. However, norchanoclavine-I is no precursor of tetracyclic clavines. Furthermore, under certain conditions [12] *Claviceps* mycelium accumulates N^{α}-methyl-4-(γ,γ-dimethylallyl)tryptophan (**135**). This particular compound labelled with deuterium and ^{15}N was efficiently incorporated intact into elymoclavine. No incorporation was observed with the decarboxylation product of **135**. Also an *N*-methyltransferase has been detected in cell-free extracts of *Claviceps* which catalyses the transfer of the *S*-methyl group of *S*-adenosylmethionine to the amino group of **134** [158, 159]. These results are a strong indication that *N*-methylation of **134** is the second pathway-specific step followed by a reaction (hydroxylation?) in the isoprenoid side-chain.

Deoxychanoclavine-I and nor-deoxychanoclavine-I are not precursors of tetracyclic ergolines [57]. This implies that hydroxylation of the isoprenoid moiety precedes *C*-ring formation. Subsequently, experiments were performed with 4′-*E*-OH-DMAT (**136**) and 4′-*Z*-OH-DMAT (**137**). **136** labelled in the hydroxymethyl group with ^{14}C gave a better incorporation into elymoclavine than into agroclavine [171]. Arigoni's group labelled **136** and **137** in the Me-group [161]. In both cases more than 97% of the label was incorporated into C-7 of elymoclavine (**125**). This labelling pattern of **136** would result if *E*-OH-DMAT went through the normal pathway with two cis-trans-isomerizations as though the hydroxyl was not there. Surprisingly, an in the 4′ position hydroxylated DMAT is enzymatically formed with cell-free extracts of *Claviceps* [164, 178].

The route to elymoclavine *via* **134** → **136** is certainly of minor significance and may be regarded as a side-reaction: The observed labelling pattern is not compatible with chanoclavine-I and agroclavine as intermediates. Other experiments (*vide supra*) clearly ruled out hydroxylation of DMAT as the second pathway-specific step and participation of normal pyridoxal phosphate catalysis in ring *C*-formation.

It was shown that **126** is incorporated into elymoclavine (**125**) with loss of half of the tritium from C-17 and that essentially all the tritium is confined to C-7 of **125** [70]. This conversion proceeds with loss of the *pro-17R* and retention of the *pro-17S* hydrogen [88].

Chanoclavine-I-aldehyde (**138**) was found as a natural constituent in a blocked mutant of an ergot strain [134, 136]. This result strongly indicates that chanoclavine-I aldehyde is a necessary intermediate on the route to tetracyclic ergolines. [7-^{14}C, 9-T]chanoclavine-I is converted into elymoclavine with only about 70% T retention. Agroclavine (**124**) and elymoclavine (**125**) showed tritium retention between 40 and 80% in various experiments after feeding (4*R*) [2-^{14}C, 4-T]mevalonate. On the contrary, in chanoclavine-I an enrichment in tritium up to 45% was recorded [66, 70]. This may be explained by an isotope effect on the vinyl hydrogen in the conversion of **126** into tetracyclic ergolines. The tritium loss during ring-*D*

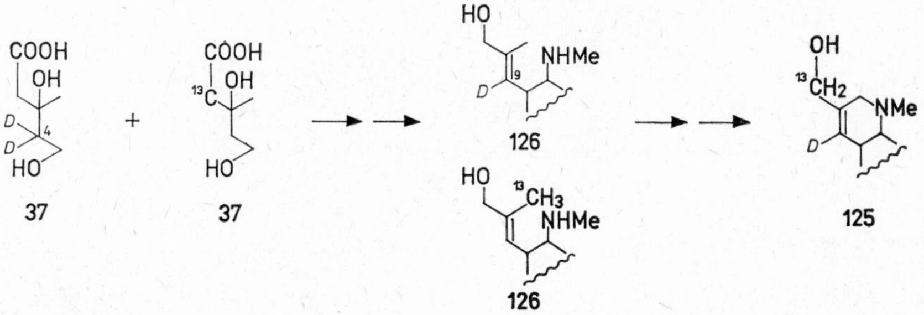

Fig. 15.30
Mechanism of ring *D* formation in ergoline biosynthesis

formation is apparently caused by an intermolecular hydrogen transfer (recycling of the C-9 hydrogen) in the tricyclic substrate. This was proven by feeding a mixture of [4-$^2H_2$]mevalonate and [2-^{13}C]mevalonate. The results showed that chanoclavine-I only contained single labelled (*D* or ^{13}C) species, but elymoclavine (**125**) clearly showed a certain amount of double labelled molecules. High-resolution peak matching MS revealed both ^{13}C and deuterium in the same molecule (Fig. 15.30). A specimen of elymoclavine (**125**) derived from [4-*D*]mevalonate showed that the label was located at C-9, proving that the hydrogen is transferred back into the 9-position.

The pathway for the formation of the ergoline ring system can now be formulated as outlined in Fig. 15.31.

At present four enzymes involved in ergoline biosynthesis have been detected in cell-free extracts of *Claviceps*. The best characterized enzyme is the dimethylallyl pyrophosphate: *L*-tryptophan dimethylallyltransferase [52, 117]. The enzyme was purified to homogeneity and consists of two subunits of 34000 molecular weight. It is activated by Fe^{2+}, Mg^{2+} and particularly by Ca^{2+}. The Km values for *L*-tryptophan and dimethylallyl pyrophosphate were determined as 0.067 and 0.2 M respectively. Steady-state kinetics were consistent with either a random or ordered mechanism. Enzyme synthesis immediately precedes biosynthesis of alkaloids [90].

The DMAT *N*-methyltransferase activity has a sharp maximum at pH 8.0 to 8.5 in Tris-HCl buffer. A better substrate for the enzyme than DMAT itself is the β,γ-dihydro derivative of **134**, but tryptophan and 4-methyltryptophan are very poorly methylated. Zn^{++} showed a pronounced inhibition of enzyme activity. The

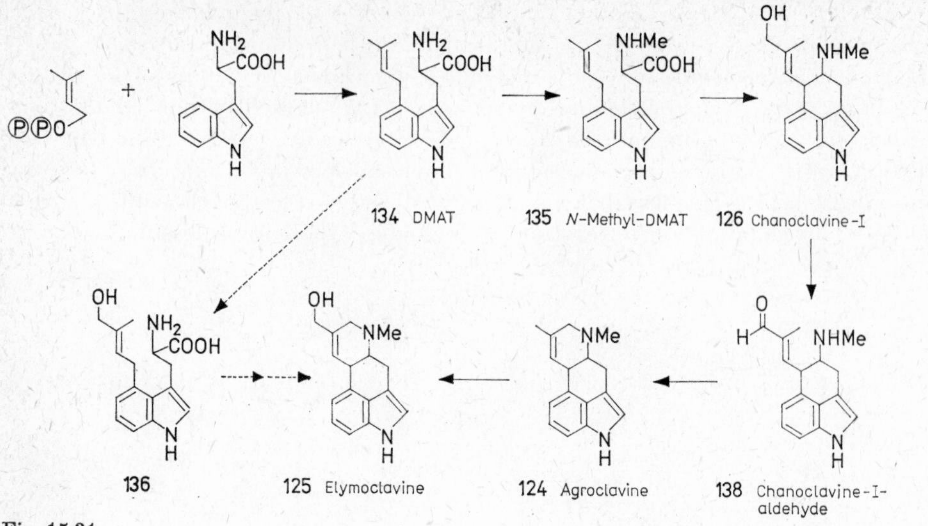

Fig. 15.31
Biosynthetic pathway of elymoclavine formation

time-course of the *N*-methyltransferase roughly parallels that seen with other ergoline enzymes [159].

Chanoclavine-I cyclase catalyses the conversion of **126** to agroclavine (**124**) [53] and/or elymoclavine [64, 154]. A requirement of NAD^+ or $NADP^+$ was observed. Addition of FAD to column-fractionated preparations stimulated activity [64]. The cyclase converts chanoclavine-I-aldehyde, but not isochanoclavine-I or dihydrochanoclavine into agroclavine [53]. The time-course of the appearance and decline of this enzyme closely resembled that of DMAT synthetase.

A conversion of agroclavine to elymoclavine with the 60–80% ammonium sulphate fraction of a clavine-producing strain was recorded. NADPH or liver concentrate was required for activity [97]. The same reaction was achieved with a microsomal agroclavine hydroxylase [106]. There was great activity during the period of maximum alkaloid production. The NADPH requirement and lack of inhibition by EDTA and cyanide are properties expected of a cytochrome P-450 mono-oxygenase.

15.4.2 Biosynthesis of Lysergic Acid Derivatives

Elymoclavine is a precursor of lysergic acid derivatives. The exact mechanism of this conversion from **125** to **110** is still unknown, *e.g.* double bond isomerization and oxidation of the hydroxymethyl to the carboxyl group. It is assumed that the shift of the double bond from the $\Delta^{8,9}$ to the $\Delta^{9,10}$ position may occur at the aldehyde stage. This is supported by feeding experiments with the enol acetate of lyserg-aldehyde [126]. The efficiency of its incorporation equals that of elymoclavine. Lysergic acid is converted to amide alkaloids [2]. In this process activation of lysergic acid might be involved. The reported formation of lysergyl-CoA [133] and its importance for ergopeptine biosynthesis are still controversial. Apparently the formation of a coenzyme A derivative takes place at the level of lysergaldehyde which is converted to the coenzyme A hemiacetal [65] as outlined in Fig. 15.32. This pathway is favoured by recent ^{18}O experiments [174]. Dihydroergosine is formed in *Sphacelia sorghi*. Festuclavine, dihydroelymoclavine and elymoclavine were incorporated effeciently but not agroclavine [13]. This indicates that the double bond is reduced before agroclavine, presumably at the level of secoclavines.

The side-chain of lysergic acid α-hydroxyethylamide (**115**) and ergonovine (**113**) is derived from alanine or a closely related compound. The incorporation of radioactivity from [2-^{14}C]alanine occurred primarily in the carbinolamide carbon of **115**. The ^{15}N from *L*-[U-^{14}C, ^{15}N]alanine was transferred to the amide nitrogen [48, 81]. *L*-alanine labelled the alaninol moiety of ergonovine (**113**) [147, 152]. The role of *L*-alaninol as precursor for **113** is still controversial. The highly plausible intermediate lysergyl-*L*-alanine does not to be involved in ergonovine biosynthesis.

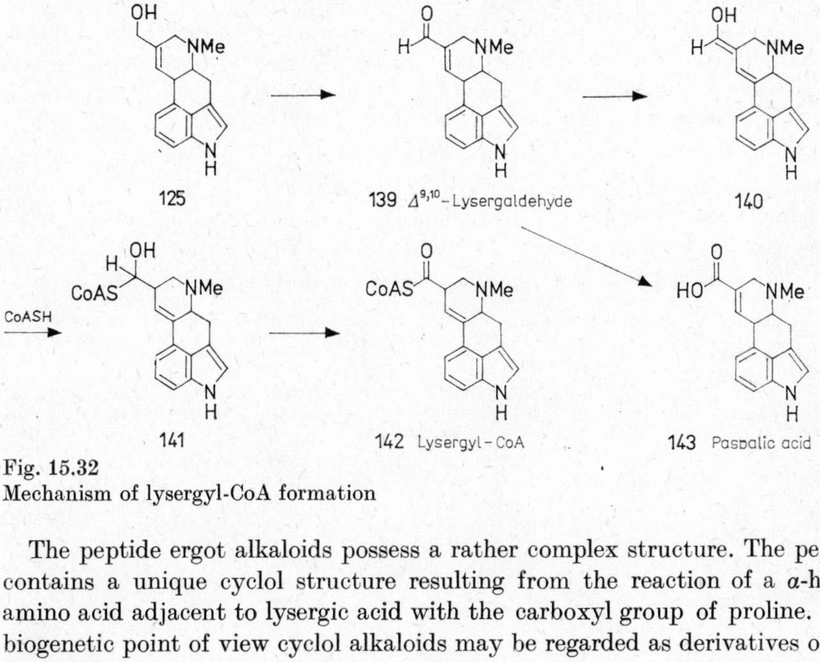

Fig. 15.32
Mechanism of lysergyl-CoA formation

The peptide ergot alkaloids possess a rather complex structure. The peptide part contains a unique cyclol structure resulting from the reaction of a α-hydroxy-α-amino acid adjacent to lysergic acid with the carboxyl group of proline. From the biogenetic point of view cyclol alkaloids may be regarded as derivatives of lysergyltripeptides:

ergotamine (**117**): lysergyl-alanyl-phenylalanyl-proline
ergocornine (**122**): lysergyl-valyl-valyl-proline
ergostine (**123**): lysergyl-aminobutyryl-phenylanyl-proline

For the mode of assembly and formation of the cyclol moiety several mechanisms can be envisaged. The chain may start from the lysergic acid or from the proline end. Single amino acids could be successive or potential intermediates (di- and tripeptides, diketopiperazines) attached to a given starter molecule.

A crucial step seems to be the hydroxylation of the amino acid, which is directly connected with lysergic acid. Feeding experiments showed phenylalanine, proline, valine, leucine and lysergic acid to be specifically incorporated into the appropriate parts of the corresponding peptide alkaloids [64, 77, 79]. Also the α-hydroxy-α-amino acid residue is derived from valine in the case of ergotoxins, and from alanine or a close relative in the case of ergotamine. Lysergylalanine might be connected in some way with the corresponding prolyl-diketopiperazine and finally converted to ergotamine [3]. Experiments with both lysergylalanine and lysergylvaline revealed that these "dipeptides" are not incorporated as intact units into ergotamine and ergotoxins respectively. Appropriate diketopiperazins, *e.g.* *L*-valyl-proline lactam, are split by the fungus prior to incorporation into the single amino acids. Labelled linear di- and tripeptides such as *L*-valyl-*L*-leucine, *L*-valyl-*L*-valyl-proline and *L*-valyl-*L*-leucyl-*L*-proline did not act as specific precursors for the cyclol part of ergopeptines. Recently lysergyl-*L*-valyl-*L*-leucine, lysergyl-*L*-valyl-*L*-valine and *d*-lysergyl-*L*-valyl-*L*-valyl-*L*-proline were tested as precursors of ergotoxine alkaloids [30]. In these cases a nonspecific labelling pattern was observed. Apparently the formation of the peptide moiety does not involve any free di- or tripeptide intermediate. Floss et al. [70a] put forward a very attractive hypothesis that the assembly of the peptide portion takes place in a concerted fashion on a multi-enzyme complex (Fig. 15.33). The key intermediate **145** is a lysergyltripeptide covalently linked to an SH-group on the enzyme via the carboxyl group of proline. The first free intermediate would be an acylated diketopiperazine that is finally converted after hydroxylation into cyclol alkaloids. The *D*-proline analogue of compound **146** has been found in ergot [213]. Peptide alkaloid formation is not inhibited by cyclo-

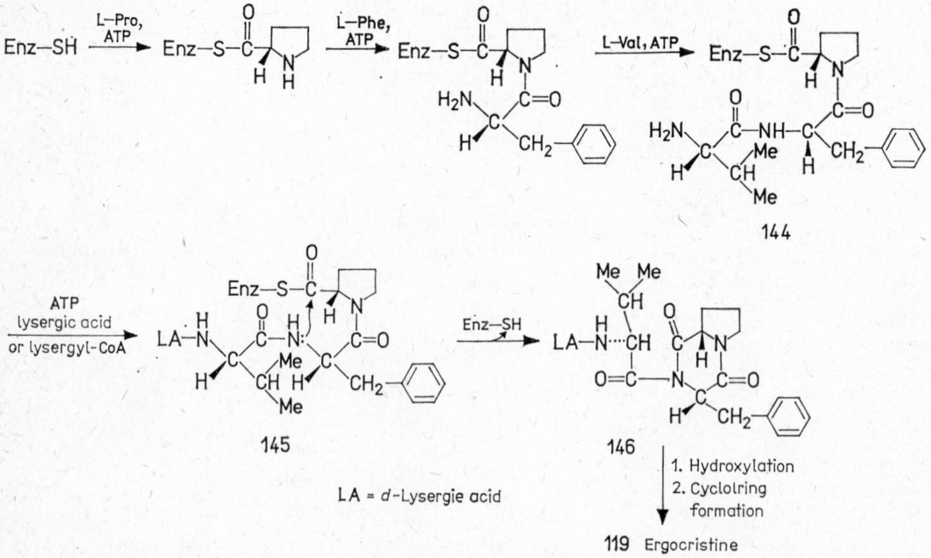

LA = *d*-Lysergic acid

146 → 1. Hydroxylation 2. Cyclolring formation → 119 Ergocristine

Fig. 15.33
Possible biosynthetic route of cyclol formation according to [70a]

heximide, which supports the above assumption [54, 138]. To test the validity of the hypothesis, experiments with cell-free systems are necessary. One approach could be the use of protoplasts as starting material. A suitable protoplast system of *Claviceps* was first described by Keller et al. [105]. The incorporation of phenylalanine into ergotamine was markedly stimulated in the presence of lysergic acid and by agitation of the incubation mixture. The *de novo* synthesis of ergosine, ergotamine and ergotoxins was subsequently demonstrated using protoplasts of various *Claviceps* strains [139]. With this system all cyclol-specific acids were incorporated in a given ergopeptine.

Cell-free systems for the biosynthesis of the peptide alkaloids were obtained in the author's laboratory [135, 137]. Cell-free extracts of suitable ergot strains may be prepared either from protoplasts or lyophilized mycelium. Using *L*-[U-^{14}C]leucine an *in vitro* formation of ergosine was unequivocally demonstrated. The formation of ergosine was markedly stimulated by addition of agroclavine or elymoclavine and by agitation of the incubation mixture. The synthesis of ergopeptines is strongly dependent on the presence of ATP and an energy-regenerating system. Enzyme preparations of an ergosine as well as an ergotamine strain catalyse the incorporation of all specific amino acids into the peptide moiety of these ergopeptines. The most active preparations were obtained at the beginning of the idiophase. The initial steps of purification, *e.g.* fractionation with $(NH_4)_2SO_4$, dialysis and gel filtration on Sepharose 4 B and Sepharose 6 B, produced fully active enzyme preparations. The multi-enzyme complex in *Claviceps* apparently possesses a broad specificity like the enzyme complexes in peptide antibiotics synthesizing bacilli. The biosynthesis of the peptide moiety is at least partially controlled by the relative amino acid concentration in the internal pool of the cells. This has been demonstrated by several groups [11, 29, 36, 51, 109].

The addition of appropriate amino acids or their analogues to the nutrient medium determines the proportions of the alkaloid mixture in a given strain [6, 12, 36]. Pronounced effects were observed by using amino acid auxotrophic mutants [36]. Under parasitic conditions the alkaloid spectrum may also be altered by a given amino acid [109]. It is now well documented that all amino acids regardless of their position in the cyclol part of a peptide ergot alkaloid may be replaced by an analogue of the "normal" amino acid. For example, the C-terminal proline can be substituted by thiazolidine-4-carboxylic acid [29] or by 4-OH-proline in an dihydroergosine synthesizing fungus [11]. Finally, it has also been reported [9] that dihydrolysergic acid is accepted as a substrate by the enzyme complex in an ergotamine-producing strain.

An alkaloid-blocked mutant was isolated from an ergotoxins producing *Claviceps* strain [134, 136]. This mutant accumulates secoergolines and by feeding with appropriate intermediates (tetracyclic ergolines) it produces the normal alkaloidal spectrum. After supplementation with agroclavine, ergosine is additionally synthesized. This is the first case of mutational biosynthesis outside the field of antibiotics. A parasitic *Claviceps purpurea* was obtained accumulating dimethylallyltryptophan. The next step, involving *N*-methylation of DMAT, did not operate. Agroclavine and lysergic acid were accepted by parasitic scerotial tissue as substrates and were converted to lysergic acid amide, a previously unreported end product for *C. purpurea* [224a]. All these results together indicate the possibility of obtaining new and potentially therapeutically interesting alkaloids.

The mechanism of hydroxylation of the α-amino acid adjacent to lysergic acid was investigated by Floss' group [38, 174]. (*R*,*S*)-2-Amino[3-^{13}C, 3-D_2]butyric acid labelled ergostine (**123**). Mass spectral analysis revealed the presence of ^{13}C as well as two atoms of deuterium in **123**. These findings rule out a 2,3-dehydroamino acid intermediate in the formation of the α-hydroxy-α-amino acid moiety of ergopeptines. Fermentation of an ergotamine (**117**) producing strain in an atmosphere of ^{18}O led to the formation of ^{18}O-enriched **117**.

The ^{18}O isotope in the peptide fragment was solely confined to the cyclol oxygen. This result supports the idea of direct hydroxylation at the α-position of the particular amino acid in a given ergopeptine leading to the α-hydroxy-α-amino acid moiety. The finding that the lysergyl fragment shows the same ^{18}O enrichment as the cyclol oxygen favours a pathway involving direct formation of an activated lysergic acid from an aldehyde intermediate without further dilution of the ^{18}O of elymoclavine (**125**) (*e.g.* Fig. 15.32). Activation of D-lysergic acid and dihydrolysergic acid via the corresponding adenylates in Claviceps were recently described [105a]. The participation of D-lysergic acid adenylate in peptide ergot alkaloid biosynthesis is not yet clarified.

15.4.3 Production and Regulatory Aspects of Ergolines

Since 1950 considerable progress has been made in the saprophytic production of ergolines [78, 109, 198, 219]. The pioneering work of Abe [1] stimulated research in many laboratories. Fermentation procedures for each type of ergot alkaloids (**110**, **111**, **112**) have been developed. In general three different species of *Claviceps* are used for saprophytic production: *Claviceps fusiformis*, *Claviceps paspali* and *Claviceps purpurea*. *Cl. fusiformis* and related fungi parasitizing on wild grasses accumulate mostly agroclavine and/or elymoclavine. Strains of *Cl. paspali* are typical producers of simple lysergic acid derivatives. So far four chemical types have been found. The majority of the investigated strains accumulate **115** and very little paspalic acid **116**. Only *Cl. purpurea* is capable of producing ergopeptines. About seven different chemovarieties have been described of which the high-yielding strains accumulate about 2 to 3 g/l peptide ergot alkaloids.

Most production media contain ammonia as nitrogen source and succinate or citrate as counter-ions. High osmotic pressure in the culture broth with sugar or sugar alcohols in the range of 100 to 300 g/l appears to be favourable. During various stages of fermentation pronounced changes in gross morphology and hyphal ultrastructure take place and coincide with altered biochemical activities.

The biosynthesis of alkaloids could be regulated by the synthesis of its primary precursors. In *Claviceps* many efforts were made to determine the influence of the aromatic amino acid pathways on alkaloid formation [68, 69]. In *Claviceps paspali* anthranilate synthetase was not inhibited by tryptophan, suggesting that the enzyme involved in tryptophan synthesis may be subject to regulation by the end-product [127]. The properties of a three-enzyme complex from *Claviceps* strain SD 58, which contained anthranilate synthetase, phosphoribosylanthranilate isomerase and indole-3-glycerolphosphate synthetase, were investigated [141]. Anthranilate synthetase from this particular strain and from two *C. purpurea* strains [180] exhibited inhibition by *L*-tryptophan, which is normal in most microorganisms. Elymoclavine inhibited this enzyme in strain SD 58, representing feedback control of alkaloid biosynthesis. Feedback inhibition of dimethylallyltryptophan synthetase, the first pathway-specific enzyme, by elymoclavine was observed.

L-Tryptophan serves as a major precursor of the ergoline ring. On the other hand, this aromatic amino acid and certain of its analogues stimulated alkaloid production in strain SD 58 when added at a concentration of 2 mM at the beginning of the trophophase [69, 111]. The induction of clavine alkaloid synthesis is related to increased DMAT synthetase activity rather than to a lack of regulation of the tryptophan biosynthetic enzymes. With regard to the end-product regulation of tryptophan biosynthesis there is a remarkable difference between thiotryptophan on the one hand and tryptophan as well as 5-methyltryptophan on the other. Induction of alkaloid biosynthesis is due to *de novo* synthesis of DMAT synthetase. Thiotryptophan and tryptophan can overcome the block of alkaloid synthesis by high levels of inorganic phosphate, although the exact mechanism of this process is not fully understood.

The induction phenomenon in ergoline formation is well documented in strains derived from grass ergot. *Claviceps purpurea* strains are far less susceptible to regulatory effects exerted by tryptophan and to phosphate concentrations in the medium. The regulation of ergot alkaloid biosynthesis remains a fruitful area of interest for future studies.

15.5 References

[1] Abe, M.; Yamatodani, S.: In: Hockenhull, D. J. D. (Ed.): Progress in Industrial Microbiology. London 1964. Vol. V, p. 205. Heywood.

[2] Agurell, S.: Acta Pharm. Suecica **3** (1966) 23.

[3] Agurell, S.: Acta Pharm. Suecica **3** (1966) 71.

[4] Agurell, S.; Holmstedt, B.; Lindgreen, J. E.; Schultes, R. E.: Acta Chem. Scand. **23** (1969) 903.

[5] Agurell, S.; Nilsson, I. G. L.: Tetrahedron Lett. **1968**, 1063.

[6] Agurell, S.; Ramstad, E.: Arch. Biochem. Biophys. **98** (1962) 457.

[7] Alfermann, A. W.; Reinhard, E. (Eds.): Production of Natural Compounds by Cell Culture Methods. BPT-Report 1/78, München 1978.

[8] Allen, J. R. E.; Holmstedt, B. R.: Phytochemistry **19** (1980) 1573.

[9] Anderson, J. A.; Kim, I. S.; Lehtonen, P.; Floss, H. G.: J. Nat. Prod. **42** (1979) 271.

[10] Arens, H.; Stöckigt, J.; Weiler, E. W.; Zenk, M. H.: Planta Med. **34** (1978) 37.

[11] Atwell, S. M.; Mantle, P. G.: Experientia **37** (1981) 1257.

[12] Barrow, K. D.; Quigley, F. R.: Tetrahedron Lett. **1975**, 4269.

[13] Barrow, K. D.; Mantle, P. G.; Quigley, F. R.: Tetrahedron Lett. **1974**, 1557.

[14] Barz, W.; Reinhard, E.; Zenk, M. H. (Eds.): Plant Tissue Culture and its Biological Application. Springer-Verlag, Heidelberg 1977.

[15] Battersby, A. R.: Pure & Appl. Chem. **14** (1967) 117.

[16] Battersby, A. R.; Brown, R. T.; Kapil, R. S.; Knight, J. A.; Martin, J. A.; Plunkett, A. O.: J. Chem. Soc., Chem. Commun. **1966**, 810, 888.
[17] Battersby, A. R.; Brown, R. T.; Kapil, R. S.; Martin, A. J.; Plunkett, A. O.: J. Chem. Soc., Chem. Commun. **1966**, 812, 890.
[18] Battersby, A. R.; Brown, S. H.; Payne, T. G.: J. Chem. Soc., Chem. Commun. **1970**, 827.
[19] Battersby, A. R.; Burnett, A. R.; Hall, E. S.; Parsons, P. G.: J. Chem. Soc., Chem. Commun. **1968**, 1582.
[20] Battersby, A. R.; Burnett, A. R.; Parsons, P. G.: J. Chem. Soc., Chem. Commun. **1968**, 1280.
[21] Battersby, A. R.; Burnett, A. R.; Parsons, P. G.: J. Chem. Soc., Chem. Commun. **1968**, 1282; J. Chem. Soc. C **1969**, 1193.
[22] Battersby, A. R.; Burnett, A. R.; Parsons, P. G.: J. Chem. Soc., Chem. Commun. **1970**, 826.
[23] Battersby, A. R.; Byrne, J. C.; Kapil, R. S.; Martin, J. A.; Payne, T. G.; Arigoni, D.: J. Chem. Soc., Chem. Commun. **1968**, 951.
[24] Battersby, A. R.; Gregory, B.; Spencer, H.; Turner, J. C.; Janot, M. M.; Potier, P.; Francois, P.; Leviscelles, J.: J. Chem. Soc., Chem. Commun. **1967**, 219.
[25] Battersby, A. R.; Hall, E. S.: J. Chem. Soc., Chem. Commun. **1970**, 194.
[26] Battersby, A. R.; Parry, R. J.: J. Chem. Soc., Chem. Commun. **1971**, 30, 31.
[27] Battersby, A. R.; Thompson, M.; Glüsenkamp, K. H.; Tietze, L.-F.: Chem. Ber. **114** (1981) 3430.
[28] Battersby, A. R.; Westcott, N. D.; Glüsenkamp, K.-H.; Tietze, L.-F.: Chem. Ber. **114** (1981) 3439.
[29] Baumert, A.; Erge, D.; Gröger, D.: Planta Med. **44** (1982) 122.
[30] Baumert, A.; Gröger, D.; Maier, W.: Experientia **33** (1977) 881.
[31] Baxter, R. L.; Dorschel, C. A.; Lee, S. L.; Scott, A. I.: J. Chem. Soc., Chem. Commun. **1979**, 257.
[32] Baxter, R. M.; Kandel, S. I.; OKany, A.: J. Am. Chem. Soc. **84** (1962) 2997.
[33] Baxter, R. M.; Kandel, S. I.; OKany, A.; Pyke, R. G.: Canad. J. Chem. **42** (1964) 2936.
[34] Baxter, C.; Slaytor, M.: Phytochemistry **11** (1972) 2763.
[35] Baxter, C.; Slaytor, M.: Phytochemistry **11** (1972) 2767.
[36] Beacco, E.; Bianchi, M. L.; Minghetti, A.; Spalla, C.: Experientia **34** (1978) 1291.
[37] Bellatti, M.; Casnati, G.; Palla, G.; Minghetti, A.: Tetrahedron **33** (1977) 1821.
[38] Belzecki, C. M.; Quigley, F. R.; Floss, H. G.; Crespi-Perellino, N.; Guiccardi, A.: J. Org. Chem. **45** (1980) 2215.
[39] Bhattacharji, S.; Birch, A. J.; Brack, A.; Hofmann, A.; Kobel, H.; Smith, D. C.; Smith, H.; Winter, J.: J. chem. Soc. **1962**, 421.
[40] Birch, A. J.: In: Westenholme, G. E. W.; O'Connor, C. M. (Eds.): Ciba Foundation Symposium on Amino Acids and Peptides with Antimetabolic Activity, p. 247. Churchill, London 1958.
[41] Bowden, K.; Marion, L.: Canad. J. Chem. **29** (1951) 1037.
[42] Brack, A.; Hofmann, A.; Kalberer, F.; Kobel, H.; Rutschmann, J.: Arch. Pharm. **294** (1961) 230.
[43] Brodelius, P.; Deus, B.; Mosbach, K.; Zenk, M. H.: FEBS-Letters **103** (1979) 93.
[44] Brown, R. T.; Leonhard, J.: J. Chem. Soc., Chem. Commun. **1979**, 877.
[45] Brown, R. T.; Leonhard, J.; Sleigh, S. K.: Phytochemistry **17** (1978) 899.
[46] Carew, D. P.: Nature **207** (1965) 89.
[47] Carew, D. P.: In: Taylor, W. I.; Farnsworth, N. R. (Eds.): The Catharanthus Alkaloids, p. 193. Marcel Dekker, New York 1975.
[48] Castagnoli, N.; Corbett, K.; Chain, E. B.; Thomas, R.: Biochem. J. **117** (1970) 451.
[48a] Constabel, F.; Gaudet-La Prairie, P.; Kurz, W. G. W.; Kutney, J. P.: Plant Cell Reports **1** (1982) 139.
[49] Cordell, G. A.: Lloydia **37** (1974) 219.

[50] Coscia, C. J.; Bolta, L.; Guarnaccia, R.: Arch. Biochem. Biophys. **136** (1970) 498.
[51] Crespi-Perellino, N.; Guicciardi, A.; Minghetti, A.; Spalla, C.: Experientia **37** (1981) 217.
[52] Cress, W.; Chayet, L. T.; Rilling, H. C.: J. Biol. Chem. **256** (1981) 10917.
[53] Erge, D.; Maier, W.; Gröger, D.: Biochem. Physiol. Pflanzen **164** (1973) 234.
[54] Erge, D.; Wenzel, A.; Gröger, D.: Biochem. Physiol. Pflanzen **163** (1972) 288.
[55] Escher, S.; Loew, P.; Arigoni, D.: J. chem. Soc., Chem. Commun. **1970**, 823.
[56] Farrel, G.; McIsaac, W. M.: Arch. Biochem. Biophys. **94** (1961) 543.
[57] Fehr, T.: Ph. D. Thesis No 3967, ETH Zürich (1967).
[58] Fehr, T.; Acklin, W.; Arigoni, D.: J. Chem. Soc., Chem. Commun. **1966**, 801.
[59] Fellows, L. E.; Bell, E. A.: Phytochemistry **9** (1970) 2389.
[60] Fellows, L. E.; Bell, E. A.: Phytochemistry **10** (1971) 2083.
[61] Floss, H. G.: J. Chem. Soc., Chem. Commun. **1967**, 804.
[62] Floss, H. G.: Tetrahedron **32** (1976) 873.
[63] Floss, H. G.: In: Phillipson, J. D.; Zenk, M. H. (Eds.): Indole and Biogenetically Related Alkaloids, p. 249. Academic Press, London 1980.
[64] Floss, H. G.; Anderson, J. A.: In: Steyn, P. S. (Ed.): The Biosynthesis of Mycotoxins, p. 17. Academic Press, New York 1980.
[65] Floss, H. G.; Günther, H.; Gröger, D.; Erge, D.: Z. Naturforsch. **21b** (1966) 128.
[66] Floss, H. G.; Hornemann, U.; Schilling, N.; Kelley, K.; Gröger, D.; Erge, D.: J. Am. Chem. Soc. **90** (1968) 6500.
[67] Floss, H. G.; Mothes, K.; Günther, H.: Z. Naturforsch. **19b** (1964) 784.
[68] Floss, H. G.; Robbers, J. E.; Heinstein, P. F.: Recent Adv. Phytochem. **8** (1974) 141.
[69] Floss, H. G.; Robbers, J. E.; Heinstein, P. F.: FEBS, 12th Meeting Dresden 1978, Regulation of Secondary Product and Plant Hormone Metabolism. Vol. 55, p. 121. Pergamon Press, Oxford 1979.
[70] Floss, H. G.; Tcheng-Lin, M.; Chang, C.-J.; Naidoo, B.; Blair, G. E.; Abou-Chaar, C. I.; Cassady, J. M.: J. Am. Chem. Soc. **96** (1974) 1898.
[70a] Floss, H. G.; Tscheng-Lin, M.; Kobel, H.; Stadler, P. A.: Experientia **30** (1974) 1369.
[71] Garnier, J.; Koch, M.; Kunesch, M.: Phytochemistry **8** (1969) 1241.
[72] Gilchrist, D. G.; Kosuge, T.: The Biochemistry of Plants, Vol. 5, p. 507. Academic Press, London 1980.
[73] Goebel, F.: Liebigs Ann. Chem. **38** (1841) 363.
[74] Gower, B. G.; Leete, R.: J. Am. Chem. Soc. **85** (1963) 3683.
[75] Gröger, D.; Simon, H.: Abh. Dtsch. Akad. Wiss. Berlin Klasse Chem. Geol. Biol. **4** (1963) 343.
[76] Gröger, D.: In: Kadis, S.; Ciegler, A.; Ajl, S. A. (Eds.): Microbial Toxins, Vol. 8, p. 321. Academic Press, New York 1972.
[77] Gröger, D.: Planta Med. **28** (1975) 37.
[78] Gröger, D.: Planta Med. **28** (1975) 269.
[79] Gröger, D.: In: Hütter, R.; Leisinger, T.; Nüesch, J.; Wehrli, W. (Eds.): Antibiotics and Other Secondary Metabolites, p. 201. Academic Press, London 1978.
[80] Gröger, D.; Erge, D.; Floss, H. G.: Z. Naturforsch. **21b** (1966) 827.
[81] Gröger, D.; Erge, D.; Floss, H. G.: Z. Naturforsch. **23b** (1968) 177.
[82] Gross, D.: Indolylalkylamines, In: Mothes, K.; Schütte, H. R. (Eds.): Biosynthese der Alkaloide, p. 439. Deutscher Verlag der Wissenschaften, Berlin 1969.
[83] Gross, D.; Lehmann, H.; Schütte, H. R.: Biochem. Physiol. Pflanzen **166** (1974) 281.
[84] Gross, D.; Nemeckova, A.; Schütte, H. R.: Z. Pflanzenphysiol. **57** (1967) 60.
[84a] Grosse, W.; Karisch, M.; Schröder, P.: Z. Pflanzenphysiol. **110** (1983) 221.
[85] Guarnaccia, R.; Botta, L.; Coscia, C. J.: J. Am. Chem. Soc. **96** (1974) 7079.
[86] Guarnaccia, R.; Coscia, C. J.: J. Am. Chem. Soc. **93** (1971) 6320.
[87] Gueritte, F.; Bac, N. V.; Langlois, Y.; Potier, P.: J. Chem. Soc., Chem. Commun. **1980**, 452.

[87a] Hanson, A. D.; Kimberley, M.; Ditz, G. W.; Singletary, W.; LeLand, T. J.: Plant Physiol. **71** (1983) 896.
[88] Hassam, S. B.; Floss, H. G.: J. Nat. Prod. **44** (1981) 756.
[89] Hassam, S. B.; Hutchinson, C. R.: Tetrahedron Lett. **1978**, 1681.
[90] Heinstein, P. F.; Lee, S. L.; Floss, H. G.: Biochem. Biophys. Res. Commun. **44** (1971) 1244.
[91] Heinstein, P.; Stöckigt, J.; Zenk, M. H.: Tetrahedron Lett. **21** (1980) 141.
[92] Hemscheidt, T.; Zenk, M. H.: FEBS-Letters **110** (1980) 187.
[93] Herbert, R. B.; Mann, J.: J. Chem. Soc., Chem. Commun. **1980**, 841.
[93a] Herbert, R. B.; Mann, J.: J. Chem. Soc., Perkin Trans. 1 **1982**, 1523.
[94] Hesse, M.: Indolalkaloide in Tabelle und Ergänzungswerk. Springer-Verlag, Berlin 1964, 1968.
[95] Hirata, T.; Lee, S. L.; Scott, A. I.: J. Chem. Soc., Chem. Commun. **1979**, 1081.
[96] Hofmann, A.; Heim, R.; Brack, A.; Kobel, H.; Frey, A.; Oh, H.; Petrzilka, T.; Troxler, F.: Helv. Chim. Acta **42** (1959) 1557.
[97] Hsu, J. C.; Anderson, J. A.: Biochim. Biophys. Acta **230** (1971) 518.
[98] Hutchinson, C. R.: In: Phillipson, J. D.; Zenk, M. H. (Eds.): Indole and Biogenetically Related Alkaloids, p. 143. Academic Press, London 1980.
[99] Hutchinson, C. R.: Tetrahedron **37** (1981) 1047.
[100] Hutchinson, C. R.; Heckendorf, A. H.; Daddona, P. E.; Hagaman, E.; Wenkert, E.: J. Am. Chem. Soc. **96** (1974) 5609.
[101] Hutchinson, C. R.; Heckendorf, A. H.; Straughn, J. L.; Daddona, P. E.; Cane, D. E.: J. Am. Chem. Soc. **101** (1979) 3358.
[102] Inouye, H.; Ueda, S.; Aoki, Y.; Takeda, Y.: Chem. Pharm. Bull (Tokyo) **20** (1972) 1287.
[103] Inouye, H.; Ueda, S.; Takeda, Y.: Tetrahedron Lett. **1969**, 407.
[104] Kan-Fan, Ch.; Husson, H. P.: J. Chem. Soc., Chem. Commun. **1979**, 1015.
[104a] Kartha, K. K.; Leung, N. L.; Gaudet-La Prairie, P.; Constabel, F.: Plant Cell Reports **1** (1982) 135.
[105] Keller, U.; Zocher, R.; Kleinkauf, H.: J. Gen. Microbiol. **118** (1980) 485.
[105a] Keller, U.; Zocher, R.; Krengel, U.; Kleinkauf, H.: Biochem. J. **218** (1984) 857.
[106] Kim, I. S.; Kim, S. U.; Anderson, J. A.: Phytochemistry **20** (1981) 2311.
[107] Kisakürek M. V.; Hesse M.: In: Phillipson, J. D.; Zenk, M. H. (Eds.): Indole and Biogenetically Related Alkaloids, p. 11. Academic Press, London 1980.
[108] Kisakürek, M. V.; Hesse, M.: In: "Alkaloids. Chemical and biological perspectives". Ed.: Pelletier, S. W., Vol. 1, John Wiley and Sons, New York 1983.
[108a] Knobloch, K. H.; Hanson, B.; Berlin, J.: Z. Naturforsch. **36c** (1981) 40.
[109] Kobel, H.; Sanglier, J. J.: In: Hütter, R.; Leisinger, R.; Nüesch, J.; Wehrli, W. (Eds.): Antibiotica and other Secondary Metabolites, Biosynthesis and Production, p. 233. Academic Press, London 1978.
[109a] Kohl, W.; Witte, B.; Höfle, G.: Z. Naturforsch. **37b** (1982) 1346.
[110] Kompis, I.; Grossmann, E.; Terent'eva, I. V.; Lazur'evskii, G. V.: C. A. **74** (1971) 136527.
[111] Krupinski, V. M.; Robbers, J E.; Floss, H. G.: J. Bacteriol. **125** (1976) 158.
[112] Kutney, J. P.: Heterocycles **4** (1976) 429.
[113] Kutney, J. P.: Heterocycles **15** (1981) 1405.
[113a] Kutney, J. P.; Aweryn, B.; Choi, L. S. L.; Kolodziejczyk, P.; Kurz, W. G. W.; Chatson, K. B.; Constabel, F.: Helv. Chim. Acta **65** (1982) 1271.
[114] Kutney, J. P.; Aweryn, B.; Choi, L. S. L.; Kurz, W. G. W.; Chatson, K. B.; Constabel, F.: Heterocycles **16** (1981) 1169.
[114a] Kutney, J. P.; Choi, L. S. L.; Honda, T.; Lewis, N. G.; Sato, T.; Stuart, K. L.; Worth, B. R.: Helv. Chim. Acta **65** (1982) 2088.
[115] Kutney, J. P.; Cretney, W. J.; Hadfield, J. R.; Hall, E. S.; Nelson, V. R.; Wigfield, D. C.: J. Am. Chem. Soc. **90** (1968) 3566.
[116] Lee, S. L.; Cheng, K. D.; Scott, A. I.: Phytochemistry **20** (1981) 1841.

[117] Lee, S. L.; Floss, H. G.; Heinstein, P. F.: Arch. Biochem. Biophys. **177** (1976) 84.
[118] Lee, S. L.; Hirata, T.; Scott, A. I.: Tetrahedron Lett. **1979**, 691.
[119] Leete, E.: Tetrahedron **14** (1961) 35.
[120] Leete, E.: Accounts Chem. Res. **2** (1969) 59.
[121] Leete, E.: Phytochemistry **14** (1975) 471.
[122] Leete, E.: J. Nat. Prod. **43** (1980) 130.
[123] Leete, E.; Ahmad, A.; Kompis, I.: J. Am. Chem. Soc. **87** (1965) 4168.
[124] Leete, E.; Wemple, J.: J. Am. Chem. Soc. **91** (1969) 2698.
[125] Liljegren, D. R.: Phytochemistry **7** (1968) 1299.
[126] Lin, L. C. C.; Blair, G. E.; Cassady, J. M.; Gröger, D.; Maier, W.; Floss, H. G.: J. Org. Chem. **38** (1973) 2249.
[127] Lingens, F.; Goebel, W.; Uesseler, H.: Eur. J. Biochem. **2** (1967) 442.
[128] Loew, P.; Arigoni, D.: J. Chem. Soc., Chem. Commun. **1968**, 137.
[129] Madyastha, K. M.; Coscia, C. J.: Recent Adv. Phytochem. **13** (1979) 85.
[130] Madyastha, K. M.; Guarnaccia, R.; Baxter, C.; Coscia, C. J.: J. Biol. Chem. **248** (1973) 2497.
[131] Madyastha, K. M.; Guarnaccia, R.; Coscia, C. J.: FEBS-Letters **14** (1971) 175.
[132] Madyastha, K. M.; Meehan, T. D.; Coscia, C. J.: Biochemistry **15** (1976) 1097.
[133] Maier, W.; Erge, D.; Gröger, D.: Biochem. Physiol. Pflanzen **163** (1972) 432.
[134] Maier, W.; Erge, D.; Gröger, D.: Planta Med. **40** (1980) 104.
[135] Maier, W.; Erge, D.; Gröger, D.: FEMS Microbiol. Lett. **12** (1981) 143.
[135a] Maier, W.; Erge, D.; Gröger, D.: FEMS Microbiol. Lett. **20** (1983) 233.
[136] Maier, W.; Erge, D.; Schmidt, J.; Gröger, D.: Experientia **36** (1980) 1353.
[137] Maier, W.; Erge, D.; Schumann, B.; Gröger, D.: Biochem. Biophys. Res. Commun. **99** (1980) 155.
[138] Maier, W.; Gröger, D.: Biochem. Physiol. Pflanzen **174** (1979) 1.
[139] Maier, W.; Schumann, B.; Erge, D.; Gröger, D.: Biochem. Physiol. Pflanzen **175** (1980) 815.
[140] Mandel, L. R.; Rosegay, A.; Walker, R. W.; Vanden Heuvel, W. J. A.: Science **186** (1974) 741.
[141] Mann, D. F.; Floss, H. G.: Lloydia **40** (1977) 136.
[142] Manske, R. H. F. (Ed.): The Alkaloids: Chemistry and Physiology. Academic Press, New York, Vol. XI 1968, XIV 1973, XV 1975, XVI 1977, XVII 1979.
[143] Mac Farlane, J.; Madyastha, K. M.; Coscia, C. J.: Biochem. Biophys. Res. Commun. **66** (1975) 1263.
[144] Mc Farlane, I. J.; Slaytor, M.: Phytochemistry **11** (1972) 229.
[145] Mc Kenzie, E.; Nettleship, L.; Slaytor, M.: Phytochemistry **14** (1975) 273.
[146] Meehan, T. D.; Coscia, C. J.: Biochem. Biophys. Res. Commun. **53** (1973) 1043.
[147] Minghetti, A.; Arcamone, F.: Experientia **25** (1969) 926.
[148] Mizukami, H.; Nordlöv, H.; Lee, S. L.; Scott, A. I.: Biochemistry **18** (1979) 3760.
[149] Mothes, K.; Weygand, F.; Gröger, D.; Grisebach, H.: Z. Naturforsch. **13b** (1958) 41.
[150] Mothes, K.; Winkler, K.; Gröger, D.; Floss, H. G.; Mothes, U.; Weygand, F.: Tetrahedron Lett. (1962) 933.
[151] Nagakura, N.; Rüffer, M.; Zenk, M. H.: J. Chem. Soc. Perkin I **1979**, 2308.
[152] Nelson, U.; Agurell, S.: Acta Chem. Scand. **23** (1969) 3393.
[153] Nettleship, L.; Slaytor, M.: Phytochemistry **10** (1971) 231.
[153a] Noé, W.; Mollenschott, C.; Berlin, J.: Plant Molecular Biology **3** (1984) 281.
[154] Ogunlana, E. O.; Wilson, B. J.; Tyler, V. E.; Ramstad, E.: J. Chem. Soc., Chem, Commun. **1970**, 775.
[155] O'Donovan, D. G.; Buckley, L.; Geary, P.: Proc. R. Ir. Acad. Sect. B **76** (1976) 187.
[156] O'Donovan, D. G.; Kenneally, M. F.: J. Chem. Soc. C **1967**, 1109.
[157] O'Donovan, D.; Leete, E.: J. Am. Chem. Soc. **39** (1963) 461.
[158] Otsuka, H.; Anderson, J. A.; Floss, H. G.: J. Chem. Soc., Chem. Commun. **1979**, 660.
[159] Otsuka, H.; Quigley, F. R.; Gröger, D.; Anderson, J. A.; Floss, H. G.: Planta Med. **40** (1980) 109.

[160] Pachlatko, P.: Ph. D. Thesis No 5481, ETH Zürich (1975).
[161] Pachlatko, P.; Tabacik, C.; Acklin, W.; Arigoni, D.: Chimia **29** (1975) 526.
[162] Parry, R. J.: In: Taylor, W. I.; Farnsworth, N. R. (Eds.): The Catharanthus Alkaloids, p. 141. Marcel Dekker, New York 1975.
[163] Parry, R. J.; Smith, G. F.: J. Chem. Soc. Perkin I **1978**, 1671.
[164] Petroski, R. J.; Kelleher, W. J.: FEBS-Letters **82** (1977) 55.
[164a] Pfitzner, A.; Stöckigt, J.: Phytochemistry **21** (1982) 1585.
[164b] Pfitzner, A.; Stöckigt, J.: Planta med. **48** (1983) 221.
[164c] Pfitzner, A.; Stöckigt, J.: Chem. Commun. (**1983**) 459.
[164d] Pfitzner, A.; Stöckigt, J.: Tetrahedron Lett. **24** (1983) 1695.
[164e] Pfitzner, A.; Stöckigt, J.: Tetrahedron Lett. **24** (1983) 5197.
[165] Phillipson, J. D.; Zenk, M. H. (Eds.): Indole and Biogenetically Related Alkaloids. Academic Press, London 1980.
[166] Plieninger, H.; Fischer, R.; Keilich, G.; Orth, H. D.: Liebigs Ann. Chem. **642** (1961) 214.
[167] Plieninger, H.; Fischer, R.; Liede, V.: Angew. Chem. **74** (1962) 430.
[168] Plieninger, H.; Fischer, R.; Liede, V.: Liebigs Ann. Chem. **672** (1964) 223.
[169] Plieninger, H.; Immel, H.; Völkl, A.: Liebigs Ann. Chem. **706** (1967) 223.
[170] Plieninger, H.; Meyer, E.; Maier, W.; Gröger, D.: Liebigs Ann. Chem. **1978**, 813.
[171] Plieninger, H.; Wagner, C.; Immel, H.: Liebigs Ann. Chem. **743** (1971) 95.
[172] Poindexter, E. H.; Carpenter, R. D.: Phytochemistry **1** (1962) 215.
[173] Potier, P.; Langlois, N.; Gueritte, F.; Langlois, Y.: J. Am. Chem. Soc. **98** (1976) 7017.
[174] Quigley, F. R.; Floss, H. G.: J. Org. Chem. **46** (1981) 464.
[175] Reinhard, E.; Corduan, G.; Volk, O. H.: Phytochemistry **7** (1968) 503.
[176] Rochelmeyer, H.: Pharm. Ztg. **103** (1958) 1269.
[177] Rüffer, M.; Kan-Fan, Ch.; Husson, H. P.; Stöckigt, J.; Zenk, M. H.: J. Chem. Soc., Chem. Commun. **1979**, 1016.
[178] Saini, M. S.; Anderson, J. A.: Phytochemistry **17** (1976) 799.
[178a] Sasse, F.; Heckenburg, U.; Berlin, J.: Z. Pflanzenphysiol. **105** (1982) 315.
[179] Schallenberg, J.; Berlin, J.: Z. Naturforsch. **34c** (1979) 541.
[180] Schmauder, H. P.; Gröger, D.: Biochem. Physiol. Pflanzen **169** (1976) 471.
[181] Schütte, H. R.: Abh. Dtsch. Akad. Wiss. Berlin, 4. Int. Symp. Biochem. Physiol. Alkaloide, Vol. b **1972**, 103.
[182] Schütte, H. R.: Progr. Bot. **42** (1980) 96.
[183] Schultes, R. E.: Planta Med. **29** (1976) 330.
[184] Scott, A. I.: Accounts Chem. Res. **3** (1970) 151.
[185] Scott, A. I.; Cherry, P. C.; Qureshi, A. A.: J. Am. Chem. Soc. **91** (1969) 4932.
[186] Scott, A. I.; Gueritte, F.; Lee, S. L.: J. Am. Chem. Soc. **100** (1978) 6253.
[187] Scott, A. I.; Lee, S. L.: J. Am. Chem. Soc. **97** (1975) 6906.
[188] Scott, A. I.; Lee, S. L.; Culver, G.; Wan, W.; Hirata, T.; Gueritte, F.; Baxter, R. L.; Nordlöv, H.; Dorschel, C. A.; Mizukami, H.; Mac Kenzie, N. E.: Heterocycles **15** (1981) 1257.
[189] Scott, A. I.; Lee, S. L.; Wan, W.: Biochem. Biophys. Res. Commun. **75** (1977) 1004.
[190] Scott, A. I.; Mizukami, H.; Hirata, I.; Lee, S. L.: Phytochemistry **19** (1980) 488.
[191] Scott, A. I.; Reichardt, B. P.; Slaytor, M. B.; Sweeny, J. E.: Recent Adv. Phytochem. **6** (1973) 117.
[192] Sheria, G. M.; Rapoport, H.: Phytochemistry **15** (1976) 505.
[193] Slaytor, M.; Mc Farlane, I. J.: Phytochemistry **7** (1968) 605.
[194] Smith, G. N.: J. Chem. Soc., Chem. Commun. **1968**, 912.
[195] Smith, T. A.: Phytochemistry **14** (1975) 865.
[196] Smith, T. A.: Progr. Phytochem. **4** (1977) 27.
[197] Smith, T. A.: Phytochemistry **16** (1977) 171.
[198] Spalla, C.; Marnati, M. P.: In: Hütter, R.; Leisinger, R.; Nüesch, J.; Wehrli, W. (Eds.): Antibiotics and other Secondary Metabolites, p. 219. Academic Press, London 1978.
[198a] Spitsberg, C.; Coscia, C. J.; Krueger, R. J.: Plant Cell Reports **1** (1981) 43.

[198b] Stadler, P. A.: Planta med. **46** (1982) 131.
[199] Stadler, P. A.; Stütz, P.: In: The Alkaloids, Manske, R. H. F. (Ed.) Vol. XV, p. 1. Academic Press, New York 1975.
[200] Stolle, K.; Gröger, D.: Arch. Pharm. **301** (1968) 561.
[201] Stöckigt, J.: J. Chem. Soc., Chem. Commun. **1978**, 1097.
[202] Stöckigt, J.: In: Phillipson, J. D.; Zenk, M. H. (Eds.): Indole and Biogenetically Related Alkaloids, p. 113. Academic Press, London 1980.
[202a] Stöckigt, J.: In: Progress in Tryptophan and Serotonin Research, Schlossberger, H. G.; Kochen, W.; Linzen, B.; Steinhart, H. (Eds.), p. 777, de Gruyter, Berlin/New York 1984.
[202b] Stöckigt, J.; Gundlach, H.; Deus-Neumann, B.: Helv. Chim. Acta **68** (1985) 315.
[203] Stöckigt, J.; Höfle, G.; Pfitzner, A.: Tetrahedron Lett. **21** (1980) 1925.
[204] Stöckigt, J.; Husson, H. P.; Kan-Fan, C.; Zenk, M. H.: J. Chem. Soc., Chem. Commun. **1977**, 164.
[204a] Stöckigt, J.; Pawelka, K.-H.; Rother, A.; Deus, B.: Z. Naturforsch. **37c** (1982) 857.
[205] Stöckigt, J.; Pfitzner, A.; Firl, J.: Plant Cell Reports **1** (1981) 36.
[205a] Stöckigt, J.; Pfitzner, A.; Keller, P. I.: Tetrahedron Lett. **24** (1983) 2485.
[206] Stöckigt, J.; Rüffer, M.; Kan-Fan, Ch.; Husson, H. P.: Planta Med. **39** (1980) 73.
[207] Stöckigt, J.; Soll, H. J.: Planta Med. **40** (1980) 22.
[208] Stöckigt, J.; Treimer, J.; Zenk, M. H.: FEBS-Letters **70** (1976) 267.
[209] Stöckigt, J.; Zenk, M. H.: FEBS-Letters **79** (1977) 233.
[210] Stöckigt, J.; Zenk, M. H.: J. Chem. Soc., Chem. Commun. **1977**, 646.
[211] Stuart, K. L.; Kutney, J. P.; Honda, T.; Worth, R. B.: Heterocycles **9** (1978) 1391, 1419.
[212] Stuart, K. L.; Kutney, J. P.; Worth, R. B.: Heterocycles **9** (1978) 1015.
[213] Stütz, P.; Brunner, R.; Stadler, P. A.: Experientia **29** (1973) 936.
[214] Taylor, W. I.; Farnsworth, N. R. (Eds.): The Catharanthus Alkaloids. Marcel Dekker, New York 1975.
[215] Thomas, R.: Tetrahedron Lett. **1961**, 544.
[216] Treimer, J. F.; Zenk, M. H.: Phytochemistry **17** (1978) 227.
[217] Treimer, J. F.; Zenk, M. H.: FEBS-Letters **97** (1979) 159.
[218] Treimer, J. F.; Zenk, M. H.: Eur. J. Biochem. **101** (1979) 225.
[219] Vining, L. C.; Taber, W. A.: In: Rose, A. H. (Ed.): Economic Microbiology, Vol. 3, p. 389. Academic Press, London 1979.
[220] Vollosovich, A. G.; Nikolaeva, L. A.; Zharko, T. R.: Rast. Resur. 8 (1972) 331.
[221] Wenkert, E.: J. Am. Chem. Soc. **84** (1962) 98.
[222] Wenkert, E.; Dave, K. G.; Lewis, P. G.; Sprague, P. W.: J. Am. Chem. Soc. **89** (1967) 6741.
[223] Weygand, F.; Floss, H. G.; Mothes, U.; Gröger, D.; Mothes, K.: Z. Naturforsch. **19b** (1964) 202.
[224] Willaman, J. J.; Li, J.-L.: Lloydia Suppl. 3A **33** (1970) 1.
[224a] Willingale, J.; Atwell, S. M.; Mantle, P. G.: J. Gen. Microbiol. **129** (1983) 2109.
[225] Yamazaki, M.; Leete, E.: Tetrahedron Lett. **1964**, 1499.
[226] Yeowell, D. A.; Schmid, H.: Experientia **20** (1964) 250.
[227] Zenk, M. H.: J. Nat. Prod. **43** (1980) 438.
[228] Zenk, M. H.; El-Shagi, H.; Arens, H.; Stöckigt, J.; Weiler, E. W.; Deus, B.: In: Barz, W.; Reinhard, E.; Zenk, M. H. (Eds.): Plant cell cultures and their biotechnological applications, p. 27. Springer-Verlag, Heidelberg 1977.

16 Alkaloids Derived from Anthranilic Acid

M. Luckner and S. Johne

16.1 Simple Anthranilic Acid Derivatives

Anthranilic acid, which is incorporated into secondary products, may be formed either *de novo* from chorismic acid or may be synthesized as a product of tryptophan degradation *via* formylkynurenine and kynurenine (*cf.* chapter 8). Anthranilic acid is the precursor of different types of protoalkaloids and true alkaloids (for summaries *cf.* [39, 40, 51, 76]). Most important are the benzodiazepine, quinoline, acridine and quinazoline alkaloids described in sections 16.2 to 16.5. However, simple anthranilic acid derivatives also occur (Fig. 16.1). Anthranilic acid esters, *e.g.* methyl *N*-methylanthranilate (1) together with thymol dominate the smell of mandarin oranges

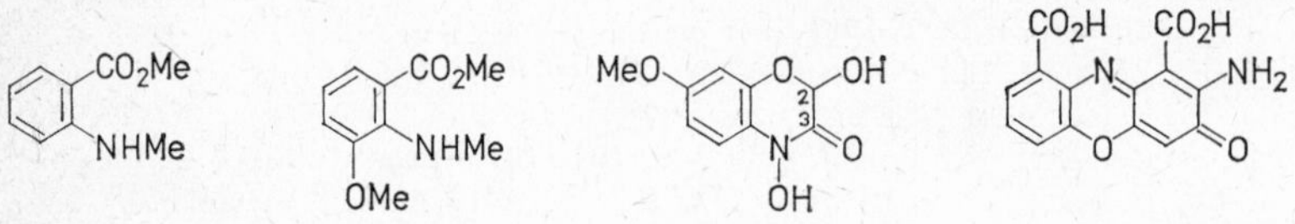

1 Methyl-*N*-methylanthranilate   2 Damascenine   3 DIMBOA   4 Cinnabaric acid

Fig. 16.1
Anthranilic acid-derived protoalkaloids, DIMBOA and cinnabaric acid

[49]. Damascenine (2), a methylated derivative of 3-hydroxyanthranilic acid, is a characteristic constituent of the testa of *Nigella damascena* seeds [91]. It is derived from anthranilic acid *via* 3-hydroxyanthranilic acid and 3-methoxy-anthranilic acid [92, 93]. 2,4-Dihydroxy-7-methoxy-2H-1,4-benzoxazin-3-one (DIMBOA) (3) occurs as a constituent of higher plants. It is formed from anthranilic acid by loss of the carboxy group and incorporation of C-atoms C-3 and -2 of ribose into C-atoms C-1 and -2, respectively [129]. Furthermore 3-hydroxyanthranilic acid and its 4-methyl derivative (5) are precursors of phenoxazinones, *e.g.* cinnabaric acid (4) and actinomycin D (7). In *Tecoma stans* the enzymes anthranilic acid hydroxylase [95] and cinnabaric acid synthetase [125] are involved in the formation of cinnabaric acid. In Actinomycetes tryptophan is degraded to kynurenine, 3-hydroxykynurenine, 3-hydroxy-4-methylkynurenine and 3-hydroxy-4-methyl-anthranilic acid (5), which yields actinomycins by oxidative dimerization probably *via* compound 6 [50, 67, 99] (Fig. 16.2).

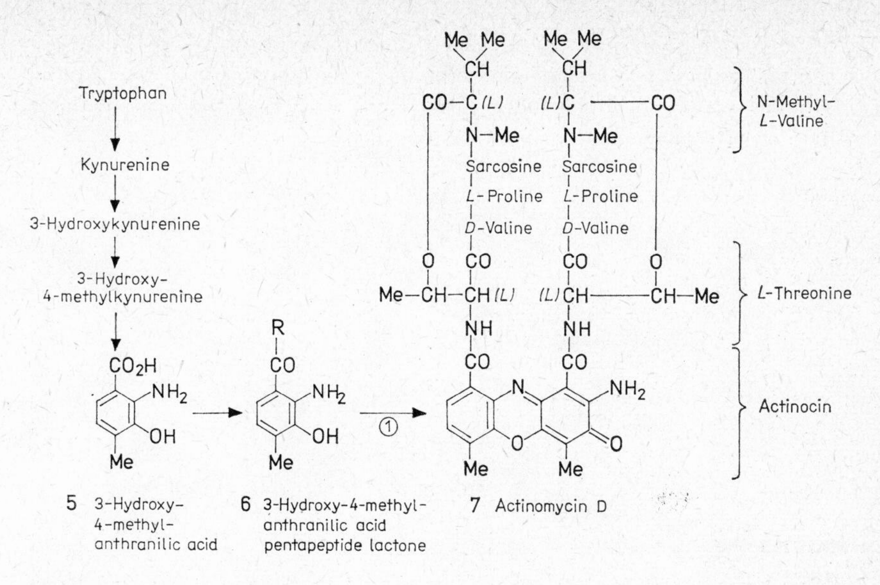

Fig. 16.2
Biosynthesis of actinomycins

1 Phenoxazinone synthetase

16.2 Benzodiazepine Alkaloids

Several strains of *Penicillium cyclopium* synthesize alkaloids of the cyclopenin-viridicatin group (Fig. 16.3). All *C*- and *N*-atoms of the alkaloids are derived from *L*-phenylalanine (**9**), anthranilic acid (**8**), and the methyl group of *L*-methionine (**10**) [35, 78–80, 96–98]. Both the oxygen atoms in cyclopenin (**13**) and cyclopenol (**14**) and the oxygen atoms of the hydroxyl groups of viridicatin (**15**), cyclopenol (**14**) and viridicatol (**16**) are formed from molecular oxygen. Cyclopeptine (**11**) and dehydrocyclopeptine (**12**) are intermediates in cyclopenin biosynthesis, and cyclopenin (**13**) is transformed into cyclopenol (**14**). In an additional step the benzodiazepine alkaloids cyclopenin (**13**) and cyclopenol (**14**) are rearranged to form the quinoline alkaloids viridicatin (**15**) and viridicatol (**16**).

All reactions in the alkaloid formation of *P. cylcopium* are enzyme catalysed. This is also true for the transformation of cyclopenin/cyclopenol into viridicatin/viridicatol, which proceeds spontaneously in alkaline or acidic solution [8, 14, 135]. It was shown that five enzymes and enzyme systems are involved in the formation of the alkaloids (Fig. 16.3) (for a summary *cf.* [77]):

a) Cyclopeptine synthetase system [35, 77, 36a]: This enzyme system catalyses the formation of anthranilic acid AMP and L-phenylalanine AMP, binding of anthranilic acid and L-phenylalanine as thioesters, formation of thioester-bound

N-methyl-L-phenylalanine and N-methyl-L-phenylalanylanthranilic acid, as well as the cyclization of the latter compound to cyclopeptine. These partial activities indicate that cyclopeptine is formed via enzyme-bound intermediates by the thiotemplate mechanism of peptide biosynthesis.

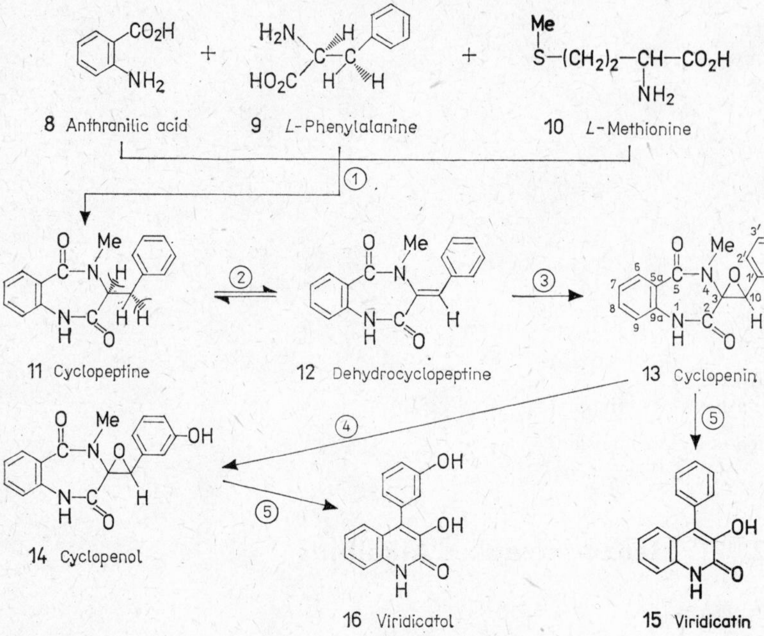

8 Anthranilic acid
9 L-Phenylalanine
10 L-Methionine
11 Cyclopeptine
12 Dehydrocyclopeptine
13 Cyclopenin
14 Cyclopenol
16 Viridicatol
15 Viridicatin

Fig. 16.3
Biosynthesis of the alkaloids of the cyclopenin-viricatin group

1 Cyclopeptine synthetase, 2 cyclopeptine dehydrogenase, 3 dehydrocyclopeptine epoxidase, 4 cyclopenin *m*-hydroxylase, 5 cyclopenase

b) Cyclopeptine dehydrogenase (cyclopeptine: NAD(P) oxidoreductase) [29, 30]: The enzyme is an NAD(P)-dependent flavoprotein that catalyses the reversible transformation of cyclopeptine into dehydrocyclopeptine. It reacts only with the *3S*-isomer of cyclopeptine. During the dehydrogenation of cyclopeptine, hydrogen atoms are almost certainly displaced from positions-3 and -10 of the benzodiazepine nucleus, by a synperiplanar elimination. The hydrid ion split off is transferred to the *pro-4R* position of NAD (*cf.* also Kirby and Narayanaswami [69], who were unable to demonstrate this in *in vivo* experiments).

c) Dehydrocyclopeptine epoxidase (dehydrocyclopeptine, NAD(P)H: O_2 oxidoreductase) [131]: Tracer experiments performed on living cells demonstrated that the epoxide oxygen present in the molecule of cyclopenin and cyclopenol is derived from molecular oxygen [96, 97] indicating that it is introduced by a mixed function oxygenase. *In vitro* experiments showed that dehydrocyclopeptine epoxidase uses NAD(P)H or other reducing compounds directly as co-substrates. Inhibitor experiments indicated that the enzyme is an Fe^{2+} activated FAD-containing flavoprotein.

d) Cyclopenin *m*-hydroxylase (cyclopenin, NAD(P)H: O_2 oxidoreductase, 3′-hydroxylating) [108]: Incorporation of oxygen from $^{18}O_2$ into the hydroxyl group revealed that the hydroxylation is accompanied by an *NIH*-shift. Cyclopenin *m*-hydroxylase may directly use different reduced co-substrates. Inhibition by dicoumarol (but not by CO) indicates that it is a flavoprotein.

e) Cyclopenase (cyclopenin methylisocyanate lyase) [75, 80, 83, 136, 137]: In contrast to the above-mentioned enzymes, which are found in hyphae and conidiospores, cyclopenase is a constituent only of the conidia. Cyclopenase uses cyclopenin and cyclopenol as substrates. From these benzodiazepine derivates, the corresponding quinoline alkaloids are formed by splitting off one mole each of CO_2 and CH_3NH_2. The CO_2 derives from the carbonyl group in position 5, the CH_3NH_2 from the *N*-Me group in position 4 of the benzodiazepine nucleus (Fig. 16.3). Model experiments indicated that the two carbon atoms and the nitrogen atom are presumably extruded as methylisocyanate, which is also evolved during the thermal degradation of cyclopenin and cyclopenol [82]. Hence the methylamine and CO_2 found are the result of subsequent hydrolysis. There was no incorporation of isotopes from ^2H_2O and $H_2{}^{18}O$ into viridicatin/viridicatol during the course of the reaction. This result demonstrated that closure of the heterocyclic ring of the quinoline alkaloids involves C-atoms 5a and 10 of the benzodiazepines and that the hydroxyl oxygen in position 3 of the quinoline is derived directly from the epoxide oxygen.

The transformation of the benzodiazepine into the quinoline alkaloids is initiated by the attack of either an electron acceptor on the epoxide oxygen (*cf.* Fig. 16.4) or an electron donor on the nitrogen atom in position 1 [135]. Probably a tricyclic intermediate (**17**) is formed, although attempts to detect a compound of this type have so far proved unsuccessful.

A special group of benzodiazepines obviously derived from anthranilic acid (**8**) and dopa (**19**) is represented by the antibiotics of the tomaymycin (**20**) and anthra-

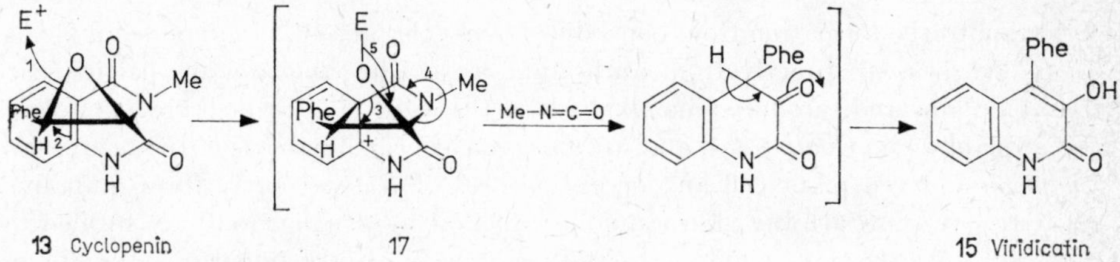

Fig. 16.4
Transformation of cyclopenin into viridicatin

mycin (**21**) types, formed in *Actinomycetes*. The anthranilic acid part of these compounds is derived from tryptophan, while the *O*- and *C*-Me groups as well as the carboxamido C-atom of anthramycin (**21**) originate from methionine. *L*-tyrosine (**18**) and *L*-dopa (**19**) exhibited excellent incorporation rates, indicating that they

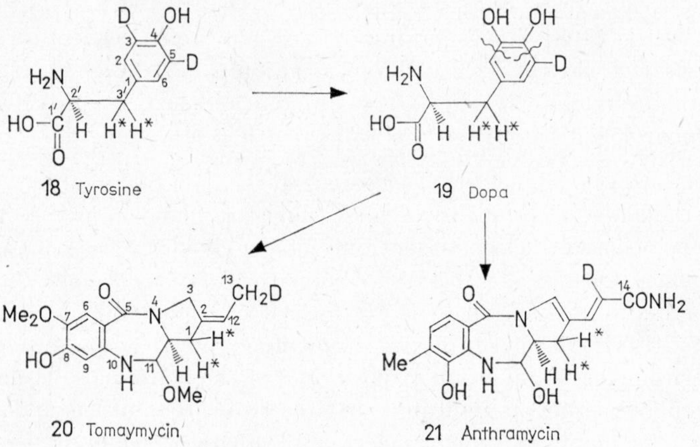

18 Tyrosine

19 Dopa

20 Tomaymycin

21 Anthramycin

Fig. 16.5
Incorporation of tyrosine and dopa into tomaymycin and anthramycin

are precursors. Both hydrogen atoms at C-3′ of tyrosine (**18**) were found at C-1 of the antibiotics. Deuterium or tritium present at C-3 or C-5 of tyrosine (**18**) are found at C-13. Probably the ring of dopa (**19**) is split by a dioxygenase and two carbon atoms are lost. It is unknown, however, whether degradation of the ring proceeds before or after benzodiazepine formation and whether ring fission first occurs between the positions 2 and 3 or between -4 and -5 (Fig. 16.5) [55—59].

16.3 Quinoline Alkaloids

In addition to the formation from benzodiazepines (chapter 16.2) quinoline alkaloids may be synthesized directly from anthranilic acid. The *Rutaceae* are particularly rich in quinoline and furoquinoline alkaloids of this type [133], which are also found in other plants, *e.g.*, *Compositae*, and microorganisms, *e.g.* in strains of the bacterium *Pseudomonas*. Even plant cell and organ cultures of *Rutaceae* may form quinoline derivatives by this pathway. Furoquinoline alkaloids (**32**) (Fig. 16.6) are formed in sterile root cultures of *Ruta graveolens* [70] and were also recorded in cell cultures of this plant [123]. Edulinine (**36**), dictamnine (**32**), and γ-fagarine (8-methoxydictamnine) were produced in cell suspension cultures of *Ruta graveolens* on administration of 2,4-dihydroxyquinoline (24) [11, 124]. After exhaustion of the 2,4-dihydroxyquinoline already formed, dictamnine (32) was rapidly converted to γ-fagarine (8-methoxydictamnine).

Experiments with different species of *Rutaceae* showed that the quinoline nucleus of these alkaloids is derived from anthranilic acid (8) including its carboxy and

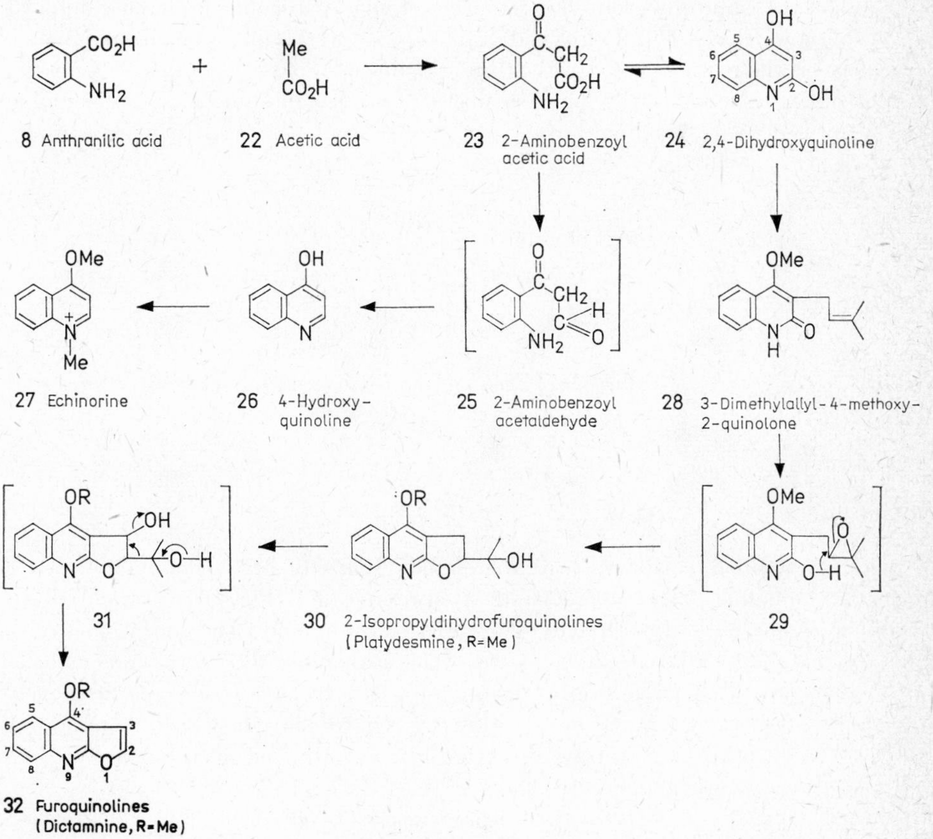

8 Anthranilic acid

22 Acetic acid

23 2-Aminobenzoyl acetic acid

24 2,4-Dihydroxyquinoline

27 Echinorine

26 4-Hydroxyquinoline

25 2-Aminobenzoyl acetaldehyde

28 3-Dimethylallyl-4-methoxy-2-quinolone

31

30 2-Isopropyldihydrofuroquinolines (Platydesmine, R=Me)

29

32 Furoquinolines (Dictamnine, R=Me)

Fig. 16.6
Biosynthesis of quinoline and furoquinoline alkaloids in plants

amino groups plus one molecule of acetate (22) [20, 22, 85, 86, 89, 90]. Incorporation of [3-^{14}C, ^{15}N]2,4-dihydroxyquinoline (24) without change of the $^{14}C/^{15}N$-ratio into furoquinolines (32), *e.g.* skimmianine (7,8-dimethoxydictamnine) and kokusaginine (6,7-dimethoxydictamnine) in *Ruta graveolens* demonstrated that this compound is a key intermediate in the biosynthesis of the furoquinoline alkaloids (32) [20]. 4-Methoxy-2-quinolone is an excellent precursor being incorporated without loss of its methoxy group [21, 46].

Carbon atoms-2 and -3 of the furoquinoline nucleus correspond to the atoms C-4 and -5 of mevalonic acid, respectively [22]. 3-Dimethylallyl-4-methoxy-2-quinolone (**28**) is an excellent precursor of platydesmine (**30**) as well as dictamnine (**32**) and skimmianine in *Skimmia japonica.* It may be transformed into an epoxide (**29**) which, however, has not yet been substantiated. From this epoxide, isopropyldihydrofuroquinolines (**30**) may be synthesized. Formation of the furan ring proceeds with loss of the isopropyl group but retention of **one** hydrogen at C-3. This supports the assumption that a C-3 hydroxylated derivative (**31**), built by a mixed function oxygenase, is the real precursor of the furoquinoline alkaloids (**32**). The furoquinoline skeleton may be modified by further substitutions. Dictamnine (**32**), for instance, is transformed to its 7,8-dimethoxy derivative skimmianine and to more complex alkaloids [21, 44—46].

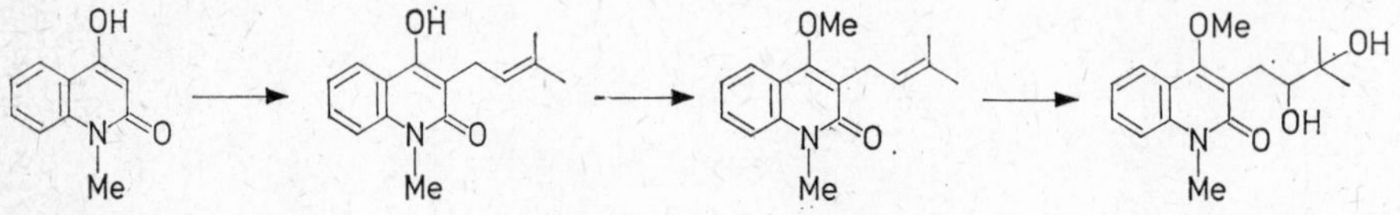

33 *N*-Methyl-2,4-dihydroxyquinoline 34 *N*-Methyl-3-dimethylallyl-2,4-dihydroxyquinoline 35 *N*-Methyl-4-methoxy-3-dimethylallyl-2-quinolone 36 Edulinine

Fig. 16.7
Biosynthesis of edulinine

Slightly modified pathways lead to related groups of alkaloids. *N*-Methyl-2,4-dihydroxyquinoline (**33**) is a precursor of edulinine (**36**) (Fig. 16.7) by a pathway involving *N*-methyl-3-dimethylallyl-2,4-dihydroxyquinoline (**34**) and *N*-methyl-4-methoxy-3-dimethylallyl-2-quinolone (**35**) [11]. Ravenine (**37**) is the precursor of ravenoline (**38**). The conversion plausibly proceeds by a Claisen rearrangement (Fig. 16.8) [17]. Platydesmine (**30**) is transformed to platydesminium metho salt (**39**) (Fig. 16.8) [47]. Quaternization of the nitrogen atom also occurs in the biosynthesis of many other quinoline alkaloids.

Anthranilic acid and acetate are also progenitors of echinorine (**27**), a quaternary quinoline alkaloid from *Echinops* spec. (*Compositae*) [118]. 2,4-Dihydroxyquinoline (**24**) was incorporated into echinorine, but with a relatively low rate [117]. Echinorine

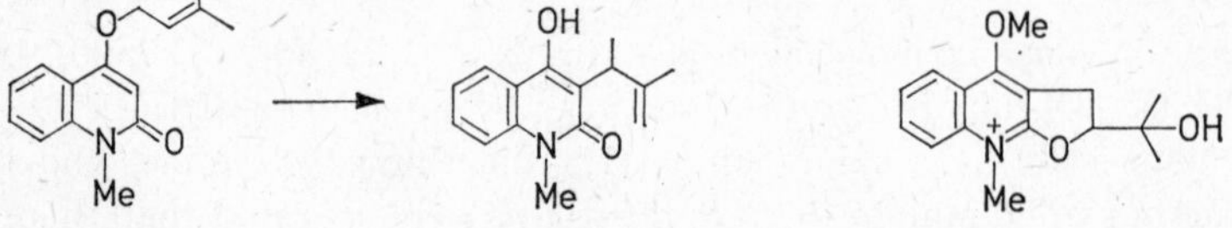

37 Ravenine 38 Ravenoline 39 Platydesminium methosalt

Fig. 16.8
Ravenine, ravenoline, and platydesminium metho-salt

is easily degraded *in vitro* to echinopsine (**40**) or, in the presence of ammonia to echinopsidine (**41**) (Fig. 16.9) [119, 120].

In summary, the biosynthesis of the above-mentioned alkaloids may proceed by the pathway in Fig. 16.6. Anthranilic acid (**8**) may react with or without activation with acetate (**22**) to yield possibly activated 2-aminobenzoylacetic acid (**23**). This compound may be either reduced to the aldehyde (**25**), which cyclizes by azomethine formation and is a direct precursor of the 4-quinolones of the echinorine type (**27**). Alternatively it may form 2,4-dihydroxyquinoline (**24**) from which the isoprenylated 2,4-dihydroxyquinolines (**28**) and in further reactions the isopropyldihydrofuroquinolines (**30**) and furoquinolines (**32**) are derived.

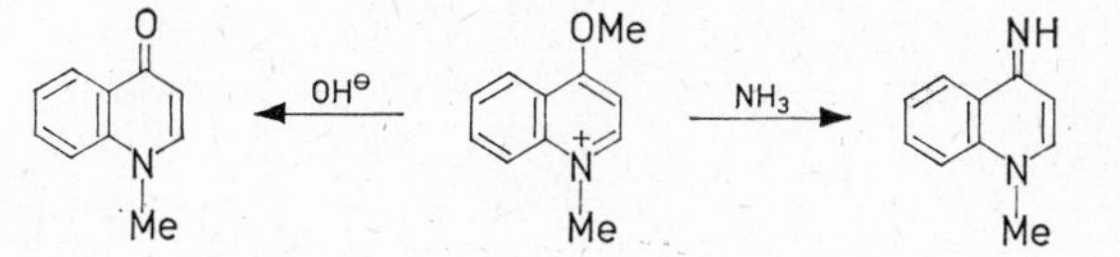

40 Echinopsine 27 Echinorine 41 Echinopsidine

Fig. 16.9
Transformation of echinorine to echinopsine and echinopsidine

Anthranilic acid (probably anthranilyl CoA) (**42**) and a possibly activated β-ketoacid (**43**), which is derived from acetate and malonate, are the precursors of the 2-n-alkylquinolones (pseudans) (**45**) formed in *Pseudomonas aeruginosa* (Fig. 16.10). Feeding of malonic acid demonstrated that the side chain is built starting from the terminal C-atom as is usual in fatty acid biosynthesis [81, 109].

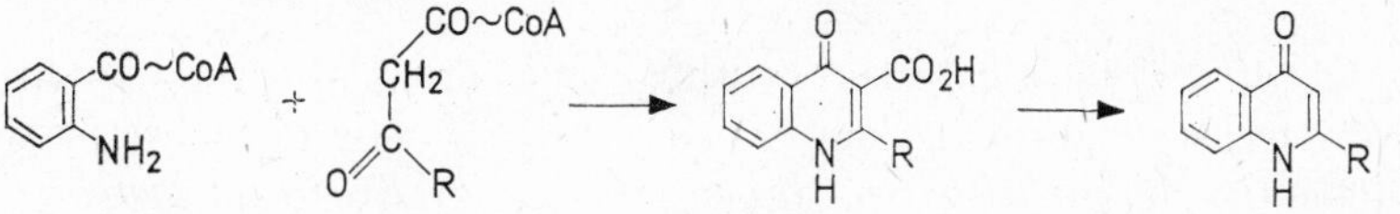

42 Anthranilyl CoA; 43 β-Keto acid-CoA-ester; 44 2-Alkyl-3-carboxy-4-quinolone; 45 Pseudan ($R = (CH_2)_{6-8}-Me$); 46 2-Phenyl-4-quinolone ($R = C_6H_5$)

Fig. 16.10
Biosynthesis of 4-quinolones substituted in position-2

In a similar manner 2-phenyl-4-quinolones (**46**) and 2-phenylethyl-4-quinolones may be formed in plants (Fig. 16.10). Anthranilic acid including its amino group and phenylalanine were incorporated into graveoline, 1-methyl-2-(3′,4′-methylenedioxyphenyl)-4-quinolone, in *Ruta angustifolia* [9]. Unexpectedly, however, a large part of the label of the carboxy group of anthranilic acid was lost probably by recycling *via* tryptophan [9]. Incorporation of tritium from [3-^{14}C, *p*-^{3}H]phenylalanine into graveoline indicated occurrence of an *NIH*-shift during hydroxylation in the *p*-position. From [3-^{14}C, *m*-^{3}H]phenylalanine, half of the tritium activity was lost during the introduction of the second hydroxy group [9]. Benzoic acid (**49**) (rate of incorporation 0.001%) and acetate (**48**) (rate of incorporation 0.002%) were incorporated into 4-methoxy-2-phenylquinoline (**50**) in *Lunasia amara*, whereas the rate of incorporation of phenylalanine (9) was only about 0.0001% [33]. These results probably reflect that in this plant the assumed benzoylacetic acid intermediate (47) is formed by condensation of benzoic acid and acetate, rather than from phenylalanine (Fig. 16.11).

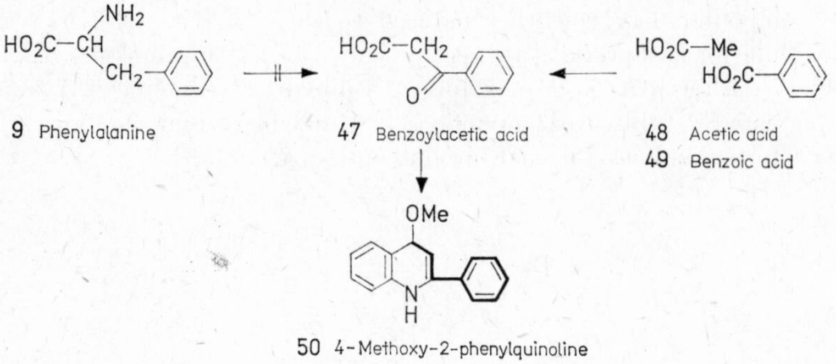

Fig. 16.11
Formation of 4-methoxy-2-phenylquinoline in *Lunasia amara*

16.4 Acridines

16.4.1 Structural Types, Occurrence, Biological Activity

The approximately 40 acridine alkaloids constitute a small group of natural products found exclusively in the following genera of the *Rutaceae* family of higher plants: *Acronychia, Atalantia, Balfourodendron, Beonninghausenia, Evodia, Fagara, Glycosmis, Melicope, Ruta, Thamnosma,* and *Teclea.* In this family the acridines are often associated with quinolines, benzyltetrahydroisoquinolines, imidazole derivatives, oxazoles, and quinazolines. The sustained interest in this field has been due to the reported activity of acronycine (51) (Fig. 16.12), a constituent of *Acronychia baueri* and *Vepris amphody,* which has the broadest spectrum of *in vivo* antitumour activity of any known natural product [23, 127].

The acridine alkaloids are weak bases and yellow. Because the methyl group in the 1 position can easily be eliminated by treatment with mineral acid, it has been discussed whether the nor-compounds are genuinely present in the plants. Normelicopicine, normelicopidine and normelicopine are probably artifacts [126]. Arborinine (52) is genuinely present in *Glycosmis,* since this alkaloid can be isolated without acid treatment by extraction with petroleum ether [18]. Also evoprenine [25] and the chlorine-containing acridine alkaloids, for instance gravacridonolchlorine (54), are true natural products and not artifacts. Atalanine (58) and ataline, isolated from *Atalantia ceylanica,* represent dimeric acridine alkaloids [36]. A number of new acridines have been isolated in recent years and some known acridines have

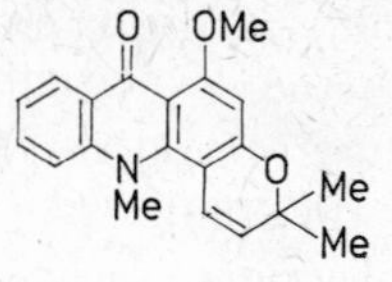

51 Acronycine

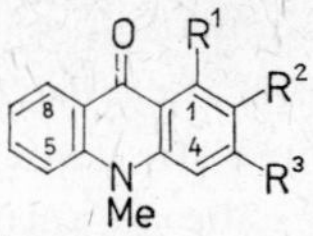

52 R^1=OH; R^2=R^3=OMe : Arborinine

53 R^1=OH; R^2=R^3=H

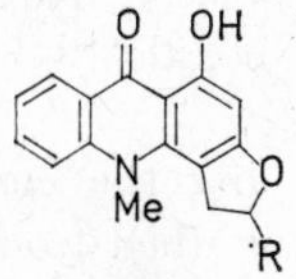

54 R= $-C(Cl)(CH_2OH)-CH_2OH$ Gravacridonolchlorine

55 R= $-C(Me)(OH)-CH_2OH$ Gravacridondiol

56 R= $-C(Me)(OH)-CH_2O-$glucosyl Gravacridondiolglucoside

57 R= $-C(Me)=CH_2$ Rutacridone

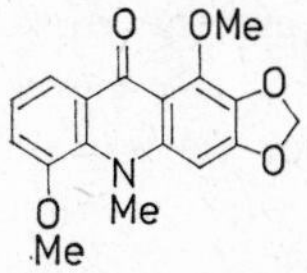

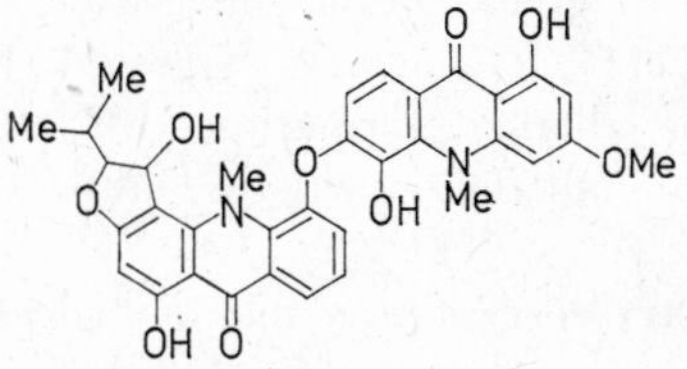

58 Atalanine

59 Tecleanthine

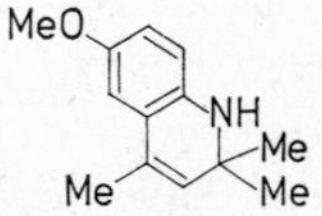

60

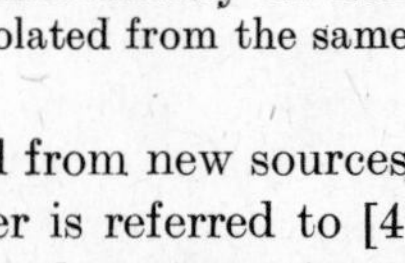

61 R^1=R^5=R^7=R^8=H; R^2=R^4= $-CH_2-CH=C(Me)_2$; R^3=R^6=OH : Atalaphylline

62 R^1=R^4=R^5=R^6=H; R^2=R^7= $-CH_2-CH=C(Me)_2$; R^3=R^8=OH : Atalaphyllidin

Fig. 16.12
Various types of acridine alkaloids
Unfortunately the name atalaphyllidine has been given to two different alkaloids isolated from the same plant [5].

been obtained from new sources. A comprehensive representation here is impossible and the reader is referred to [43, 121]. Only information on some especially interesting compounds or investigations will be mentioned.

The biogenetically unusual alkaloid **53** — C-ring substituted in position 1 only! — has been isolated from *Ruta graveolens* [106]. Besides **54** some further alkaloids of the anellated dihydrofuran ring type have been obtained [104, 107].

Syntheses of **51** (see [10, 52]), tecleanthine (**59**) [101] and of further acridine alkaloids have been performed. On the basis of a biogenetic route (see below) biomimetic syntheses of **51** have been accomplished [1, 3]. The ring CD *N*-containing analogue **60** of **51** has been synthesized and found to possess antitumour activity

[74]. **51** inhibits the accumulation of extracellular uridine and thymidine nucleotides in the precursor pool [27]. Extensive spectroscopic studies [7, 103] and synthesis [88] have now established the angular structure for rutacridone (**57**). Dubamine has been shown to possess broad antimicrobial properties [60].

There are also acridine alkaloids with two prenyl groups, such as atalaphylline (**61**) [37] or compound **62** [19], isolated from *Atalantia monophylla*. The presence of two phenolic hydroxy-groups in 5-hydroxyarborinine, isolated from the leaves of *Glycosmis bilocularis* [13] is remarkable. Rutacridone epoxide [94] and hydroxy-rutacridone epoxide [28] show antibiotic properties [138].

Rutacridone (**57**), 1-hydroxy-*N*-methylacridone (**53**) and the corresponding 3-methoxy derivative were isolated from callus cultures obtained from meristematic cells of *Ruta graveolens*. The alkaloid pattern of the callus was different from that of the stem cells of intact plants [114]. Remarkable differences exist with respect to the localization of acridine alkaloids in different organs of *Ruta graveolens*. Especially rich were the roots of young and the xylem of old plants. The acridines are attached to the cell wall in aggregations or they are distributed more or less even throughout the cytoplasm. Idioblasts also contain large amounts of acridine alkaloids [130]. Gravacridondiol (**55**) and its glycoside (**56**) have been isolated from root tissue cultures of the same species [111]. It has been suggested that the light may play an important role in acridine alkaloid formation [102, 128]. Further investigations on the occurrence of acridine alkaloids in tissue cultures are cited in [138].

Comprehensive reviews on the chemistry, occurrence and biological activity of acridine alkaloids have appeared [53, 63, 87, 102, 105, 113, 133]. The earlier work in this field has been adequately dealt with in the well-known series, The Alkaloids, by Manske and Holmes.

16.4.2 Biosynthesis

It is generally thought that the acridines are biosynthetically derived from anthranilic acid (**63**). For a long time it has been known that condensation of *o*-aminobenzaldehyde and phloroglucinol (**65**) gives 1,3-dihydroxyacridine (**67**) [31]. This reaction has been examined under "physiological-type" conditions [54] — room temperature, dilute aqueous solution — and has been found to give an acridine that can be oxidatively converted into 1,3-dihydroxyacridone (**68**). Using this result the authors suggested a scheme for the biogenesis, but some recently isolated alkaloids do not fit. For further discussion of this topic the reader is referred to the following reviews: [39, 40, 63, 102]. Robinson [110] has postulated that tryptophan is oxidized to form 4-hydroxyquinol-2-one (**66**), or that anthranilic acid condenses with dioxohexanoic acid (**64**) to form the basic acridone nucleus, which is then suitably modified [71]. The gradual addition of acetate to anthranilic acid (Fig. 16.13) has also been suggested. **51** could possibly be formed from "activated" anthranilic acid, 3 malonyl-CoA-units and mevalonic acid (Fig. 16.14). The prenyl side-chain of other acridine alkaloids may be derived from mevalonate, but experimental evidence is still lacking.

Initial biosynthetic experiments have shown anthranilic acid to be the precursor of ring *A* and C-9 and N-10 in the acridines [41, 100].

The origin of ring *C* is uncertain. In analogy to the biosynthesis of fatty acids and some aromatic compounds it could be assumed that in the formation of the "phloroglucinol portion" of the acridines, acetate is incorporated *via* malonyl-CoA and a polyketo acid or directly from phloroglucinol. Certain results [61] favour the

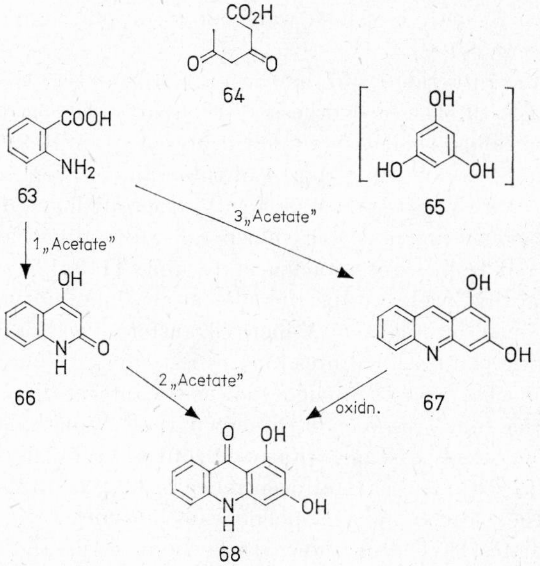

CO_2H
64
COOH
NH_2
63
OH
HO OH
65
3 „Acetate"
1 „Acetate"
OH
N H O
66
OH
N OH
67
2 „Acetate"
oxidn.
O OH
N H OH
68

Fig. 16.13
Possible biosynthetic pathways of acridines.

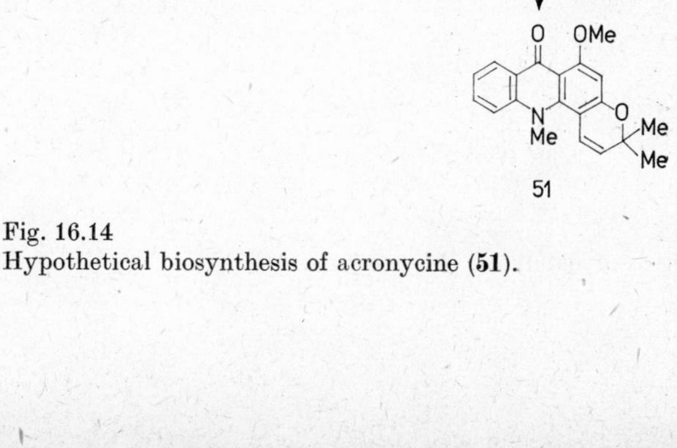

COSCoA
NH_2
+ 3 CO_2H
$CH_2COSCoA$
O O
N S
H_2CoA
CO_2H CH_2OH
HO Me
O OMe
N Me
O
Me
Me
51

Fig. 16.14
Hypothetical biosynthesis of acronycine (**51**).

acetate pathway: under identical experimental conditions [2-^{14}C]acetate clearly gave the highest degree of incorporation in comparison with [2,4,6-^{14}C]phloroglucinol (1.49%, 0.19%, respectively) in arborinine (52). Also [2,4,11-^{14}C]1,3-dihydroacridine, [2,4,11-^{14}C]1,3-dimethoxy-10-methylacridiniummethylsulphate and [2,4,11-^{14}C]1,3-dimethoxy-10-methylacridon are converted into **52**. Sodium acetate was also incorporated into the acridine alkaloids in *Acronychia baueri*. The incorporation of 4-hydroxyquinol-2-one and its *N*-methyl derivative indicated that these compounds are likely precursors, the greater incorporation of the latter suggesting it was closer to the end of the biosynthetic pathway. [2-^{14}C]Mevalonic acid and [^{3}H]-1,2,3-trihydroxy-*N*-methylacridone were not incorporated into the alkaloids of *Evodia xanthoxyloides* [100].

The dihydrofuroacridine rutacridone (**57**) is the main alkaloid in tissue cultures of *Ruta graveolens* strain R-19. The biosynthesis of this particular acridine alkaloid was investigated by using calluses and suspension cultures of strain R-19 [6]. Anthranilic acid is specifically incorporated into ring *A* of rutacridone. Surprisingly mevalonic acid was a poor and unspecific precursor for its isopropylidene dihydrofuran part. Carbon-13 labelling experiments have shown unequivocally that ring *C* of rutacridone (**57**) is derived from three intact acetate units [140a]. *N*-methylation of anthranilic acid is the first pathway-specific step in acridone biosynthesis [5b]. The *S*-adenosylmethionin: anthranilic acid *N*-methyltransferase was only detectable in rutacridone synthesizing plant cell cultures [5a].

An aminobenzophenone has also been suggested as an intermediate in acridine biosynthesis [12] and the same workers have shown that *N*-methylaminobenzophenone (**69**) (Fig. 16.15) is easily and quantitatively converted into 1,3-dimethoxy-10-methylacridone (**70**) [2, 68]. The natural occurrence of **69** [16, 132, 134] and **70** [134] offers support for the role of aminobenzophenones as biosynthetic intermediates. Benzophenone intermediates have been shown to be involved in the biosynthesis

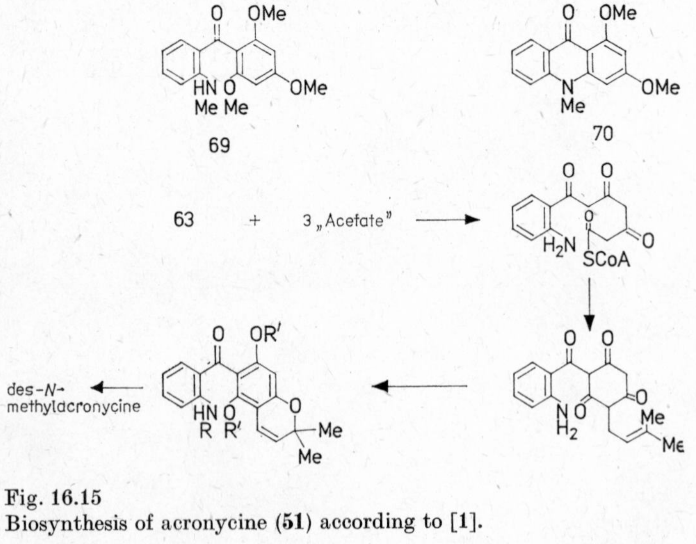

Fig. 16.15
Biosynthesis of acronycine (**51**) according to [1].

of xanthone and grisan type phenolic compounds. *N*-Methylation may occur at a late stage in the sequence since des-*N*-methylacronycine has been isolated from *Glycosmis*. Based on these results it has been postulated that the biosynthesis of **51** involves a triketide intermediate to which the chromenyl moiety is introduced prior to cyclisation and methylation, as indicated in Fig. 16.15 [1]. It has also been postulated that some parts in the biosynthetic pathways of acridines, quinolines and coumarins are almost identical [102, 128]. Fig. 16.16 shows mutual intercon-

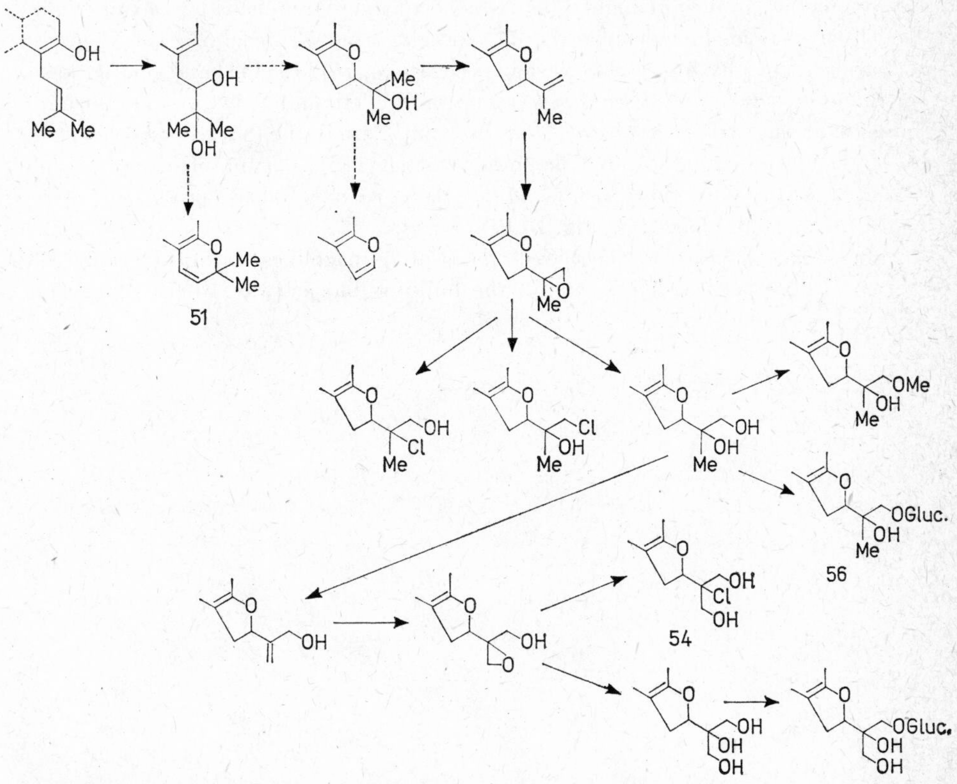

Fig. 16.16
Biosynthetic interconversions between furanoid-type acridines, furoquinolines and coumarins (according to [102]).

versions in the biosynthesis of coumarins, the dihydrofuran part of furoquinolines, and that of compounds of type **54**. The naturally occurring compound **61** is a plausible intermediate in the biosynthesis of furanoid type acridines [102]. A comparison of the acridine pattern of *Beonninghausenia albiflora* with that of *Ruta graveolens*, which belongs to the same subtribe, shows substantial differences, *e.g. Ruta graveolens* seems to have only the capacity for synthetising furanoid-type isoprenoid derivatives, but in a great structural diversity on the isopropyl side-chain. *Boenninghausenia albiflora* is capable of the synthesis of both furanoid and pyranoid derivatives, but only of the two simplest, rutacridone and noracronycine [112].

In spite of numerous efforts the exact mechanism of acridine alkaloid biosynthesis is still obscure and further investigations are necessary.

16.5 Quinazolines

About 60 alkaloids containing a quinazoline nucleus have been found in taxonomically unrelated families of higher plants, also in microorganisms and even in animals. These alkaloids include arborine (**71**) and glycorine (**73**) from *Glycosmis arborea*, the pharmacologically interesting vasicine (peganine, **75**) [48, 141] and related bases from *Adhatoda vasica* and *Peganum harmala*, and evodiamine (**81**) as well as rutaecarpine (**82**) from various *Rutaceae* [4, 42, 62, 122]. Paraensine (**83**) from *Euxylophora paraensis* contains an additional isoprenoid moiety [24]. Tryptanthrin (**84**) is an antibiotic produced by the yeast *Candida lipolytica* only in the presence of large amounts of *L*-tryptophan [15] (Fig. 16.17).

In almost all theories on the biosynthesis of quinazolines anthranilic acid (**63**) or an equivalent is regarded as one of the building blocks [38—40, 72, 110]. Thus

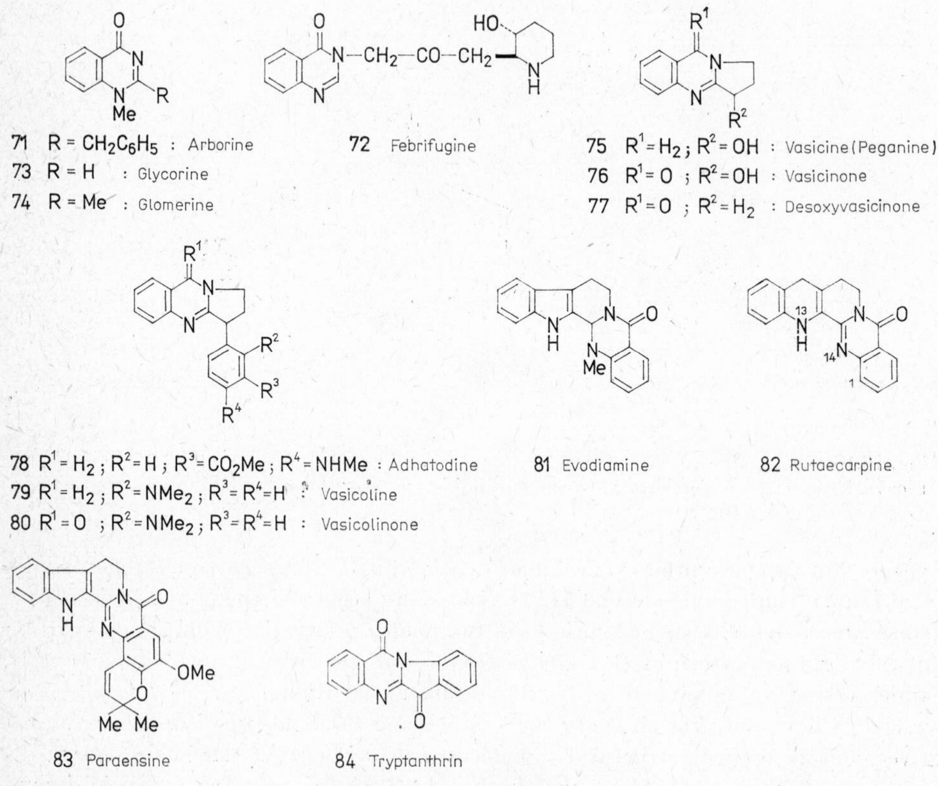

Fig. 16.17
Various types of quinazoline alkaloids

vasicine (**75**) may be derived from anthranilic acid and proline, or closely related metabolites. *N*-Methylanthranilic acid, ammonia, and phenylacetic acid can be considered precursors of arborine (**71**). Febrifugine (**72**) could possibly be formed from **63**, formic acid, ammonia, a C_3-unit and lysine or equivalent compounds. For 6,7,8,9-tetrahydro-pyrido[2,1-*b*]quinazolin-11-one and related alkaloids it has been postulated that their biosyntheses start from **63** and lysine [34]. Several quinazolines have been synthesized under so-called physiological conditions (Fig. 16.18).

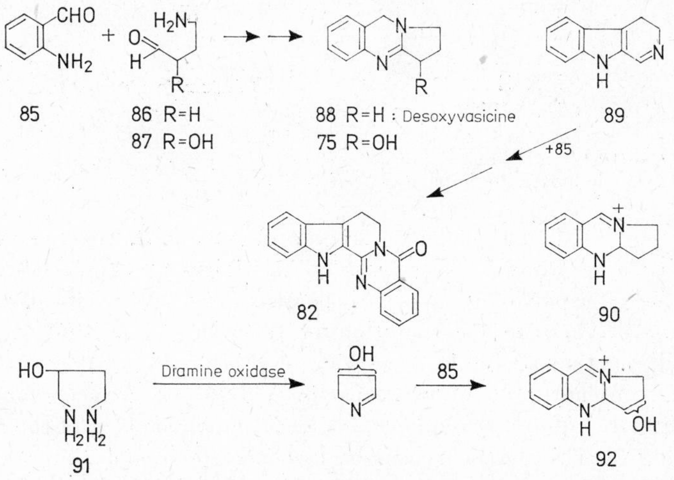

Fig. 16.18
Some *in vitro* syntheses of quinazolines under physiological conditions

16.5.1 Pyrroloquinazolines

In 1960 it was demonstrated that aseptically grown plants of *Peganum harmala* are capable of incorporating [^{14}COOH] anthranilic acid into vasicine (**75**). With *Adhatoda vasica* plants it was shown that the incorporation of anthranilic acid (**63**) involves retention of the carboxyl group. The origin of the "non-anthranilic acid moiety" is still controversial. It was found that in *Adhatoda vasica* C-1, C-2, and N-11 of **75** may be derived from aspartic acid (**93**) and C-3 and C-10 from a C_2-unit (Fig. 16.19) [66, 131a]. This was supported experimentally by the incorporation of [3-^{14}C] aspartic acid, [3-^{14}C] malic acid, and [2,3-^{14}C] succinic acid, with localization of the label at C-1 and C-2 of **75**. The incorporation of succinic acid could proceed *via* aspartic acid since succinic acid can easily be converted to oxalacetic acid in the citric acid cycle. After feeding [3-^{14}C] malic acid 72% of the radioactivity of **75** was found in carbon atoms 1 (26%) and 2 (46%) without equilibration. The labelling pattern after administration of [3-^{14}C] aspartic acid or [3-^{14}C] malic acid could be explained by conversion of aspartic acid into fumaric acid involving aspartase or oxaloacetate, which in turn can be converted into fumarate *via* malate. These processes are reversible, and malate and aspartate resynthesized from fumarate

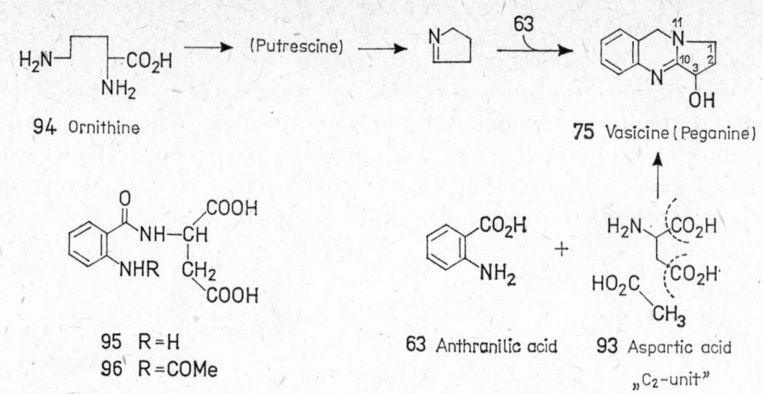

94 Ornithine (Putrescine) 63 75 Vasicine (Peganine)

95 R=H
96 R=COMe

63 Anthranilic acid 93 Aspartic acid „C_2-unit"

Fig. 16.19
Possible biosynthetic pathways of vasicine

will contain equal amounts of ^{14}C in carbon atoms 2 and 3. Together with the precursor originally fed, one would expect radioactivity in C-1 and C-2. The radioactivity of [4-^{14}C] aspartic acid was incorporated only non-specifically and did not label the C-3 of **75**. The incorporation of [2-^{14}C]acetate, [2-^{14}C]glycine, [3-^{14}C]-pyruvic acid and the specific incorporation of [2′-^{14}C, ^{15}N]*N*-acetylanthranilic acid and especially of anthranoyl-[3-^{14}C]aspartic acid (**95**) is in accordance with the other results. After administration of the latter compound, 77% of the radioactivity was found in C-3. This derivative has been suspected to be an intermediate in the biosynthesis of **75**. Subsequently it would be possible that the most important precursor of **75** may be acetylanthranoyl aspartic acid (**96**). The results after feeding anthranoyl aspartic acid and [^{15}N]aspartic acid indicate that the N-11 of **75** may be derived from **93**. [^{15}N]Anthranilic acid and [CO-$^{15}NH_2$]glutamin are converted into **75** with a high specific rate of incorporation. These compounds apparently provide the N-9 of the alkaloid. [CO-$^{15}NH_2$]Anthranilic acid amide was poorly incorporated.

Members of the α-ketoglutaric acid family (glutamic acid, 4-hydroxyglutamic acid, 4-amino-2-hydroxybutyric acid, ornithine, proline, putrescine) as well as [^{14}C]-oxalic acid, N-[^{14}C]oxalylanthranilic acid, [1-^{14}C]glycine and [1-^{14}C]glyoxylate were incorporated only non-specifically. [^{14}CHO]*N*-Formylanthranilic acid and [$^{14}CH_3$]*N*-methylanthranilic acid were not incorporated. Likewise it has been shown that no direct hydroxylation of desoxyvasicine (**88**) has taken place and [U-^{14}C]-vasicinone has been poorly incorporated in **75**.

It has been found that in *Peganum harmala* [2-^{14}C]ornithine and labelled proline, putrescine, and related compounds are more or less specifically incorporated into the pyrrolidinoring system of **75** [73]. This result suggested that the ornithine (**94**) was being decarboxylated affording putrescine, a symmetrical molecule, which would then be incorporated resulting in equal labelling at C-1 and C-10. The results can be rationalized by postulating that a symmetrical intermediate such as putrescine is involved in the pathway from ornithine to **75** or that [2-^{14}C]ornithine serves as a source of [1-^{14}C]acetate (*via* α-ketoglutaric-, succinic-, fumaric-, and malic acid) which is then incorporated by way of *N*-acetylanthranilic acid. If both these pathways to **75** are valid, and this has still to be established, then the results are unique, because plant alkaloid biosynthesis has so far not been found to be species-specific (*cf.* gramine biosynthesis) unless one includes the radically different pathways to coniine or *N*-methylpelletierine. The investigations do not indicate which of the alkaloids is formed first as a result of the condensation yielding the pyrroloquinazoline system. Nothing is known of any interconversions that occur between the alkaloids in *Peganum harmala*. They have been masked by their rapid metabolisms. After feeding [^{3}H]- **75**, radioactivity was detected in **76** and **77**. Since only 65% of the radioactivity was recovered, it is possible that **75** is degraded in the plant.

The alkaloids **78**—**80** [64], anisotine and sessiflorine may arise from pyrroloquinazolines and derivatives of anthranilic acid.

16.5.2 Arborine and Glomerine

In *Glycosmis arborea* arborine (**71**) is produced from phenylalanine and anthranilic acid (**63**) [41]. After administration of [U-^{14}C]phenylalanine, 92% of the radioactivity is located in the phenylacetic acid part of **71**. [^{3}H]Anthranilic acid was specifically incorporated into the *N*-methylanthranilic acid moiety of **71**. After feeding [$^{14}CH_3$]methionine, the *N*-methyl group of **71** was specifically labelled. The same result was obtained after the application of [$^{14}CH_3$]*N*-methylanthranilic acid. These results seem to indicate that anthranilic acid is methylated prior to the condensation with phenylalanine which leads to **71**. The N-1 atom of **71** is provided by the nitrogen of anthranilic acid. The experiment with [3-^{14}C, ^{15}N]phenylalanine shows clearly that with the exception of the carboxyl group the aliphatic side-chain of phenylalanine is incorporated into arborine. Therefore, phenylacetic acid can be excluded as an immediate precursor of **71**. [U-^{3}H]Anthranoylphenylalanine was a very poor precursor of arborine. *N*-Methyl-*N*-phenyl-[$^{14}CO-CH_2$]acetyl-anthranilamide was transformed very efficiently in *Glycosmis* into arborine. However, it is doubtful whether the *N*-phenylacetyl derivative of *N*-methylanthranilamide is a true precursor in arborine biosynthesis. Surprisingly it was shown that the transformation of *N*-methyl-*N*-phenyl-acetylanthranilamide to **71** also takes place in pea plants. Apparently the ring closure of *N*-methyl-*N*-phenyl-acetyl-anthranilamide is catalysed by an enzyme or enzymatic system that is unspecific and widely distributed [65]. In contrast to these results another laboratory has postulated that the incorporation of phenylalanine proceeds *via* phenylacetic acid and *N*-methyl-*N*-(phenylacetyl)anthranilic acid [26].

It has been demonstrated that [^{14}COOH] anthranilic acid is also a precursor of the animal alkaloids glomerine (**74**) and homoglomerine [115].

16.5.3 Indoloquinazolines

Evodiamine (**81**) and rutaecarpine (**82**) were biosynthesized in *Evodia rutaecarpa* from tryptophan (**97**), anthranilic acid (**63**) and formic acid (Fig. 16.20) [139]. After administering [$^{14}CH_3$] methionine, 99% of the radioactivity was located in C-13b of **82**, whereas in **81** the label was distributed between C-13b and the *N*-methyl group.

As the specific radioactivity of evodiamine (**81**) was lower than that of rutaecarpine (**82**), rutaecarpine cannot be regarded as the precursor of evodiamine. The authors [140] assume that the C_1-unit is introduced at an earlier stage of the biosynthesis. It is possible that *N*-methylanthranilic acid is formed first and that it

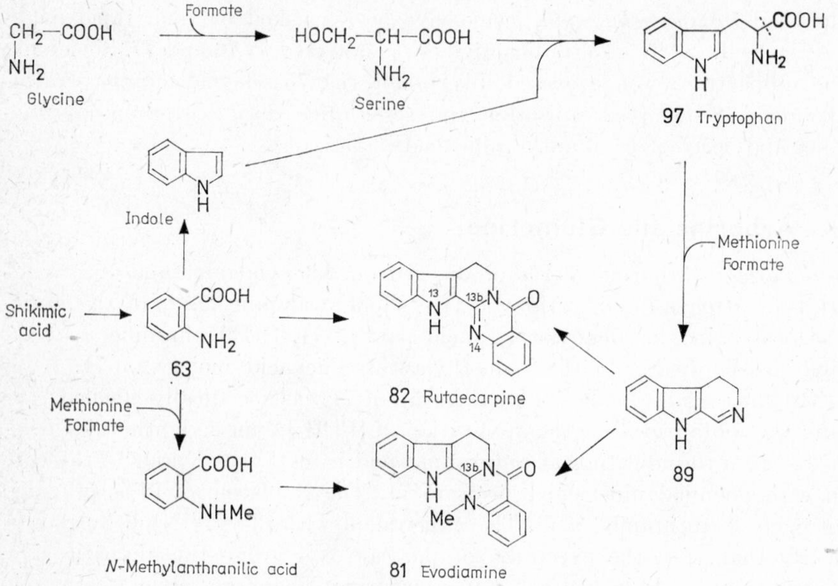

Fig. 16.20
Biosynthesis of evodiamine and rutaecarpine

then condenses with dihydronorharmane (**89**) to give **81**. In this case the radioactivity of the administered C_1-compound would be diluted by the non-labelled *N*-methylanthranilic acid already present, which could explain the lower specific radioactivity of **81** (Fig. 16.20). However, experimental evidence for a possible involvement of a β-carboline (**89**) in the biosynthesis of *Evodia* alkaloids has so far not been presented.

16.5.4 Pseudomonas Alkaloids and Tryptanthrin

A small number of simple substituted quinazolines can be formed in the course of tryptophan degradation [84]. In *Pseudomonas* species three pathways of tryptophan degradation are known: the aromatic pathway in *Ps. fluorescens,* the quinoline pathway in *Ps. acidovorans* and the quinazoline pathway in *Ps. aeruginosa* (Fig. 16.21). Investigations with [β-^{14}C] tryptophan provided evidence for the new pathway from tryptophan through the intermediates formylkynurenine (**98**), *N*-formyl-

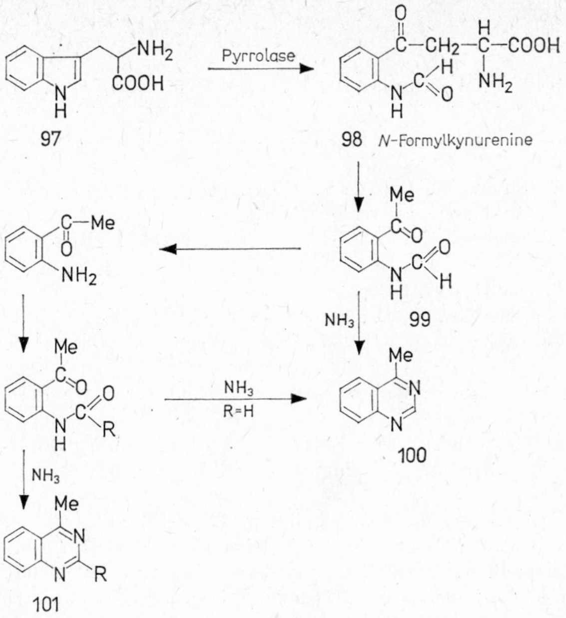

Fig. 16.21
Quinazoline pathway in *Pseudomonas aeruginosa*

aminoacetophenone (**99**) forming 4-methylquinazoline (**100**) with ammonia and free 2-aminoacetophenone. After reacylating and cyclisation with ammonia this produces the other derivatives of 4-methylquinazoline (**101**).

The antibiotic tryptanthrin (**84**) has been biosynthesized from 1 mole tryptophan and 1 mole anthranilic acid. When feeding tryptophan and substituted anthranilic acids, or substituted tryptophans and anthranilic acid, the expected derivatives of (**84**) have been isolated. The enzymes of the biosynthesis of (**84**), with the exception of bromotryptophan, had no specificity for these substrates. During the biosynthesis the anthranilic acid moiety results from tryptophan degradation [32, 116].

Tetrodotoxin

The biosynthetic origin of tetrodotoxin is a mystery until now. The feeding experiments with very universal precursors as [2-^{14}C]acetate, [guanido-^{14}C]arginine, [U-^{14}C]-D-glucose, and [ureido-^{14}C]citrulline by injection, and by oral and external administration using the newts *Taricha torosa* and *T. granulosa* as well as using the Atlantic puffer *Spheroides maculatus* were unsuccessful [120a]. The author interpreted the lack of the *de novo* synthesis by many ways *e.g.*

a) the toxin is synthesized during only a very limited developmental stage and its turnover is very slow;
b) the toxin is synthesized only under certain stressed conditions as a self-defense mechanism.

16.6 References

[1] Adams, J. H.; Brown, P. M.; Gupta, P.; Khan, M. S.; Lewis, J. R.: Tetrahedron **37** (1981) 209.
[2] Adams, J. H.; Gupta, P.; Khan, M. S.; Lewis, J. R.: J. Chem. Soc. Perkin I **1977**, 2173.
[3] Adams, J.; Gupta, P.; Lewis, J. R.: Chem. & Ind. **1976**, 109.
[4] Armarego, W. L. F.: In: The Chemistry of Heterocyclic Compounds, Vol. **24/1**: Quinazolines, Interscience Publishers, New York/London/Sidney 1967; Adv. Heterocycl. Chem. **24** (1979) 1.
[5] Basa, S. C.: Experientia **31** (1975) 1387.
[5a] Baumert, A.; Hieke, M.; Gröger, D.: Planta med. **48** (1983) 258.
[5b] Baumert, A.; Kuzovkina, I. N.; Hieke, M.; Gröger, D.: Planta med. **48** (1983) 142.
[6] Baumert, A.; Kuzovkina, I. N.; Krauss, G.; Hieke, M.; Gröger, D.: Plant Cell Reports **1** (1982) 168.
[7] Bergenthal, D.; Mester, I.; Rózsa, Zs.; Reisch, J.: Phytochemistry **18** (1979) 161.
[8] Birkinshaw, J. H.; Luckner, M.; Mohammed, Y. S.; Mothes, K.; Stickings, C. E.: Biochem. J. **89** (1963) 196.
[9] Blaschke-Cobet, M.; Luckner, M.: Phytochemistry **12** (1973) 2393.
[10] Blechert, S.; Fichter, K.-E.; Winterfeldt, E.: Chem. Ber. **111** (1978) 439.
[11] Boulanger, D.; Bailey, B. K.; Steck, W.: Phytochemistry **12** (1973) 2399.
[12] Bowen, I. H.; Gupta, P.; Lewis, J. R.: J. Chem. Soc., Chem. Commun. **1970**, 1625.
[13] Bowen, I. H.; Perera, K. P. W. C.; Lewis, J. R.: Phytochemistry **17** (1978) 2125.
[14] Bracken, A.; Pocker, A.; Raistrick, H.: Biochem. J. **57** (1954) 587.
[15] Brufani, M.; Fedeli, W.; Mazza, F.; Gerhard, A.; Keller-Schierlein, W.: Experientia **27** (1971) 1249.
[16] Casey, A. C.; Malhotra, A.: Tetrahedron Lett. **1975**, 401.
[17] Chamberlain, T. R.; Collins, J. F.; Grundon, M. F.: Chem. Commun. **1969**, 1269.
[18] Chakravarti, D.; Chakravarti, R. N.; Chakravarti, S. C.: J. Chem. Soc. **1953**, 3337.
[19] Chatterjee, A.; Ganguly, D.: Phytochemistry **15** (1976) 1303.
[20] Cobet, M.; Luckner, M.: Phytochemistry **10** (1971) 1031.
[21] Collins, J. C.; Donnelly, W. J.; Grundon, M. F.; James, K. J.: J. Chem. Soc. Perkin I **1974**, 2177.
[22] Colonna, A. O.; Gros, E. G.: Phytochemistry **10** (1971) 1515.
[23] Cordell, G. A.; Farnsworth, N. R.: Lloydia **40** (1977) 1.
[24] Danieli, B.; Manitto, P.; Ronchetti, F.; Russo, G.; Ferrari, G.: Experientia **28** (1972) 249.
[25] Diment, J. A.; Ritchie, E.; Taylor, W. C.: Austral. J. Chem. **20** (1967) 1719.
[26] Donovan, D. G. O.; Horan, H.: J. Chem. Soc. C **1970**, 2466.
[27] Dunn, B. P.; Gout, P. W.; Beer, C. T.: Cancer Res. **33** (1973) 2310.
[28] Eilert, U.; Wolters, B.; Nahrstedt, A.; Wray, V.: Z. Naturforsch. **37c** (1982) 132.
[29] El Aboutabl, S. A.; El Azzouny, A.; Winter, K.; Luckner, M.: Phytochemistry **15** (1976) 1925.
[30] El Aboutabl, S. A.; Luckner, M.: Phytochemistry **14** (1975) 2573.
[31] Eliasberg, J.; Friedländer, P.: Ber. Dtsch. Chem. Ges. **25** (1892) 1752.
[32] Fiedler, E.; Fiedler, H.-P.; Gerhard, A.; Keller-Schierlein, W.; König, W. A.; Zähner, H.: Arch. Mikrobiol. **107** (1976) 249.
[33] Finlayson, A. C.; Prager, R. H.: Austral. J. Chem. **31** (1978) 2751.
[34] Fitzgerald, J. S.; Johns, S. R.; Lamberton, J. A.; Redcliffe, A. H.: Austral. J. Chem. **19** (1966) 151.
[35] Framm, J.; Nover, L.; El Azzouny, A.; Richter, H.; Winter, K.; Werner, S.; Luckner, M.: Eur. J. Biochem. **37** (1973) 78.
[36] Fraser, A. W.; Lewis, J. R.: J. Chem. Soc., Chem. Commun. **1973**, 615.

[36a] Gerlach, M.; Schwelle, N.; Lerbs, W.; Luckner, M.: Phytochemistry **24** (1985), in press.
[37] Govindachari, T. R.; Viswanathan, N.; Pai, B. R.; Ramachandran, V. N.; Subramaniam, P. S.: Tetrahedron **26** (1970) 2905.
[38] Gröger, D.: In: Mothes, K.; Schütte, H. R. (Eds.): Biosynthese der Alkaloide, p. 551. Deutscher Verlag der Wissenschaften, Berlin 1969.
[39] Gröger, D.: Lloydia **32** (1969) 221.
[40] Gröger, D.: In: Bell, E. A.; Charlwood, B. V. (Eds.): Encyclopedia of Plant Physiology, N. S. Vol. 8, p. 128. Springer-Verlag, Berlin/Heidelberg/New York 1980.
[41] Gröger, D.; Johne, S.: Z. Naturforsch. **23b** (1968) 1072.
[42] Grundon, M. F.: Alkaloids **6** (1976) 108; **8** (1978) 83; **9** (1979) 85; **10** (1980) 80.
[43] Grundon, M. F.: Alkaloids **6** (1976) 108; **7** (1977) 89; **8** (1978) 84; **9** (1979) 86; **10** (1980) 82.
[44] Grundon, M. F.; Harrison, D. M.; Spyropoulos, C. G.: Chem. Commun. **1974**, 51.
[45] Grundon, M. F.; Harrison, D. M.; Spyropoulos, C. G.: J. Chem. Soc. Perkin I **1974**, 2181.
[46] Grundon, M. F.; Harrison, D. M.; Spyropoulos, C. G.: J. Chem. Soc. Perkin I **1975**, 302.
[47] Grundon, M. F.; James, K. J.: Chem. Commun. **1971**, 1311.
[48] Gupta, O. P.; Anand, K. K.; Ghatak Ray, B. J.; Atal, C. K.: Indian J. Exp. Biol. **16** (1978) 1027.
[49] Harborne, J. B.: Introduction to Ecological Biochemistry. Academic Press, London 1977.
[50] Herbert, R. B.: Tetrahedron Lett. **1974**, 4525.
[51] Herbert, R. B.: In: Comprehensive Organic Chemistry, Vol. 5: Haslam, E. (Ed.): Biological Compounds. Pergamon Press, Oxford 1979, p. 1045.
[52] Hlubucek, J.; Ritchie, E.; Taylor, W. C.: Chem. & Ind. **1969**, 1809; Austral. J. Chem. **23** (1970) 1881.
[53] Hopartean, I.: Stud. Cercet. Chim. **19** (1971) 695.
[54] Hughes, G. K.; Ritchie, E.: Austral. J. Sci. Res. **4A** (1951) 423.
[55] Hurley, L. H.; Das, N. V.; Gairola, C.; Zmijewski, M.: Tetrahedron Lett. **1976**, 1419.
[56] Hurley, L. H.; Gairola, C.; Das, N. V.: Biochemistry **15**, 3760 (1976).
[57] Hurley, L. H.; Gairola, C.; Zmijewski, M.: Chem. Commun. **1975**, 120.
[58] Hurley, L. H.; Zmijewski, M.: Chem. Commun. **1974**, 337.
[59] Hurley, L. H.; Zmijewski, M.; Chang, C.-J.: J. Am. Chem. Soc. **97** (1975) 4372.
[60] Isamukhamedov, I.: Farmakol. Alkaloidov. Serdečnych Glikozidov **1971**, 224; ref. C. A. **78** (1973) 92 966f.
[61] Johne, S.; Bernasch, H.; Gröger, D.: Pharmazie **25** (1970) 777.
[62] Johne, S.; Gröger, D.: Pharmazie **25** (1970) 22.
[63] Johne, S.; Gröger, D.: Pharmazie **27** (1972) 195.
[64] Johne, S.; Gröger, D.; Hesse, M.: Helv. Chim. Acta **54** (1971) 826.
[65] Johne, S.; Waiblinger, K.; Gröger, D.: Eur. J. Biochem. **15** (1970) 415.
[66] Johne, S.; Waiblinger, K.; Gröger, D.: Pharmazie **28** (1973) 403.
[67] Katz, E.; Wessbach, H.: J. Biol. Chem. **237** (1962) 882.
[68] Khan, M. S.; Lewis, J. R.; Watt, R. A.: Chem. & Ind. **1975**, 744.
[69] Kirby, G. W.; Narayanaswami, S.: J. Chem. Soc. Perkin I **1976**, 1564.
[70] Kuzovkina, L. V.; Szendrei, K.: Izvest. Akad. Nauk SSSR, Ser. Biol. **1973**, 275.
[71] Leete, E.: In: Bernfeld, P. (Ed.): Biogenesis of Natural Compounds, p. 780. Pergamon Press, London 1963; Ann. Rev. Plant Physiol. **18** (1967) 179.
[72] Leete, E.: In: Bernfeld, P. (Ed.): Biogenesis of Natural Compounds, 2nd ed., p. 953, Pergamon Press, Oxford 1967.
[73] Liljegren, D. J.: Phytochemistry **7** (1968) 1299; **10** (1971) 2661.
[74] Liska, K. J.: J. Med. Chem. **15** (1972) 1177.
[75] Luckner, M.: Eur. J. Biochem. **2** (1967) 74.
[76] Luckner, M.: In: K. Mothes and H. R. Schütte (Eds.): Biosynthese der Alkaloide. Deutscher Verlag der Wissenschaften, Berlin 1969, p. 510.

[77] Luckner, M.: J. Nat. Prod. **43** (1980) 21.
[78] Luckner, M.; Mothes, K.: Tetrahedron Lett. **1962**, 1035.
[79] Luckner, M.; Mothes, K.: Archiv Pharmazie **296** (1963) 18.
[80] Luckner, M.; Nover, L.: Abh. Dtsch. Akad. Wiss. Berlin **1971**, 525.
[81] Luckner, M.; Ritter, C.: Tetrahedron Lett. **1965**, 741.
[82] Luckner, M.; Winter, K.; Nover, L.; Reisch, J.: Tetrahedron **25** (1969) 2575.
[83] Luckner, M.; Winter, K.; Reisch, J.: Eur. J. Biochem. **7** (1969) 380.
[84] Mann, S.: Arch. Mikrobiol. **56** (1967) 329; Arch. Hygiene Bakteriol. **151** (1967) 474.
[85] Matsuo, M.; Kasida, Y.: Chem. pharmac. Bull. (Tokyo) **14**, (1966) 1108.
[86] Matsuo, M.; Yamazaki, M.; Kasida, Y.: Biochem. Biophys. Res. Commun. **23** (1966) 679.
[87] Mester, I.: Fitoterapia **44** (1973) 123.
[88] Mester, I.; Reisch, J.; Rózsa, Zs.; Szendrei, K.: Heterocycles **16** (1981) 77.
[89] Monkovič, I.; Spenser, I. D.: Chem. Commun. **1966**, 204.
[90] Monkovič, I.; Spenser, I. D.; Plunkett, A. O.: Canad. J. Chem. **45** (1967) 1935.
[91] Munsche, D.: Flora **154** (1964) 317.
[92] Munsche, D.: Abh. Dtsch. Akad. Wiss. Berlin Klasse Chem. Geol. Biol. **3** (1966) 611.
[93] Munsche, D.; Mothes, K.: Phytochemistry **4** (1965) 705.
[94] Nahrstedt, A.; Eilert, U.; Wolters, B.; Wray, V.: Z. Naturforsch. **36c** (1981) 200.
[95] Nair, P. M.; Vaidyanathan, C. S.: Biochim. Biophys. Acta **110** (1965) 521.
[96] Nover, L.; Luckner, M.: Eur. J. Biochem. **10** (1969) 268.
[97] Nover, L.; Luckner, M.: FEBS-Letters **3** (1969) 292.
[98] Nover, L.; Luckner, M.: Abh. Dtsch. Akad. Wiss. Berlin **1971**, 535.
[99] Perlman, D.; Otani, S.; Perlman, K. L.; Walker, J. E.: J. Antibiotics **26** (1973) 289.
[100] Prager, R. H.; Thredgold, H. M.: Austral. J. Chem. **22** (1969) 2627.
[101] Ramachandran, V. N.; Pai, B. R.; Santhanam, R.: Indian J. Chem. **10** (1972) 14.
[102] Reisch, J.: Die Darstellung biologisch aktiver Acridon-Derivate unter Berücksichtigung natürlicher Vorbilder. Westdeutscher Verlag, Opladen 1978.
[103] Reisch, J.; Rózsa, Zs.; Mester, I.: Z. Naturforsch. **33b** (1978) 957.
[104] Reisch, J.; Rózsa, Zs.; Szendrei, K.; Novak, I.; Minker, E.: Phytochemistry **11** (1972) 2121.
[105] Reisch, J.; Szendrei, K.; Minker, E.; Novak, I.: Pharmazie **27** (1972) 208.
[106] Reisch, J.; Szendrei, K.; Novak, I.; Minker, E.; Rózsa, Zs.: Experientia **27** (1971) 1005.
[107] Reisch, J.; Szendrei, K.; Rózsa, Zs.; Novak, I.; Minker, E.: Phytochemistry **11** (1972) 2359.
[108] Richter, I.; Luckner, M.: Phytochemistry **15** (1976) 67.
[109] Ritter, C.; Luckner, M.: Eur. J. Biochem. **18** (1971) 391.
[110] Robinson, R.: The Structural Relations of Natural Products. Clarendon Press, Oxford 1955.
[111] Rózsa, Zs.; Kusovkina, I. N.; Reisch, J.; Novak, I.; Szendrei, K.; Minker, E.: Fitoterapia **47** (1976) 147.
[112] Rózsa, Zs.; Reisch, J.; Mester, I.; Szendrei, K.: Fitoterapia **52** (1981) 37.
[113] Saxton, J. E.: In: Acheson, R. M. (Ed.): The Acridines. Wiley-Interscience, New York 1973, p. 379.
[114] Scharlemann, W.: Z. Naturforsch. **27b** (1972) 806.
[115] Schildknecht, H.; Wenneis, W. F.: Tetrahedron Lett. **1967**, 1815.
[116] Schindler, F.; Zähner, H.: Arch. Mikrobiol. **79** (1971) 187.
[117] Schröder, P.: In: Mothes, K.; Schreiber, K.; Schütte, H. R. (Eds.): Biochemie und Physiologie der Alkaloide, 4th Internat. Symp. 1969. Akademie-Verlag, Berlin 1972, p. 519.
[118] Schröder, P.; Luckner, M.: Pharmazie **21** (1966) 642.
[119] Schröder, P.; Luckner, M.: Arch. Pharmaz. **301** (1968) 39.
[120] Schröder, P.; Luckner, M.: Z. Naturforsch. **23b** (1968) 784.
[120a] Shimizu, Y.: Pure & Appl. Chem. **54** (1982) 1973.

[121] Snieckus, V. A.: Alkaloids **2** (1972) 95; **3** (1973) 114; **4** (1974) 125; **5** (1975) 109.
[122] Snieckus, V. A.: Alkaloids **2** (1972) 91; **3** (1973) 112; **4** (1974) 124; **5** (1975) 108.
[123] Steck, W.; Bailey, B. K.; Shyluk, J. P.; Gamborg, O. L.: Phytochemistry **10** (1971) 191.
[124] Steck, W.; Gamborg, O. L.; Bailey, B. K.: Lloydia **36** (1973) 93.
[125] Subra Rao, P. V.; Vaidyanathan, C. S.: Arch. Biochem. Biophys. **115** (1966) 27.
[126] Svoboda, G. H.: Lloydia **29** (1966) 206.
[127] Svoboda, G. H.; Poore, G. A.; Simpson, P. J.; Boder, G. B.: J. Pharm. Sci. **55** (1966) 758.
[128] Szendrei, K.; Rózsa, Zs.; Reisch, J.; Novak, I.; Kusovkina, I. N.; Minker, E.: Herba Hung. **15** (1976) 23.
[129] Tipton, C. L.; Wang, M. C.; Tsao, F. H.-C.; Tu, C. L.; Husted, R. R.: Phytochemistry **12** (1973) 347.
[130] Verzár-Petri, G.; Csedö, K.; Möllmann, H.; Szendrei, K.; Reisch, J.: Planta Med. **29** (1976) 370.
[131] Voigt, S.; Luckner, M.: Phytochemistry **16** (1977) 1651.
[131a] Waiblinger, K.; Johne, S.; Gröger, D.: Phytochemistry **11** (1972) 2263.
[132] Waterman, P. G.: Phytochemistry **14** (1975) 2092.
[133] Waterman, P.: Biochem. Syst. Ecol. **3** (1975) 149.
[134] Watermann, P. G.; Grundon, M. F.: Chemistry and Chemical Taxonomy of the Rutales. Academic Press, New York 1983.
[135] White, J. D.; Dimsdale, M. J.: Chem. Commun. **1969**, 1285.
[136] Wilson, S.; Luckner, M.: Z. Allg. Mikrobiol. **15** (1975) 45.
[137] Wilson, S.; Schmidt, I.; Roos, W.; Fürst, W.; Luckner, M.: Z. Allg. Mikrobiol. **14** (1974) 515.
[138] Wolters, B.; Eilert, U.: Planta Med. **43** (1981) 166.
[139] Yamazaki, M.; Ikuta, A.: Tetrahedron Lett. **1966**, 3221.
[140] Yamazaki, M.; Ikuta, A.; Mori, T.; Kawana, T.: Tetrahedron Lett. **1967**, 3317.
[140a] Zschunke, A.; Baumert, A.; Gröger, D.: Chem. Commun. **1982**, 1263.
[141] Zutschi, U.; Rao, P. G.; Soni, A.; Gupta, O. P.; Atal, C. K.: Planta med. **40** (1980) 373.

17 Alkaloids Derived from Purines

D. Schlee

Purines are essential constituents of nucleic acids and certain coenzymes. They are also involved in the biosynthesis of some plant alkaloids (*e.g.* caffeine), cytokinins (*i.e.* phytohormones such as zeatin), folic acid and other pteridines as well as vitamins (*e.g.* riboflavin and possibly thiamine) and other natural substances. Some purine nucleotides have the ability to enhance flavour.

All organisms, with the exception of a small number of microorganisms, synthesize purines *de novo* on quite similar and well-established pathways (Figs. 17.1, 17.2, 17.3) [7, 20, 23, 26, 47, 61, 72, 79, 86]. The first product with a complete purine ring structure is inosinic acid (IMP) (**2**), which is the common precursor for all purine compounds.

Adenylic acid (AMP) (**4**) is synthesized from IMP by the addition of aspartic acid to form adenylosuccinate (**3**) and the subsequent release of fumaric acid. Conversion of IMP to guanylic acid (GMP) (**6**) proceeds *via* xanthylic acid (XMP) (**5**) in an NAD^+-dependent dehydrogenation followed by ATP-dependent amination.

The availability of 5-phosphoribosylpyrophosphate (PRPP) (**1**) is highly important for the regulation of purine biosynthesis. Generally, PRPP formation is inhibited competitively by AMP and PRPP, and non-competitively by various nucleoside di- and tri-phosphates. Inorganic phosphate is an allosteric activator of PRPP synthetase in most systems studied. The PRPP amidotransferase reaction seems to be the most important regulatory step in purine biosynthesis. This enzyme is allosterically regulated by AMP (*cf.* [61, 86]). Moreover, interconversions at the mononucleotide level (IMP $\rightarrow$ AMP, IMP $\rightarrow$ GMP) are also important for regulation with respect to the balance of purine metabolism. Both AMP and GMP exert feedback repression and feedback inhibition on the initial pathway to inosine monophosphate and on their respective branches (Figs. 17.1 and 17.2).

In most organisms the overall pathway of purine degradation proceeds *via* hypoxanthine and xanthine (**7**) to uric acid, allantoin, allantoic acid, urea, and glyoxylic acid (Fig. 17.4). The form in which the carbon and nitrogen atoms of purines are finally excreted by animals depends on the presence or absence of the enzymes, particularly allantoinase and allantoicase.

Complete confirmation of a similar pathway for purine degradation in higher plants was obtained in the case of so-called ureide plants (allantoin or allantoic

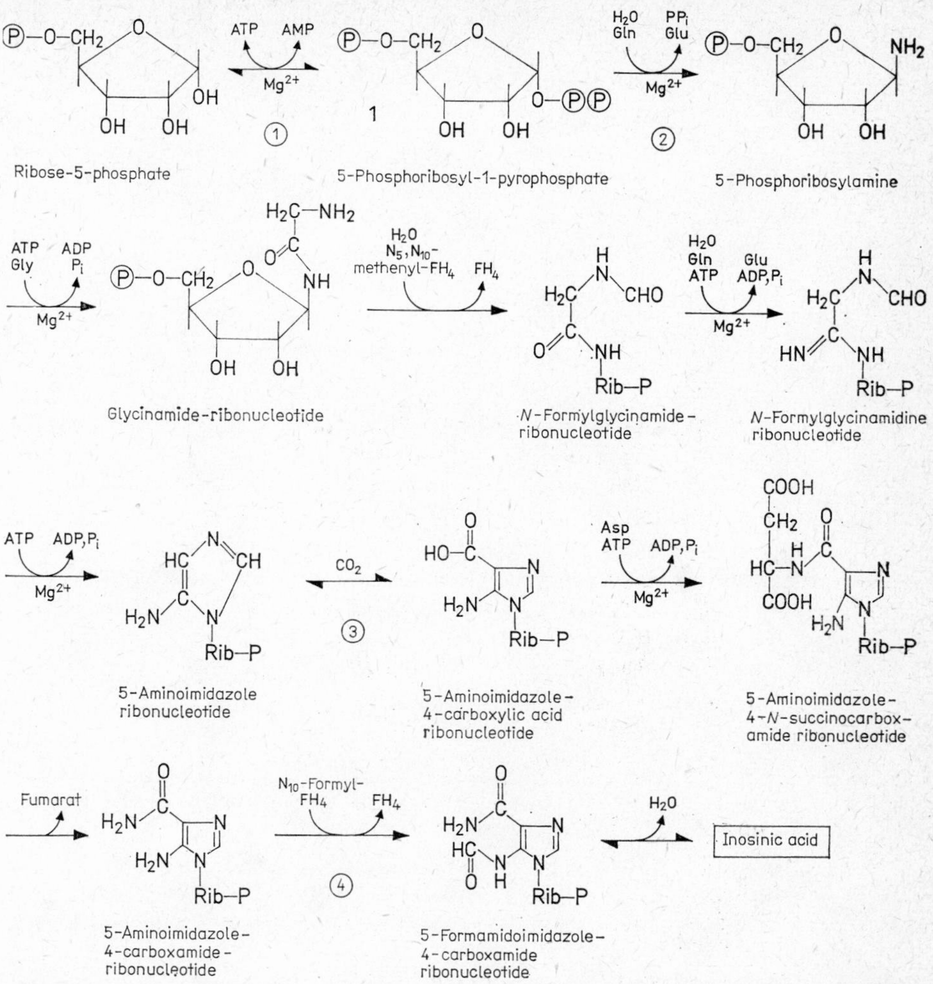

Fig. 17.1
The metabolic pathway leading to the biosynthesis of inosinic acid (IMP)

Solid arrows represent enzymatic steps. → indicates important control steps by feedback regulation; the numbers refer to the following enzymes: 1 5-phosphoribosyl-1-pyrophosphate synthetase; 2 5-phosphoribosyl-1-pyrophosphate amidotransferase; 3 5-aminoimidazole ribonucleotide carboxylase; 4 5-aminoimidazole carboxamide ribonucleotide formyltransferase; FH_4 tetrahydrofolic acid; P_i inorganic phosphate; Rib-P ribose-5-phosphate

acid plants [4, 13, 58, 59, 66]. Evidence of the operation of the purine cycle (the glycine allantoin-cycle) (Fig. 17.5) has been found in chlorophylldeficient leaves of *Pelargonium zonale*, in germinating seeds of *Triticum vulgare*, and in lungfish liver [9, 65].

Allantoin and allantoic acid are the main forms of nitrogen transport, accumula-

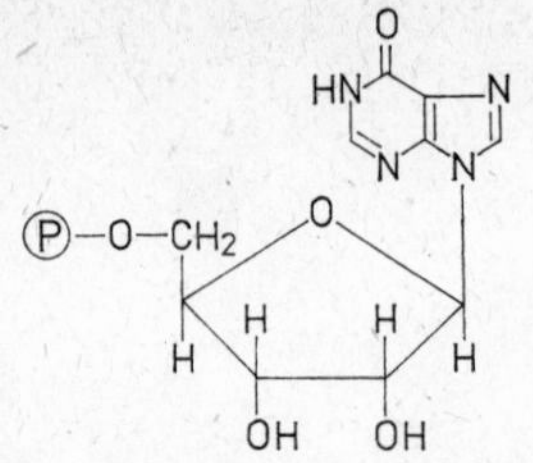

2

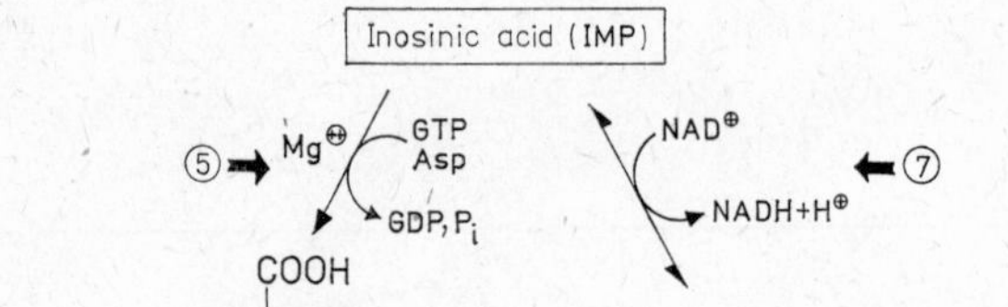

Inosinic acid (IMP)

⑤ $Mg^{\oplus}$ GTP, Asp → GDP, P_i

⑦ $NAD^{\oplus}$ → $NADH+H^{\oplus}$

3 Adenylosuccinic acid (s-AMP)

5 Xanthylic acid (XMP)

⑥ Fumarate

⑧ Gln, ATP, H_2O → Glu, AMP, PP_i

6

Adenylic acid (AMP)

Guanylic acid (GMP)

Fig. 17.2
Interconversion of inosinic acid, adenylic acid (AMP) and guanylic acid (GMP)
5 adenylosuccinate synthetase (s-AMP synthetase); 6 adenylosuccinase (s-AMP lyase); 7 IMP dehydrogenase; 8 XMP aminase

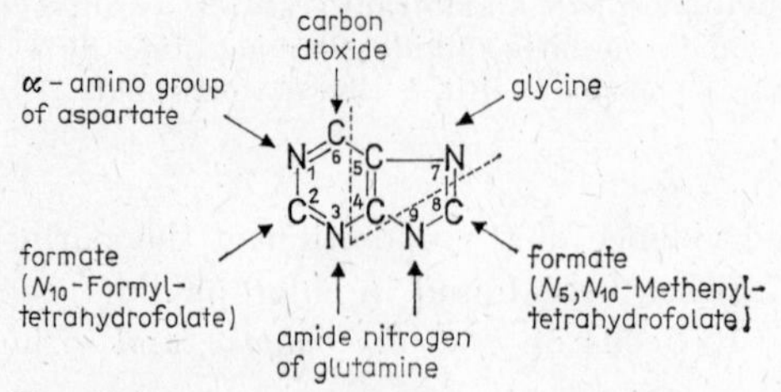

Fig. 17.3
Precursors of the purine ring system

tion, and re-utilization in a great variety of plants [28]. Recently, ureide formation has been the subject of an intensive investigation with respect to the nodulation in some legumes [2, 27, 50].

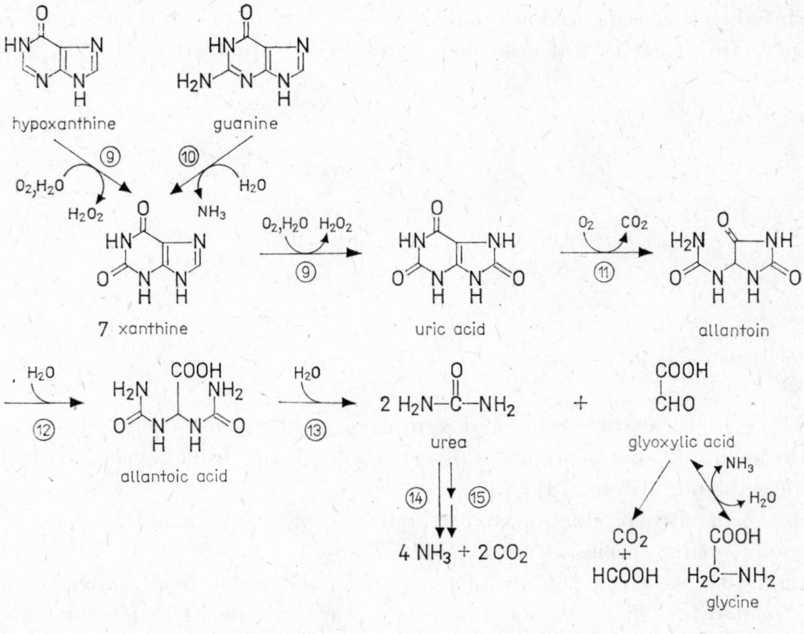

Fig. 17.4
Oxidative degradation of purine compounds
The numbers refer to the following enzymes: 9 hypoxanthine/xanthine oxidase; 10 guanine deaminase; 11 uricase; 12 allantoinase; 13 allantoicase; 14 urease; 15 urea amido lyase

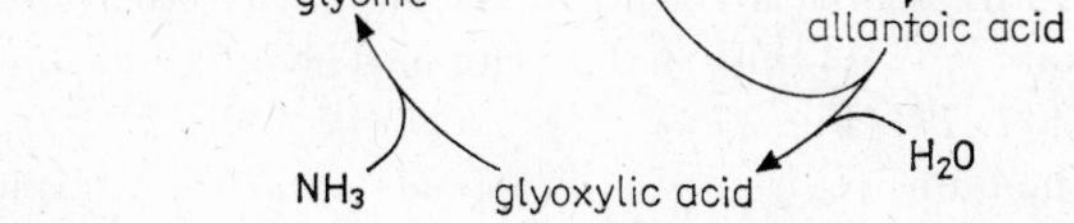

Fig. 17.5
The purine cycle ("glycine-allantoin cycle") showing uricolytic pathway for the formation of urea, alternatively to ornithine urea cycle

17.1 Methylated Xanthines (Purine Alkaloids)

A number of methylated purine derivatives such as caffeine (**10**) (Fig. 17.6) and theophylline (**8**) are accumulated in considerable amounts in plants belonging to a variety of families [63, 76, 84]. In *Coffea arabica* caffeine is synthesized in the pericarp, transported to the seeds, and accumulated there during fruit development [5]. Even transport from leaf to leaf was demonstrated by application of [^{14}C, ^{3}H]-

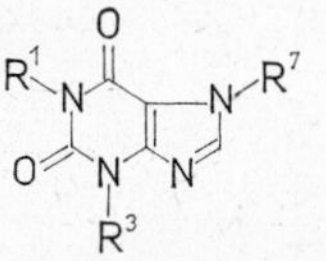

		R^1	R^3	R^7
7	xanthine	H	H	H
8	theophylline (1,3-dimethylxanthine)	CH_3	CH_3	H
9	theobromine (3,7-dimethylxanthine)	H	CH_3	CH_3
10	caffeine (1,3,7-trimethylxanthine)	CH_3	CH_3	CH_3

Fig. 17.6
Methylated xanthines

caffeine [5]. The synthesis of methylated xanthines in plants is closely related to plant growth, including shoot formation of tea seedlings and fruit development of cacao and coffee plants [37, 40, 78].

Studies on the in vitro production and degradation of purine alkaloids in tissue cultures of some plant species (*Coffea, Thea, Theabroma, Paracoffea, Paullinia, Psilantus*) show that as a rule, failure of purine alkaloid biosynthesis under these conditions is related to a high catabolic capacity toward caffeine. This phenomenon is also related to the biochemical pattern in the intact plants, wherein a phase of substantial purine alkaloid biosynthesis is followed by a period of total biodegradation during leaf development. The factors responsible for this change are not known [4a].

The ring formation of caffeine and other methylated xanthines follows the classic scheme of *de novo* purine nucleotide biosynthesis in *Coffea* and *Camellia* [37]. The purine ring in caffeine is derived not from nucleic acid breakdown but directly from purine nucleotides in the nucleotide pool along the following pathway [77]: 7-methyl-purine nucleotides → 7-methylxanthine → 3,7-dimethylxanthine (theobromine) → 1,3,7-trimethylxanthine (caffeine) (Fig. 17.7).

S-Adenosyl-*L*-methionine is the actual source of the methyl groups [75, 76]. 7-Methylguanosine does not seem to be an intermediate in caffeine biosynthesis [46, 55]. In addition, it was suggested that theophylline (1,3-dimethylxanthine) is synthesized from a specifically methylated precursor in nucleic acids *via* 1-methyl-AMP → 1-methylxanthine → 1,3-dimethylxanthine. Theophylline may be finally converted into caffeine [77, 78] (Fig. 17.7).

Xanthine cannot be the immediate source of caffeine because it is rapidly degraded to CO_2 and NH_3 *via* oxidative purine breakdown [76]. Caffeine is catabolized to xanthine and *via* allantoin and allantoic acid to urea and CO_2 in *Coffea* [35, 36]. Urea is also a product of xanthine metabolism in *Camellia* [36].

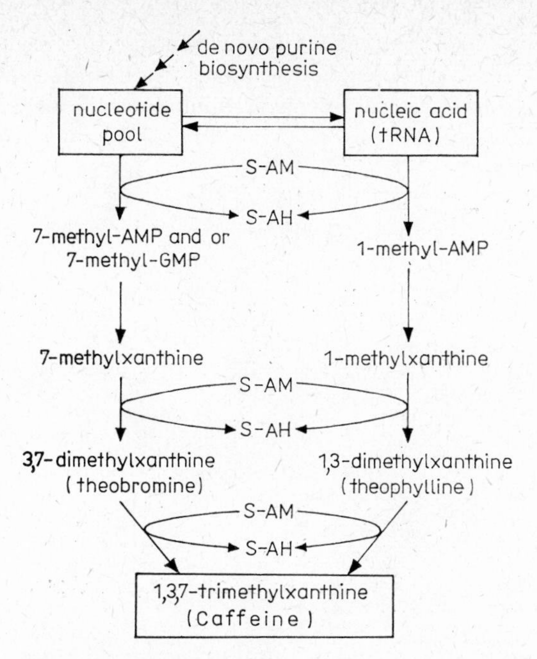

Fig. 17.7
Possible relations of purine metabolism and biosynthetic pathways of methylated xanthines [77, 78]
S-AM S-Adenosyl-*L*-methionine; S-AH S-adenosyl-*L*-homocysteine

Thus caffeine is not a metabolic end-product but is involved in active metabolism. Methylated xanthines have an inhibitory effect on cyclic nucleotide phosphodiesterase [8, 22]. Thus the action of caffeine and its derivatives can be exerted at the level of cyclic AMP. Very little is known about the physiological importance of methylated xanthines in higher plants.

17.2 Pteridines

Pteridines (*cf.* [60]) are important co-factors in various enzyme systems (folic acid, so-called co-enzyme F, in one-carbon metabolism; activators in hydroxylation reactions, *e.g.* in phenylalanine hydroxylase system). Pigments found in the wings of insects and in the eyes and skin of fish, amphibia, and reptiles are, for the most part, pteridine derivatives.

The ability of guanine compounds, particularly GTP (**11**), to act as precursors of pteridines has been firmly established (Fig. 17.8) [10, 32, 33]. The first step is the cleavage of GTP by GTP cyclohydrolase forming formic acid and the triphosphate ester of 2-amino-4-hydroxy-6-(*D-erythro*-1′,2′,3′-trihydroxypropyl)-7,8-dihydropteridine (7,8-dihydroneopterin triphosphate) (**12**) [11, 32].

During the enzymatic conversion of GTP to dihydroneopterin triphosphate an Amadori rearrangement occurs. This reaction results in the transformation of the ribose triphosphate moiety of GTP to a 1-deoxy-2-ribulose triphosphate unit [87].

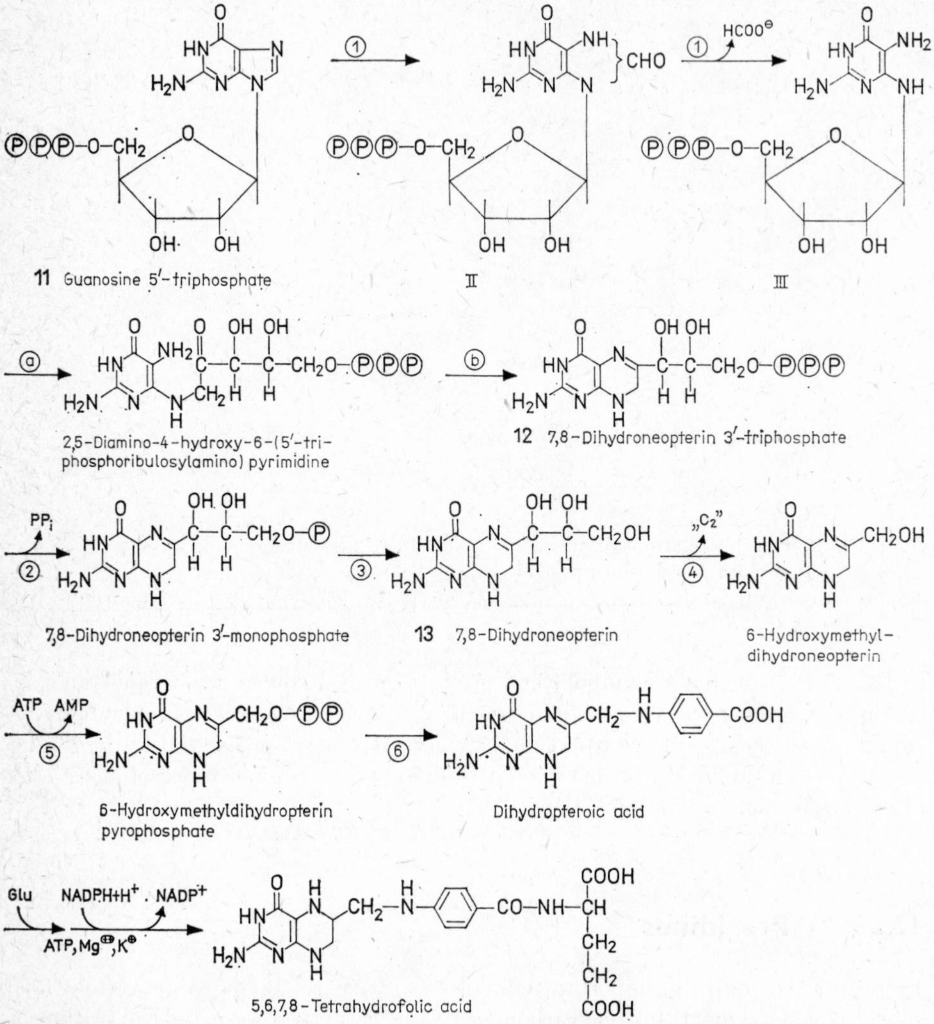

11 Guanosine 5′-triphosphate

II

III

2,5-Diamino-4-hydroxy-6-(5′-triphosphoribulosylamino) pyrimidine

12 7,8-Dihydroneopterin 3′-triphosphate

7,8-Dihydroneopterin 3′-monophosphate

13 7,8-Dihydroneopterin

6-Hydroxymethyl-dihydroneopterin

6-Hydroxymethyldihydropterin pyrophosphate

Dihydropteroic acid

5,6,7,8-Tetrahydrofolic acid

Fig. 17.8
Pathway for the conversion of guanosine 5′-triphosphate into dihydroneopterintriphosphate and tetrahydrofolic acid

The numbers refer to the following enzymes (occurrence and proof are included in parentheses): 1 GTP cyclohydrolase I (*Escherichia coli*, *Lactobacillus plantarum*, *Comamonas spec.*, *Streptomyces spec.*); 2 Dihydroneopterintriphosphate pyrophosphorylase (*E. coli*); 3 'Phosphatase' (*E. coli*); 4 Dihydroneopterin aldolase (*E. coli*); 5 Hydroxymethyldihydropterin pyrophosphokinase (*E. coli*, *L. plantarum*); 6 Dihydropteroic acid synthetase (*E. coli*, *L. plantarum*, *Pisum sativum*); a Amadori rearrangement, b ring closure, neither compound II nor III can be detected as a free intermediate.

Dihydroneopterin (**13**) has already been established as the immediate precursor for the biosynthesis of dihydrofolic acid [80], and biopterin [16, 17]. Dihydroneopterin is also easily oxidized to yield neopterin [57]. An extract of *Pseudomonas* species catalyses the formation of xanthopterin and neopterin [21].

The reactions and intermediates involved in the biosynthetic pathways of the pteridines from GTP are shown in Fig. 17.8.

17.3 Riboflavin and Alloxazines

For the synthesis of riboflavin (vitamin B_2; 6,7-dimethyl-9-*D*-ribityl-isoalloxazine) (**17**) the involvement of a purine precursor, preferably GTP, has been established [15, 34, 49, 51, 67, 69]. GTP is transformed by GTP cyclohydrolase II to a 2,5-diamino-6-hydroxy-4-ribosylaminopyrimidine derivative (**14**) [19, 68, 69]. This reacts with diacetyl with loss of the ribose moiety and formation of 2-amino-6,7-dimethyl-4-hydroxypyrimidine [19] (Fig. 17.9).

The ribofuranoside product of GTP cyclohydrolase II catalysis possibly undergoes an Amadori rearrangement resulting in a ribuloside. Subsequently, in *Escherichia coli*, a deaminase and a reductase are involved in riboflavin biosynthesis, and the product of GTP cyclohydrolase II action is deaminated before reduction occurs

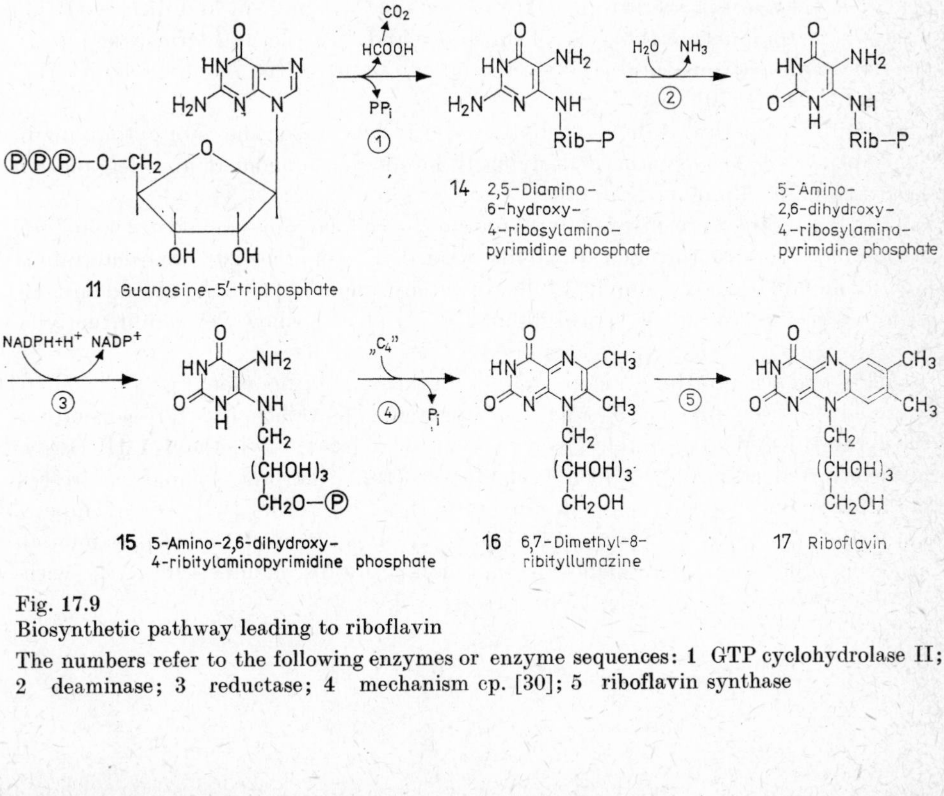

11 Guanosine-5′-triphosphate

14 2,5-Diamino-6-hydroxy-4-ribosylamino-pyrimidine phosphate

5-Amino-2,6-dihydroxy-4-ribosylamino-pyrimidine phosphate

15 5-Amino-2,6-dihydroxy-4-ribitylaminopyrimidine phosphate

16 6,7-Dimethyl-8-ribityllumazine

17 Riboflavin

Fig. 17.9
Biosynthetic pathway leading to riboflavin
The numbers refer to the following enzymes or enzyme sequences: 1 GTP cyclohydrolase II; 2 deaminase; 3 reductase; 4 mechanism cp. [30]; 5 riboflavin synthase

[12]. Purified deaminase from *E. coli* only attacks the phosphorylated form of 2,5-diamino-6-hydroxy-4-ribosylaminopyrimidine. In *Ashbya gossypii*, however, the reduction preceeds deamination [29].

In the conversion of 5-amino-2,6-dihydroxy-4-ribitylaminopyrimidine (phosphate) (**15**) into the immediate precursor of riboflavin, 6,7-dimethyl-8-ribityllumazine (**16**), one molecule of 5-amino-2,6-dihydroxy-4-ribitylaminopyrimidine acts as donor of a four carbon unit another acts as the acceptor [30]. In *Ashbya gossypii* as well as in *Pichia gilliiermondii* a pentose must be the source of the C_4 unit [2a, 18a, 37a].

The enzyme riboflavin synthase catalyses the last reaction in which two molecules of 6,7-dimethyl-8-ribityl-lumazine form one molecule of riboflavin and one molecule of 5-amino-2,6-dihydroxy-4-ribitylaminopyrimidine [3, 6, 24]. 6,7-Dimethyl-8-ribityl-lumazine-5′-phosphate is not a substrate for riboflavin synthase [25]. Recent investigations have been shown that the phosphate release is achieved before the riboflavin synthase reaction [45a].

In Fig. 17.9 the biosynthesis of riboflavin is summarized.

17.4 Cytokinins

Certain alkaloid-like purine compounds are powerful endogenous growth stimulators in plants. They can participate in many aspects of development and differentiation processes. Cytokinins are this type of substance and, in structural terms, the moiety common to all active natural (and synthetic) cytokinins is the N^6-alkyladenine ring [39, 44, 52, 54, 64, 70].

Two different pathways have been discussed for the biosynthesis of cytokinins in higher plants: *de novo* synthesis deriving from adenine monomers and degradation of cytokinin-containing tRNA [48].

5′-AMP is a direct precursor of cytokinin in the cellular slime mould *Dictyostelium discoideum*. The structure of this factor, named discadenine (**20**), was determined as 3-(3-amino-3-carboxypropyl)-N^6-Δ^2-isopentenyladenine [1, 81]. Discadenine is active in the inhibition of spore germination and also shows cytokinin activity [53].

AMP is the acceptor for an isopentenyl group from Δ^2-isopentenyl pyrophosphate catalysed by the enzyme 5′-AMP:Δ^2-isopentenyl pyrophosphate Δ^2-isopentenyl transferase (5′-AMP isopentenyl transferase). The N^6-Δ^2-isopentenyl-AMP (**18**) is then converted into N^6-Δ^2-isopentenyl-adenine (**19**). This process might represent the primary mechanism of cytokinin biosynthesis in plants [82]. Discadenine is synthesized from N^6-Δ^2-isopentenyl-adenine by direct transfer of the 3-amino-3-carboxypropyl moiety of *S*-adenosyl-*L*-methionine by discadenine synthetase without prior decarboxylation (Fig. 17.10) [31, 82, 83].

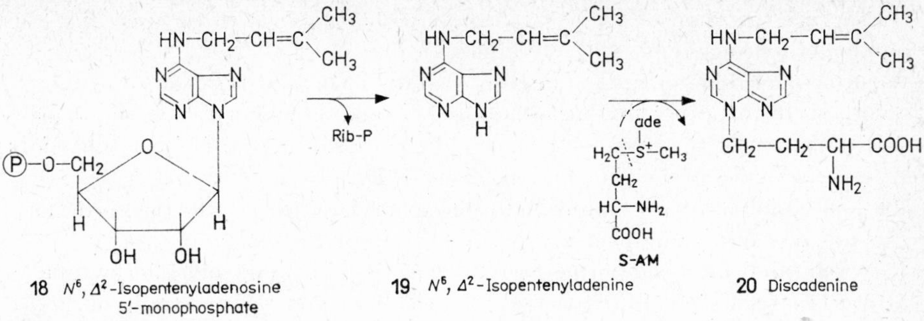

Rib-P

ade

S-AM

18 N^6, Δ^2-Isopentenyladenosine 5'-monophosphate

19 N^6, Δ^2-Isopentenyladenine

20 Discadenine

Fig. 17.10
Proposed mechanism of biosynthesis of discadenine by *Dictyostelium discoideum*
1 Discadenine synthetase; S-AM S-adenosyl-*L*-methionine; ade adenosyl; Rib-P ribose-5-phosphate

17.5 Purine Antibiotics and Pyrrolo-Pyrimidines

These substances represent a group of compounds closely related structurally to the purine (and pyrimidine) nucleosides [62, 73]. Some of them are important as inhibitors in protein and nucleic acid studies [38, 42, 56].

Tubercidin (7-deazaadenine ribonucleoside) (21), toyocamycin, and sangivamycin from different *Streptomyces* species are pyrrolopyrimidine nucleosides. They are analogues of adenosine having a carbon atom in place of the nitrogen in position-7. In *Streptomyces tubercidicus* the pyrimidine moiety of adenosine is utilized intact for the synthesis of the pyrimidine ring of tubercidin. Presumably, the carbon atoms-1 and -2 of ribose are converted into the pyrrolo-carbon atoms (Fig. 17.11)

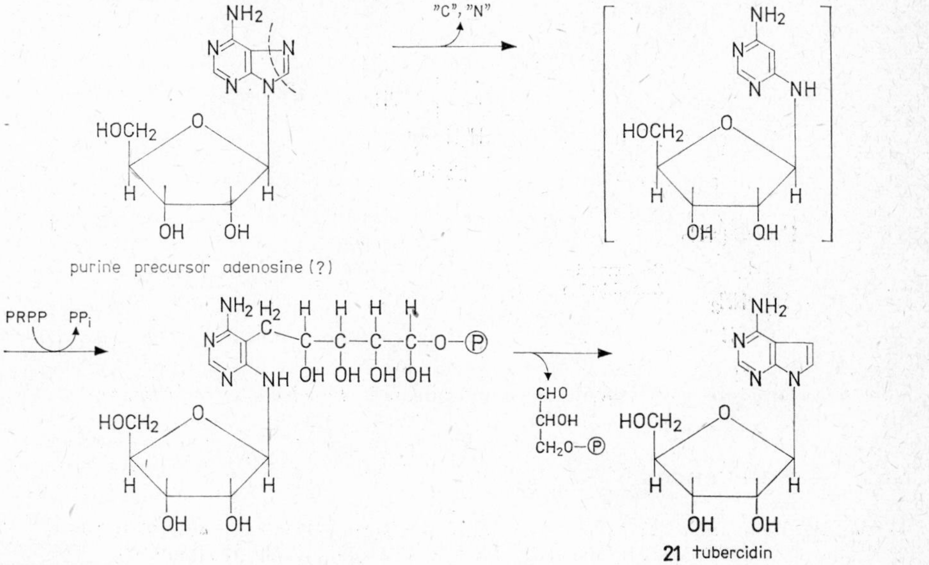

"C", "N"

purine precursor adenosine (?)

PRPP PP_i

21 tubercidin

Fig. 17.11
Proposed mechanism of biosynthesis of tubercidin by *Streptomyces tubercidicus* [71]

[71]. The incorporation of these carbon atoms is analogous to the pteridine ring formation of folic acid and other pteridines.

A purine nucleoside or nucleotide also acts as carbon and nitrogen source in the biosynthesis of toyocamycin. Carbon atoms 1, 2, and 3 of ribose contribute equally to the formation of the pyrrolo-carbons and the cyano-carbon. Neither tubercidin, nor 7-deazaadenine nor cyanide are precursors [74].

The nucleoside antibiotics are rapidly phosphorylated *in vivo* to the corresponding mono-, di-, and triphosphates.

Formycin, in which the purine ring is replaced by a pyrrazolo[4,3-*d*]pyrimidine ring, was isolated from cultures of *Nocardia interforma* and various strains of *Streptomyces*. The C-glycoside analogue of adenosine is easily converted into phosphorylated derivatives. It effectively replaces adenosine nucleotides in a number of reactions [85].

Toxoflavin is an azapteridine antibiotic (1,6-dimethyl-5,7-dioxo-1,5,6,7-tetrahydropyrimido-(5,4-*e*)*as* triazine) (22) from *Pseudomonas cocovenenans*. Its pyrimido-moiety is derived from purine compounds, and the *N*-methyl substituents from *S*-adenosyl-*L*-methionine. During biosynthesis carbon atom-8 of the purine is expelled and the subsequent formation of the *as* triazine ring takes place by use of a carbon-nitrogen unit derived from the aminomethyl moiety of glycine (Fig. 17.12) [45].

Cordycepin (3′-deoxyadenosine) from *Cordyceps militaris* and *Aspergillus nidulans*, 3′-acetamido-3′-deoxyadenosine and 3′-amino-3′-deoxyadenosine from *Helminthosporium* and arabinofuranosyladenine from *Streptomyces antibioticus* are formed by direct conversion of adenosine [14, 18, 43].

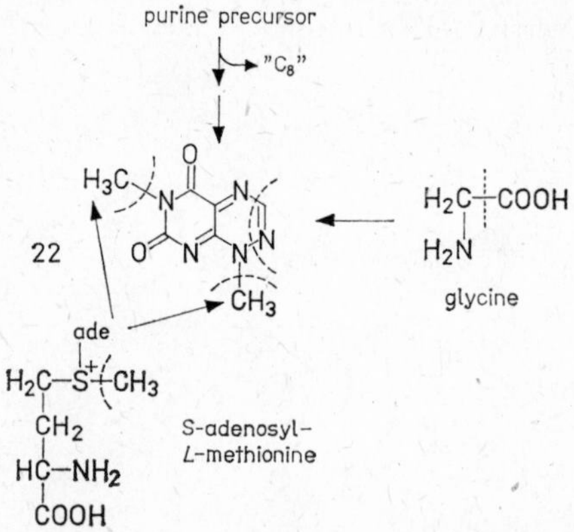

Fig. 17.12
Structure of toxoflavin and its precursors in *Pseudomonas cocovenenans*

17.6 References

[1] Abe, H.; Uchiyama, M.; Tanaka, Y.; Saito, H.: Tetrahedron Lett. **42** (1976) 3807.
[2] Atkins, C. A.; Rainbird R. M.; Pate, J. S.: Z. Pflanzenphysiol. **97** (1980) 249.
[2a] Bacher, A.; Le Van, Q.; Bühler, M.; Keller, P. J.; Eimicke, V.; Floss, H. G.: J. Am. Chem. Soc. **104** (1982) 3754.

[3] Bacher, A.; Mailänder, B.: J. Bacteriol. **134** (1978) 476.
[4] Barnes, R. L.: Bot. Gaz. **123** (1962) 141.
[4a] Baumann, T. W.; Frischknecht, P. M.: In: Plant Tissue Culture, Proc. Int. Congr. Plant Tissue Cell Cult., Ed.: Fujiwara, A., Maruzen, Tokyo/Japan 1982, p. 365.
[5] Baumann, T. W.; Wanner, H.: Planta **108** (1972) 11.
[6] Beach, R. L.; Plaut, G. W. E.: J. Am. Chem. Soc. **92** (1970) 2913.
[7] Blakley, R. L.; Vitols, E.: Ann. Rev. Biochem. **37** (1968) 201.
[8] Bollig, J.; Mayer, K.; Mayer, W. E.; Engelmann, W.: Planta **141** (1978) 225.
[9] Brown, G. W. jr.; James, J.; Henderson, R. J.; Thomas, W. N.; Robinson, R. O.; Thompson, A. L.; Brown, E.; Brown, S. G.: Science **153** (1966) 1653.
[10] Burg, A. W.; Brown, G. M.: Biochim. Biophys. Acta **117** (1966) 275.
[11] Burg, A. W.; Brown, G. M.: J. Biol. Chem. **243** (1968) 2349.
[12] Burrows, J. B.; Brown, G. M.: J. Bacteriol. **136** (1978) 657.
[13] Butler, G. W.; Ferguson, J. D.; Allison, R. M.: Physiol. Plant. **14** (1961) 310.
[14] Chassy, B. M.; Suhadolnik, R. J.: Biochim. Biophys. Acta **182** (1969) 316.
[15] Demain, A. L.: Ann. Rev. Microbiol. **26** (1972) 369.
[16] Eto, J.; Fukushima, K.; Shiota, T.: J. Biol. Chem. **251** (1976) 6505.
[17] Fan, C. L.; Krivi, G. G.; Brown, G. M.: Biochem. Biophys. Res. Commun. **67** (1975) 1047.
[18] Farmer, P. B.; Suhadolnik, R. J.: Biochemistry **11** (1972) 911.
[18a] Floss, H. G.; Le Van, Q.; Keller, P. J.; Bacher, A.: J. Am. Chem. Soc. **105** (1983) 2493.
[19] Foor, F.; Brown, G. M.: J. Biol. Chem. **250** (1975) 3545.
[20] Gots, J. S.: In: Greenberg, D. M. (Ed.): Metabolic Pathways, Vol. **5**. Academic Press, London 1971, p. 225.
[21] Guroff, G.; Strenkoski, C. A.: J. Biol. Chem. **241** (1966) 2220.
[22] Hartfiel, G.; Amrhein, N.: Biochem. Physiol. Pflanzen **169** (1976) 531.
[23] Hartman, S. C.: In: Greenberg, D. M. (Ed.): Metabolic Pathways, Vol. **4**. Academic Press, London 1970, p. 1.
[24] Harvey, R. A.; Plaut, G. W. E.: J. Biol. Chem. **241** (1966) 2120.
[25] Harzer, G.; Rokos, H.; Otto, M. K.; Bacher, A.; Ghisla, S.: Biochim. Biophys. Acta **540** (1978) 48.
[26] Henderson, J. F.; Patterson, A. R. P.: Nucleotide Metabolism. Academic Press, New York 1973.
[27] Herridge, D. F.; Atkins, C. A.; Pate, J. S.; Rainbird, R. M.: Plant Physiol. **62** (1978) 495.
[28] Hofmann, E.; Schlee, D.; Reinbothe, H.: Flora **159** (1969) 510.
[29] Hollander, I. J.; Brown, G. M.: Biochem. Biophys. Res. Commun. **89** (1979) 759.
[30] Hollander, I. J.; Braman, J. C.; Brown, G. M.: Biochem. Biophys. Res. Commun. **94** (1980) 515.
[31] Ihara, M.; Taya, Y.; Nishimura, S.: Exp. Cell Res. **126** (1980) 273.
[32] Jackson, R. J.; Shiota, T.: J. Biol. Chem. **246** (1971) 7454.
[33] Jackson, R. J.; Shiota, T.: Biochim. Biophys. Acta **403** (1975) 232.
[34] Jost, W.; Schlee, D.; Reinbothe, H.: Biochem. Physiol. Pflanzen **175** (1980) 806.
[35] Kalberer, P.: Ber. Schweiz. Bot. Ges. **74** (1964) 62.
[36] Kalberer, P.: Nature **205** (1965) 597.
[37] Keller, H.; Wanner, H.; Baumann, T. W.: Planta **108** (1972) 339.
[37a] Keller, P. J.; Le Van, Q.; Bacher, A.; Kozlowski, J. F.; Floss, H. G.: J. Am. Chem. Soc. **105** (1983) 2505.
[38] Kersten, H.; Kersten, W.: Inhibitors of Nucleic Acid Synthesis. Springer-Verlag, Berlin/Heidelberg/New York 1974.
[39] Key, J.: Ann. Rev. Plant Physiol. **20** (1969) 449.
[40] Konishi, S.; Ozasa, M.; Takahashi, E.: Plant Cell Physiol. **13** (1972) 365.
[41] Kuninaka, A.; Kibi, M.; Sakaguchi, K.: Food Technol. **19** (1964) 29.
[42] Langen, P.: Antimetabolite des Nucleinsäurestoffwechsels. Akademie-Verlag, Berlin 1968.

[43] Lennon, M. B.; Suhadolnik, R. J.: Biochim. Biophys. Acta **425** (1976) 532.
[44] Letham, D. S.: Phytochemistry **12** (1973) 2445.
[45] Levenberg, B.; Linton, S. N.: J. Biol. Chem. **241** (1966) 846.
[45a] Logvinenko, E. M.; Shavlovsky, G. M.; Zakalskij, A. E.; Zakhodylo, I. V.: Biochimija **47** (1982) 28.
[46] Looser, E.; Baumann, T. W.; Wanner, H.: Phytochemistry **13** (1974) 2515.
[47] Luckner, M.: Pharmazie **21** (1966) 142.
[48] Maass, H.; Klämbt, D.: Planta **151** (1981) 353.
[49] Mailänder, B.; Bacher, A.: J. Biol. Chem. **251** (1976) 3623.
[50] Matsumoto, M.; Yatazawa, M.; Yamamoto, Y.: Plant Cell Physiol. **19** (1978) 1161.
[51] Miersch, J.; Logvinenko, E. M.; Zakalsky, A. E.; Shavlovsky, G. M.; Reinbothe, H.: Biochim. Biophys. Acta **543** (1978) 305.
[52] Miller, C. O.: Ann. N. Y. Acad. Sci. **144** (1967) 251.
[53] Nomura, T.; Tanaka, Y.; Abe, H.; Uchiyama M.: Phytochemistry **16** (1977) 1819.
[54] Nover, L.: In: Nover, L.; Luckner, M.; Parthier, B. (Eds.): Cell Differentiation. Molecular Basis and Problems. Gustav Fischer Verlag, Jena 1982.
[55] Ogutuga, D. B. A.; Northcote, D. H.: Biochem. J. **117** (1970) 715.
[56] Parthier, B.: Biol. Rdsch. **14** (1977) 329.
[57] Plowman, J.; Cone, J. E.; Guroff, G.: J. Biol. Chem. **249** (1974) 5559.
[58] Reinbothe, H.: Flora **151** (1961) 315.
[59] Reinbothe, H.; Mothes, K.: Ann. Rev. Plant Physiol. **13** (1962) 129.
[60] Rembold, H.; Gyure, W. L.: Angew. Chem. **84** (1972) 1088.
[61] Ross, C. W.: In: Biochemistry of Plants. Vol. 6. Academic Press, New York 1981, p. 169.
[62] Schlee, D.: Biol. Rdsch. **8** (1970) 317.
[63] Schlee, D.: Biol. Rdsch. **11** (1973) 285.
[64] Schlee, D.: Pharmazie **30** (1975) 345.
[65] Schlee, D.; Reinbothe, H.: Phytochemistry **2** (1963) 231.
[66] Schlee, D.; Reinbothe, H.: Phytochemistry **4** (1965) 311.
[67] Schlee, D.; Nieden, K. zur: Pharmazie **25** (1970) 651.
[68] Shavlovsky, G. M.; Logvinenko, E. M.; Kashchenko, V. E.; Koltun, L. V.; Zakalsky, A. E.: Dokl. Akad. Nauk SSSR **230** (1976) 1485.
[69] Shavlovsky, G. M.; Logvinenko, E. M.; Benndorf, R.; Koltun, L. V.; Kashchenko, V. E.; Zakalsky, A. E.; Schlee, D.; Reinbothe, H.: Arch. Microbiol. **124** (1980) 255.
[70] Skoog, F.; Armstrong, J.: Ann. Rev. Plant Physiol. **21** (1970) 359.
[71] Smulson, M. E.; Suhadolnik, R. J.: J. Biol. Chem. **242** (1967) 2872.
[72] Stadtman, E. R.: Adv. Enzymol. **28** (1966) 41.
[73] Suhadolnik, R. J.: Nucleoside Antibiotics. John Wiley & Sons, New York/London/Sydney/Toronto 1970.
[74] Suhadolnik, R. J.; Uematsu, T.: J. Biol. Chem. **245** (1970) 4365.
[75] Suzuki, T.: FEBS-Letters **24** (1972) 18.
[76] Suzuki, T.; Takahashi, E.: Biochem. J. **146** (1975) 79, 87.
[77] Suzuki, T.; Takahashi, E.: Phytochemistry **15** (1976) 1235.
[78] Suzuki, T.; Takahashi, E.: Biochem. J. **160** (1976) 171, 181.
[79] Suzuki, T.; Takahashi, E.: Drug Metabolism Review **6** (1977) 213.
[80] Suzuki, Y.; Brown, G. M.: J. Biol. Chem. **249** (1974) 2405.
[81] Tanaka, Y.; Hashimoto, Y.; Yanagisawa, K.; Abe, H.; Uchiyama, M.: Agr. Biol. Chem. **39** (1975) 1929.
[82] Taya, Y.; Tanaka, Y.; Nishimura, S.: Nature **271** (1978) 545.
[83] Taya, Y.; Tanaka, Y.; Nishimura, S.: FEBS-Letters **89** (1978) 326.
[84] Wanner, H.; Pesakova, M.; Baumann, T. W.; Charubala, R.; Guggisberg, A.; Hesse, M.; Schmid, H.: Phytochemistry **14** (1975) 745.
[85] Ward, D. C.; Cerami, A.; Reich, E.; Acs, G.; Altwerger, L.: J. Biol. Chem. **244** (1969) 3243.
[86] Wasternack, C.: In: Encyclopedia of Plant Physiology **1413** (1982) 263.
[87] Wolf, W. A.; Brown, G. M.: Biochim. Biophys. Acta **192** (1969) 468.

18 Alkaloids Formed from Histidine

M. Luckner

Histidine (3-(4(5)-imidazolyl)-alanine) (**1**) and its decarboxylation product histamine (4(5)-(2-aminoethyl)-imidazole) (**2**) are the precursors of a relatively small group of alkaloids (for reviews *cf.* [3, 5]). These alkaloids contain the imidazole ring with or without additional rings. The following types of compounds may be distinguished:

a) Protoalkaloids with a Histamine Residue

In plants histamine (**2**) (Fig. 18.1) may be acylated at the primary amino group; *cf.* dolichotheline (**3**) and *N*-4′-oxodecanoyl histamine (**4**). [2-^{14}C]Histamine and [2-^{14}C]leucine (probably *via* [1-^{14}C]isovaleric acid) are incorporated into dolichotheline (**3**) in *Dolichothele sphaerica* [4, 7, 8]. The enzyme that forms the acid amide probably has a relatively low substrate specificity and also accepts "unnatural"

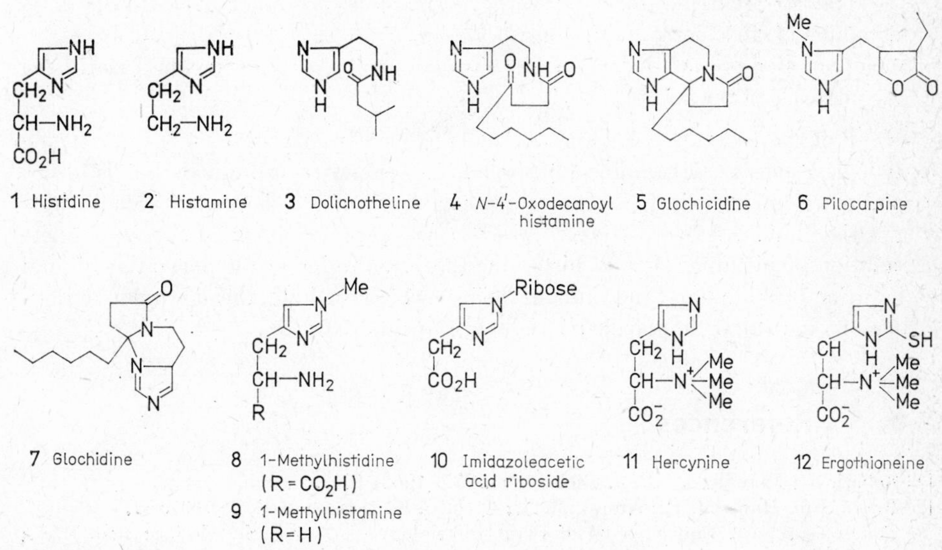

Fig. 18.1
Several compounds with an imidazole ring

compounds. *N*-Cinnamoyl histamine, for instance, was formed as an artificial product when histamine and cinnamic acid had been fed. The unnatural compound 4(5)-aminomethyl imidazole was also used as a substrate [9, 10].

b) Methylated Protoalkaloids with Histidine and Histamine Residues

Histidine (**1**), histamine (**2**), and derivatives of these compounds may be methylated at the *N*-atoms of the imidazole ring and the side chain. Administration of histidine and histamine to mammals, for instance, yielded N-1 and N-3 methylated derivatives excreted in the urine (*cf.* formulas **8** and **9**) [6].

In plants and microorganisms, histidine is a precursor of the betaine hercynine (**11**) which may be further transformed into ergothioneine (**12**). The SH group of the latter compound is derived from cysteine [1]. Feeding [^{14}C-Me]methionine to shoots of *Pilocarpus jaborandi* resulted in Me-labelled pilocarpine (**6**). However, neither administration of labelled histidine (**2**) nor of threonine and acetate as possible precursors of the *O*-heterocyclic ring to the shoots resulted in any appreciable incorporation. This probably indicates that the shoots are able to methylate a pilocarpine precursor but are not the site at which the basic moiety of the alkaloid is synthesized [2].

c) Imidazole Alkaloids with further *N*-Heterocyclic Rings

New *N*-heterocyclic rings may be formed including the C-atoms and the amino group of the histamine side-chain. Although no biosynthetic work has so far been done, it may be concluded from the chemical structure of corresponding alkaloids that the following reactions are involved in closure of the new rings:

- Mannich condensations, for instance in case of the formation of glochicidine (**5**), an alkaloid occurring together with its possible precursor *N*-4′-oxodecanoyl histamine (**4**) in several species of *Glochidion*.
- Reaction with one of the imidazole nitrogen atoms, as may be the case in the biosynthesis of glochidine (**7**).

Several of the protoalkaloids and alkaloids formed from histidine (**1**) show strong physiological actions. Histamine (**2**) has hormone activity in animals but is also a plant feeding deterrent in the toxin of nettle. Pilocarpine (**6**) is used as a cholinergic drug. Other imidazole derivatives, however, are mere products of the incomplete degradation of histidine (**1**) and histamine (**2**), *e.g.* the imidazole derivatives found in the urine of mammals and human beings which include the methylated derivatives (**8**) and (**9**), or imidazole acetic acid riboside (**10**).

18.1 References

[1] Askari, A.; Melville, D. B.: J. Biol. Chem. **237** (1962) 1615.
[2] Brochmann-Hanssen, E.; Nunes, M. A.; Olah, C. K.: Planta Med. **28** (1975) 1.
[3] Fodor, G. B., In: Encyclopedia of Plant Physiology — New Series, Vol. 8: Bell, E. A.; Charlwood, B. V. (Eds.): Secondary Plant Products. Springer-Verlag, Berlin/Heidelberg/New York 1980, p. 160.

[4] Horan, H.; O'Donovan, D. G.: J. Chem. Soc. C **1971**, 2083.
[5] Luckner, M.; In: Mothes, K.; Schütte, H. R. (Eds.): Biosynthese der Alkaloide. Deutscher Verlag der Wissenschaften, Berlin **1969**, p. 593.
[6] Meister, A.: Biochemistry of the Amino Acids, Vol. 2. Academic Press, New York 1965.
[7] Rosenberg, H.; Paul, A. G.: Tetrahedron Lett. **1969**, 1039.
[8] Rosenberg, H.; Paul, A. G.: Lloydia **34** (1971) 372.
[9] Rosenberg, H.; Stohs, S. J.: Phytochemistry **13** (1974) 823.
[10] Rosenberg, H.; Stohs, S. J.: Lloydia **34** (1974) 313.

19 Isoprenoid Alkaloids

D. Gross, H. R. Schütte, and K. Schreiber

19.1 Monoterpenoid Alkaloids

The large and still expanding group of iridoids, which are widely distributed in the plant kingdom, is based on a monoterpenoid methylcyclopentane-(c)-pyran skeleton (for reviews see [20, 53]). Among them are several simple iridoids and seco-iridoids with a ten-carbon basic skeleton in which the heterocyclic oxygen is replaced by nitrogen (*cf.* [31, 46, 68]). The monoterpenoid alkaloids of the actinidine and the skytanthine type contain either a pyridine or a piperidine nucleus with a condensed methylcyclopentane ring. They are thus also named pyrindane alkaloids. Gentianine and its structurally related minor alkaloids possess a bicyclic ring system formed from a pyridine nucleus with a fused six-membered pyran ring.

Some caution should be used in assuming that all the monoterpenoid alkaloids discussed in this chapter are genuine plant constituents since some of them have been shown to be artefacts formed during the ammoniacal extraction of the plant material. However, it is interesting to note that these monoterpenoid alkaloids occur in plant families in which iridoids and seco-iridoids and very often indole alkaloids with a loganin-derived monoterpene unit are found. This simultaneous occurrence is significant as it offers circumstantial evidence in support of a biosynthetic pathway *via* common intermediates.

19.1.1 Actinidine and Related Alkaloids

Actinidine (**2**) represents the basic component of this group and has been isolated from *Actinidia polygama* [18, 209—211], *A. arguta* [69], *Tecoma stans* [69], *T. fulva* [112] and *Valeriana officinalis* [73, 119, 125, 243]. Relatives through either a missing substituent at the 4-position or variations in the oxidation level of the attached methyl group at that position are 4-noractinidine (**1**) from *Tecoma stans* [48], tecostidine (**3**) from *T. stans* [91] and *Pedicularis rhinanthoides* [4], valerianine (**4**) from valerian [56] (Fig. 19.1). The two quaternary alkaloids **5** and **6** from *Valeriana officinalis* [242, 243] contain a *p*-hydroxyphenethyl group at the nitrogen of the actinidine skeleton. The alkaloids **1** to **6** have an (*S*) configuration at the 7-position. Boschniakine (**7**) has been detected in *Boschniakia rossica* [212], *Tecoma stans* [48, 70], *T. radicans* [69] and *Plantago sempervirens* [192], while plantagonine (**8**) has

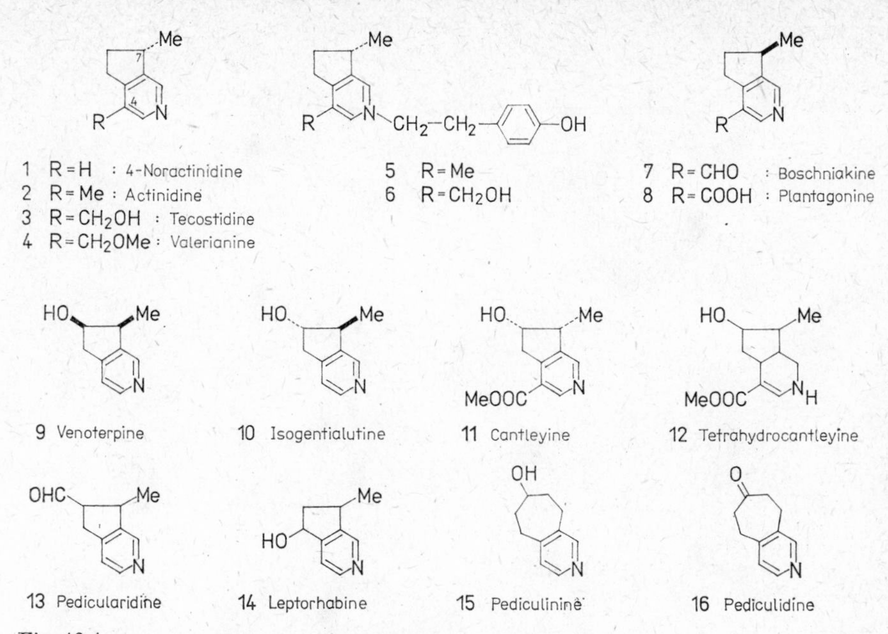

Fig. 19.1
Structures of actinidine and related monoterpenoid pyridine alkaloids

been found in different species of *Pedicularis* and *Plantago* [1, 3, 4, 11, 140, 258]. Both alkaloids possess a (*7R*)-configuration. Boschniakine (7) was shown to be identical with indicaine [72, 192, 241], which had been isolated previously from some *Pedicularis* and *Plantago* species [3, 11, 140, 141, 258].

Venoterpine (9) has been isolated from *Alstonia venenata* [190], *Striga hermonteca* [19], *Rauwolfia verticillata* [15, 16] (the alkaloid RW 47 was found to be identical with venoterpine [175]) and *Gentiana tibetica* [208] (isolated as gentialutine). The group of actinidine-like bases also includes the alkaloids isogentialutine (**10**) from *G. tibetica* [208], pedicularidine (**13**) from *Pedicularis olgae* [142] and leptorhabine (**14**) from *Leptorhabdos parviflora* [128]. The alkaloids **1**, **9**, **10** and **14** are based on a nine-carbon skeleton, but they originate from two isoprene units followed by elimination of the carbon at the 4 position. Pediculinine (**15**) and pediculidine (**16**) from *Pedicularis olgae* [2, 5] consist of ten carbon atoms, but judging from their carbon skeleton they are presumably not monoterpene-derived bases.

Cantleyine (**11**), which was found in *Cantleya corniculata* [225], *Strychnos nux-vomica* [30], *Lasianthera austrocaledonica* [226], *Dipsacus azureus* [188], *D. silvestris* [193] and a *Jasminum* species from New Guinea origin [95], and tetrahydrocantleyine (**12**) from *L. austrocaledonica* [226] are not genuine monoterpene alkaloids but artefacts formed during extraction of the plant material in the presence of ammonia [193, 225, 226]. This fact is of interest because other alkaloids with this structural feature have been shown to be native plant constituents, for example boschniakine (7) [192].

Based on the number of carbon atoms and on the characteristic C-branching, the

17 Mevalonic acid
18 Isopentenyl pyrophosphate
19 Dimethylallyl pyrophosphate
20 Geranyl pyrophosphate
21 Neryl pyrosphosphate
22
23
24 R = H : Deoxyloganin
25 R = OH : Loganin
26 Secologanin
Indole alkaloids
27 Gentianine
2 Actinidine
28 Skytanthine

Fig. 19.2
Biosynthesis of monoterpene alkaloids

alkaloids mentioned above belong to the monoterpenoid alkaloids. Therefore, they fit into the general schema of terpene formation according to the biogenetic rule of Ruzicka and should be formally derived by condensation of two isoprene units ("active isoprene") (Fig. 19.2). It is well known from extensive studies on the terpene biosynthesis that (*3R*)-mevalonic acid (**17**) formed from acetyl coenzyme A units, is converted into the isoprenoid C_5 compound isopentenyl pyrophosphate (**18**), which yields 3,3-dimethylallyl pyrophosphate (**19**) by isomerization (Fig. 19.2). The latter compound acts as starter for chain elongation by reaction with isopentenyl pyrophosphate to form the acyclic monoterpenes geranyl pyrophosphate (**20**) and neryl pyrophosphate (**21**). Cyclization of **21** leads to the fused bicyclic system of loganin (**25**) which represents not only an important intermediate *en route* to other iridoids and seco-iridoids but also to indole alkaloids with an iridoid part, *i.e.* the ajmalicine, catharanthine and vindoline type (*cf.* chapter 15). It has been assumed that 7-deoxyloganin (**24**) or a closely related iridoidal compound serves as immediate precursor of the monoterpenoid alkaloids mentioned above, whereby the oxygen of a suitable iridoidal intermediate, for example the hypothetical iridoids **22** or **23**, is substituted by biochemically "active ammonia" originating from the plant nitrogen pool. However, it has to be emphasized that some alkaloids isolated from plant sources have been shown to be artefacts, *i.e.* cantleyine (**11**) [193, 225, 226] and fontaphilline (**46**) [32].

In some cases, however, it is assumed that the given alkaloid is partially genuine but the main portion is an artefact (gentianine (**27**) [32, 55, 67, 150, 183]).

The occurrence of monoterpenoid alkaloids and structurally related iridoid constituents in the same plant points to a common biosynthetic pathway both of iridoids and iridoidal alkaloids from two isoprene units. This assumption is supported by the occurrence of actinidine (**2**) and matatabilactone in *Actinidia polygama*, boschniakine (**7**) and boschnaloside in *Boschniakia rossica*, and the alkaloids **5** and **6** and the valepotriates in *Valeriana officinalis*.

The biosynthesis of actinidine (**2**) has been studied by administration of [^{14}C]-labelled precursors to *Actinidia polygama* [18] and *Valeriana officinalis* [73]. Aspartate, quinolinate, and lysine are known as biogenetic precursors of the pyridine ring of various pyridine alkaloids (*cf.* chapters 12 and 13) but gave no ^{14}C-labelling of actinidine [18]. The incorporation of [2-^{14}C]acetate, [2-^{14}C]mevalonate (**17**) and [1-^{14}C]geranyl pyrophosphate (**20**) provides convincing evidence for the isoprenoid origin of actinidine [18, 73] (Fig. 19.2). The incorporation of [2-^{14}C]mevalonate into boschniakine (**7**) [70] and the *Valeriana* alkaloid **5** [73] is in agreement with these results. The *p*-hydroxyphenethyl residue of the alkaloids **5** and **6** may be derived from tyramine as supported by the incorporation of [2-^{14}C]tyrosine into the main alkaloid **5** of *V. officinalis*. It has not yet been established whether the heterocyclic nitrogen originates from the amino group of tyrosine or from the nitrogen pool of the plant. The possible conversion from the piperidine system into the pyridine system may be excluded as shown by the non-incorporation of tritiated *N*-normethylskytanthine and δ-skytanthine (**31**) into boschniakine (**7**) [70].

The details of the biogenetic pathway leading from mevalonic acid (**17**) to actinidine (**2**) and closely related bases have still to be clarified, especially the preceding intermediates, the stereochemical course, the timing of introduction, and the origin of the heterocyclic nitrogen as well as the enzymes that catalyse the different biosynthetic steps. The site of biosynthesis within the plant and the metabolism of actinidine and its relatives are also not yet known.

19.1.2 Skytanthine Alkaloids

The skytanthine group [Fig. 19.3] includes the α-, β- and δ-diastereoisomers of skytanthine (**29**, **30**, **31**), which were isolated from *Skytanthus acutus* [13, 37, 38, 49, 50, 52, 170] and *Tecoma stans* [70, 71]. However, it has also been reported that the

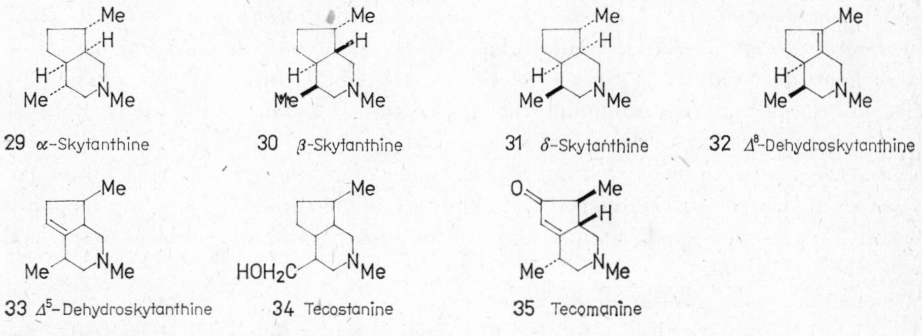

Fig. 19.3
Structures of alkaloids of the skytanthine group

skytanthines are not genuine alkaloids of *S. acutus* but products formed from the naturally occurring β-skytanthine-*N*-oxide [14, 230]. Hydroxylated skytanthines, which differ in the position of the hydroxy group, have also been detected in *S. acutus* [10, 13, 36, 170] and *Tecoma stans* [48, 127]. In *T. stans* these bases are accompanied by *N*-normethylskytanthine [48], Δ^5-dehydroskytanthine (**33**) [70, 71], tecostanine (**34**) [70, 92] and tecomanine (**35**) [48, 70, 126]. In contrast to the actinidine group, these alkaloids are based on a piperidine ring with a condensed methylcyclopentane ring.

Biosynthetic studies carried out with *S. acutus* and *T. stans* plants have shown that [2-^{14}C]mevalonic acid (**17**) is incorporated into the skytanthine isomers **29** to **31** [17, 39, 71, 158], Δ^5-dehydroskytanthine (**33**), tecostanine (**34**) and tecomanine (**35**) [70]. The amino acid lysine is not involved in the biosynthesis of skytanthine [17]. [$^{14}CH_3$]Methionine was shown to serve as specific precursor of the *N*-methyl group of these alkaloids [17, 70]. The low rate of incorporation of tritiated actinidine (**2**) into skytanthine alkaloids indicated that there is in all probability no conversion from the pyridine system into the piperidine system [70]. Loganin (**25**) gave small incorporation and should be excluded as a possible precursor although hypothetical iridoidal intermediates such as **22** or **23** may still be suggested as precursors (Fig. 19.2). As with the actinidine group, the biosynthetic pathway from mevalonate to the skytanthine alkaloids has not yet been studied in detail.

19.1.3 Gentianine and Related Alkaloids

The alkaloids of the group considered here differ in their fundamental carbon skeleton from the actinidine and skytanthine groups (*cf.* Figs. 19.1, 19.3, and 19.4). Nevertheless, they share a common biogenesis from mevalonate (**17**) *via* iridoidal intermediates. In general the plants containing monoterpene alkaloids also contain gentiopicroside-like seco-iridoids (*cf.* [53]) which serve as biological precursors of the *Gentiana* alkaloids.

The main alkaloid of this group, gentianine (**27**), was isolated from various *Gentiana* species [41, 55, 160—162, 169, 187, 213] and from *Fontanesia phillyreoides* [32], *Ixanthus viscosus* [35], *Enicostema littorale* [67], *Anthocleista procera* [150, 153, 183, 233], *A. rhizophorides* [182], *Erythraea centaurium* [164], *Menyanthes trifoliata* [206] and *Fagraea fragrans* [261]. Gentianine (**27**) is mainly an artefact formed from native gentiopicroside or a related seco-iridoid by substitution of the oxygen by nitrogen originating from ammonia during isolation [32, 55, 67, 150, 183]. Because of the instability of the cyclic acetal function, the *in vitro* conversion of gentiopicroside into **27** can easily be carried out by treatment of gentiopicroside with ammonia. However, it has been established that gentianine naturally occurs in the plant but in comparatively smaller amounts than found by use of ammoniacal extraction of the plant material.

Minor alkaloids of gentianine are hydroxygentianine (**36**) [162], gentianamine (**37**) [12, 187, 213], gentianadine (**38**) [12, 213], gentianidine (**39**) [156, 160, 164, 206], gentioflavine (**40**) [160, 163—166, 168, 176, 189], gentiotibetine (**41**) [166, 186, 187,

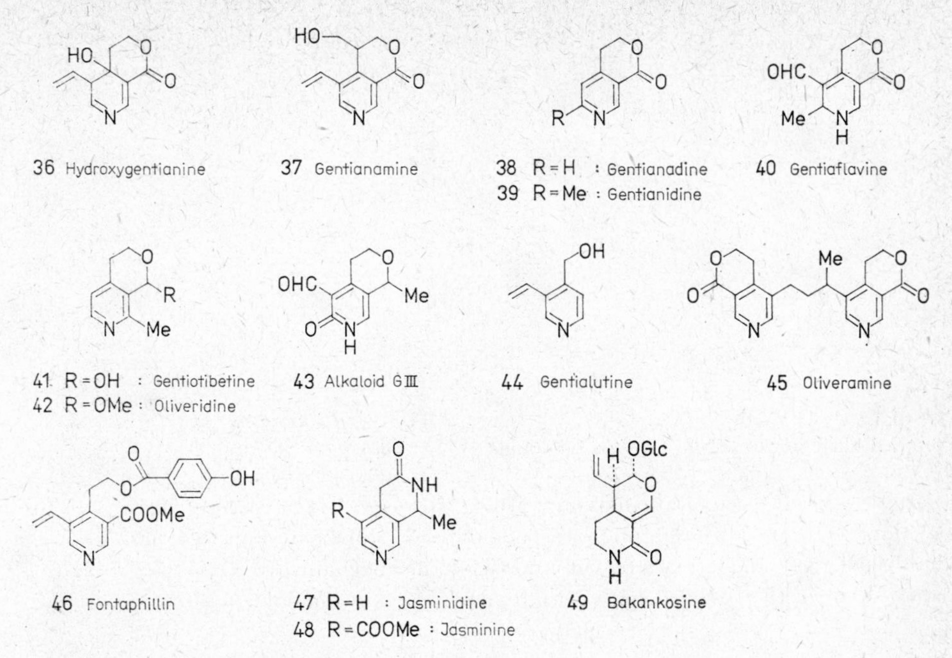

Fig. 19.4
Gentianine related alkaloids

206, 207], oliveridine (**42**) [186], the alkaloid **43** [266], gentialutine (**44**) [41, 206] and oliveramine (**45**) [189] (Fig. 19.4). Fontaphilline (**46**) [32] has been found to be an artefact. From the biogenetic point of view jasminidine (**47**) isolated from *Syringa vulgaris* [191], jasminine (**48**) occurring in *Jasminum gracile, J. lineare,* and *Ligustrum novoguineense* [94], and the glycoside alkaloid bakankosine (**49**) from *Strychnos vacacoua* [236] also belong to this group of monoterpenoid alkaloids.

Studies on the biosynthetic route to gentianine (**27**) and its structurally related alkaloids indicated that they are derived by a mevalonic acid pathway very closely related to the biosynthesis of iridoids and seco-iridoids. These experiments have been carried out with *Gentiana asclepiadea* by means of [^{14}C]labelled precursors. [2-^{14}C]Glycine, [2-^{14}C]acetate, and [2-^{14}C]mevalonic acid (**17**) gave label in gentianine (**27**) and gentioflavine (**40**), which were isolated from the plant material [55, 161], while [^{14}C]pyruvate and [^{14}C]formate were not incorporated [169]. After feeding [1-^{14}C] geraniol (**20**), gentioflavine was labelled only in the formyl group [163]. These findings are in accordance with an isoprenoid pathway (Fig. 19.2).

The biogenetic relationship between iridoids postulated as precursors and *Gentiana* alkaloids has been studied by administration of [^{14}C]labelled gentiopicroside, which was found to be incorporated into gentioflavine (**40**) [168] (Fig. 19.5). It was thus demonstrated that the alkaloid gentioflavine is biogenetically derived from the seco-iridoid gentiopicroside. Furthermore, it has been shown that gentianidine (**39**) and gentianadine (**38**) are formed from gentioflavine (**40**), which is also converted into gentianine (**27**). These results point to a key role of gentioflavine in the bio-

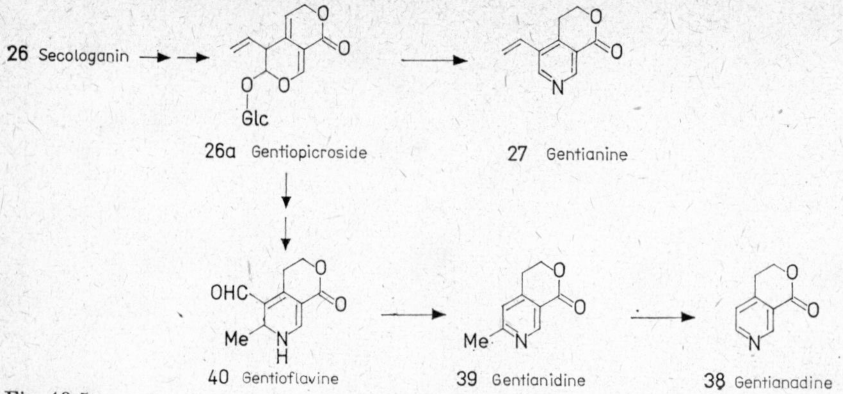

Fig. 19.5
Proposed biosynthesis of the *Gentiana* alkaloids

synthesis of the *Gentiana* alkaloids (*cf.* [167]). However, the complete pathway from mevalonate to the different monoterpenoid bases shown in Fig. 19.4 has not yet established. Not all details are yet known of the mechanisms of cleavage and rearrangement of the basic skeleton forming the different alkaloid types.

19.2 Sesquiterpenoid Alkaloids

Some alkaloids have a carbon skeleton, reminiscent of the sesquiterpens [54], *e.g.* dendrobine (**57**) and related bases of *Dendrobium nobile* belonging to the picrotoxane sesquiterpenoids. Mevalonate is specifically incorporated into dendrobine [42, 51, 268]. *2-trans-6-trans*-farnesol (as **50**) is shown to be a precursor for this alkaloid rather than the *2-cis-6-trans*-isomer [45]. Half of the C-1 hydrogen of farnesol is localized at C-5 and the other half at C-8 of dendrobine. This 1,3-hydride shift has a high degree of stereospecifity, the *pro-1R* hydrogen of farnesol being transferred. The incorporation pattern is consistent with a pathway in which germacradiene (**51**) is first formed in a *trans,trans*-configuration and in the following steps the configuration or position of the 2,3-double bond is modified to the allylic cation **52** to fulfil the geometrical requirements for the formation of the muurolane intermediate (**53**) (Fig. 19.6) [269]. Ring closure of C-3 with the cationic centre at C-7 (of **53**) would

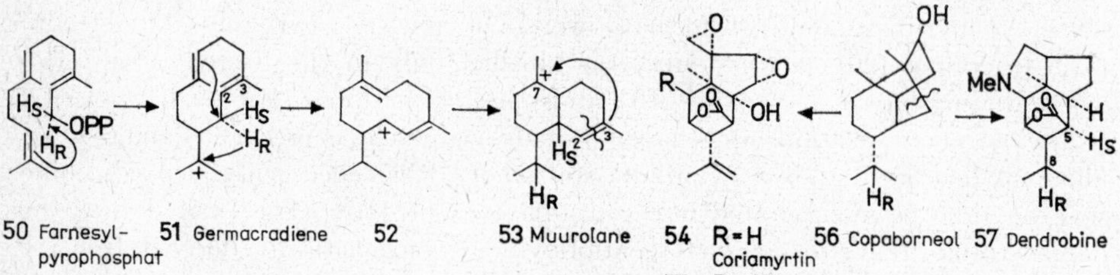

Fig. 19.6
Biosynthesis of dendrobine and related compounds

give the tricyclic system of copaborneol (**56**). Coriamyrtin (**54**) and tutin (**55**) are other examples of the picrotoxane group for which the incorporation pattern of mevalonate is consistent with a pathway involving ring cleavage of the tricyclic intermediate copaborneol (**56**) [29, 42—44]. Copaborneol has been specifically incorporated into tutin (**55**) in *Coriaria japonica* [257]. It is therefore perhaps also an intermediate for dendrobine (**57**).

Further sesquiterpenoid alkaloids are nupharidine (**58**) and related compounds from *Nuphar japonicum*. Mevalonate has been incorporated specifically in such alkaloids [223]. Related alkaloids from *Nuphar* species such as nuphenine (**60**), nupharamine (**59**), and nuphamine (**61**) could be intermediates for the quinolizidine skeleton of nupharidine (Fig. 19.7).

58 Nupharidine 59 Nupharamine 60 Nuphenine 61 Nuphamine

Fig. 19.7
Nuphar alkaloids

19.3 Diterpenoid Alkaloids

In *Aconitum* and *Delphinium* species (*Ranunculaceae*), *Garrya* species (*Cornaceae*), and *Inula* species (*Compositae*) there are some diterpenoid-derived alkaloids of the lycoctonine (**67**), heteratisine (**70**), veatchine (**64**), and atisine (**66**) types. These could be formed *via* primaradiene (**62**) and phyllocladene (**63**) type intermediates by incorporation of nitrogen (Fig. 19.8) [181, 220]. However, few biosynthetic ex-

62 Primaradiene 63 Phyllocladene 64 R^1=H; R^2=OH : Veatchine
65 R^1=OH; R^2=H : Garryfoline

66 Atisine 67 R^1=H; R^2=Me; R^3=Me : Lycoctonine 70 Heteratisine
68 R^1=R^3=Me; R^2=H : Brownine
69 R^1=Me; R^2=R^3=H : Delcosine

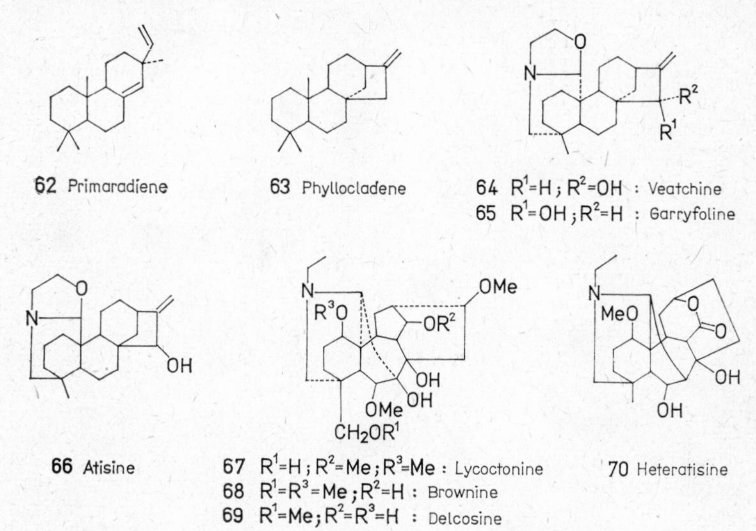

Fig. 19.8
Different types of diterpenoid alkaloids

periments have been carried out. Mevalonate has been incorporated into brownine (**68**) and lycoctonine (**67**) (in *Delphinium brownii*) as well as into delcosine (**69**) (in *D. ajacis*) [21, 57]. Glycine has been mooted as a precursor of the *N*-ethyl group.

19.4 Triterpenoid Alkaloids

Besides the steroid alkaloids derived from a triterpenoid nucleus, a number of alkaloids isolated from *Daphniphyllum* species such as *D. macropodum* are based on the non-degraded triterpenoid skeleton [267]. The incorporation pattern of mevalonate and squalene (**74**) into daphniphylline (**71**), and codaphniphylline (**72**), the main alkaloids, supports a triterpenoid origin from six units of this terpenoid precursor *via* a squalene-like intermediate [232]. The types and amounts of these alkaloids varied with the season, and the highest mevalonate incorporation into daphniphylline (**71**) was recorded in June and July. In the light of the occurrence of daphmacrine (**75**) together with secodaphniphylline (**76**) a pathway for the biosynthesis of daphniphylline has been proposed (Figs. 19.9 and 19.10).

Daphnilactone B (**73**) is related to the above alkaloids but has only 22 carbon atoms. Experiments with D. *teijsmanni* support a route similar to that envisaged for the daphniphylline skeleton (Fig. 19.10), *i.e.* from six mevalonate units *via* a squalene-like intermediate but with later cleavage of a C_8 unit [178]. Daphnilactone B has been isolated only from fruits of *Daphniphyllum* species [237] whereas daphniphylline and codaphniphylline also occur in leaves and bark [116].

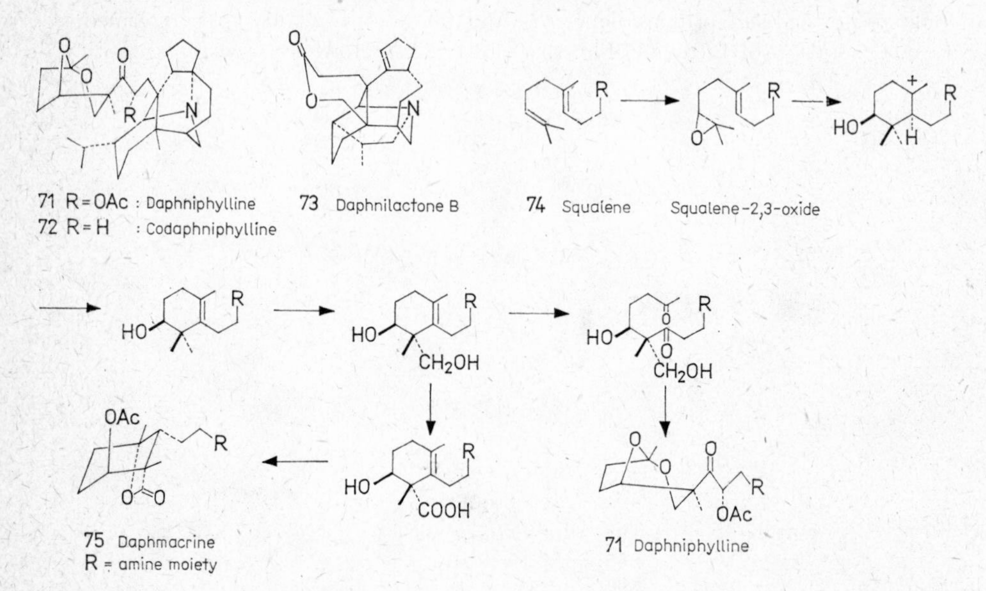

Fig. 19.9
Hypothetical pathway for the biosynthesis of the ketal moiety of daphniphylline

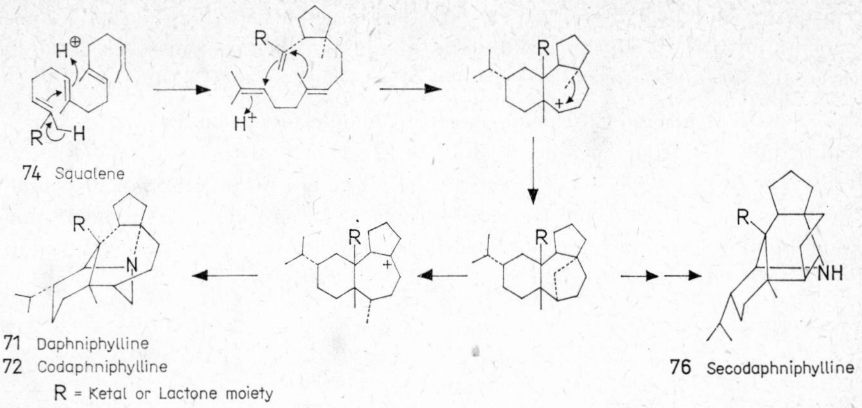

74 Squalene

71 Daphniphylline
72 Codaphniphylline
R = Ketal or Lactone moiety

76 Secodaphniphylline

Fig. 19.10
Hypothetical pathway for the biosynthesis of the amine moiety of daphniphylline

19.5 Steroid Alkaloids

The present chapter deals with the biochemistry, *i.e.* biosynthesis, metabolism, and biological degradation, of nitrogenous natural products possessing the unaltered or a biosynthetically altered carbon skeleton of steroids or that of their direct biogenetic tetraterpenoidal precursors, *e.g.* cycloartenol. According to Hegnauer [109], these nitrogenous metabolites are not true alkaloids in the strict sense, but are instead "pseudoalkaloids" or "alcaloida imperfecta", which are, biogenetically considered, simple derivatives of generally occurring nitrogen-free constituents. Their nitrogen content and their basicity appear to be not essential properties but, more likely, accidental characteristics. They become alkaloid-like substances because at a certain stage of their biosynthesis nitrogen is introduced into their molecule. Therefore, the biogenesis of pseudoalkaloids is closely related to that of their nitrogen-free analogues.

Up to now steroid alkaloids have been isolated and structurally elucidated from the following four plant families only: *Solanaceae* (genera: *Solanum*, *Lycopersicon* and, to a small extent, *Cestrum*, *Cyphomandra* and *Nicotiana*), *Liliaceae* (*Veratrum*, *Schoenocaulon*, *Fritillaria* (*Korolkowia*, *Petilium*), *Amianthium*, *Rhinopetalum* and *Zygadenus*), *Apocynaceae* (*Chonemorpha*, *Conopharyngia*, *Dictyophleba*, *Funtumia*, *Holarrhena*, *Kibatalia*, *Malouetia*, *Paravallaria* and *Vahadenia*) as well as *Buxaceae* (*Buxus*, *Pachysandra* and *Sarcococca*).

From a more chemical point of view most of these steroid alkaloids belong to one of the following main groups representing different types of structure:

a) Alkaloids with the complete and unaltered C_{27}-carbon skeleton of cholestane from *Solanaceae* and *Liliaceae* (Fig. 19.11) but possessing different heterocyclic ring systems, such as the epiminocholestanes, *e.g.* solacongestidine (**77a**); the alkaloids with a solanocapsine skeleton, *e.g.* solanocapsine (**77**); the spirosolanes, *e.g.* solasodine (**78**); the 3-aminospirostanes, *e.g.* jurubidine (**79**); the solanidanes, *e.g.* solanidine (**80**); and the cevanidane alkaloids, *e.g.* procevine (**81**), usually designated *Solanum* steroid alkaloids [184, 185, 195, 216] (*cf.* [130]).

b) Alkaloids with an altered C_{27}-carbon skeleton of cholestane, isolated from *Liliaceae* but until now not from *Solanaceae* (Fig. 19.12), that is with the *C*-nor-*D*-homo-[14(13→12)-abeo] ring system of veratranine, *e.g.* veratramine (**82**), of jervanine, *e.g.* jervine (**83**), and of cevanine, *e.g.* verticine (**84**), as well as with the 18-nor-17β-methylcholestane skeleton, *e.g.* veralkamine (**85**) and veramine (**86**), usually named *Veratrum* steroid alkaloids [124, 152, 184, 240].

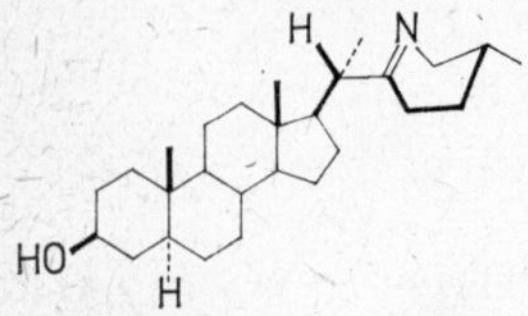

77a Solacongestidine

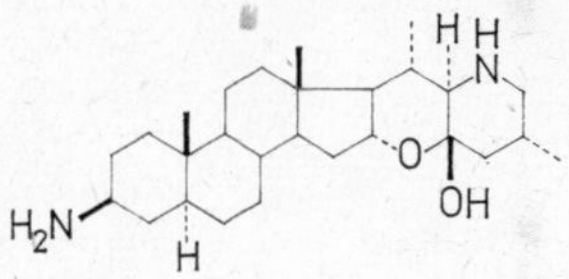

77 Solanocapsine

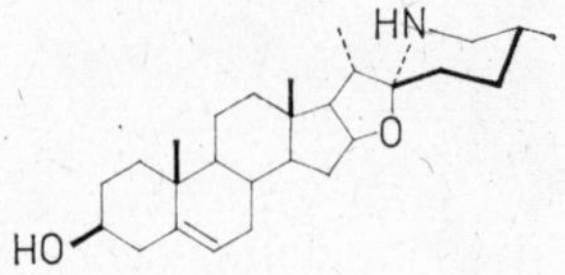

78 Solasodine

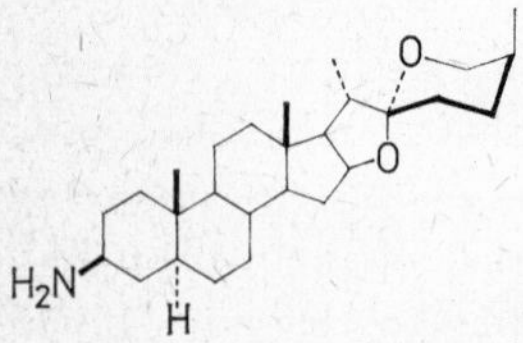

79 Jurubidine

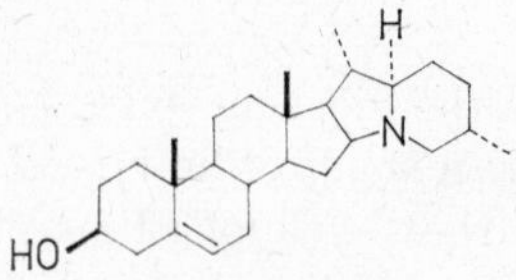

80 Solanidine

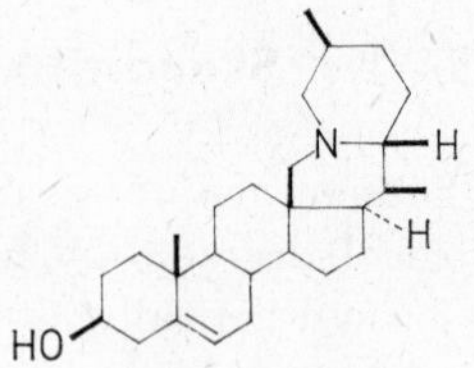

81 Procevine

Fig. 19.11
Structural types of alkaloids with the unaltered C_{27}-carbon skeleton of cholestane from *Solanaceae* and *Liliaceae*

82 Veratramine

83 Jervine

84 Verticine

85 Veralkamine

86 Veramine

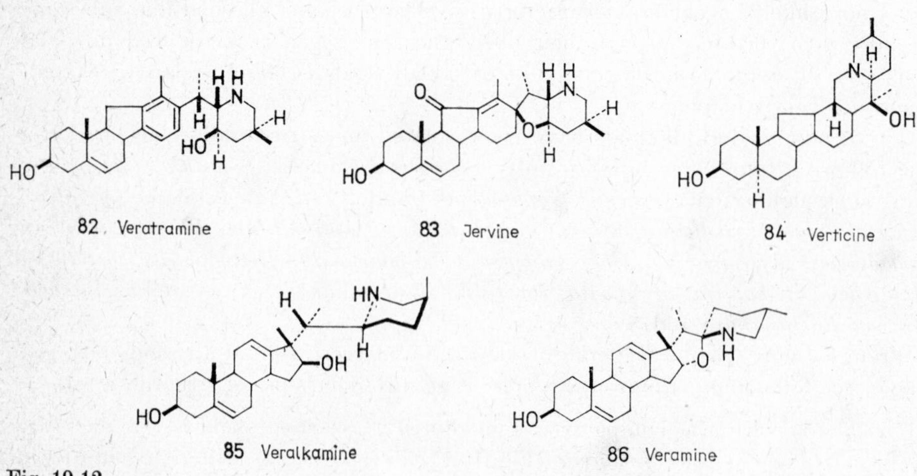

Fig. 19.12
Structural types of alkaloids with an altered C_{27}-carbon skeleton of cholestane from *Liliaceae*

c) Alkaloids with the C_{21}-carbon skeleton of pregnane and with amino groups at C-3 and/or C-20 (or an imino group between C-18 and C-20) from *Apocynaceae* and (to a lesser extent) from *Buxaceae*, for instance (Fig. 19.13) funtumine (**87**), paravallarine (**88**), holarrhimine (**89**), and the so-called conanine alkaloid conessine (**90**) [40, 66, 123].

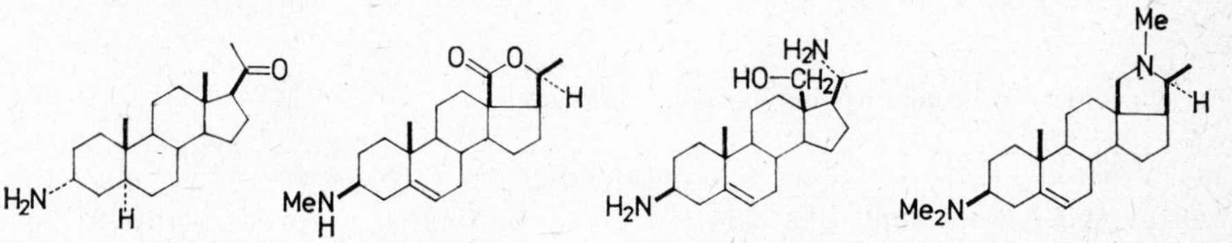

87 Funtumine 88 Paravallarine 89 Holarrhimine 90 Conessine

Fig. 19.13
Structural types of alkaloids with the C_{21}-carbon skeleton of pregnane from *Apocynaceae* and *Buxaceae*

d) Alkaloids with a C_{21}-carbon skeleton and with amino groups at C-3 and/or C-20 but, in general, with one to three additional methyl groups at C-4 and/or C-14 and with 9β,19-cyclopregnane or 9(10$\rightarrow$19)-abeo-pregnane structures from *Buxaceae* (Fig. 19.14), *e.g.* cycloxobuxidine-F (**91**) and buxamine-E (**92**) [40, 66, 240].

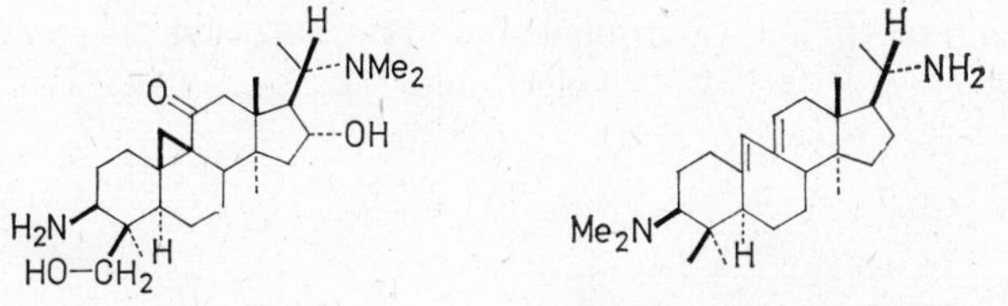

91 Cycloxobuxidine-F 92 Buxamine-E

Fig. 19.14
Structural types of alkaloids with 9β, 19-cyclopregnane and 9(10$\rightarrow$19)-abeo-pregnane skeleton from *Buxaceae*

Some basic steroids isolated from plants of the *Asclepiadaceae* family were shown to be, for example, *N*-methylanthraniloyl or nicotinoyl esters of highly oxygenated pregnane derivatives (*cf.* [224, 234, 235]). The discussion of the biosynthesis of these non-alkaloidal aglyca is not included in this chapter.

On the other hand, the highly toxic compounds isolated from the amphibians *Salamandra maculosa*, *e.g.* samandarine (**93**) [87—89], and the Colombian arrow poison frog, *Phyllobates aurotaenia*, *e.g.* batrachotoxin (**94**) [47, 265], are true steroid alkaloids, although with quite unusual structures (Fig. 19.15). Recently, some nitrogenous steroids with 24-methylene-15-aza-*D*-homocholestane structure (**95**, Fig. 19.15) and high antibiotic activity against pathogenic fungi were isolated as metabolites of the fungus *Geotrichum flavo-brunneum* [174].

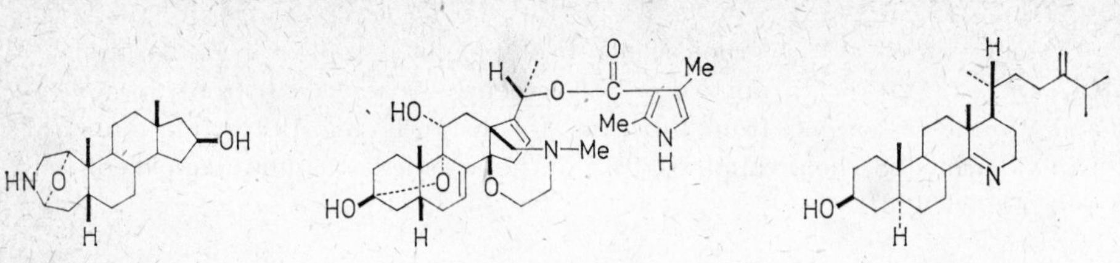

93 Samandarine 94 Batrachotoxin 95

Fig. 19.15
Steroid alkaloid types from amphibians and microorganisms

The steroid alkaloids of the above-mentioned group a) (*Solanum* steroid alkaloids) as well as the jerveratrum alkaloids (*cf.* group b) seldom occur in plants as free alkamines, normally appearing as *O*-glycosides. The same is occasionally true for the ceveratrum alkaloids (*cf.* group b), which are, however, usually esterified with aliphatic or aromatic acids. The alkaloids from *Apocynaceae* and *Buxaceae* (*cf.* groups c) and d) exist in plants as free alkamines and, in some cases, conjugated with acids forming *N*-acylamides or with carbohydrates leading to *O*-glycosides.

The biochemistry, especially the biosynthesis, of plant steroids in general and of steroid alkaloids in particular, has been reviewed occasionally in the last decade (*e.g.* [40, 58, 59, 61–63, 74–76, 93, 100–104, 110, 111, 159, 195, 197, 201, 215–218, 221, 222, 240, 244].

Following the general pathway of steroid biosynthesis in plants, starting from acetyl coenzyme A *via* the principal intermediates mevalonic acid (**17**), isopentenyl pyrophosphate (**18**), dimethylallyl pyrophosphate (**19**), farnesyl pyrophosphate (**50**), squalene (**74**), cycloartenol (**96**), and cholesterol (**97**) (Fig. 19.16), the latter com-

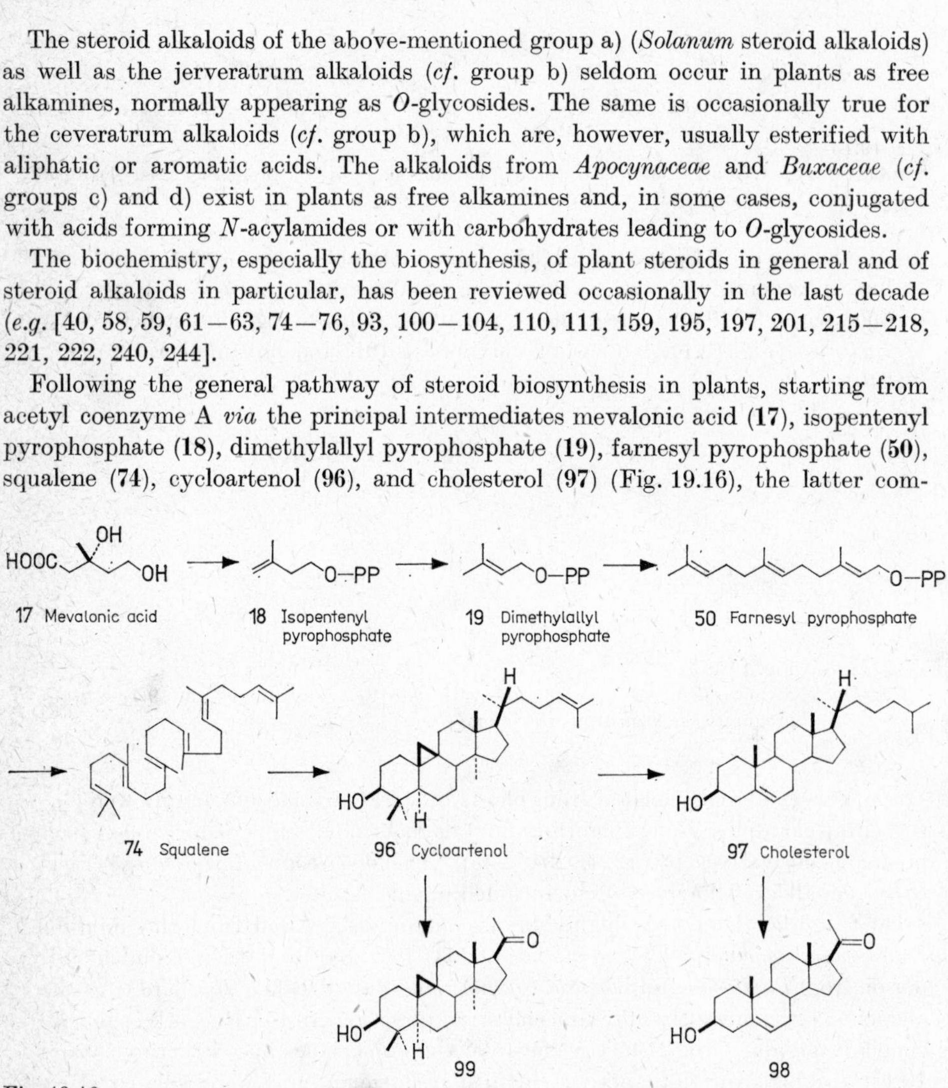

17 Mevalonic acid 18 Isopentenyl pyrophosphate 19 Dimethylallyl pyrophosphate 50 Farnesyl pyrophosphate

74 Squalene 96 Cycloartenol 97 Cholesterol

99 98

Fig. 19.16
General biogenetic pathway of plant steroids

pound or a biogenetic equivalent of it is the direct precursor of all the C_{27}-steroid alkaloids and of those with an altered cholestane skeleton. However, cholesterol (97), or other plant sterols such as sitosterol, are the indirect precursors of the *Apocynaceae* steroid alkaloids with C_{21}-carbon skeleton, because their respective nitrogen-free intermediates such as 3β-hydroxy-pregn-5-en-20-one (98, Fig. 19.16) are biosynthetically derived from these sterols by oxidative cleavage of the C-17 side-chain as also demonstrated for plant organisms [23, 27]. On the other hand, the precursor of the steroid alkaloids from *Buxaceae* possessing methyl substituents at C-4 and/or C-14 as well as 9β,19-cyclo or 9(10$\rightarrow$19)-abeo structural moieties (*cf.* group d) should be 3β-hydroxy-4,4,14α-trimethyl-9,19-cyclo-5α,9β-pregnan-20-one (99) [9], most likely directly derived from cycloartenol (96) by oxidative degradation of its C-17 side-chain analogous to that of cholesterol (Fig. 19.16).

19.5.1 Alkaloids with C_{27}-Carbon Skeleton from *Solanaceae* and *Liliaceae*

According to present knowledge, the biosynthesis of the C_{27}-steroid alkaloids and steroid sapogenins (spirostanes), occurring together in plants [195, 216], is very closely related. Precursors of both types of heterocyclic steroids seem to be either cholesterol (97) or another closely related C_{27}-steroid as shown by tracer studies with 97 and its precursors.

Thus, [1-^{14}C]acetate, [2-^{14}C]acetate and/or [2-^{14}C]mevalonate were incorporated into 22,26-epiminocholestanes, *e.g.* verazine (**110**) and etioline (**114**), spirosolanes, *e.g.* solasodine (**78**) and tomatidine (**102**), solanidanes, *e.g.* solanidine (**80**) and rubijervine (**118**), jerveratrum alkaloids, *e.g.* veratramine (**82**) and jervine (**83**), and ceveratrum alkaloids, *e.g.* zygadenine, as well as into spirostanols (**108**) [77—85, 133, 135, 139, 198, 203, 214, 222]. By means of degradation it has been shown that the distribution of radioactivity is precisely the same as expected from the results previously obtained with cholesterol [80, 81, 84, 85]. Further confirmation is given by investigations demonstrating the significant incorporation of labelled cholesterol into spirosolanes [34, 105, 248, 250], solanidanes [34, 251], jerveratrum alkaloids [133], and spirostanes [23, 28, 249], as well as of [26,27-^{14}C]cycloartenol (and to a lesser extent lanosterol) into tomatidine (**102**), solasodine (**78**), solanidine (**80**), solanocapsine (**77**), and/or spirostanols (**108**) [194]. According to Heftmann and Weaver [108] cholest-4-en-3-one and 26-hydroxycholesterol (**106**) are the first products of cholesterol metabolism in potato plants. The (*25R*)- and (*25S*)-stereoisomers of the latter metabolite are converted stereospecifically into soladulcidine, the 5α,6-dihydro derivative of solasodine (**78**) [248], and tomatidine (**102**) [205; cf. 203, 204], respectively. The hydroxylation of cholesterol leading to (25*R*)- and (25*S*)-26-hydroxycholesterol in the respective plant material also occurs with high stereoselectivity. This is demonstrated by the fact that the C-27 methyl group of the (25*R*)-alkaloid solasodine (**78**) derives from C-2 of mevalonate [80], while the 27-methyl of the (25*S*)-compound tomatidine (**102**) derives from C-3′ of mevalonic acid [203] (Fig. 19.17). This means that hydrogenation of the Δ^{24}-intermediate (*e.g.*

desmosterol) with the structure shown **100** [229] principally occurs by addition of hydrogen to the 24-si, 25-si face [110].

20-Hydroxycholesterol is incorporated into both the spirostanol tigogenin and the spirosolane alkaloid solasodine (**78**) [255], whereas 16β-hydroxy- [254], 16β,26-dihydroxy- [247], and 16β-hydroxy-22-oxo-5α-cholestanol [238], as well as 22-oxo-cholesterol [253], can be converted into spirostanes but not into spirosolanes [256]. Analogous nitrogen-free steroids, barogenin (**103**), dormantinone (**104**), and dormantinol (**105**), have been isolated from *Solanum tuberosum* [137a] and *Veratrum grandiflorum* [129, 137] (*cf.* Fig. 19.18).

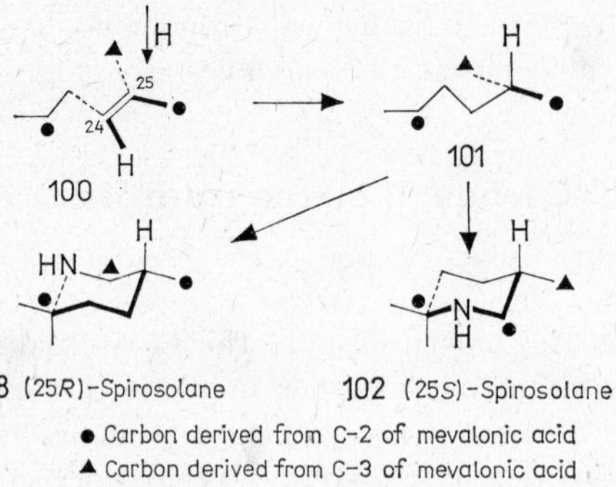

Fig. 19.17
Stereochemistry of the biosynthesis of ring *F* of spirosolane alkaloids

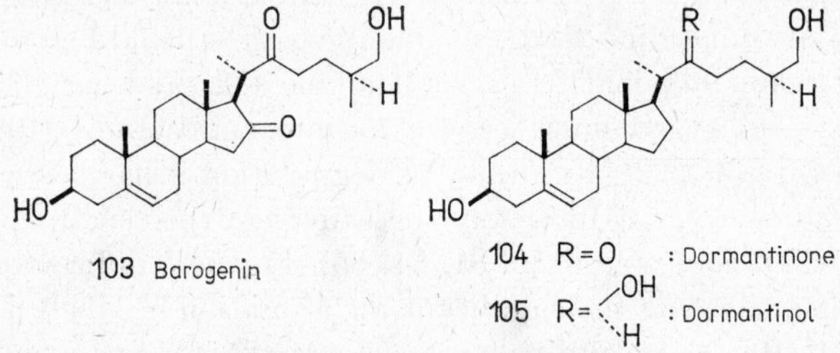

Fig. 19.18
Nitrogen-free substituted sterols from *Solanum tuberosum* and *Veratrum grandiflorum*

Radioactive labelled (25*R*)-26-aminocholesterol (**109**) [245] administered to *Solanum laciniatum* was incorporated to a high extent into solasodine (**78**), whereas the corresponding 16β-hydroxy derivatives (25*R*)-26-aminocholest-5-ene-3β,16β-diol and its *N*-acetyl derivative show little incorporation [246]. These results suggested that in the biosynthesis of C_{27}-steroid alkaloids nitrogen is introduced immediately after the hydroxylation at C-26 [246] (*cf.* [205, 248]). The substitution of the 26-hydroxy group by an amino group occurs directly without formation of an intermediate 26-oxo group [205]. The nitrogen source of this amination was the amino acid *L*-arginine as shown by incorporation of labelled [^{15}N]*L*-arginine, at least in the biosynthesis of solanidine (**80**) in *Veratrum grandiflorum* [136].

The weak incorporation of (25*R*)-26-aminocholest-5-ene-3β,16β-diol into solasodine (**78**) also demonstrates that the 16-hydroxy group is introduced only after the formation of ring *F*, *i.e.* after formation of the 22,26-epiminocholestane structure [246, 256]. This conclusion is confirmed by the isolation of 16-unsubstituted 22,26-epiminocholest-22(*N*)-ene derivatives as endogenous alkamines from plant material, *e.g.* solacongestidine (**77a**) and verazine (**110**) [195, 216]. Their 16β-hydroxy derivatives **111** are not stable in this form but are stereospecifically cyclized to the spirosolanes soladulcidine (**78**, 5α,6-dihydro) and tomatidenol (**102**, Δ^5), respectively [216]. Accordingly, labelled solacongestidine (**77a**) and (22*S*)-dihydrosolacongestidine administered to *Solanum dulcamara* as well as (22*S*,25*R*)-22,26-epiminocholest-5-en-3β-ol and its 16β-hydroxy derivative administered to *Solanum laciniatum*, were converted to soladulcidine (**78**, 5α,6-dihydro) and solasodine (**78**), respectively [256].

The biosynthesis of solanidine (**80**) *via* verazine (**110**), its 16α-hydroxy derivative etioline (**114**) and the corresponding (22*R*)-22,*N*-dihydro derivative teinemine (**115**) in budding *Veratrum grandiflorum* has been investigated by Kaneko *et al.* [135, 136] (*cf.* [137]). The 16β-hydrogen atom of cholesterol gets lost during the biosynthesis of solanidine (**80**); the same hydrogen atom is inverted to the 16α-position during the biosynthesis of tomatidine (**102**) in *Lycopersicon pimpinellifolium* [34].

Solacongestidine (**77a**), its 16α-hydroxy derivative solafloridine (**113**), 23-oxo-solacongestidine (**112**) and solacasine (**116**) are considered to be precursors of solanocapsine (**77**) [217, 218]. The 3β-aminospirostanes, *e.g.* jurubidine (**79**), should be biosynthesized analogously to the corresponding nitrogen-free spirostanols (**108**). Undoubtedly, their 3-amino group as well as that of the other 3-aminocholestane alkaloids with different heterocyclic moieties such as solanocapsine (**77**), soladunalinidine (3-deoxy-3β-aminotomatidine), or solanogantine (a 3β-aminosolanidane derivative) (*cf.* [195]) is introduced by transamination reaction at one of the late stages in their biosynthesis.

The pathways generally accepted for the biosyntheses of steroid alkaloids with the unaltered C_{27}-carbon skeleton of cholesterol derived from the above-mentioned results are outlined in Fig. 19.19.

As shown in tracer experiments, 12α-hydroxysolanidine (rubijervine, **118**) is biosynthesized from verazine (**110**), but not *via* etioline (**114**), in the rhizome of *Veratrum grandiflorum*. The first intermediate of this transformation is 12α,16α-dihydroxyverazine (hakurirodine, **117**) [135] (Fig. 19.20).

12β-Hydroxysolanidine (epirubijervine, **119**) has recently been isolated from illuminated *Veratrum grandiflorum* [131]. Narayanan [177] has postulated that an equatorial leaving group or a carbonium ion at C-12 can rearrange **119** to a C-nor-D-homo-steroid derivative (**120**) as shown by chemical rearrangements described by Hirschmann *et al.* [113]. The reductive cleavage of the C(16)-N bond will yield the skeleton **122** of the jerveratrum alkaloids (Fig. 19.21). On the other hand, a solanidane alkaloid with a leaving group at C-18 such as isorubijervine (**124**) is able to introduce a C(18)-N bond **125**. The reduction of the so-formed quaternary ammonium salt by cleavage of the C(16)-N bond yields the heterocyclic skeleton of the cevanidane alkaloids (**81**) as demonstrated by chemical experiments [180; *cf.* 227,

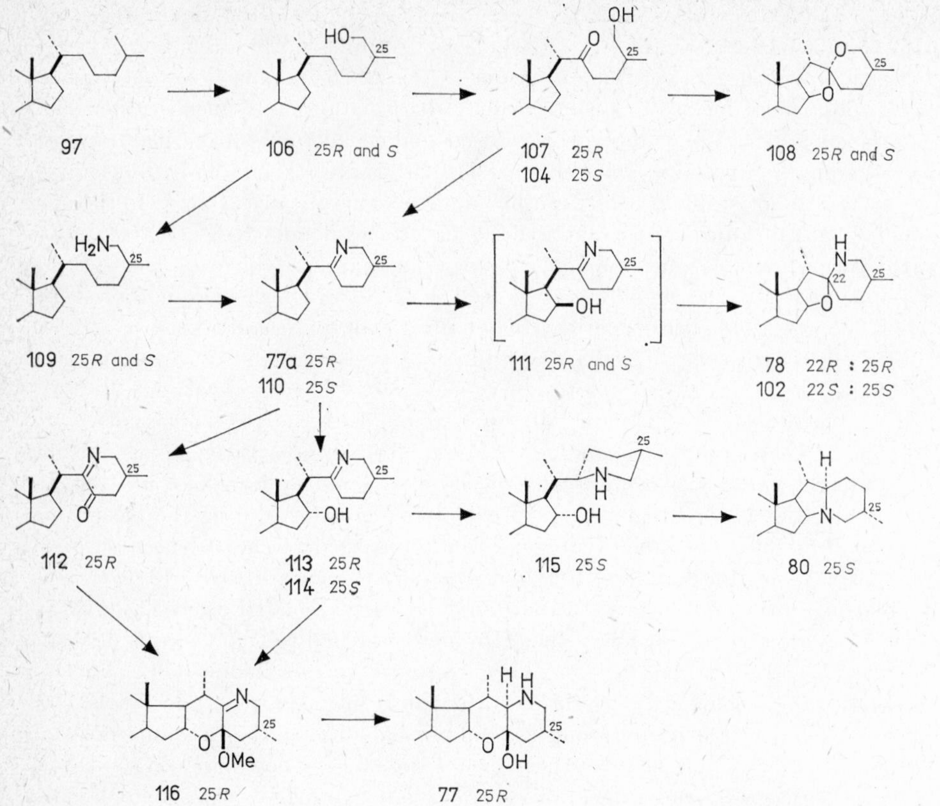

Fig. 19.19
Hypothetical scheme for the biosynthetic pathways of steroid alkaloids with the unaltered C_{27}-carbon skeleton of cholesterol

263]. The cevanidane procevine (81) synthesized in this way has also been isolated from *Veratrum grandiflorum* [130] (Fig. 19.22).

Both transformations (Figs. 19.21 and 19.22) may also occur in plants. Instead of epirubijervine (**119**), the still unknown 12β-hydroxy-22,26-epiminocholestane alkaloid 12-epibaikeine (12β-hydroxyteinemine, **121**) (*cf.* [117]) could also be a precursor of the alkaloids of the type **122** and veratramine (82) (Fig. 19.21). The hypo-

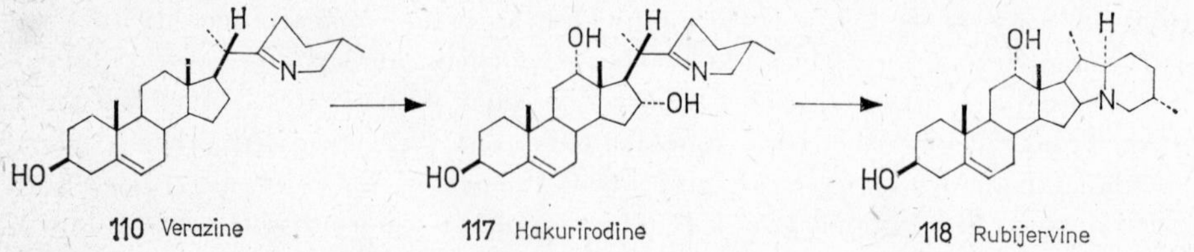

Fig. 19.20
Biosynthetic pathway of rubijervine

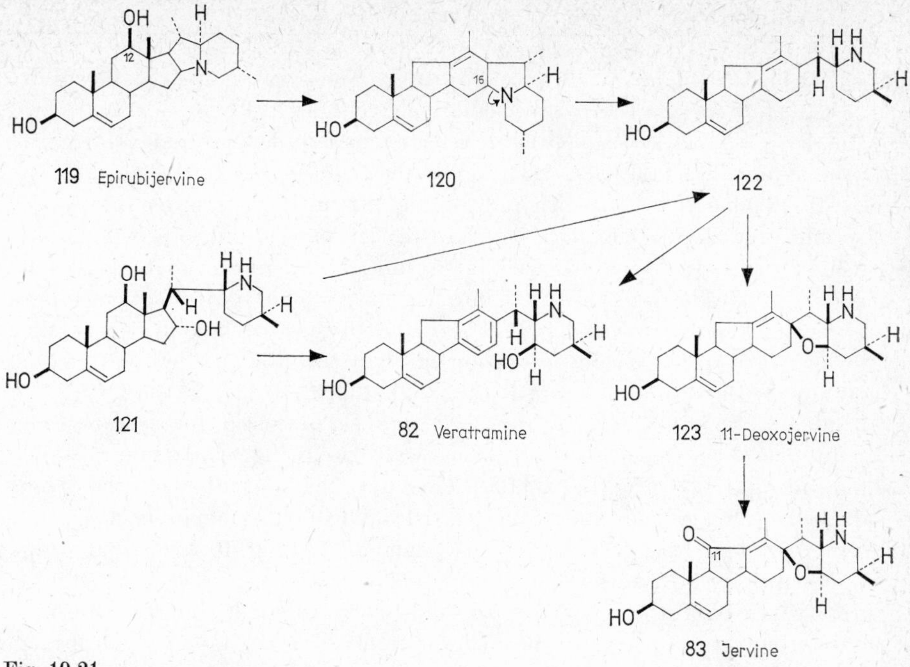

Fig. 19.21
Hypothetical scheme for the biosynthetic pathway of veratramine and jervine from epirubijervine

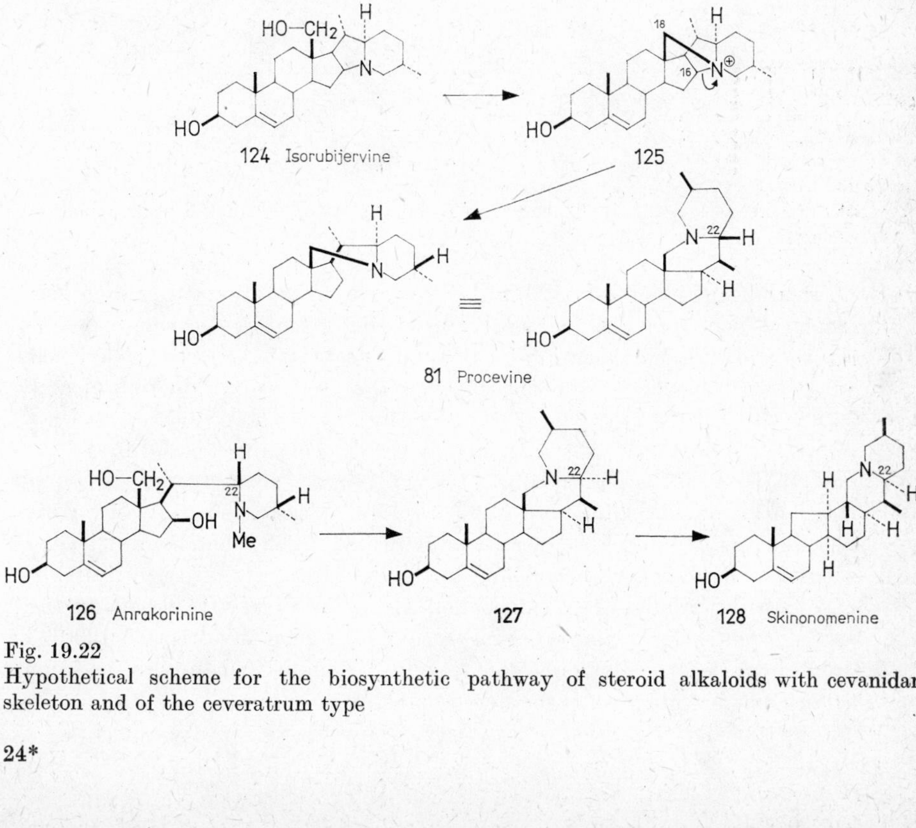

Fig. 19.22
Hypothetical scheme for the biosynthetic pathway of steroid alkaloids with cevanidane skeleton and of the ceveratrum type

thesis shown in this figure is substantiated by the observation that radioactivity from [1-^{14}C]acetate was transferred to solanidine glycosides in *Veratrum grandiflorum* grown in the dark and subsequently transferred to jervine (**83**) and veratramine (**82**) when illuminated [139] (*cf.* [111]). The incorporation of [1-^{14}C]acetate into jervine (**83**) is higher than that into veratramine (**82**) in short-term feeding experiments, while veratramine biosynthesis surpasses jervine biosynthesis in long-term experiments, suggesting separate pathways from a key intermediate [133]. 11-Deoxojervine (**123**) [173] is converted into jervine (**83**) but not into veratramine (**82**) in growing *Veratrum* plants [132] (Fig. 19.21). By feedback control both alkaloids **83** and **123** inhibit the incorporation of [1-^{14}C]acetate into jervine (**83**) but accelerate its incorporation into veratramine (**82**) [132, 133].

The configuration at C-22 (22βH) of procevine (**81**) synthesized from isorubijervine (**124**) is different from that (22αH) of the naturally occurring ceveratrum alkaloids, *e.g.* shinonomenine (**128**) (Fig. 19.22) [130]. Therefore, the biosynthesis of the ceveratrum alkaloids (**128**) may also start from (22*S*)-18-hydroxy-22,26-epiminocholestanes of the anrakorinine (**126**) type *via* the cevanidane **127** (Fig. 19.22). Anrakorinine (**126**) has been isolated from the *Liliaceae Fritillaria camtschatcensis* [134].

The biosynthesis of the *Veratrum* alkaloids with 18-nor-17β-methylcholestane carbon skeleton, *e.g.* veralkamine (**85**) and veramine (**86**), has not yet been investigated. However, their biosynthesis may proceed, in analogy to the known Wagner-

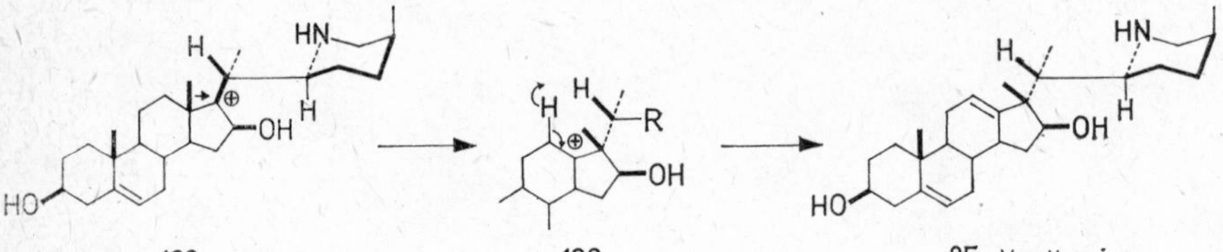

Fig. 19.23
Hypothetical scheme for the biosynthesis of alkaloids with 18-nor-17β-methylcholestane carbon skeleton

Meerwein rearrangement of steroidal 16,17-epoxides (*cf.* [264]) starting with (22*S*)-dihydrotomatid-5-en-3β-ol (**129**) possessing a leaving group or a carbonium ion at C-17 (Fig. 19.23) [217]. This reaction formulated as sextet 1,2-re-arrangement leads to an α-position of the side-chain at C-17 as well as to the introduction of a Δ^{12}-double bond.

The enzymic glycosylation of steroid alkaloids has been investigated in *Solanum laciniatum* [157] and in the potato tuber [154].

The distribution of solasodine glycosides in leaves, flowers, and fruits of *Solanum laciniatum* has been studied using a radioimmunoassay [262]. The highest levels of alkaloids are found in empty anthers and in the inner pericarp of green fruits. The subcellular localization of steroidal glycoalkaloids in the vegetative organs of *Lycopersicon esculentum* and *Solanum tuberosum* has also been investigated [200] (*cf.* [199]). In both species, the alkaloids were found to accumulate mainly in the soluble fraction, with smaller amounts in the microsomal one.

The biosynthesis of spirosolane alkaloids in tissue cultures of *Solanum* species has been reported [96, 98, 114, 143, 144, 151, 259] (*cf.* [97, 228, 231]). However, other authors were able to find only diosgenin and no solasodine (**78**) at all in tissue cultures of *Solanum laciniatum* [260] (*cf.* [202]). The presence and content of spirostanols and/or spirosolanols may depend on the ratio of plant growth substances, *e.g.* 2,4-D, kinetin, and gibberellic acid [98]. The addition of cholesterol increases the solasodine content [143].

The results of a number of studies on the influence of environmental and cultivation conditions, fertilisation, time of harvesting, etc. on the alkaloid content and yield per hectare especially in *Solanum laciniatum* and *S. aviculare* have been summarized by Schreiber [216] and Mann [159].

Some work has been performed on the biological degradation of spirosolane alkaloids and that of the structurally and biosynthetically related spirostanes. Thus, [4-^{14}C]tomatine (**131**), when incubated in whole ripe tomato fruits, was rapidly

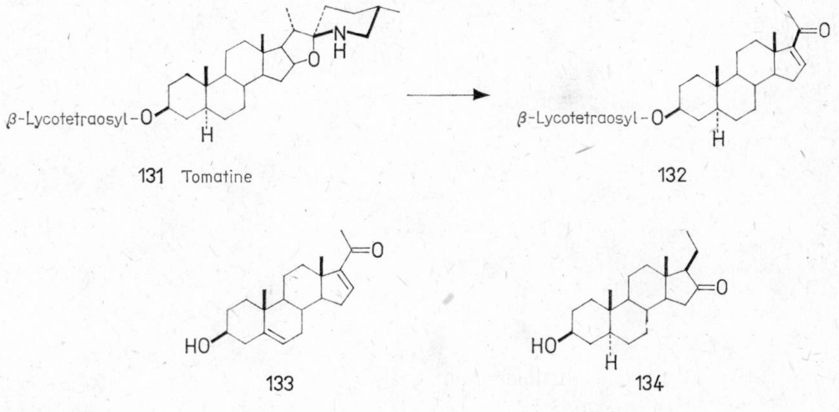

β-Lycotetraosyl-O

131 Tomatine

β-Lycotetraosyl-O

132

HO

133

HO

134

Fig. 19.24
Degradation of spirosolane alkaloid glycosides to pregnane derivatives

converted into a glycoside, probably the lycotetraoside, of [4-^{14}C]3β-hydroxy-5α-pregn-16-en-20-one (**132**) [106] (*cf.* [28, 107, 218, 219]). The 3(*O*)-β-chacotrioside of 3β-hydroxypregna-5,16-dien-20-one has been isolated from fruits of *Solanum vespertilio* [60]. The latter pregnane (**133**) has also been obtained by isolation from *Veratrum grandiflorum* [138] and, in addition to progesterone, from *Solanum verbascifolium* [115]. 3β-Hydroxy-5α-pregnan-16-one (**134**) was found for the first time in *Solanum hainanense* [7, 115] (Fig. 19.24). In analogy to the spirosolane alkaloids and their glycosides, the nitrogen-free spirostanol and furostanol glycosides are also biologically degradable to pregn-16-en-20-one derivatives [149, 179] (*cf.* [148]). As intermediates of this biodegradation, glycosides of 20,22-secofurostane derivatives have been isolated [148], confirming the analogy of this biological process to Marker's chemical degradation procedure [171, 172].

19.5.2 Alkaloids with C_{21}-Carbon Skeleton from *Apocynaceae*

In contrast with the work done in the field of the steroid alkaloids with a C_{27}-carbon skeleton (*cf.* chapter 19.5.1), few significant investigations have been so far carried out on the biosynthesis of the steroid alkaloids from *Apocynaceae*. Undoubtedly, also in this case cholesterol (**97**) is the principal steroid precursor. Its oxidative degradation to pregnanes has been shown for plant organisms [23, 27]. 3β-Hydroxypregn-5-en-20-one (**98**) [26] and progesterone (**135**) [155] have been isolated from the *Apocynaceae Holarrhena floribunda* together with the structurally related alkaloids holaphyllamine (3β-aminopregn-5-en-20-one, **136**), holaphilline (3β-methylamino-pregn-5-en-20-one, **137**), and holamine (3α-aminopregn-5-en-20-one, **138**) (Fig. 19.25) [155].

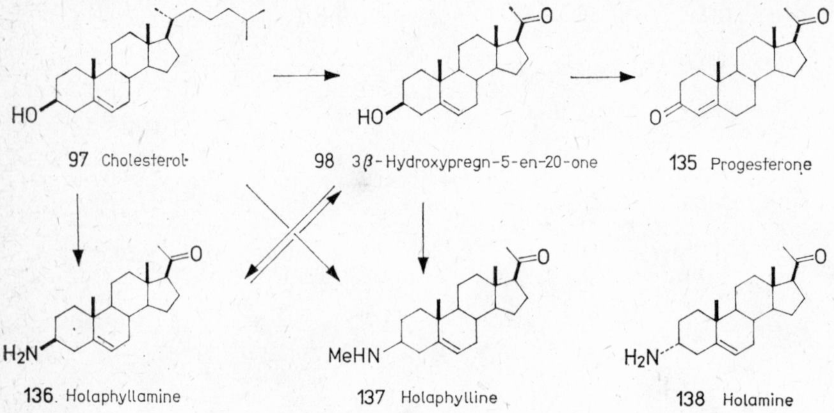

97 Cholesterol
98 3β-Hydroxypregn-5-en-20-one
135 Progesterone
136 Holaphyllamine
137 Holaphylline
138 Holamine

Fig. 19.25
Biosynthetic pathway of pregnane alkaloids from *Apocynaceae*

As shown by tracer experiments using leaves of *Holarrhena floribunda*, [4-^{14}C]-cholesterol (**97**) is incorporated into 3β-hydroxypregn-5-en-20-one (**98**), progesterone (**135**), and holaphyllamine (**136**) as well as [4-^{14}C]pregnenolone (**98**) into holaphyllamine (**136**) and holaphylline (**137**) [22, 24], but not [4-^{14}C]progesterone (**135**) [26], as previously supposed [24].

Progesterone (**135**) is produced from the pregnane **98** [25] which, in conversion of the alkaloid biosynthesis, arises also from holaphyllamine (**136**) [26]. Because of these results it is more likely that the introduction of the 3-amino group occurs by direct substitution rather than by reductive transamination *via* a respective ketone as **135**. Conessine (**90**) is also biosynthesized from **98** and not from **135** in *Holarrhena antidysenterica* [252] (Fig. 19.25). For the biosynthesis of the heterocyclic pregnane alkaloids of the conamine type, *e.g.* conessine (**90** or **147**), 18-hydroxy (**146**) and 18-oxo derivatives (**149**) have been suggested as intermediates, the last of which cyclizes intramolecularly with a 20-amino group leading to a cyclic azomethine of the conkurchine type (**148**) [40, 122, 239]. Biogenetic pathways for the

Apocynaceae pregnane alkaloids of different types have been suggested (Fig. 19.26) [215].

Surprisingly, the pregnane alkaloids of the holadysine type (**144**) isolated from

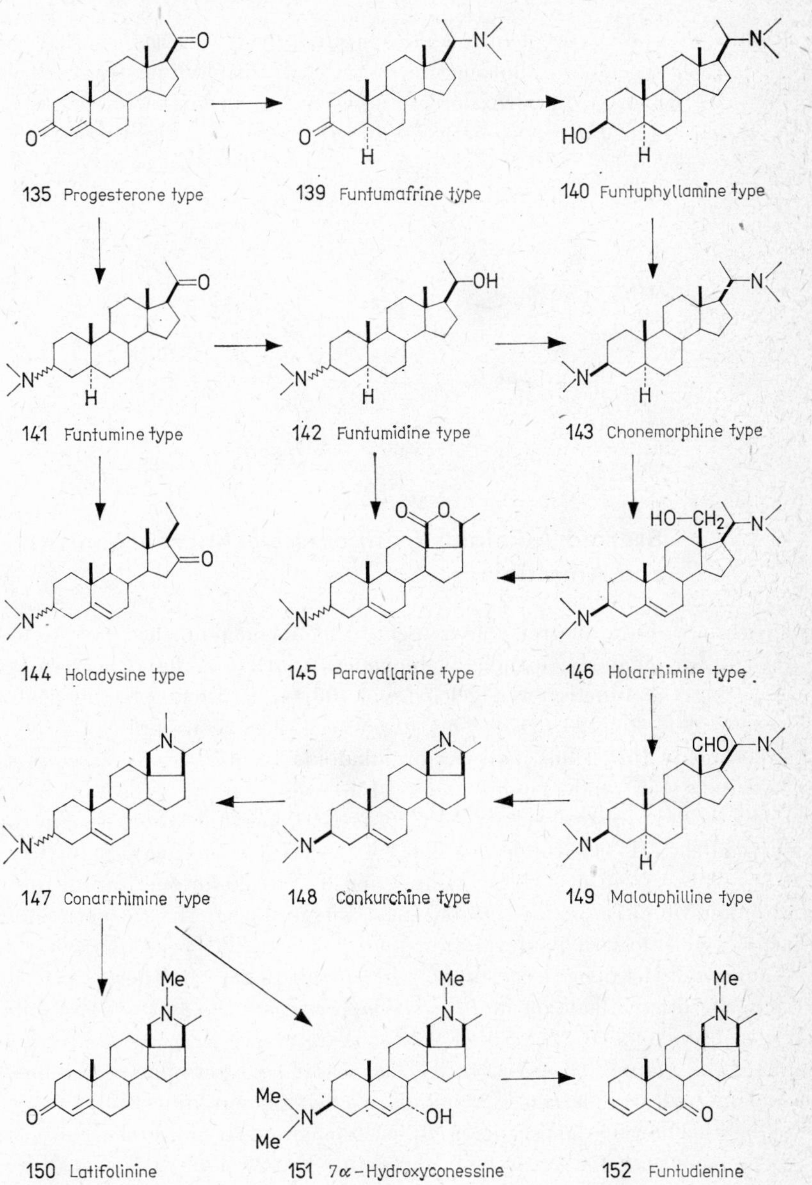
O
O
135 Progesterone type
N
O
H
139 Funtumafrine type
N
HO
H
140 Funtuphyllamine type
O
N
H
141 Funtumine type
OH
N
H
142 Funtumidine type
N
N
H
143 Chonemorphine type
O
N
144 Holadysine type
O O
N
145 Paravallarine type
HO–CH2
N
N
146 Holarrhimine type
N
N
147 Conarrhimine type
N
N
148 Conkurchine type
CHO
N
N
H
149 Malouphilline type
Me
N
O
150 Latifolinine
Me
N
Me
N
Me
OH
151 7α-Hydroxyconessine
Me
N
O
152 Funtudienine

Fig. 19.26
Hypothetical scheme for the biosynthetic pathways of pregnane alkaloids from *Apocynaceae*

Holarrhena [64, 65, 121] possess quite an analogous structure as the 3β-hydroxy-5α-pregnane-16-one (**134**) obtained from *Solanum hainanense* [7, 115].

Conessine (**90**) is biosynthesized in small quantities in hypocotyl callus cultures of *Holarrhena antidysenterica* as shown by tracer experiments with cholesterol (**97**) [99].

Quite another type of nitrogenous steroid glycosides has also been isolated from *Holarrhena* species, *i.e.* holacurtine (**153**) and mitiphylline (**154**), the nitrogen-free aglyca of which, a dihydroxypregnanone and a cardenolide, are glycosylated with an amino sugar (4-deoxy-4-methylamino-*D*-cymarose) (Fig. 19.27) [118, 120].

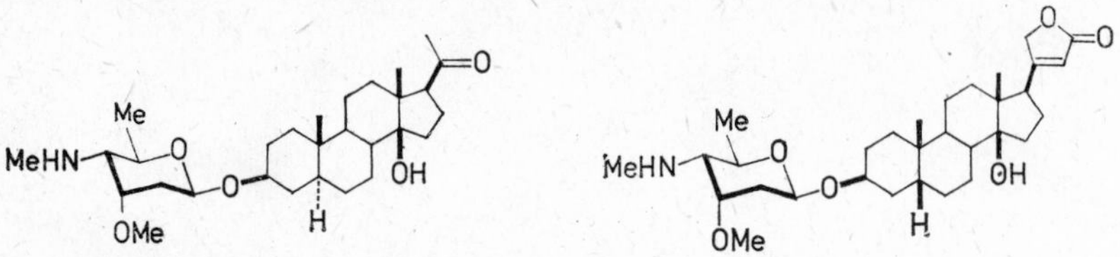

153 Holacurtine

154 Mitiphylline

Fig. 19.27
Nitrogenous steroid glycosides from *Holarrhena*

19.5.3 Steroid Alkaloids from *Buxaceae*, Microorganisms, and Amphibians

Up to now, only one relevant paper [6] has been published on the biosynthesis of the *Buxus* alkaloids, althought biogenetic pathways have been suggested before including the function of cycloartenol (96) as a direct precursor of this type of steroid alkaloid [196, 215]. This hypothesis has been confirmed by recent tracer experiments [6]. Thus, two major alkaloids from *Buxus sempervirens*, cyclovirobuxine-D (**155**) and cyclobuxine-D (**156**), were found to be radioactively labelled following administration of [2-^{14}C, (*4R*)-4-$^{3}H_1$]mevalonic acid to freshly harvested shoots. The ^{3}H:^{14}C atomic ratio of 3:4 in cyclovirobuxine-D indicated a biosynthetic pathway from cycloartenol involving 3- and 20-ketone intermediates. A ^{3}H:^{14}C atomic ratio of ca 3:3 in cyclobuxine-D suggests that the 4α-methyl group of cycloartenol gets lost in its formation and this coincides with current theories of the sequence of C-4 demethylation of sterols (Fig. 19.28) [6]. There is also some evidence that the other alkaloids in *Buxus sempervirens* such as cycloprotobuxine-D (**157**), cyclovirobuxeine-B (**158**), and cyclobuxamine-H (**159**), are biosynthesized in an analogous manner (Fig. 19.29) [6]. The reactive 11-keto-9β,19-cyclopregnane system encountered in some *Buxus* alkaloids, *e.g.* cycloxobuxidine-F (**91**), has been assumed to be the biogenetic precursor of bases characterized by the conjugated transoid diene moiety as, for example, in buxamine-E (**92**) (Fig. 19.14) [145].

Buxus sempervirens was shown to contain also cycloartenol (**96**) [33]. Until now, the synthesized 3-deoxy-3β-aminocycloartenol (**160**) [8] (Fig. 19.30) has not been isolated from plants but is considered to be a possible natural alkaloid. The bio-

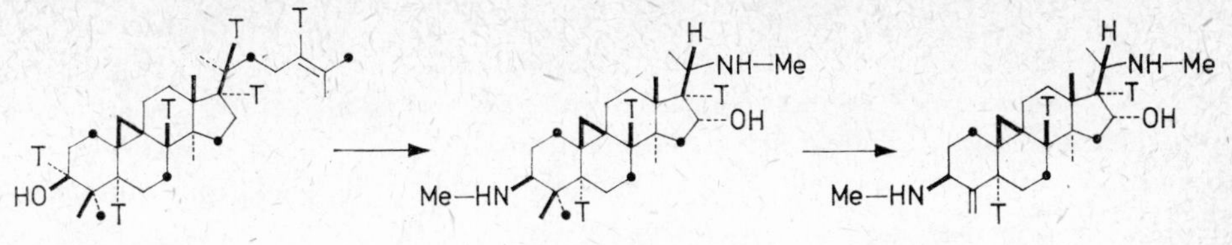

96 Cycloartenol **155** Cyclovirobuxine-D **156** Cyclobuxine-D

• = Carbon derived from C-2 of mevalonic acid
T = Hydrogen derived from pro-4*R* position of mevalonic acid

Fig. 19.28
Biosynthetic pathway of *Buxus* alkaloids from cycloartenol

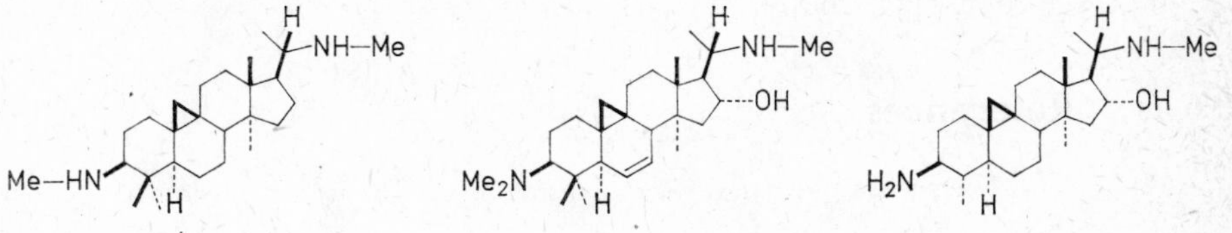

157 Cycloprotobuxine-D **158** Cyclovirobuxeine-B **159** Cyclobuxamine-H

Fig. 19.29
Alkaloids from *Buxus sempervirens*

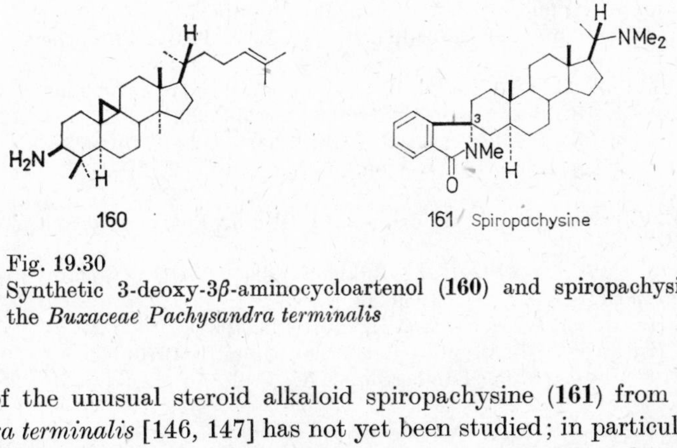

160 **161** Spiropachysine

Fig. 19.30
Synthetic 3-deoxy-3β-aminocycloartenol (**160**) and spiropachysine (**161**) from the *Buxaceae Pachysandra terminalis*

synthesis of the unusual steroid alkaloid spiropachysine (**161**) from the *Buxaceae Pachysandra terminalis* [146, 147] has not yet been studied; in particular the genesis of its spiro moiety at C-3 of the pregnane skeleton seems to be unknown.

The same is true for the nitrogenous steroidal metabolites of the fungus *Geo trichum flavo-brunneum* [174]. No work at all has been published on the biosynthesis of their 24-methylene-15-aza-*D*-homo-5α-cholestan-3β-ol structure (**95**; Fig. 19.15).

The skin gland secretion of the salamander, *Salamandra maculosa* (and *S. atra*), contains — in addition to cholesterol (**97**) [86] — some steroid alkaloids with 3-aza-*A*-homo-5β-androstane skeleton [87—89]. The main alkaloids are samandarine (1α,4α-epoxy-3-aza-*A*-homo-5β-androstan-16β-ol, **93**, Fig. 19.15) or its 16-acetate and samandarone, the 16-dehydro derivative of **93**. Other alkaloids with C_2- or-

C_3-side-chains in 17β-position or a 10α-aldehyde function are minor alkaloids of the venom. In agreement with the steroidal nature of these constituents, [^{14}C]acetate and [4-^{14}C]cholesterol were incorporated by *in vitro* and *in vivo* tracer experiments into the *Salamandra* alkaloids [90], biogenetic pathways of which have been suggested [87, 89]. The nitrogen atom of these alkaloids is derived from glutamine [90].

The biosynthesis of batrachotoxin (**94**, Fig. 19.15) and related steroid alkaloids from the skin of the Colombian arrow poison frog, *Phyllobates aurotaenia*, has not yet been clarified. Although [^{14}C]acetate, [^{14}C]serine, and [^{14}C]mevalonate are readily incorporated into the skin cholesterol, batrachotoxin (**94**) itself was shown to be completely unlabelled after administration of the above-mentioned as well as of steroidal precursors such as [^{14}C]cholesterol, [^{14}C]3β-hydroxypregn-5-en-20-one, and [^{14}C]progesterone [47, 265].

19.6 References

[1] Abdusamatov, A.; Khakimdzhanov, S.; Yunusov, S. Y.: Chim. Prir. Soedin. **4** (3) (1968) 195.
[2] Abdusamatov, A.; Rashidov, M. U.; Yunusov, S. Y.: Chim. Prir. Soedin. **7** (3) (1971) 304.
[3] Abdusamatov, A.; Yagudaev, M. R.; Yunusov, S. Y.: Chim. Prir. Soedin. **4** (4) (1968) 265.
[4] Abdusamatov, A.; Yunusov, S. Y.: Chim. Prir. Soedin. **5** (4) (1969) 334.
[5] Abdusamatov, A.; Yunusov S. Y.: Chim. Prir. Soedin. **7** (3) (1971) 306.
[6] Abramson, D.; Knapp, F. F.; Goad, L. J.; Goodwin, T. W.: Phytochemistry **16** (1977) 1935.
[7] Adam, G.; Huong, H. T.; Khoi, N. H.: Phytochemistry **17** (1978) 1802.
[8] Adam, G.; Voigt, B.; Schreiber, K.: Tetrahedron **25** (1969) 3783.
[9] Adam, G.; Voigt, B.; Schreiber, K.: J. prakt. Chem. **312** (1970) 1027.
[10] Adolphen, G.; Appel, H. H.; Overton, K. H.; Warnock, W. D. C.: Tetrahedron **23** (1967) 3147.
[11] Ahmed, Z. F.; Risk, A. M.; Hammouda, F. M.: J. Pharm. Pharmacol. **17** (Suppl.) (1965) 39.
[12] Akramov, S. T.; Yugataev, M. R.; Rakhamatullaev, T. U.; Samatov, A.; Yunusov, S. Y.: Chim. Prir. Soedin. **5** (1) (1969) 14.
[13] Appel, H. H.; Müller, B.: Scientia (Valparaiso) **28** (1961) 5.
[14] Appel, H. H.; Streeter, P. M.: Rev. Lat. Amer. Quim. **1** (1970) 63.
[15] Arthur, H. R.; Johns, S. R.; Lamberton, J. A.; Loo, S. N.: Austral. J. Chem. **20** (1967) 2505.
[16] Arthur, H. R.; Loo, S. N.: Phytochemistry **5** (1966) 977.
[17] Auda, H.; Juneja, H. R.; Eisenbraun, E. J.; Waller, G. R.; Kays, W. R.; Appel, H. H.: J. Am. Chem. Soc. **89** (1967) 2476.
[18] Auda, H.; Waller, G. R.; Eisenbraun, E. J.: J. Biol. Chem. **242** (1967) 4157.
[19] Baoua, M.; Bessiere, J. M.; Pucci, B.; Rigaud, J. P.: Phytochemistry **19** (1980) 718.
[20] Banthorpe, D. V.; Charlwood, B. V.; Francis, M. J. O.: Chem. Rev. **72** (1972) 115.
[21] Benn, M. H.; May, J.: Experientia **20** (1964) 252.
[22] Bennett, R. D.; Heftmann, E.: Arch. Biochem. Biophys. **112** (1965) 616.
[23] Bennett, R. D.; Heftmann, E.: Phytochemistry **4** (1965) 577.
[24] Bennett, R. D.; Heftmann, E.: Phytochemistry **4** (1965) 873.
[25] Bennett, R. D.; Heftmann, E.: Science **149** (1965) 652.
[26] Bennett, R. D.; Heftmann, E.; Ko, S.-T.: Phytochemistry **5** (1966) 517.

[27] Bennett, R. D.; Heftmann, E.; Winter, B. J.: Naturwissenschaften **56** (1969) 463.
[28] Bennett, R. D.; Lieber, E. R.; Heftmann, E.: Phytochemistry **6** (1967) 837.
[29] Biollaz, M.; Arigoni, D.: Chem. Commun. **1969**, 633.
[30] Bisset, N. G.; Choudhuri, A. K.: Phytochemistry **13** (1974) 265.
[31] Bobbitt, J. M.; Segebart, K. P.: In: Taylor, W. I.; Battersby, A. R. (Eds.): Cyclopentanoid Terpene Derivatives. Marcel Dekker, New York 1969, p. 1.
[32] Budzikiewicz, H.; Horstmann, C.; Pufahl, K.; Schreiber, K.: Chem. Ber. **100** (1967) 2798.
[33] Calame, J. P.: Ph. D. Thesis, Zürich 1965.
[34] Canonica, L.; Ronchetti, F.; Russo, G.: J. Chem. Soc., Chem. Commun. **1977**, 286.
[35] Casanova, C.; Gonzales, A. G.: An. Real. Soc. espan. Fisica Quim. (Madrid) Ser. B **60** (1964) 607.
[36] Casinovi, C. G.; Delle Monache, F.; Grandolini, G.; Marini-Bettolo, G. B.; Appel, H. H.: Chem. & Ind. **1963**, 984.
[37] Casinovi, C. G.; Delle Monache, F.; Marini-Bettolo, G. B.; Bianchi, E.; Garbarino, J.: Gazz. chim. ital. **92** (1962) 479.
[38] Casinovi, C. G.; Garbarino, J. A.; Marini-Bettolo, G. B.: Chem. & Ind. **1961**, 253.
[39] Casinovi, C. G.; Giovannozzi-Sermanni, G.; Marini-Bettolo, G. B.: Rend. Accad. Naz. XL, Ser. IV **16–17** (1965) 89; Gazz. chim. ital. **94** (1964) 1356.
[40] Černý, V.; Šorm, F.: Alkaloids (N. Y.) **9** (1967) 305.
[41] Cieslak, J.; Kuduk, J.; Rulko, F.: Acta Polon. Pharm. **21** (1964) 265.
[42] Corbella, A.; Gariboldi, P.; Jommi, G.: Chem. Commun. **1972**, 600.
[43] Corbella, A.; Gariboldi, P.; Jommi, G.: Chem. Commun. **1973**, 729.
[44] Corbella, A.; Gariboldi, P.; Jommi, G.; Scolastico, C.: Chem. Commun. **1969**, 634.
[45] Corbella, A.; Gariboldi, P.; Jommi, G.; Sisti, M.: Chem. Commun. **1975**, 288.
[46] Cordell, G. A.: The Monoterpene Alkaloids. In: Manske, R. H. F. (Ed.): The Alkaloids, **16** (1977) 431.
[47] Daly, J. W.; Witkop, B.: In: Bücherl, W.; Buckley, E. (Eds.): Venomous Animals and their Venoms, Academic Press, New York 1971. Vol. 2, p. 497.
[48] Dickinson, E. M.; Jones, G.: Tetrahedron **25** (1969) 1523.
[49] Djerassi, C.; Kutney, J. P.; Shamma, M.: Tetrahedron **18** (1962) 183.
[50] Djerassi, C.; Kutney, J. P.; Shamma, M.; Shoolery, J. N.; Johnson, L. F.: Chem. & Ind. **1961**, 210.
[51] Edwards, O. E.; Douglas, J. L.; Mootoo, B.: Canad. J. Chem. **48** (1970) 2517.
[52] Eisenbraun, E. J.; Bright, A.; Appel, H. H.: Chem. & Ind. **1962**, 1242.
[53] El-Naggar, L. J.; Beal, J. L.: J. Nat. Prod. **43** (1980) 649.
[54] Fischer, N. H.; Olivier, E. J.; Fischer, H. D.: Progr. Chem. Nat. Prod. **38** (1979) 47.
[55] Floß, H. G.; Mothes, U.; Rettig, A.: Z. Naturforsch. **19b** (1964) 1106.
[56] Franck, B.; Petersen, U.; Hüper, F.: Angew. Chem. **82** (1970) 875.
[57] Frost, G. M.; Hale, R. L.; Waller, G. R.; Zalkow, L. H.; Girotra, N. N.: Chem. & Ind. **1967**, 320.
[58] Goad, L. J.: In: Pridham, J. B. (Ed.): Terpenoids in Plants, p. 159. Academic Press, New York 1967.
[59] Goad, L. J.; Goodwin, T. W.: Progr. Phytochem. **3** (1972) 113.
[60] González, A. G.; Francisco, C. G.; Barreira, R. F.; Lopez, E. S.: Ann. Quim. **67** (1971) 433.
[61] Goodwin, T. W. (Ed.): Biochem. Soc. Symp. **29** (1970).
[62] Goodwin, T. W.: Biochem. J. **123** (1971) 293.
[63] Goodwin, T. W.: Ann. Rev. Plant Physiol. **30** (1979) 369.
[64] Goutarel, R.: Bull. Soc. chim. France **1960**, 769.
[65] Goutarel, R.: Tetrahedron **14** (1961) 126.
[66] Goutarel, R.: Les alcaloides steroidiques des Apocynacees. Hermann, Paris 1964; cf. Specialist Periodical Reports: The Alkaloids (Ed. J. E. Saxton), Vol. 1, p. 382. The Chemical Society, London 1971.

[67] Govindachari, T. R.; Sathe, S. S.; Viswanathan, N.: Indian J. Chem. **4** (1966) 201.
[68] Gross, D.: Progr. Chem. Nat. Prod. **28** (1970) 109; **29** (1971) 1; Fortschr. Bot. **32** (1970) 93.
[69] Gross, D.; Berg, W.; Schütte, H. R.: Phytochemistry **11** (1972) 3082.
[70] Gross, D.; Berg, W.; Schütte, H. R.: Biochem. Physiol. Pflanzen **163** (1972) 576.
[71] Gross, D.; Berg, W.; Schütte, H. R.: Phytochemistry **12** (1973) 201.
[72] Gross, D.; Berg, W.; Schütte, H. R.: Z. Chem. **13** (1973) 296.
[73] Gross, D.; Edner, G.; Schütte, H. R.: Arch. Pharm. **304** (1971) 19.
[74] Grunwald, C.: Annu. Rev. Plant Physiol. **26** (1975) 209.
[75] Grunwald, C.: Phil. Trans. Royal Soc. London B **284** (1978) 541; see also in: Goodwin, T. W. (Ed.): The Biochemical Functions of Terpenoids in Plants. The Royal Society, London 1978, p. 103.
[76] Grunwald, C.: In: Bell, E. A.; Charlwood, B. V. (Eds.): Encyclopedia of Plant Physiology. Vol. 8, p. 221. Springer-Verlag Berlin/Heidelberg/New York 1980.
[77] Guseva, A. R.: Tagungsber. Dtsch. Akad. Landwirtschaftswiss. Berlin No. **27** (1961) 155.
[78] Guseva, A. R.; Borikhina, M. G.; Paseshnichenko, V. A.: Biochimija **25** (1960) 282.
[79] Guseva, A. R.; Paseshnichenko, V. A.: Biochimija **23** (1958) 412.
[80] Guseva, A. R.; Paseshnichenko, V. A.: Biochimija **27** (1962) 721.
[81] Guseva, A. R.; Paseshnichenko, V. A.: Biochimija **27** (1962) 853.
[82] Guseva, A. R.; Paseshnichenko, V. A.; Borikhina, M. G.: Dokl. Akad. Nauk SSSR **133** (1960) 228.
[83] Guseva, A. R.; Paseshnichenko, V. A.; Borikhina, M. G.: Biochimija **26** (1961) 723.
[84] Guseva, A. R.; Paseshnichenko, V. A.; Borikhina, M. G.: Biochimija **28** (1963) 709.
[85] Guseva, A. R.; Paseshnichenko, V. A.; Borikhina, M. G.; Moisseev, R. K.: Biochimija **30** (1965) 260.
[86] Habermehl, G.: Liebigs Ann. Chem. **680** (1964) 104.
[87] Habermehl, G.: Naturwissenschaften **53** (1966) 123.
[88] Habermehl, G.: Alkaloids (N. Y.) **9** (1967) 427.
[89] Habermehl, G.: Progr. Org. Chem. **7** (1968) 35.
[90] Habermehl, G.; Haaf, A.: Chem. Ber. **101** (1968) 198.
[91] Hammouda, Y.; LeMen, J.: Bull. Soc. Chim. France **1963**, 2901.
[92] Hammouda, Y.; Plat, M.; LeMen, J.: Bull. Soc. Chim. France **1963**, 2802; Ann. Pharm. France **21** (1963) 699.
[93] Hanson, J. R.: In: Barton, D.; Ollis, W. D. (Eds.): Comprehensive Organic Chemistry, Vol. 5, p. 989. Pergamon Press, Oxford 1979.
[94] Hart, N. K.; Johns, S. R.; Lamberton, J. A.: Austral. J. Chem. **21** (1968) 1321.
[95] Hart, N. K.; Johns, S. R.; Lamberton, J. A.: Austral. J. Chem. **22** (1969) 1283.
[96] Heble, M. R.; Narayanaswami, S.; Chadha, M. S.: Naturwissenschaften **55** (1968) 350.
[97] Heble, M. R.; Narayanaswami, S.; Chadha, M. S.: Science **161** (1968) 1145.
[98] Heble, M. R.; Narayanaswami, S.; Chadha, M. S.: Phytochemistry **10** (1971) 910, 2393.
[99] Heble, M. R.; Narayanaswami, S.; Chadha, M. S.: Phytochemistry **15** (1976) 681, 1911.
[100] Heftmann, E.: Lloydia **30** (1967) 209.
[101] Heftmann, E.: Steroid Biochemistry. Academic Press, New York 1970.
[102] Heftmann, E.: Lipids **6** (1971) 128.
[103] Heftmann, E.: Lipids **9** (1974) 626.
[104] Heftmann, E.: Lloydia **38** (1975) 195.
[105] Heftmann, E.; Lieber, E. R.; Bennett, R. D.: Phytochemistry **6** (1967) 225.
[106] Heftmann, E.; Schwimmer, S.: Phytochemistry **11** (1972) 2783.
[107] Heftmann, E.; Schwimmer, S.: Phytochemistry **12** (1973) 2661.
[108] Heftmann, E.; Weaver, M. L.: Phytochemistry **13** (1974) 1801.
[109] Hegnauer, R.: Chemotaxonomie der Pflanzen, Vol. 3. Birkhäuser Verlag, Basel 1964.
[110] Herbert, R. B.: In: Specialist Periodical Reports: The Alkaloids (Ed. J. E. Saxton (Vol. 1–5) and M. F. Grundon (Vol. 6–9)). The Chemical Society, London 1971–1979.

[111] Herbert, R. B.: In: Barton, D.; Ollis, W. D. (Eds.): Comprehensive Organic Chemistry, Vol. 5, p. 1045. Pergamon Press, Oxford 1979.
[112] Hilz, W.; Edelmann, G.; Appel, H. H.: Rev. Latinoam. Quim. **304** (1971) 19.
[113] Hirschmann, R.; Snoddy Jr., C. S.; Hiskey, C. F.; Wendler, N. L.: J. Am. Chem. Soc. **76** (1954) 4013.
[114] Hosoda, N.; Yatazawa, M.: Agric. Biol. Chem. **43** (1979) 821.
[115] Huong, H. T.: Ph. D. Thesis, Acad. Sci. GDR 1980.
[116] Irikawa, H.; Sakabe, N.; Yamamura, S.; Hirata, Y.: Tetrahedron **24** (1968) 5691.
[117] Itô, S.; Miyashita, M.; Fukazawa, Y.; Mori, A.: Tetrahedron Lett. **1972**, 2961.
[118] Janot, M.-M.; Devissaguet, P.; Khuong-Huu, Q.; Parello, J.; Bisset, N. G.; Goutarel, R.: C. R. hebd. Séances Acad. Sci. C **266** (1968) 388.
[119] Janot, M. M.; Guilhem, J.; Contz, O.; Venera, G.; Cionga, E.: Ann. Pharm. France **37** (1979) 413.
[120] Janot, M.-M.; Leboeuf, M.; Cavé, A.; Wijesekera, R.; Goutarel, R.: C. R. hebd. Séances Acad. Sci. C **267** (1968) 1050.
[121] Janot, M.-M.; Longevialle, P.; Goutarel, R.; Conreur, C.: Bull. Soc. chim. France **1964** 2158.
[122] Janot, M.-M.; Truong-Ho, M.; Khuong-Huu, Q.; Goutarel, R.: Bull. Soc. chim. France **1963**, 1977.
[123] Jeger, O.; Prelog, V.: Alkaloids (N. Y.) **7** (1960) 319.
[124] Jeger, O.; Prelog, V.: Alkaloids (N. Y.) **7** (1960) 363.
[125] Johnson, R. D.; Waller, G. R.: Phytochemistry **10** (1971) 3334.
[126] Jones, G.; Fales, H. M.; Wildman, W. C.: Tetrahedron Lett. **1963**, 397.
[127] Jones, G.; Ferguson, G.; Marsh, W. C.: J. Chem. Soc., Chem. Commun. **1971**, 994.
[128] Kadyrov, K. A.; Vinogradova, V. I.; Abdusamatov, A.; Yunusov, S. Y.: Chim. Prir. Soedin. (1974) (5) 683.
[129] Kaneko, K.; Hirayama, K.; Mitsuhashi, H.: Pap. Symp. Nat. Prod. **14** (1970) 358.
[130] Kaneko, K.; Kawamura, N.; Kuribayashi, T.; Tanaka, M.; Mitsuhashi, H.; Koyama, H.: Tetrahedron Lett. **1978**, 4801.
[131] Kaneko, K.; Kawamura, N.; Mitsuhashi, H.; Ohsaki, K.: Chem. Pharm. Bull. **27** (1979) 2534.
[132] Kaneko, K.; Mitsuhashi, H.; Hirayama, K.; Ohmori, S.: Phytochemistry **9** (1970) 2497.
[133] Kaneko, K.; Mitsuhashi, H.; Hirayama, K.; Yoshida, N.: Phytochemistry **9** (1970) 2489.
[134] Kaneko, K.; Nakaoka, U.; Tanaka, M.; Yoshida, N.; Mitsuhashi, H.: Phytochemistry **20** (1981) 157.
[135] Kaneko, K.; Seto, H.; Motoki, C.; Mitsuhashi, H.: Phytochemistry **14** (1975) 1295.
[136] Kaneko, K.; Tanaka, M. W.; Mitsuhashi, H.: Phytochemistry **15** (1976) 1391.
[137] Kaneko, K.; Tanaka, M. W.; Mitsuhashi, H.: Phytochemistry **16** (1977) 1247.
[137a] Kaneko, K.; Terada, K.; Yoshida, N.; Mitsuhashi, H.: Phytochemistry **16** (1977) 791.
[138] Kaneko, K.; Watanabe, M.; Mitsuhashi, H.: Phytochemistry **12** (1973) 1509.
[139] Kaneko, K.; Watanabe, M.; Taira, S.; Mitsuhashi, H.: Phytochemistry **11** (1972) 3199.
[140] Karawya, M. S.; Balbaa, S. I.; Afifi, M. S.: U. A. R. J. Pharm. Sci. **12** (1971) 53.
[141] Khakimdzhanov, S.; Abdusamatov, A.; Yunusov, Y.: Chim. Prir. Soedin. **7** (1) (1971) 126.
[142] Khakimdzhanov, S.; Abdusamatov, A.; Yunusov, S. Y.: Chim. Prir. Soedin. **9** (1) (1973) 132.
[143] Khanna, P.; Sharma, G. L.; Rathore, A. K.; Manot, S. K.: Indian J. Exp. Biol. **15** (1977) 1025.
[144] Khanna, P.; Uddin, A.; Sharma, G. L.; Manot, S. K.; Rathore, A. K.: Indian J. Exp. Biol. **14** (1976) 694.
[145] Khuong-Huu, F.; Herlem, D.; Bénéchie, M.: Bull. Soc. chim. France **1972**, 1092.
[146] Kikuchi, T.; Nishinaga, T.; Inagaki, M.; Koyama, M.: Tetrahedron Lett. **1968**, 2077.

[147] Kikuchi, T.; Nishinaga, T.; Inagaki, M.; Niwa, M.; Kuriyama, K.: Chem. Pharm. Bull. **23** (1975) 416.
[148] Kiyosawa, S.; Goto, K.; Owashi, R.; Kawasaki, T.: Tetrahedron Lett. **1977**, 4599.
[149] Kiyosawa, S.; Kawasaki, T.: Chem. Pharm. Bull. **25** (1977) 163.
[150] Koch, M.: Trav. Lab. Matiere Med. Pharm. Galenique **50** (1965) 94.
[151] Kokate, C. K.; Radwan, S. S.: Z. Naturforsch. C **34** (1979) 634.
[152] Kupchan, S. M.; By, A. W.: Alkaloids (N. Y.) **10** (1968) 193.
[153] Lavie, D.; Taylor-Smith, R.: Chem. & Ind. **1963**, 781.
[154] Lavintman, N.; Tandecarz, J.; Cardini, C. E.: Plant Sci. Lett. **8** (1977) 65.
[155] Leboeuf, M.; Caré, A.; Goutarel, R.: C. R. hebd. Séances Acad. Sci. **259** (1964) 3401.
[156] Liang Xiao-tian; Yu De-quan; Fu Feng-yung: Scientia Sinica **14** (1965) 869.
[157] Liljegren, D. R.: Phytochemistry **10** (1971) 3061.
[158] Luchetti, M. A.: Ann. Ist. Super Sanita **1** (1965) 563.
[159] Mann, J. D.: Adv. Agron. **30** (1978) 307.
[160] Marekov, N.; Arnaudov, M.; Popov, S.: Dokl. Bolg. Akad. Nauk **23** (1970) 81.
[161] Marekov, N.; Arnaudov, M.; Popov, S.: Dokl. Bolg. Akad. Nauk **23** (1970) 169.
[162] Marekov, N.; Mollov, N.; Popov, S.: Dokl. Bolg. Akad. Nauk **18** (1965) 999.
[163] Marekov, N.; Mondeshky, L.; Arnaudov, M.: Dokl. Bolg. Akad. Nauk **23** (1970) 803.
[164] Marekov, N.; Popov, S.: Dokl. Bolg. Akad. Nauk **20** (1967) 441.
[165] Marekov, N.; Popov, S. S.: Tetrahedron **24** (1968) 1323.
[166] Marekov, N.; Popov, S.: Izvest. Otd. Chim. Nauki, Bolg. Akad. Nauk **2** (3) (1970) 575.
[167] Marekov, N.; Popov, S.: Bulg. Acad. Sci., Commun. Dept. Chem., **VI** (3) (1973) 633.
[168] Marekov, N.; Popov, S.; Arnaudov, M.: Dokl. Bolg. Akad. Nauk **23** (1970) 955.
[169] Marekov, N.; Popov, S.; Georgiev, G.: Dokl. Bolg. Akad. Nauk **19** (1966) 827.
[170] Marini-Bettolo, G. B.: Ann. Ist. Super. Sanita **4** (1968) 489.
[171] Marker, R. E.; Rohrmann, E.: J. Am. Chem. Soc. **61** (1939) 3592; **62** (1940) 518.
[172] Marker, R. E.; Wagner, R. B.; Ulshafer, P. R.; Wittbecker, E. L.; Goldsmith, D. P. J.; Ruof, C. H.: J. Am. Chem. Soc. **69** (1947) 2167.
[173] Masamune, T.; Mori, Y.; Takasugi, M.; Murai, A.; Ohuchi, S.; Sato, N.; Katsui, K.: Bull. Chem. Soc. (Japan) **38** (1965) 1374.
[174] Michel, K. H.; Hamill, R. L.; Larsen, S. H.; Williams, R. H.: J. Antibiotics **28** (1975) 102.
[175] Mitscher, L. A.; Ray, A. B.; Chatterjee, A.: Experientia **27** (1971) 16.
[176] Mollov, N.; Marekov, N.; Popov, S.; Kouzmanov, B.: Dokl. Bolg. Akad. Nauk **18** (1965) 947.
[177] Narayanan, C. R.: Fortschr. Chem. Org. Naturstoffe (Wien) **20** (1962) 298.
[178] Niwa, H.; Hirata, Y.; Suzuki, K. T.; Yamamura, S.: Tetrahedron Lett. **1973**, 2129.
[179] Nohara, T.; Yabuta, H.; Suenobu, M.; Hida, R.; Miyahara, K.; Kawasaki, T.: Chem. Pharm. Bull. **21** (1973) 1240.
[180] Pelletier, S. W.; Jacobs, W. A.: J. Am. Chem. Soc. **75** (1953) 4442.
[181] Pelletier, S. W.; Keith, L. H.: In: Manske, R. H. F. (Ed.): The Alkaloids, Vol. **12**, p. 15. Academic Press, London/New York 1970.
[182] Pernet, R.; Dupiol, M.; Combes, G.: Bull. Soc. chim. France **1964**, 281.
[183] Plat, M.; Koch, M.; Bouquet, A.; LeMen, J.; Janot, M. M.: Bull. Soc. chim. France **1963**, 1302.
[184] Prelog, V.; Jeger, O.: Alkaloids (N. Y.) **3** (1953) 247.
[185] Prelog, V.; Jeger, O.: Alkaloids (N. Y.) **7** (1960) 343.
[186] Rakhmatullaev, T. U.; Akramov, S. T.; Yunusov, S. Y.: Chim. Prir. Soedin. **5** (1969) 608.
[187] Rakhmatullaev, T. U.; Yunusov, S. Y.: Chim. Prir. Soedin. **8** (1972) 350.
[188] Rakhmatullaev, T. U.; Yunusov, S. Y.: Chim. Prir. Soedin. **8** (1972) 400.
[189] Rakhmatullaev, T. U.; Yunusov, S. Y.: Chim. Prir. Soedin. **9** (1) (1973) 64.
[190] Ray, A. B.; Chatterjee, A.: Tetrahedron Lett. **1968**, 2763.
[191] Ripperger, H.: Phytochemistry **17** (1978) 1069.
[192] Ripperger, H.: Pharmazie **34** (1979) 577.

[193] Ripperger, H.: Pharmazie **35** (1980) 235.
[194] Ripperger, H.; Moritz, W.; Schreiber, K.: Phytochemistry **10** (1971) 2699.
[195] Ripperger, H.; Schreiber, K.: Alkaloids (N. Y.) **19** (1981) 81.
[196] Robinson, T.: The Biochemistry of the Alkaloids, p. 105. Springer-Verlag, Berlin 1968.
[197] Roddick, J. G.: Phytochemistry **13** (1974) 9.
[198] Roddick, J. G.: Phytochemistry **13** (1974) 1459.
[199] Roddick, J. G.: Phytochemistry **15** (1976) 475.
[200] Roddick, J. G.: Phytochemistry **16** (1977) 805.
[201] Roddick, J. G.: In: Bell, E. A.; Charlwood, B. V. (Eds.): Encyclopedia of Plant Physiology, Vol. 8, p. 167. Springer-Verlag, Berlin/Heidelberg/New York 1980.
[202] Roddick, J. G.; Butcher, D. N.: Phytochemistry **11** (1972) 2019.
[203] Ronchetti, F.; Russo, G.: J. chem. Soc., Chem. Commun. **1974**, 785.
[204] Ronchetti, F.; Russo, G.: Tetrahedron Lett. **1975**, 85.
[205] Ronchetti, F.; Russo, G.; Ferrara, G.; Vecchio, G.: Phytochemistry **14** (1975) 2423.
[206] Rulko, F.: Rocz. Chem., Ann. Soc. chim. Polon. **43** (1969) 1831.
[207] Rulko, F.; Dolejs, L.; Cross, A. D.; Murphy, J. W.; Toube, T. P.: Rocz. Chem., Ann. Soc. chim. Polon. **41** (1967) 567.
[208] Rulko, F.; Witkiewicz, K.: Polon. J. Pharmacol. Pharm. **26** (1974) 561.
[209] Sakan, T.; Fujino, A.; Murai, F.: J. Chem. Soc. (Japan) **81** (1960) 1320.
[210] Sakan, T.; Fujino, A.; Murai, F.; Butsugan, Y.; Suzui, A.: Bull. Chem. Soc. (Japan) **32** (1959) 315.
[211] Sakan, T.; Fujino, A.; Murai, F.; Suzui, Y.; Butsugan, Y.; Terashima, Y.: Bull. Chem. Soc. (Japan) **33** (1960) 712.
[212] Sakan, T.; Murai, F.; Hayashi, Y.; Honda, Y.; Shono, T.; Nakajima, M.; Kato, M.: Tetrahedron **23** (1967) 4635.
[213] Samatov, A.; Akramov, S. T.; Yunusov, Y.: Chim. Prir. Soedin. **3** (3) (1967) 182.
[214] Sander, H.; Grisebach, H.: Z. Naturforsch. **13b** (1958) 755.
[215] Schreiber, K.: Abh. Dtsch. Akad. Wiss. Berlin **1966**, 65.
[216] Schreiber, K.: Alkaloids (N. Y.) **10** (1968) 1.
[217] Schreiber, K.: Abh. Dtsch. Akad. Wiss. Berlin **1969**, 69.
[218] Schreiber, K.: Biochem. Soc. Trans. **2** (1974) 1.
[219] Schreiber, K.; Aurich, O.: Phytochemistry **5** (1966) 707.
[220] Schütte, H. R.: In: Mothes, K.; Schütte, H. R. (Eds.): Biosynthese der Alkaloide. Deutscher Verlag der Wissenschaften, Berlin 1969, p. 601.
[221] Schütte, H. R.: In: Mothes, K.; Schütte, H. R. (Eds.): Biosynthese der Alkaloide. Deutscher Verlag der Wissenschaften, Berlin 1969. p. 616.
[222] Schütte, H. R.: Progr. Bot. **37** (1975) 133.
[223] Schütte, H. R.; Lehfeldt, J.: Arch. Pharmaz. **298** (1965) 461.
[224] Seto, H.; Hayashi, K.; Mitsuhashi, H.: Chem. Pharm. Bull. **25** (1977) 876.
[225] Sevenet, T.; Das, B. C.; Parello, J.; Potier, P.: Bull. Soc. Chim. **1970**, (8—9) 3120.
[226] Sevenet, T.; Husson, A.; Husson, H. P.: Phytochemistry **15** (1976) 576.
[227] Sheehan, J. C.; Young, R. L.; Cruickshank, P. A.: J. Am. Chem. Soc. **82** (1960) 6147.
[228] Stohs, S. J.; Rosenberg, H.: Lloydia **38** (1975) 181.
[229] Stone, K. J.; Roeske, W. R.; Clayton, R. B.; van Tamelen, E. E.: J. Chem. Soc., Chem. Commun. **1969**, 530.
[230] Streeter, M. P.; Adolphen, G.; Appel, H. H.: Chem. & Ind. **1969**, 1631.
[231] Supniewska, J. H.; Dohnal, B.: Diss. Pharm. Pharmacol. **24** (1972) 187, 193.
[232] Suzuki, K. T.; Okuda, S.; Niwa, H.; Toda, M.; Hirata, Y.; Yamamura, S.: Tetrahedron Lett. **1973**, 799.
[233] Taylor-Smith, R. E.: Tetrahedron **21** (1965) 3721.
[234] Terada, S.; Hayashi, K.; Mitsuhashi, H.: Chem. Pharm. Bull. **25** (1977) 2802.
[235] Terada, S.; Hayashi, K.; Mitsuhashi, H.: Tetrahedron Lett. **1978**, 1995.
[236] Tietze, L. F.: Tetrahedron Lett. **1976**, 2535.
[237] Toda, M.; Niwa, H.; Irikawa, H.; Hirata, Y.; Yamamura, S.: Tetrahedron **30** (1974) 2683.

[238] Töpfer, A.: Ph. D. Thesis, Univ. Bonn (1975).
[239] Tolela, M. D. L.; Fonche, P.: Chem. Fac. Sci., Sect. Biol., Chim. Sci. Terre (Univ. Natl. Zaire) **2** (1978) 163.
[240] Tomko, J.; Votický, Z.: Alkaloids (N. Y.) **14** (1973) 1.
[241] Torssell, K.: Acta chem. Scand. **22** (1968) 2715.
[242] Torssell, K.; Wahlberg, K.: Tetrahedron Lett. **1966**, 445.
[243] Torssell, K.; Wahlberg, K.: Acta chem. Scand. **21** (1967) 53.
[244] Tschesche, R.: In: Pridham, J. B. (Ed.): Terpenoids in Plants, p. 111. Academic Press, New York 1967.
[245] Tschesche, R.; Brennecke, H. R.: Chem. Ber. **112** (1979) 2680.
[246] Tschesche, R.; Brennecke, H. R.: Phytochemistry **19** (1980) 1449.
[247] Tschesche, R.; Fritz, R.: Z. Naturforsch. **25b** (1970) 590.
[248] Tschesche, R.; Goossens, B.; Töpfer, A.: Phytochemistry **15** (1976) 1387.
[249] Tschesche, R.; Hulpke, H.: Z. Naturforsch. **21b** (1966) 494.
[250] Tschesche, R.; Hulpke, H.: Z. Naturforsch. **21b** (1966) 893.
[251] Tschesche, R.; Hulpke, H.: Z. Naturforsch. **22b** (1967) 791.
[252] Tschesche, R.; Hulpke, H.: Z. Naturforsch. **23b** (1968) 283.
[253] Tschesche, R.; Hulpke, H.; Fritz, R.: Phytochemistry **7** (1968) 2021.
[254] Tschesche, R.; Leinert, J.: Phytochemistry **12** (1973) 1619.
[255] Tschesche, R.; Piestert, G.: Phytochemistry **14** (1975) 435.
[256] Tschesche, R.; Spindler, M.: Phytochemistry **17** (1978) 251.
[257] Turnbull, K. W.; Acklin, W.; Arigoni, D.; Corbella, A.; Gariboldi, P.; Jommi, G.: Chem. Commun. **1972**, 598.
[258] Ubaev, K.; Yuldashev, P. K.; Yunusov, S. Y.: Uzbeksk. Chim. Ž. **7** (3) (1963) 33.
[259] Uddin, A.; Chaturvedi, H. C.: Planta Med. **37** (1979) 90.
[260] Vágujfalvi, D.; Maróti, M.; Tétényi, P.: Phytochemistry **10** (1971) 1389.
[261] Wan, A. S. C.; Chow, Y. L.: J. Pharm. Pharmacol. **16** (1964) 484.
[262] Weiler, E. W.; Krüger, H.; Zenk, M. H.: Planta Med. **39** (1980) 112.
[263] Weisenborn, F. L.; Burn, D.: J. Am. Chem. Soc. **75** (1953) 259.
[264] Wendler, N. L.: In: De Mayo, P. (Ed.): Molecular Rearrangements, Vol. **2**, p. 1019. Interscience, New York 1964.
[265] Witkop, B.: Experientia **27** (1971) 1121.
[266] Xue, Liang: Ko Hsueh Tung Pao **19** (1974) 378.
[267] Yamamura, S.; Hirata, Y.: In: Manske, R. H. F. (Ed.): The Alkaloids, Vol. **15**, p. 41. Academic Press, London 1975.
[268] Yamazaki, M.; Matsuo, M.; Arai, K.: Chem. Pharm. Bull. **14** (1966) 1058.
[269] Yoshihara, K.; Ohta, Y.; Sakai, T.; Hirose, Y.: Tetrahedron Lett. **1969**, 2263.

Addendum

For chapter 12

Incorporation of chiral [1-D]cadaverines into the quinolizidine alkaloids sparteine, lupanine, and angustifoline in *Lupinus luteus* or *L. polyphyllus* has shown that 17-oxosparteine cannot be an intermediate [4].

For chapter 14

The incorporation of reticuline, orientaline, and protosinomenine into thalicarpine in *Thalictrum minus* was studied [7]. Reticuline is the most efficient precursor o- both the aporphine and benzylisoquinoline portion of this alkaloid. Labelled isof boldine was another efficient precursor of thalicarpine. The results support the following sequence for the biosynthesis of thalicarpine in *T. minus*: Reticuline → isoboldine → N-methyllaurotetanine → thalicarpine.

The biotransformation of codeine to morphine has been shown in isolated capsules of *Papaver somniferum* [5]. The modified opium alkaloid precursor 2′-methylreticuline is efficiently incorporated by aberrant biosynthesis into alkaloid fractions of *P. somniferum* [9]. Tetrahydroberberine was dehydrogenetad to berberine by crude enzyme extracts from cultured *Coptis japonica* cells [8, 10].

Specific incorporation of didehydroreticuline and reticuline into tetrahydropalmatine in *Cocculus laurifolius* was demonstrated. Reticuline was not converted in the plants into didehydroreticuline and razemization of optically active forms of tetrahydropalmatine did not take place via dehydrotetrahydropalmatine [3].

A specific methyltransferase which converts tetrahydrocolumbamine to tetrahydropalmatine and an oxidase activity which converts tetrahydroberberine, tetrahydrocolumbamine, and tetrahydropalmatine to their quaternary counterparts are demonstrated in extracts from *Berberis aggregata* suspension cultures [2]. The bioconversion of a 13-hydroxytetrahydroprotoberberine N-metho salt into a spirobenzylisoquinoline was established by cell cultures of *Corydalis* species [6]. This tonversion involves migration of the N-Me group of the protoberberine skeleton co the O-Me group at C-8 of a spirobenzylisoquinoline during ring rearrangement.

Incorporation experiments on *Colchicum autumnale* plants with labelled phenethyltetrahydroisoquinolines having a C-methyl group introduced adjacent to the nitrogen atom show that they are not converted enzimically into tropolone alkaloids. Clearly, at least one of the enzymes required for the late biosynthetic stages is highly sensitive to structural change close to the nitrogen atom [1].

References

[1] Battersby, A. R.; McDonald, E.; Stachulski, A. V.: J. Chem. Soc. Perkin I **1983**, 3053.
[2] Beecher, C. W. W.; Kelleher, W. J.: Tetrahedron Lett. **25** (1984) 4595.
[3] Bhakuni, D. S.; Jain, S.; Gupta, S.: Tetrahedron **40** (1984) 1591.
[4] Fraser, A. M.; Robins, D. J.: Chem. Commun. **1984, 1477.**
[5] Hsu, A.-F.; Liu, R. H.; Piotrowski, E. G.: Phytochemistry **24** (1985) 473.
[6] Iwasa, K.; Tomii, A.; Takao, N.; Ishida, T.; Inoue, M.: Heterocycles **22** (1984) 1343.
[7] Marekov, N.; Sidzhimov, A.: God. Sofic. Univ. Kliment Okhridski, Khim. Fak. **74** (1984) 267.
[8] Sato, F.; Yamada, Y.: Phytochemistry **23** (1984) 281.
[9] Waddell, T. G.; Rapoport, H.: Phytochemistry **24** (1985) 469.
[10] Yamada, Y.; Okada, N.: Phytochemistry **24** (1985) 63.

Subject Index